Reproductive and Developmental Effects of Contaminants in Oviparous Vertebrates

Other titles from the Society of Environmental Toxicology and Chemistry (SETAC):

Multiple Stressors in Ecological Risk and Impact Assessment
Foran and Ferenc, editors
1999

Linkage of Effects to Tissue Residues: Development of a Comprehensive Database for Aquatic Organisms Exposed to Inorganic and Organic Chemicals
Jarvinen and Ankley
1999

Ecotoxicology and Risk Assessment for Wetlands
Lewis, Mayer, Powell, Nelson, Klaine, Henry, Dickson, editors
1999

Uncertainty Analysis in Ecological Risk Assessment
Warren-Hicks and Moore, editors
1998

Ecotoxicological Risk Assessment of the Chlorinated Organic Chemicals
Carey, Cook, Giesy, Hodson, Muir, Owens, Solomon, editors
1998

Sustainable Environmental Management
Barnthouse, Biddinger, Cooper, Fava, Gillett, Holland, Yosie, editors
1998

Ecological Risk Assessment Decision-Support System: A Conceptual Design
Reinert, Bartell, Biddinger, editors
1998

Principles and Processes for Evaluating Endocrine Disruption in Wildlife
Kendall, Dickerson, Giesy, Suk, editors
1998

Quantitative Structure-Activity Relationships in Environmental Science-VII
Chen and Schüürmann, editors
1997

Atmospheric Deposition of Contaminants to the Great Lakes and Coastal Waters
Baker, editor
1997

For information about any SETAC publication, including SETAC's international journal,
Environmental Toxicology and Chemistry,
contact the SETAC Office, 1010 N. 12th Avenue, Pensacola, Florida, USA 32501-3367
T 850 469 1500   F 850 469 9778   E setac@setac.org   http://www.setac.org

Reproductive and Developmental Effects of Contaminants in Oviparous Vertebrates

Edited by
Richard T. Di Giulio
Donald E. Tillitt

Proceedings from the Pellston Workshop on Reproductive and Developmental
Effects of Contaminants in Oviparous Vertebrates
13–18 July 1997
Fairmont Hot Springs Resort
Anaconda, Montana

SETAC Special Publications Series

SETAC Liason
Greg Schiefer
SETAC/SETAC Foundation

Current Coordinating Editor of SETAC Books
C.G. Ingersoll
U.S. Geological Survey

Publication sponsored by the Society of Environmental Toxicology and Chemistry
(SETAC) and the SETAC Foundation for Environmental Education

SETAC PRESS

Cover by Michael Kenney Graphic Design and Advertising
Indexing by IRIS

Library of Congress Cataloging-in-Publication Data

Pellston Workshop on Reproductive and Developmental Effects of Contaminants in Oviparous Vertebrates
(1997 : Fairmont Hot Springs Resort, Anaconda, Mont.)
   Reproductive and developmental effects of contaminants in oviparous vertebrates : proceedings from the
Pellston Workshop on Reproductive and Developmental Effects of Contaminants in Oviparous Vertebrates. 13–18
July 1997. Fairmont Hot Springs Resort, Anaconda, Montana / edited by Richard Di Giulio, Donald Tillitt.
         p. cm. -- (SETAC special publications series)
   "Publication sponsored by the Society of Environmental Toxicology and Chemistry (SETAC) and the SETAC
   Foundation for Enviornmental Education."
   Includes bibliograpical references and index.
   ISBN 1-880611-37-6
   1. Vertebrates--Effect of chemicals on--Congresses.   2. Vertebrates--Eggs--Congresses.   3. Vertebrates--
Reproduction--Congresses   4. Vertebrates--Development--Congresses   I. Di Giulio, Richard T. (Richard
Thomas), 1950–   . II. Tillitt, Donald (Donald Edward), 1957–   . III. SETAC (Society) IV. SETAC Foundation
for Environmental Education.   V. Series.
QL605.A1P45   1997
571.8' 16--dc21   98–56080
                                                                                                         CIP

International Standard Book Number 1-880611-37-6
Printed in the United States of America
06 05 04 03 02 01 00 99   10 9 8 7 6 5 4 3 2 1

Referemce Listing: Di Giulio RT, Tillitt DE, editors. 1999. Reproductive and developmental effects of contaminants in oviparous vertebrates. SETAC Pellston Workshop on Reproductive and Developmental Effects of
Contaminants in Oviparous Vertebrates; 13–18 July 1997; Fairmont Hot Springs, Anaconda, Montana. Published
by the Society of Environmental Toxicology and Chemistry (SETAC). Pensacola, Florida, USA. 458 p.

The SETAC Special Publications Series

The SETAC Special Publications Series was established by the Society of Environmental Toxicology and Chemistry (SETAC) to provide in-depth reviews and critical appraisals of scientific subjects relevant to understanding the impact of chemicals and technology on the environment. The series consists of books on topics reviewed and recommended by the SETAC Board of Directors for their importance, timeliness, and contribution to multidisciplinary approaches to solving environmental problems. The diversity and breadth of subjects covered in the series reflect the wide range of disciplines encompassed by environmental toxicology, environmental chemistry, and hazard and risk assessment. Despite this diversity, these volumes share the goal of presenting the reader with authoritative coverage of the literature. In addition, paradigms, methodologies and controversies, research needs, and new developments specific to the featured topics are presented in these volumes. All books in the series are peer reviewed for SETAC by acknowledged experts.

The SETAC Special Publications are useful to environmental scientists in research, research management, chemical manufacturing, regulation, and education, as well as to students considering careers in these areas. The series provides information for keeping abreast of recent developments in familiar subject areas and for rapid introduction to principles and approaches in new subject areas.

SETAC would like to recognize the past SETAC Special Publications Series editors:

T.W. La Point, Institute of Applied Sciences,
University of North Texas, Denton, TX

B.T. Walton, U.S. Environmental Protection Agency,
Research Triangle Park, NC

C.H. Ward, Department of Environmental Sciences and Engineering,
Rice University, Houston, TX

Contents

List of Figures

List of Tables

Preface

This book presents the proceedings of a workshop held 13-18 July 1997 at Fairmont Hot Springs Resort, Anaconda, Montana. The workshop was designed to provide a state-of-the-science summary of what is known concerning chemical impacts on reproduction and development in egg-laying wildlife, to identify key information gaps, and to discuss research strategies for filling such gaps. Workshop participants from academia, government, and industry were selected for their expertise in specific fields including environmental chemistry and toxicology, animal physiology, ecology, and risk assessment. The major issues participants deliberated included 1) pathways by which exposures to contaminants having reproductive and developmental impacts are likely to occur in oviparous vertebrates; 2) critical physiological aspects of oviparity that may be sensitive to perturbations by environmental contaminants; 3) key mechanisms by which chemicals are likely to impact reproduction and development in oviparous vertebrates; 4) reproductive and developmental effects of contaminants in these animals that are of greatest ecological significance; and 5) ecological risk assessments oriented to exposures and effects of reproductive/developmental toxicants in oviparous vertebrates. The organizers and participants of this workshop hope that this volume is useful and challenging and will contribute to the wise stewardship of chemical releases into the environment.

Richard T. Di Giulio
Donald E. Tillitt

Acknowledgments

The Workshop on Reproductive and Developmental Effects of Contaminants in Oviparous Vetebrates was made possible through the financial support of the following organizations:

- American Crop Protection,
- American Industrial Health Council,
- BASF Corporation,
- Chemical Manufacturers Association,
- Dow Chemical Company,
- Eastman Kodak Company,
- Exxon Corporation,
- Novartis Crop Protection,
- The Procter & Gamble Company,
- Rhône-Poulenc,
- Rohm and Haas Company,
- The Soap and Detergent Association,
- U.S. Environmental Protection Agency, and
- U.S. Fish and Wildlife Service.

The content of this publication does not necessarily reflect the position or policy of any of these organizations or of the United States government, and an official endorsement should not be inferred.

The organizers of this workshop sincerely commend the steering committee, discussion group leaders, and participants for contributing to lively and fruitful discussions at the workshop and for their commitment and perseverance in the preparation of this volume. We also deeply thank John Stein, National Oceanic and Atmospheric Administration, and Mary Walker, University of New Mexico, for their careful and thorough reviews of this work. The efforts of SETAC/SETAC Foundation staff for skillful assistance with workshop planning and on-site implementation are most appreciated. In particular, we thank Greg Schiefer, Rod Parrish, Linda Longsworth, and Leslie Long. We also thank the staff of the Fairmont Hot Springs Resort for their assistance and hospitality. Finally, we deeply thank Stacey Hagman and James Blackwell for their excellent work in preparing this manuscript for publication.

About the Editors

Richard T. Di Giulio is Professor and Chair of the Environmental Toxicology, Chemistry, and Risk Assessment Program at the Nicholas School of the Environment, Duke University, Durham, North Carolina. He received a B.A. in literature from the University of Texas at Austin, an M.S. in wildlife management from Louisiana State University, and a Ph.D. in wildlife biology/environmental toxicology from Virginia Polytechnic Institute and State University. Dr. Di Giulio serves as an advisor for the Scientific Advisory Board of the USEPA, for the Canadian Network of Toxicology Centres, and previously as a member of the Board of Directors of SETAC. His research interests include mechanisms of chemical toxicity and adaptation in aquatic organisms, genetic consequences, and developmental effects.

Donald E. Tillitt is leader of the Biochemistry Section at the Columbia Environmental Research Center of the U.S. Geological Survey, Columbia, Missouri. He received B.S. degrees in fisheries and wildlife and agricultural biochemistry (1981) from Michigan State University. He worked for the Michigan Department of Natural Resources until returning to Michigan State University for a an M.S. in fisheries and wildlife (1986) and a Ph.D. in environmental toxicology (1989). He has been an active member of SETAC and the Ozark-Prairie SETAC Chapter. He leads an active research program on the fate and effects of chemicals on fish and wildlife with an emphasis on reproductive effects of organochlorines on early development.

List of Attendees and Participants*

Gerald T. Ankley[2]
USEPA
Duluth, MN

Joel E. Baker
University of Maryland
Solomons, MD

Lawrence W. Barnthouse
McLaren/Hart
Oak Ridge, TN

William H. Benson[1]
University of Mississippi
University, MS

Linda Birnbaum[1]
USEPA
Research Triangle Park, NC

Stephen P. Bradbury
USEPA
Duluth, MN

Larry Brewer[2]
Ecotoxicology & Biosystems Assoc., Inc.
Sisters, OR

James R. Clark[2]
Exxon Biomedical Sciences, Inc.
East Millstone, NJ

Peter L. de Fur[1]
Virginia Commonwealth University
Richmond, VA

Kenneth L. Dickson[2]
University of North Texas
Denton, TX

Richard T. Di Giulio[1]
Duke University
Durham, NC

Anne Fairbrother[2]
ecological planning & toxicology
Corvallis, OR

Glen A. Fox
Canadian Wildlife Service
Hull, QC, Canada

Bettina Francis
University of Illinois
Urbana, IL

Nicholas W. Gard
PTI Environmental Services
Bellevue, WA

John P. Giesy
Michigan State University
East Lansing, MI

Frank Gobas
Simon Fraser University
Burnaby, BC, Canada

Jay W. Gooch[1]
The Procter & Gamble Company
Cincinnati, OH

---

*Affiliations were current at the time of the workshop
[1]Steering Committee member
[2]Workgroup Chair

Earl Gray
USEPA
Research Triangle Park, NC

David Hinton
University of California–Davis
Davis, CA

Lyndal L. Johnson
NOAA–NW Fisheries Science Center
Seattle, WA

Kevin M. Kleinow[2]
Louisiana State University
Baton Rouge, LA

Vince Kramer
Rohm and Haas Company
Spring House, PA

Robert T. Lackey
USEPA
Corvallis, OR

Paul Mastradone
USEPA
Washington, DC

Monte A. Mayes
The Dow Chemical Company
Indianapolis, IN

Anne McNabb[2]
Virginia Tech University
Blacksburg, VA

Mary Mendonca
Auburn University
Auburn, AL

Ellen M. Mihaich[1]
Rhône-Poulenc
Research Triangle Park, NC

George Monteverdi
Duke University
Durham, NC

Kelly R. Munkittrick
Environment Canada
Burlington, ON, Canada

John W. Nichols
USEPA
Duluth, MN

Thomas F. Parkerton
Exxon Biomedical Sciences, Inc.
East Millstone, NJ

Richard Peterson
University of Wisconsin–Madison
Madison, WI

Kenneth A. Rose[2]
Oak Ridge National Laboratories
Oak Ridge, TN

Carl Schreck[2]
Oregon State University
Corvallis, OR

Kyle W. Selcer
Duquesne University
Pittsburg, PA

Jennifer Specker
University of Rhode Island
Narragansett, RI

Ralph G. Stahl, Jr.
E.I. Du Pont de Nemours & Co.
Wilmington, DE

Peter Thomas
University of Texas
Port Aransas, TX

Donald E. Tillitt[1]
USGS–Biological Resources Division
Columbia, MO

Charles Tyler[1]
Brunel University
Uxbridge, Middlesex, UK

Glen Van Der Kraak
University of Guelph
Guelph, ON, Canada

Laurie Vitt
Oklahoma Museum of Natural History
Norman, OK

Maurice G. Zeeman
USEPA
Washington, DC

CHAPTER 1

Reproductive and Developmental Effects of Contaminants in Oviparous Vertebrates: Workshop Introduction

Ellen M. Mihaich, Jay Gooch, Peter deFur, William H. Benson, Charles Tyler,
Linda Birnbaum, Richard T. Di Giulio, Donald E. Tillitt

Concern has arisen in recent years about the potential for effects on reproduction and development in both human and wildlife populations. As a consequence, considerable attention has been given to various aspects of this broad issue by research scientists, environmental managers, and the general public. This workshop brought together experts from the fields of reproductive and developmental biology, environmental toxicology and chemistry, ecology, and risk assessment in order to discuss and evaluate a key subset of this issue that has great ecological significance—effects of chemicals on both terrestrial and aquatic oviparous (i.e., egg-laying) vertebrates. The workshop focused on contaminant exposure, mechanisms of effects on reproduction and development, organismal outcomes, ecological ramifications, and a discussion of ecological risk assessment focused on oviparous vertebrates. While environmental stressors other than chemical contamination also can impair reproduction, this workshop was focused on contaminant interactions and effects in oviparous vertebrates because they have unique sensitivities to contaminants as a result of their life-history strategy. For example, maternal transfer of contaminants to eggs and the subsequent effects of these contaminants on development have provided hallmark examples of ecological risk from this mechanism of toxicity and route of exposure. It was also our intention that these discussions would provide for a clearer focus on reproductive and developmental effects of contaminants on those species that hold the greatest promise for use in prospective and retrospective ecological risk assessments, including species commonly used in today's toxicity testing methods. This

CHAPTER PREVIEW

Reproductive and Developmental Effects of Contaminants in Oviparous Vertebrates. Richard T. Di Giulio and Donald E. Tillitt, editors.
©1999 Society of Environmental Toxicology and Chemistry (SETAC). ISBN 1-880611-37-6

focus brought additional clarity to areas in which research could provide the tools needed to better protect living resources.

Environmental perturbations can cause adverse effects in vertebrate wildlife populations that range from short-term mortality to long-term changes in reproductive function, development, and growth. Chemical contaminants in particular may cause effects such as acute neurological damage that leads to death in hours or days, impairment of reproductive function, or alterations in the development of the egg or embryo over many months or across generations. Environmental stressors other than chemical contamination, such as disease, thermal stress, and habitat degradation, also can impair reproduction, growth and development. These stressors usually do not occur alone in real-world conditions, where in many perturbed environments, combinations of chemical and non-chemical stressors are common. The effects of these stressors on reproduction and development are especially problematic because of the profound influence that reproductive function and outcome have on the survival and sustainability of wildlife populations.

Scientific and Historical Context for Reproductive and Developmental Effects

Environmental toxicologists, particularly those conducting prospective risk assessments for new chemicals, have long utilized measures of reproductive and developmental effects as critical elements of hazard assessments. Comparing environmental exposure concentrations to concentrations known to affect these endpoints is a central element in ensuring the safe utilization of industrial, consumer, and agricultural chemicals. Understanding the nature of potential reproductive and developmental effects, and the exposure concentrations at which they occur, is a cornerstone of linking laboratory-based hazard assessments to protection of populations, communities, and ecosystems. The emphasis on population-level effects, which is the focus of the current ecological risk assessment paradigm, has its foundation in individual-level effects. Chemical safety testing protocols and subsequent regulatory decisions designed to protect the environment are based on exposure and effects measured in test groups that are presumed to be representative of a population. Reproductive and developmental endpoints are measured in avian reproduction tests and in fish early-life-stage and life-cycle tests. Results from these tests are commonly used in risk assessments to estimate risk for population-level effects.

With much the same rationale, ecotoxicologists and field biologists investigating adverse effects on individuals and populations in the environment have often focused on evaluating reproductive and developmental effects. A number of populations of oviparous vertebrates have been adversely affected by environmental perturbations, ranging from chemical contamination to habitat destruction (see

Colborn and Clement 1992). In several avian species, the effects of environmental contamination were first observed in wild populations (for reviews, see Fox 1992; Giesy et al. 1994; Feyk and Giesy 1998). These effects were subsequently traced to chemical contaminants acting at low concentrations on sensitive life stages, with severe consequences for several bird populations. For example, in an attempt to understand the significance of organochlorine contaminants in the Great Lakes, numerous studies have been conducted on the developmental effects of dioxins and coplanar polychlorinated biphenyls in cormorants, bald eagles, Caspian terns, Forster's terns, and black-crowned night herons (Heinz et al. 1985; Heinz 1998). In addition, the now classic studies of DDT/DDE-induced eggshell thinning in birds (Cooke 1973), particularly raptors, stand as an excellent example of the utilization of these measures, as well as the potential population-level consequences of repro-ductive effects. Extensive studies have focused on evaluating the role of organochlo-rines in the reproductive performance of Great Lakes salmonid populations (Gilbertson 1984; Cook et al. 1997; Wright and Tillitt 1999). These studies have attempted to explain, through effects on reproduction and development, both the historical decline of lake trout populations and the poor recruitment from current spawning stocks (Leatherland et al. 1982; Marcquenski and Brown 1998; Cook et al. 1999). One outcome of this type of retrospective reproductive assessment has been recognition of the significance of effects from toxic chemicals that act by altering early stages of development and result in a cascading sequence of biological pro-cesses ultimately leading to the symptomatic effect.

Biology/Physiology of Oviparous Vertebrates

The reproductive physiology of birds, fishes, reptiles, and amphibians is very diverse ranging from the production of externally fertilized, yolk-laden eggs to the production of small eggs lacking yolk that are fertilized internally and require maternally supplied nutrition prior to parturition. Oviparous vertebrates are those animals that produce eggs that develop external to the female's body and are nourished by the maternal nutrients provided prior to fertilization.

Life cycles and life-history strategies have been described for many species of birds, fish, reptiles and amphibians, detailing unique and adaptive features in these animals. One focus of this workshop was to ask the following questions: Are some species more vulnerable to environmental perturbations because of their life-history strategy? Are particular life stages more important or more sensitive than others? Some oviparous vertebrates display unique developmental and growth patterns and events, such as metamorphosis with a consequent shift from water to bimodal breathing, or changes in reproductive status as a function of size and age. These life history strategies and developmental patterns are adaptations permitting species to survive a range of ambient conditions or to thrive in specific habitats.

The sequence of life stages (e.g., egg to larva to adult), the pattern of development, and the manner of reproduction are important and often critical aspects of the biology of oviparous vertebrates. While the life cycle (the sequence of stages of development) determines the basic sequence of growth and maturation, including sexual development, the life-history strategy describes the ecological and evolutionary adaptations of a population or species. Life-history strategies often vary considerably among groups and even among closely related species of oviparous vertebrates. This variation can take several forms, including extended larval stages and density-dependent reproductive performance (e.g., mating, egg laying).

Workshop Objectives

The purpose of this book is to disseminate the critical discussions on reproductive and developmental effects in oviparous vertebrates that occurred during the SETAC Pellston Workshop in Fairmont, Montana in July 1997. The workshop brought together experts from the fields of reproductive and developmental biology, environmental toxicology and chemistry, ecology, and risk assessment. These individuals were selected to achieve balance among academia, government, and the private sector.

In the context of oviparous vertebrates, the workshop focused on the routes of contaminant exposure and the mechanisms of effect on reproduction and development, organismal outcomes, and ecological ramifications. The workshop included discussions of laboratory and field approaches to link exposure and effect and to integrate this information into ecological risk assessments focused on contaminants in which reproductive and developmental effects are the primary concern.

There are numerous reasons why the workshop focused on oviparous vertebrates. The oviparous lifestyle has particular nuances that may predispose such species to early life-stage exposures and to developmental effects. These effects may occur during larval development or during the juvenile or adult stage, depending upon the species. Exposures at early-life stages may lead to alterations in key developmental processes, as well as increased susceptibility to chemical insults as adults. Exposures at maturity could also disrupt normal reproductive physiology.

Workshop Plan and Organization

The workshop focused on the effects of chemical contaminants on individual animals, populations, and, where possible, ecological communities. The organizers planned for the participants to consider a wide range of subjects covering the biology of oviparous vertebrates and the chemical properties of contaminants. Participants were charged with accomplishing the following goals:

- describe the state of the science in the areas of reproductive and developmental biology of oviparous vertebrates in aquatic and terrestrial ecosystems;
- identify major information gaps and unknowns in the state of the science, especially those gaps that are critical for environmental decision-making; and
- suggest improvements in approaches to understand the impacts on and assessment of risks to reproduction and development in this group of animals.

The workshop organizers recognized the need to address specific aspects of the biology of oviparous vertebrates and the environmental characteristics of the contaminants. The workshop highlighted those characteristics distinct for oviparous vertebrates. The participants addressed the following topics:

- critical molecular, cellular, biochemical, and physiological aspects of reproduction and development, including comparative features that suggest sensitivities to environmental contaminants;
- key mechanisms by which environmental contaminants are likely to impact reproduction and development in oviparous vertebrates;
- reproductive and developmental responses to contaminants that are of the most significant ecological concern;
- major pathways by which exposures occur or are likely to occur for chemicals affecting these biological processes; and
- features of the effects of contaminants on oviparous vertebrates that are of specific or unique concern for ecological risk assessment.

Participants were selected on the basis of their expertise and knowledge in the above areas and were asked to work in one of the following workgroups:

- Chemistry: The participants in this workgroup were environmental and analytical chemists, biochemists, and fate modelers. They were asked to deliberate on exposure, target-organ dosimetry, mechanism of maternal transfer in oviparous vertebrates, abiotic mechanisms of chemical activation, and the utility of tissue residues and biomarkers in assessing exposure of fish and wildlife to reproductive/developmental toxicants.
- Physiology: This group was comprised primarily of vertebrate reproductive and developmental biologists, endocrinologists, and comparative physiologists. They were asked to consider a number of topics, including molecular aspects of gamete and early life stage development, comparative vertebrate reproductive endocrinology, reproductive and early life-stage behavior, identification of processes and endpoints that can/cannot be extrapolated across species, and stressor-sensitive aspects of reproduction and development.
- Toxicology: The participants in this workgroup were primarily molecular biologists, biochemists, and biochemical and environmental toxicologists. They were charged with exploring various mechanisms by which toxicants

may impact reproductive and developmental processes of oviparous vertebrates, including, but not limited to, endocrine disruption, apoptosis,
neurotoxicity, interactions with other stressors (e.g., nutrition), and in vitro
and in vivo approaches for elucidating such mechanisms.

- Ecology: This group was composed primarily of population and community
ecologists, ecotoxicologists, and ecological modelers. They were asked to
consider topics such as gamete health, early life-stage development and
survival, reproductive and early life-stage behavior, population effects and the
utility of biomarkers, ecological indices, mathematical models, and other
methods for assessing the health of populations. They were also asked to
identify approaches for translating effects across levels of biological organization.

- Risk assessment: This final workgroup was composed primarily of
ecotoxicologists, ecological modelers, ecological risk experts, and environmental managers. They were asked to consider aspects of oviparity that might
need to be handled uniquely in risk assessment. They were to consider and
discuss eco-epidemiology, chemical-specific approaches, the identification of
species or endpoints that hold the most promise for obtaining time- and cost-
efficient data, and the use of risk assessment models.

Conclusion

This monograph provides insights to scientists, regulatory decision-makers, and the
chemical industry concerning critical biological and ecological endpoints, techniques for their measurement, and strategies for ecological risk assessments in and
with oviparous vertebrates. The workshop organizers and participants recognized
the importance of providing sound and reliable information to various stakeholder
groups, including decision-makers, concerning the effects on reproduction and
development in this important group of animals.

References

Colborn T, Clement C, editors. 1992. Chemically-induced alterations in sexual and functional
    development: the wildlife/human connection. Advances in Modern Environmental
    Toxicology Vol. XXI. Princeton NJ: Princeton Scientific Pub. Co., Inc. 403 p.

Cook PM, Endicott DD, Robins J, Marquis P, Berini C, Libal J, Kizlauskis A, Walker MK,
    Zabel EW, Guiney PD, Peterson RE. Effects of chemicals with an Ah receptor-mediated
    mode of early-life-stage toxicity on lake trout reproduction and survival in Lake Ontario:
    retrospective and prospective risk assessments. *Environ Sci Technol* (In preparation).

Cook PM, Zabel EW, Peterson RE. 1997. The TCDD toxicity equivalence approach for
    characterizing risks for early life-stage mortality in trout. In: Rolland RM, Gilbertson M,

Peterson RE, editors. Chemically induced alterations in functional development and reproduction in fishes. Pensacola FL: SETAC. p 9–27.

Cooke AS. 1973. Shell thinning in avian eggs by environmental pollutants. *Environ Pollut* 4:85–152.

Feyk LA, Giesy JA. 1998. Xenobiotic modulation of endocrine function in birds. In: Kendall R, Dickerson R, Giesy R, Suk W, editors. Principles and processes for evaluating endocrine disruption in wildlife. Pensacola FL: SETAC. p 121–140.

Fox G. 1992. Epidemiological and pathobiological evidence of contaminant-induced alterations in sexual development in free-living wildlife. In: Colborn T, Clement C, editors. Chemically-induced alterations in sexual and functional development: the wildlife/human connection. Advances in Modern Environmental Toxicology Vol. XXI. Princeton NJ: Princeton Scientific Pub. Co., Inc. p 147–158.

Giesy JP, Ludwig JP, Tillitt DE. 1994. Deformities in birds of the Great Lakes region: assigning causality. *Environ Sci Tech* 28(3):128A–135A.

Gilbertson M. 1984. Need for development of epidemiology for chemically induced diseases in fish in Canada. *Can J Fish Aquat Sci* 41:1534–1540.

Heinz GH. 1998. Contaminant effects on Great lakes fish-eating birds: a population perspective. In: Rolland RM, Gilberston M, Peterson RE, eds. Chemically induced alterations in functional development and reproduction of fishes. Pensacola FL: SETAC. p 141–154.

Heinz GH, Erdman TC, Haseltine SD, Stafford C. 1985. Contaminant levels in colonial waterbirds from Green Bay and Lake Michigan, 1975–1980. *Environ Monitor Assess* 5:223–236.

Leatherland JF, Copeland P, Sumpter JP, Sonstegard RA. 1982. Hormonal control of gonadal maturation and development of secondary sexual characteristics in coho salmon, *Oncorhynchus kisutch*, from lakes Ontario, Erie and Michigan. *Gen Comp Endocrinol* 48:196–204.

Marcquenski SV, Brown SB. 1998. Early mortality syndrome in salmonid fishes from the Great Lakes. In: Rolland RM, Gilberston M, Peterson RE, editors. Chemically induced alterations in functional development and reproduction of fishes. Pensacola FL: SETAC. p 135–152.

Wright PI, Tillitt DE. 1999. Embryotoxicity of Great Lakes lake trout extracts to developing rainbow trout. *Aquat Toxicol* (In press).

CHAPTER 2

Exposure, Uptake, and Disposition of Chemicals in Reproductive and Developmental Stages of Selected Oviparous Vertebrates

Kevin Kleinow, Joel Baker, John Nichols, Frank Gobas, Thomas Parkerton, Derek Muir, George Monteverdi, Paul Mastrodone

Chemical exposure is defined here as the delivery of the chemical from the surrounding environment into the organism and transport to the active site. The actual time-variable dose delivered to the active site depends in complex ways on the physicochemical speciation of the chemical outside and inside the organism and upon the behavior, bioenergetics, physiology, and life stage of the organism. Once a chemical enters an organism, it may be presented to the active site, excreted, or sequestered within tissue away from the active site. Whether a stored chemical is eventually redistributed to the active site (and completes the exposure pathway) or is metabolized, excreted, or passed to offspring depends largely upon the physiology of the organism. Such storage within the organism may serve to delay exposure, resulting in a temporal (and perhaps spatial) disconnection between external exposure and observed effect.

In another light, exposure can also be described as the result of a organism's being physically located in an area that contains the chemical in the appropriate (e.g., bioavailable) form. Taken this way, quantifying exposure then becomes a probability exercise based on the spatial and temporal distribution of bioavailable chemical and the behavior and physiology of the organism. This approach allows for time-variable exposure calculations. For example, exposure to a persistent, globally produced organochlorine such as p,p'-DDE is likely to be low and fairly uniform spatially and temporally. In contrast, exposure to a nonpersistent, high-use chemical such as an organophosphorus agrochemical may be high but short-lived in a restricted region. The ultimate impact of a chronic, continual, low-level exposure versus a large but episodic exposure is an interesting ecotoxicological question.

CHAPTER PREVIEW

Reproductive and Developmental Effects of Contaminants in Oviparous Vertebrates. Richard T. Di Giulio and Donald E. Tillitt, editors.
©1999 Society of Environmental Toxicology and Chemistry (SETAC). ISBN 1-880611-37-6

Exposure assessment often begins with the characterization of chemical release during production, transport, processing, end-use, and disposal. The extent of release will generally depend upon such factors as the production volume and physicochemical properties of the chemical, the nature of industrial application (i.e., open or closed system), availability, type and operational features of pollution control equipment, and end-use and ultimate disposal patterns of products that contain the chemical. Natural emission sources may constitute an important component of the overall emission inventory of certain chemicals, and therefore also should be considered.

Upon release into the environment, contaminants are subjected to various environmental transport and transformation processes that determine the spatial and temporal concentrations in primary (air, water, soil, sediment) and secondary (plants, animals) compartments. The concentrations in these media, when linked to species and life-stage-specific exposure factors of oviparous organisms (i.e., inhalation rates, ingestion rates, exposure patterns), dictate the total administered dose to a given population of receptors from multiple exposure pathways. The efficiency in which the administered dose is absorbed across the respiratory, gut, or skin interface determines the absorbed dose. Further consideration of toxicokinetic behavior of the chemical determines tissue residues and the effective dose at the internal target site which, in turn, mediates adverse biological effects.

For most organisms, early developmental stages are among the most sensitive to the toxic effects of chemical contaminants. Success in terms of reproduction and development represents the integration and ontogenetic development of many critical biochemical events. Chemical interference with these events may lead to toxicological effects that influence the survival of the individual and possibly populations. In addition to sensitive sites for toxicity, the process of ontogeny brings about changes in how the organism is exposed to chemicals, how significant the exposure is, and how the organism deals with the chemicals once they are absorbed. In a true sense the organism undergoes an exposure and dispositional ontogeny. All of these events are temporally spaced, presenting windows of opportunity for xenobiotics to more capably elicit their toxic effects. Because of differences in the relative importance of exposure pathways for the various life stages, extrapolation of exposure estimates from adults of even the same species are of little value.

Oviparous species present a number of unique considerations with regard to chemical exposure, uptake, and disposition in early life stages. By definition these animals undertake their early development in eggs outside of the maternal body. The egg, more or less a self-contained unit, stores yolk nutrients, provided from maternal sources, that serve to fuel growth and development. This confinement of early development in the egg modulates chemical exposure from the outside but also subjects the embryo to chemicals that may be preferentially mobilized or concentrated from maternal sources. Furthermore, the early life stages are free of maternal influences with regard to xenobiotic disposition. Low biotransformational activities and incomplete excretory pathways dictate unusual toxicological conditions in developing stages of oviparous organisms.

In this chapter we explore the major pathways resulting in exposure of a wide variety of chemicals to oviparous vertebrates, with emphasis on critical life stages for reproduction and development. First, we discuss "external processes"—those pathways that deliver chemicals to the organism. Then we discuss "internal processes"—the partitioning, redistribution, and disposition of chemicals within the organism, including maternal transfer. Finally, we discuss modeling approaches to xenobiotics in oviparous species. While fundamental processes influencing chemical speciation and bioavailability apply to all chemical classes, we have based our discussions mainly on persistent organic chemicals and metals. Similarly, chemical exposure, uptake, and disposition is discussed predominately for fish and birds. Such an emphasis is not a reflection of perceived importance of these groups; rather, it reflects the existing literature base and the limitations dictated by the format of the workshop from which this chapter arises. By design this chapter is chemically oriented, leaving toxicological considerations to later discussions in this volume.

External Factors Controlling the Magnitude and Variability of Exposure

The challenge of quantifying exposure of a wide variety of chemicals

The extremely wide variety of chemicals produced and used in many different applications compromises our ability to generalize about the processes controlling their exposure to oviparous organisms. To illustrate this variability here, we review the production, use, and physical properties of several classes of environmentally relevant chemicals.

Types and quantities of releases to the environment

In order for a chemical to exert a significant risk to a population, it must be present at the receptor site at high enough levels and for a long enough duration to cause an effect. Therefore, one way to rank the relative "threat" of various chemicals is to quantify 3 independent parameters: production, persistence, and potency. "Production" is defined here as the gross rate at which the chemical is manufactured, modified by the fraction of production that is released into the environment. Production of chemicals of environmental interest range from those produced in extremely small quantities (i.e., 2,3,7,8-tetrachlorodibenzodioxin [TCDD]) to those produced in extremely large quantities (petroleum products, surfactants, structural polymers). Likewise, the fraction of gross production released to the environment ranges from near zero (i.e., fuel grade plutonium) to 100% (agrochemicals). Production and release rates, however, are insufficient to assess potential risk, as chemicals released into the environment vary considerably in their persistence. Only those chemicals that resist abiotic and biotic degradation processes and are mobile can be transported to potentially vulnerable populations. Again, the wide variety of chemicals that have been hypothesized to cause harm to oviparous organisms have an incredible range of environmental half-lives, ranging from less than 1 day to decades or centuries. Finally, one must consider the potency of the chemical on the target receptor species. While a detailed discussion of the large number of possible "effects" mechanisms is beyond the scope of this chapter, suffice it to say that chemicals display a very wide range of modes of action and of potencies. The large variety in production, persistence, and potency across the chemicals, which may impact oviparous organisms, is evidence that one cannot generalize the environmental behaviors and exposure pathways of these chemicals. Developing quantitative models with suitable plasticity to describe such widely varying behavior is a major challenge.

Uses, properties, and environmental persistence of the chemicals

A wide range of chemicals, both anthropogenic and naturally occurring, have the potential to cause reproductive and developmental effects. In order to address the questions about both the external and internal exposure concentrations to these chemicals, we first need to know what chemicals we are dealing with, their physico-chemical properties, chemical structures, rates of degradation in the environment and elimination rates in biota. We need similar information for possible biologically active degradation products. Table 2-1 contains data information for assessment of external and internal doses for 42 chemicals. These chemicals are of potential toxicological importance for reproduction and development as determined by recent reviews (Soto et al. 1995; GEA 1996; USEPA 1997) and papers contained within this volume. This list, which is by no means comprehensive, contains chemicals that are very diverse in their properties, and hence the potential sinks of active compounds range widely. Chemicals in Table 2-1 range from relatively hydrophilic (log K_{ow} < 1: amitrole) to hydrophobic (log K_{ow} = 6.8: TCDD). Structurally these chemicals range from low-molecular-weight, cyclic aliphatics (e.g., amitrole, MW = 84 Daltons) to oligomers (e.g., nonylphenol polyethoxylates). The properties of selected chemicals, important to show the breadth of chemicals involved or important to subsequent topics, are discussed in further detail in the following text. By overlaying the compounds' fate and exposure profile with known reproductive and developmental endpoints, an assessor may begin to determine potential exposure to various life stages of organisms.

Persistent organochlorines

Persistent organochlorines include many subclasses of compounds; however, the overall environmental fate characteristics are similar. Basically the fate characteristics of organochlorines are predominantly described by the persistence of the parent or parent-like compound and by their hydrophobic tendencies. Because of the ubiquitousness of the individual organochlorines throughout terrestrial and aquatic

Table 2-1 Physicochemical properties and chemical fate data for chemicals with reputed reproductive or developmental effects

Compound	Class	Use[a]	Known effects	Quantity used annually in U.S. (T)[b]	MW	Log K_{ow}[c]	K_{oc}[d] ($\times 10^2$)	pKa	Soil/sediment[e] DT50 (days)	Water/sewage[f] DT50 (days)	Microbial[g] biodegradability	Biologically active deg'n product?	Reference
Trifluralin	Dinitroaniline	H	mitosis inhibitor	$9–11\times10^3$	335.3	3.97	3.8	Neutral	60 (60–132)	< 1 (photolysis)	+++	Aniline dealkylation pdt	1,2,3
Atrazine	Chlorotriazine	H	—	$28–30\times10^3$	215.7	2.5	0.15	1.7	60(45–119)	> 100	+	De-ethyl atrazine	1,2,3
Amitrol	Aminotriazole	H	thyroid tumors	—	84.1	<1	0.1	Neutral	14(4–23)	~70	++		1,2
Benomyl	Carbamate	F	tetratogenicity	4.8×10^2	290.4	1.1	1.9	Neutral	67(10–356)	< 1	+++	5-OH-benzimidazole carbamate	1,2,3
Iprodione	Carboxamide	F	developmental toxicity at high doses	4.3×10^2	330.2	3	0.7	Neutral	14(<7–160)	?	+++	Similar to vinclozolin	1,3
Mancozeb	Ethylene bisdithio-carbamate	F	thyroid tumors	$1.8–3.2\times10^3$	polymer	?	> 2	Neutral	70(7–139)	1–2	+++	Ethylene thiourea	1,3
Ethylene thiourea	Thiourea	F	thyroid tumors	—	102	-0.66	-1.0	Neutral	14(7–28)	?	++		1
Metiram	Dithiocarb-amate-Zn-dithione	F	thyroid tumors	4.2×10^2	1089	2	500	Neutral	20	?	+++		1,3
Tributyl tin oxide	Organo-metallic	F	imposex	—	289.7	3.8	600	pKOH ~7	> 300	7(2–14)	+	mono-, dibutyltins	4

Table 2-1 continued

Compound	Class	Use[a]	Known effects	Quantity used annually in U.S. (T)[b]	MW	Log K_{ow}[c]	K_{oc}^d (×10⁴)	pKa	Soil/sediment [e]DT50 (days)	Water/ sewage[f] DT50 (days)	Microbial[g] biodegradability	Biologically active deg'n product?	Reference
Vinclozolin	Dicarboximide	F	deg'n pdt antiandrogenic	54	286.1	3	0.1	Neutral	20(3–75)	?	+++	Hydrolysis deg'n pdt	3
Diflubenzuron	Benzamide	I	chitin growth inhibitor	35	310.7	3.89	10	Neutral	10(3–60)	?	++++	OH-metabolite	1,3
Azadirachtin	Tetranortriterpenoid	I	ecdysone blocker	—	720.7	1.09		Neutral	<10	<10	++++		5
Fenoxycarb	Phenoxycarbamate	I	molt inhibitor	—	301.3	4.07	1	Neutral	1(1–31)	?	++++	OH-metabolite	1,3
Carbaryl	Naphthylcarbamate	I	AchE inhibitor	0.9–1.8×10³	201.2	2.36	0.38	Neutral	10(4–22)	12–30	+++	1-naphthol & OH-metabolites	1,3
Methomyl	Thioacetimidate	I	AchE inhibitor	7.8×10²	162.2	0.2	0.16	Neutral	30(8–45)	<30?	++		1,3
Parathion	Nitrophenyl phosphorothioate	I	AchE inhibitor	1.8–3.2×10³	291.3	3.9	1.5–15	Neutral	14(7–35)	<7	++	p-nitrophenol, aminoparathion	1,3

Table 2-1 continued

Compound	Class	Use[a]	Known effects	Quantity used annually in U.S. (T)[b]	MW	Log K_{ow}[c]	K_{oc}[d] ($\times 10^3$)	pKa	Soil/sediment[e] DT50 (days)	Water/sewage[f] DT50 (days)	Microbial[g] biodegradability	Biologically active deg'n product?	Reference
Dicofol	Chlorophenyl trichloroethanol	I	weak estrogen	4.9×10^2	370.5	4.7	5	Neutral	45(40–50)	15–93	++	Dichloro-benzophenone	1,3
Dieldrin/aldrin	Chlorinated cyclic aliphatic	I,M	weak estrogen	NP	380.9	5.4	7.4–12	Neutral	>1000	>10	–		1,5
Endosulfan	Chlorinated cyclic aliphatic	I	weak estrogen	8.0×10^2	406.9	3.8	2.9–6.8	Neutral	50(10–200)	<7	+	Endosulfan sulfate	1,3
Toxaphene	Chlorinated bornane/camphene	I	weak estrogen	NP	414	6.4	210	Neutral	>500	>500	–	Dechlorination	1,5
Methoxychlor	Methoxy-phenol trichloroethane	I	weak estrogen	40	345.7	3.9	80–100	Neutral	120(7–210)	~46	+	O-dealkylation to phenols & OH-	1,3
p,p'-DDE	Chlorophenyl dichloroethane	M	anti-androgen	NP	318	5.7	200	Neutral	>1000		–	deg'n pdt	1,5

Table 2-1 continued

Compound	Class	Use[a]	Known effects	Quantity used annually in U.S. (T)[b]	MW	Log K_{ow}[c]	K_{oc}^{d} (×10³)	pKa	Soil/sediment [e]DT50 (days)	Water/sewage[f]DT50 (days)	Microbial[g] biodegradability	Biologically active deg'n product?	Reference
o,p'-DDT	Chlorophenyl dichloroethane	I,B	weak estrogen	NP	354.5	6	410	Neutral	>1000		+	o,p'-DDE	1,5
Chlordecone	Chlorinated cyclic aliphatic	I	weak estrogen	NP	490.7	4.5	13	Neutral	>1000	>5	-		1,5
Tetrachloro-biphenyls	Chlorinated biphenyl	FR,P	adreno-corticoid, thyroid & estrogenic effects	NP	292	5.6–6.5	163–1300	Neutral	>1000	500–900	-		6
Trichloro biphenyls	Chlorinated biphenyl	FR,P	adreno-corticoid, thyroid & estrogenic effects	NP	257.5	5.5–5.9	130–330	Neutral	>1000	500–900	-		6
Hydroxy-tetrachloro biphenyls	Chlorinated biphenylol	M	weak estrogens, thyroid hormone mimics	NP	308	4.9–5.8	33–260	?	?	?	+++	deg'n pdt in fish, birds, and mammals	
Hydroxy-trichloro Biphenyls	Chlorinated biphenylol	M	weak estrogens, thyroid hormone mimics	NP	274.5	4.8–5.2	26–65	?	?	?	+++	deg'n pdt in fish, birds, and mammals	
2,3,7,8-TCDD	Chlorinated dibenzo-p-dioxin	B	adreno-corticoid, thyroid & estrogenic effects	NP	322	6.8	>100	Neutral	>1000	<2	-		7

Table 2-1 continued

Compound	Class	Use[a]	Known effects	Quantity used annually in U.S. (T)[b]	MW	Log K_{ow}[c]	K_{oc}[d] (×10³)	pKa	Soil/sediment[e] DT50 (days)	Water/sewage[f] DT50 (days)	Microbial[g] biodegradability	Biologically active deg'n product?	Reference
p-tertbutyl hydroxy anisole	Phenolic	A	weak estrogen	—	164	3.8	2.6	Neutral				Butyl phenol	8
p-tertbutyl-phenol	Phenolic	M	weak estrogen	—	150	3.3	0.8	?					8,9
p-nonylphenol	Phenolic	S,M	weak estrogen	—	220	4.5, 5.8	13, 260	?	>7	<7	++		9, 10
p-octyl phenol	Phenolic	S,M	weak estrogen	—	206	5.4	100	?	>7	<7	++	Nonyl phenol	9
Nonylphenyl ethoxylate (EO=9)	Ethoxylate oligomer	S	weak estrogen	>242×10³	598	−5.9, −7.2	330, 6500	Neutral	>7	<7	+++	Nonyl phenol	
Nonylphenol ethoxylate dimer (EO=2)	Ethoxylate dimer	S	weak estrogen	—	304	4.4, 5.6	100, 163	Neutral	>7	'7–21	++	Nonyl phenol	10, 11
Nonylphenol carboxylate	NPE deg'n pdt	M	weak estrogen	—	256	3.5	130	?		<7	++		
Bis-phenol A	Hydroxy diphenyl-propane	M	weak estrogen	—	228.3	3.32	0.08	?	>300	<4(<1–28)	+	isopropyl phenol	9

Table 2-1 continued

Compound	Class	Use[a]	Known effects	Quantity used annually in U.S. (T)[b]	MW	Log K_{ow}[c]	K_{oc}^d (×10³)	pKa	Soil/sediment[e] DT50 (days)	Water/sewage[f] DT50 (days)	Microbial[g] biodegradability	Biologically active deg'n product?	Reference
4-hydroxy-biphenyl	Phenolic	S,M	weak estrogen	NP	206	3.2	0.65	?	1-7	1-7	+++		9, 12
Dibutyl phthalate	Phthalate ester	P	weak estrogen	—	278.3	4.72	6.8	Neutral	~180	<5(2-12)	+++	Phthalic acid	8, 12
Butylbenzyl phthalate	Phthalate ester	P	weak estrogen	>50×10³	312.4	4.91	0.35	Neutral	?	<2<1-7	++++	phthalic acid, benzyl alcohol	8, 12
Ethynylestra-diol	Phenol	SH	synthetic estrogen	—	296	3.67	1.9	?	?	>5	+++		9, 13
Beta-sitosterol	Phenol	PY	phyto-estrogen	—	414.7	>5	>40	?	?				9

[a]H=herbicide; F=fungicide; I=insecticide; M=metabolite; B=by product; FR=flame retardant; P=plasticizer; A=antioxidant; S=surfactant; SH=synthetic hormone; PY=phytoestrogen

[b]Dash indicates no annual production figures published in the open literature. Pesticide use information from USGS pesticide National Synthesis Project (http://water.wr.usgs.gov/pnsp/use 92). Represents approximate quantities annually used in period. NP = not deliberately produced or no longer produced in U.S. or western Europe, but residues remain in soils and sediments.

[c]K_{ow} from Howard 1991; Howard et al. 1991; Hansch et al. 1995; or calculated from fragment constants using structurally related compounds (e.g., OH-PCBs from PCBs).

[d]K_{oc} calculated from $0.41 \ast K_{ow}$ (Karickhoff 1981) when measured values not available.

[e]"Soil/sediment DT50" = 50% disappearance time in field studies (pesticides) or soil/sediment biodegradation tests (industrial chemicals).

[f]"Water/sewage DT50" =50% disappearance time (if available) or range of reported results for water column in field or lab studies (pesticides) or sewage sludge incubations (mainly industrial compounds).

[g]Microbial degradability: qualitative ranking from non-degradable (−) to highly degradable (++++)

References: 1) Wauchope et al. 1992; 2) Howard 1991; 3) Tomlin 1994; 4) Stewart and Mora 1990; 5) Thompson 1992; 6) Augustijn-Beckers et al. 1994; 7) Mackay et al. 1992a; 8) Mackay et al. 1992b; 9) Howard et al. 1991; 10) Hansch et al. 1995; 11) Ahel et al. 1994; 12) Kvestak and Ahel 1995 13) Budavari et al. 1996 environments, the organism appears to be subjected to assault throughout all life stages.

Polychlorinated biphenyls (PCBs): The physical properties of selected chlorobiphenyl homologues are given in Table 2-1. Most polychlorinated biphenyl (PCB) congeners, particularly those lacking adjacent unsubstituted positions on the biphenyl rings (e.g., 2,4,5-, 2,3,5-, or 2,3,6- substituted on both rings), are extremely persistent in the environment and are essentially non-biodegradable in aerobic soils or sediments (Mackay et al. 1992a). Highly chlorinated PCBs have been shown to be dechlorinated in anaerobic sediments, but only where present at relatively high concentrations (> 10 mg/g dry weight) (Brown and Bedard 1987; Rhee and Sokol 1993). PCBs also have extremely long half-lives in adult fishes. For example, an 8-year study of eels found that the half-life of PCB153 was > 10 years (de Boer et al. 1994). A large survey of freshwater fishes in U.S. rivers and lakes found PCBs were the most prominent organochlorine contaminants, with median concentrations of 209 ng/g wet weight (USEPA 1991). Penta- and tetrachlorobiphenyls were the predominant homologue groups, with median concentrations of 72 and 23 ng/g wet weight, respectively.

Polychlorinated dibenzo-*p*-dioxins and dibenzofurans: While polychlorinated dibenzo-*p*-dioxins and dibenzofurans (PCDD/Fs) are rapidly photodegraded in air, in water, and on surfaces (Buser 1988), they are extremely hydrophobic and resistant to biodegradation in soils and sediments. Historical profiles of PCDD/Fs in sediment cores from large lakes show no evidence of transformation of congeners (such as anaerobic dechlorination) over time (Hites 1990). Dechlorination appears to be a major route of degradation of 2,3,7,8-tetrachlorodibenzodioxin (TCDD) and -TCDF in sunlight in natural waters (Dung and O'Keefe 1994). The 2,3,7,8-substituted PCDD/F congeners are known to bioaccumulate in fish and invertebrates; however, non-2,3,7,8-substituted congeners (which predominate in combustion sources) are readily degraded by vertebrates (Opperhuizen and Sijm 1990). Data on the bioavailability of TCDD from sediments and water, as well as on the pharmacokinetics in fish, are available. TCDD has a relatively long half-life in adult fishes (~ 1 year) (Kuehl et al. 1989). Surveys of PCDD/Fs in fish collected in 1986 from 388 locations in the U.S. (freshwater only) found 2,3,7,8-TCDD was detectable at 70% of all locations at median concentrations of 1.4 pg/g wet weight with maximum concentrations of 204 pg/g (USEPA 1991). Current levels may be lower because highest TCDD and 2,3,7,8-TCDF levels were found in fish near pulp and paper mills using chlorine; since then, these mills have substantially reduced TCDD emissions. The USEPA surveys found 2,3,4,7,8-pentachlorodibenzofuran (PnCDF), 1,2,3,6,7,8-hexachlorodibenzodioxin (HxCDD) and 1,2,3,4,6,7,8-heptachlorodibenzodioxin (HpCDD) were more prominent in fish tissue than was 2,3,7,8-TCDD at non-paper mill sites (median concentrations in the low pg/g wet weight range).

Toxaphene: This complex mixture of polychlorobornanes and camphenes was widely used in the U.S. on cotton crops. Toxaphene is produced by the chlorination of technical camphene or α-pinene and can consist of over 300 congeners, mainly bornanes and camphenes substituted with 6 to 10 chlorines, with an average composition of $C_{10}H_{10}Cl_8$. Determining the environmental fate of a mixture of compounds such as toxaphene is difficult, as each structurally different compound in the mixture will have a specific set of chemical properties.

Use of toxaphene peaked between 1972 and 1975. Manufacturing was banned in the U.S. in 1982 and use ceased in 1986 (Voldner and Li 1993). Similar products have been, and may continue to be, used in Mexico, Central America, eastern Europe, and the former Soviet Union. Toxaphene is extremely persistent in soils following pest control application, with half-lives ranging from 1 to 14 years (Howard et al. 1991) (Table 2-1). Losses from soil are mainly through volatilization and runoff (Glotfelty and Taylor 1984). Toxaphene degrades mainly through dechlorination in sediments and its dechlorination products are bioavailable in lakes treated with toxaphene 30 years later (Miskimmin et al. 1995). A half-life for toxaphene of 63 days was reported for juvenile lake trout (Mayer et al. 1977) and up to 1 year in injected adult fishes (Delorme et al. 1993). Glassmeyer et al. (1997) found mean concentrations of toxaphene in lake trout in the 5 Great Lakes ranging from 140 to 3500 ng/g wet weight. Lowest concentrations were found in lake trout from Lake Erie and highest in samples from Lake Superior.

DDT:-1,1,1-trichloro-2,2-Bis (p-chlorophenyl) ethane: The technical product consists of 4,4'-DDT (or p, p'-substituted) and its *o,p*-DDT isomer, as well as their dechlorinated analogs [*p,p*'- and *o,p*'- 1,1-dichloro-2,2'-bis (p-chlorophenyl) ethane (DDD)]. Its use has been restricted in Canada, the U.S., and western Europe for nearly 2 decades; however, it is used in pest control programs in southern Asia, Africa, Mexico, and Central and South America (Voldner and Ellenton 1987) and may be used in China and Russia. DDT, especially its metabolite 1,1-dichloro-2,2' bis (p-chloro-phenyl) ethylene (*p,p*'-DDE), is extremely persistent in soils and sediments and has a long half-life in biota. *p,p*'-DDE is probably the most common individual organochlorine contaminant in aquatic and terrestrial biota, and it can be present at low mg/g wet weight levels in tissues of top predators such as fish-eating birds. A survey of fishes from 388 locations in the U.S. found *p,p*'-DDE in 98% of all samples at median concentrations of 58 ng/g wet weight with maximum levels of 14 mg/g wet weight (USEPA 1991). Levels of DDT and its principal metabolite, DDE, have decreased in fish and wildlife of western Europe, North America, and Japan in the past 15 years due to bans on use.

Current use pesticides

Carbamate/thiocarbamate pesticides: Table 2-1 includes a structurally diverse group of insecticides and fungicides reported to have developmental effects in invertebrates, fish and birds (USEPA 1997). Several are designed as inhibitors of various stages of insect development, e.g., fenoxycarb, diflubenzuron, azadirachtin. All are carbamates, substituted ureas, or thiocarbamates and therefore are readily hydrolyzed chemically or microbially in soils and in vivo in organisms. All of these compounds are expected to have short half-lives in fish and other aquatic organisms and therefore would not biomagnify in food chains of piscivorous fish or birds. Their relatively low log K_{ow} (1 to 4) suggest they would be accumulated mainly from water. For example, diflubenzuron has been found to have a half-life of < 2 d in bullheads and < 1 d in sunfish (Niimi 1987). Direct exposure of oviparous organisms to these relatively nonpersistent chemicals could occur through spray drift or runoff from treated fields soon after application. Thus quantities used and type of application (e.g., aerial versus ground rig) may be critical aspects of exposure.

Tributyltin: Tributyltin (TBT) is a broad-spectrum algicide, miticide, fungicide, and insecticide (Stewart and de Mora 1990). TBT and other organotin compounds were first used in agriculture; subsequently, TBT has had wide application as a marine antifoulant starting in the 1960s. Its most important entry route to the sea is directly from ships, aquaculture pens, moorings, and industrial cooling pipes to which products containing it have been applied. It may also enter the sea in runoff from agricultural areas, from boat repair yards, and through municipal wastewater and sewage sludge. TBT is found to provide effective protection for boat hulls at release rates < 4 mg/cm^2 day and has been a popular antifoulant because it maintains its efficacy for up to 5 years, compared to about 3 years for other conventional applications. Once released to the water, TBT is degraded by sequential debutylation to dibutyltin (DBT), monobutyltin (MBT), and eventually to relatively nontoxic inorganic tin compounds. The degradation time in water is short, with half-lives reported from days to a few weeks (Stewart and de Mora 1990; Dowson et al. 1993). TBT is strongly particle-reactive, with partition coefficients reported to be as high as 10^3 to 10^4 (Langston and Pope 1995). The breakdown of TBT in anaerobic sediments is much slower than that in water (Clark EA et al. 1988). Therefore, contaminated sediments are potentially an important environmental reservoir for TBT that can continue to provide a source long after the industrial use of TBT has been curtailed. TBT is moderately lipophilic and will, therefore, bioaccumulate in the marine environment.

Surfactants

Alkyl phenol ethoxylates (APEs): These surfactants (usually nonylphenol ethoxylate or octylphenol ethoxylate) are used in industrial detergents, such as those used for wool washing and metal finishing; domestic detergents, such as clothes washing liquids; some shampoos, shaving foams, and other cosmetics; laboratory detergents, including Triton X-100; and pesticide formulations. APEs are being phased out by the European Union countries to be replaced by alcohol ethyoxylates, but in North America APEs continue to be used, especially in liquid detergents.

When alkyl phenol ethoxylates break down in sewage treatment or a river, they produce 3 main groups of alkyl phenolic compounds: alkyl phenol ethoxylates with fewer ethoxylate groups, alkyl phenoxy carboxylic acids, and alkyl phenols. Studies in Switzerland have shown that these compounds persist in rivers and their sediments and in groundwater (e.g., Ahel et al. 1994, 1996). Nonylphenol di-ethoxylate (NP2EO) was found to be the most persistent degradation product (Ahel et al. 1994). Concentrations of NP2EO ranged from 2 to 8 mg/L in the River Glatt in Switzerland (Ahel et al. 1994) downstream of Zurich. Nonylphenol ethoxycarboxylate, the carboxylate analog of NP2EO, ranged in concentration from < 0.4 to 11.8 mg/L in the Fox River in Wisconsin, with highest concentrations downstream of pulp mills and municipal sewage treatment plants (Field and Reed 1996).

Alkyl phenols: Alkyl phenols (usually nonylphenol or octylphenol) are industrial products as well as degradation products of APEs. Alkyl phenols are used as antioxidants in some clear plastics, to prevent yellowing, in the form of tris-nonylphenol phosphite. They also are formed by degradation of triaryl phosphate lubricants (e.g., t-butylphenol, diphenyl phosphate) (Muir 1984). Concentrations of nonylphenol in receiving waters in the River Glatt in Switzerland, an industrial area, were found to be 2 to 4 mg/L (Ahel et al. 1994). Similar concentrations of nonylphenol (NP) have been reported in U.S. rivers (Naylor et al. 1992). Nonylphenol was rapidly accumulated from water by rainbow trout and eliminated with a half-life of 19 h (Lewis and Lech 1996). McLeese et al. (1981) determined a half-life of 4 d in salmon. They also found that depuration half-lives of alkyl phenols varied with chain branching and length. Longest half-lives in salmon were found for dodecyl phenol and shortest for p-sec-butylphenol. Analyses of fish from a river in Switzerland, near municipal and industrial sources of NP and nonylphenol ethoxylates (NP1EO), showed that NP, NP1EO and NP2EO were present in algae and fish. Highest concentrations of all 3 compounds were found in the gut, liver, and gill tissue; lowest concentrations were found in muscle. Bioconcentration factors (BCFs), (wet weight concentration in muscle/water) ranged from 50 to 100 for NP and from 5 to 250 for NP2EO (Ahel et al. 1993). McLeese et al. (1981) reported an equilibrium BCF of 280 for NP in salmon in flow-through experiments.

Modeling fate and transport and speciation of chemicals in the environment

Oviparous organisms include both aquatic and terrestrial representatives. Depending on the specific species and life stage, direct exposure to chemicals in primary compartments (i.e., air, water, soil, sediment) may occur through inhalation, ingestion, or dermal routes. Indirect exposure to concentrations in secondary compartments (i.e., plants, biota) may also occur via the dietary route. Table 2-2 illustrates the various environmental compartments and potential exposure routes to be considered. Maternal transfer to the egg represents a unique exposure pathway for oviparous organisms from development to hatching.

Table 2-2 Summary of potential exposure pathways

Exposure route	Environment compartment	Inhalation route	Ingestion route	Dermal route
Direct	Water (dissolved)	A	A*,T	A
	Water (particulate)	X	A,T	X
	Air (vapor)	A,T	X	X
	Air (aerosol)	A,T	X	X
	Soil (particulate)	X	T	T
	Sediment (particulate)	X	A	A
Indirect	Plant tissue	X	A,T	X
	Animal tissue	X	A,T	X

A = Aquatic, marine and estuarine species
T = Terrestrial species
X = Assumed to be insignificant pathway
*Ingestion of water may be a potential exposure route for andromanous species

Exposure to primary compartment concentrations

Primary compartment concentrations depend upon

- the nature of emissions, i.e., both the amount released and the release scenario
- the characteristics of the environment (e.g., advective flows, compartment sizes, and organic carbon content);
- the physicochemical properties of the contaminant that determine partitioning behavior within and between compartments; and
- the chemical/compartment-specific degradation half-lives (Mackay 1991). Emission estimates often are characterized poorly, although crude estimates for regional scale assessments may be obtained based on production volume, physicochemical properties, and use patterns of the chemical (van der Poel et al. 1995).

A generic "unit world" can be used to define environmental properties or, alternatively, site-specific information can be used if appropriate. The key physicochemical properties dictating the multimedia distribution of organic chemicals are the air-water (K_{aw}), octanol-water (K_{ow}), and octanol-air (K_{oa}) partition coefficients. Research over the last few decades has provided considerable experimental data and quantitative structure property relationships (QSPRs) for determining the K_{aw} and K_{ow} for many chemical classes. In contrast, only limited information is available on K_{oa}, although recent work for predicting this property is promising (Finizio et al. 1997). Abiotic degradation rates in air and water can be determined experimentally or estimated using QSPRs (Karickhoff et al. 1991; Kwok and Atkinson 1995; Lyman et al. 1995; Meylan and Howard 1995). While standardized protocols for determining biodegradation are available (OECD 1992) and a number of biodegradation QSPRs have been published (Degnen et al. 1993; Boethling et al. 1994; Klopman et al. 1995), a serious limitation in applying multimedia fate models is the difficulty in parameterizing compartment-specific biodegradation rates (Boethling et al. 1995;

Hales et al. 1996; Federle et al. 1997). The stochastic nature of the factors governing primary exposure concentrations is also significant. For example, the variation of emissions in time and space and the temperature dependence of partition coefficients and degradation half-lives are expected to cause variability in exposure concentrations.

Exposure to secondary compartment concentrations

A number of models have been used to predict concentrations in secondary compartments (i.e., biota) from measured or estimated primary compartment concentrations. These models range in complexity from simple QSPRs that estimate bioaccumulation based on physicochemical properties such as K_{ow} or vapor pressure (e.g., Mackay 1982; Garten and Trabalka 1983; Travis and Arms 1988; Bintein et al. 1993; McKone 1993; Tolls and McLachlan 1994; Kraaij and Connell 1997) to mechanistic models that integrate such factors as organism bioenergetics, life history, food-chain structure, and contaminant-specific toxicokinetic information (e.g., Thomann 1989; Clark T et al. 1988; Fordham and Reagan 1991; Thomann et al. 1992; Gobas 1993). A major drawback of these models is that most of them have been developed based on data from a very limited class of contaminants (e.g., poorly metabolized compounds). Consequently, these models cannot be extrapolated generically to many chemical classes. In order to avoid misapplication of these tools in exposure assessments, such limitations must be recognized (Tell and Parkerton 1997). Future work is needed to extend such model frameworks to other contaminant classes. This effort should include an explicit description of biotransformation processes in the model framework.

Use of modeling tools in exposure assessment

Despite the uncertainties and limitations of multimedia and bioaccumulation models, these tools provide important insights for assessing exposure to oviparous vertebrates. For new chemicals in which field measurements are not possible, this may be the only approach. In the case of existing chemicals, model results, if coupled with information on exposure factors (e.g., inhalation rates, ingestion rates, surface area for dermal uptake) for the receptor species of interest, can be used to prioritize which environmental exposure pathways are most important (i.e., what is the relative magnitude of the administered dose via different pathways?). This information can then be used to prioritize field monitoring efforts to characterize concentrations in the most critical environmental compartments. These calculations can also provide guidance in identifying the most relevant route of exposure to investigate in toxicity studies. For example, if the dominant pathway appears to be through the diet, priority should not be given to toxicity studies that examine exposure to the contaminant by air or water. An excellent compilation of exposure factors for wildlife that can be used for such calculations has been published by the USEPA (1993).

The above discussion provides a first step in understanding how the environmental fate properties of a contaminant influence potential exposure. However, if the ultimate objective is to link exposure estimates to a toxicological effect, the ability of the administered dose to be absorbed by the organism and then transported to the target site of action must be considered (Figure 2-1). While recognizing the important distinction between administered and absorbed dose, the extent to which such information can be considered may be constrained by the nature of the available toxicity information. If, for example, the only dose-response data that is available is

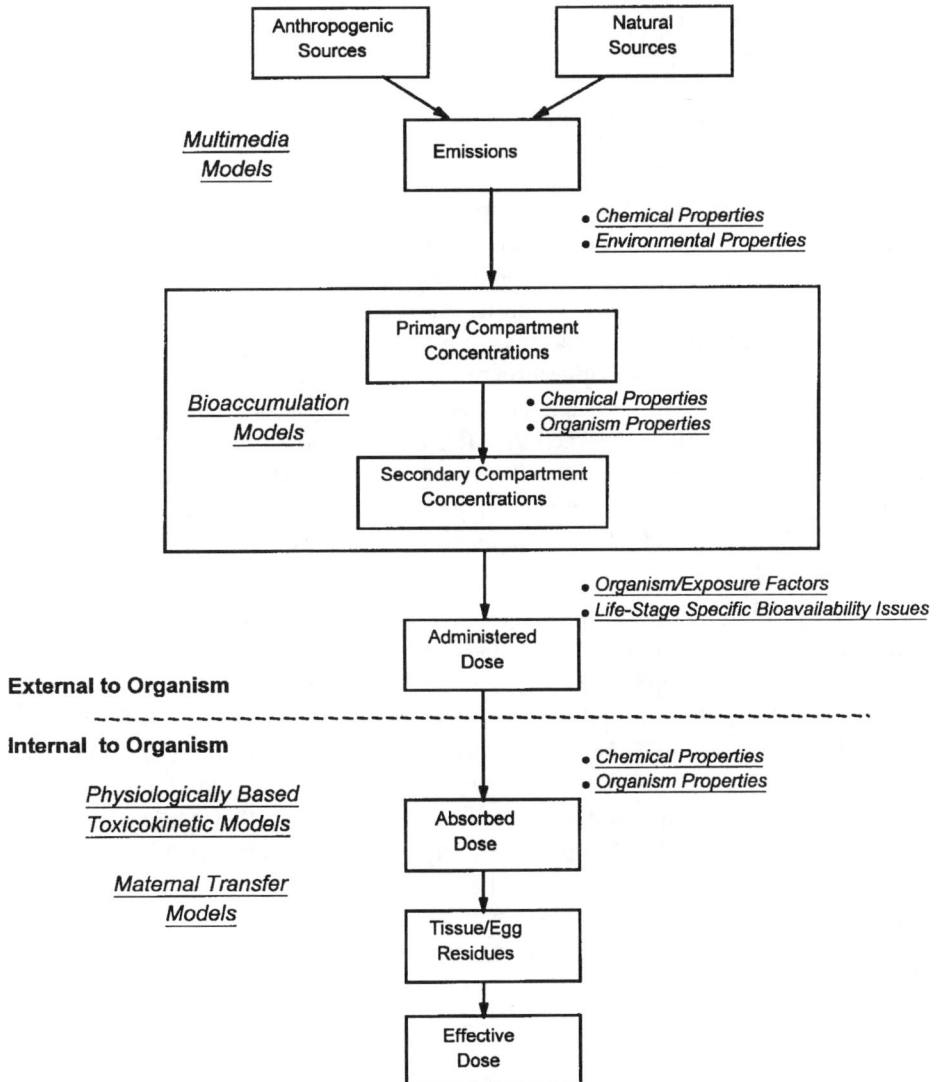

Figure 2-1 Key components influencing contaminant exposure

based on administered dose without additional information on tissue dosimetry it may not be possible to progress the exposure assessment beyond this point. Herein lies one of the main advantages of linking adverse effects of a contaminant to either whole body or specific target organ residues since concentrations in the organism provide an indirect measure of the absorbed dose. While a clear linkage between residue concentrations and reproductive or developmental effects has been shown for certain contaminants (e.g., DDT, organometalics, dioxin equivalents) such relationships may not exist for many other classes of contaminants particularly if they are readily metabolized. Conclusions reached in Ingersoll et al. (1997) also identified that the toxicological relevance of tissue residue data is an important research priority.

Quantifying external exposure

Defining bioavailability

The ability of a contaminant to be transported from the external environment to the target site within the organism depends upon: 1) the form (e.g., particulate, complexed, or freely dissolved) and chemical properties of the contaminant; 2) the characteristics of the organism that influences the efficiency by which the contaminant is extracted from the external environment (e.g., surface areas of respiratory, skin, and gut; ventilation and ingestion rates; enzymatic activities) via a given exposure route; and 3) the toxicokinetic behavior of the contaminant within the organism that determines the fraction of the absorbed dose that actually reaches the target site (i.e., the effective dose). Differences in contaminant toxicodynamics between species often form the mechanistic basis for explaining differences in toxicological sensitivity and thus extrapolation between species (Lawrence and Gobas 1997).

Confusion in the use of the term "bioavailability" arises as a result of differences in opinion regarding which of the above elements should be included in the definition (Dickson et al. 1994; Peijenburg et al. 1997). From an environmental chemist's perspective, the term bioavailability is usually limited to chemical speciation external to the organism. For example, truly dissolved concentrations of a nonpolar organic chemical are assumed to be bioavailable, while contaminant bound to dissolved organic carbon is not bioavailable (Landrum et al. 1985; McCarthy and Jiminez 1985; Black and McCarthy 1988). Thus, according to this view, the bioavailability of a chemical is independent of the organism. In contrast, ecotoxicologists and pharmacologists usually define bioavailability in terms of an absorbed dose so that the effect of the contaminant on the organism becomes implicit in the definition. For example, if the absorption efficiency of a chemical is 80% for a clam but only 60% for a trout, according to the latter definition, bioavailability in the clam is higher than in the trout.

Since the former definition limits bioavailability to chemical speciation, this restriction is problematic when applied to dietary contaminants. For example, hydrophobic organic contaminants will be primarily associated with the lipids in food and organic carbon in soils and sediments. Because the external chemical speciation is similar, the bioavailability of contaminants from different diets and soils/sediments is expected to be similar. However, experimental evidence contradicts this simple view (Vetter 1983; Landrum et al. 1992; Parkerton 1993; Alexander 1995). If instead bioavailability is defined in terms of an absorbed dose, a quantitative estimate of bioavailability can, in principle, be experimentally determined for different diets or soils/sediments. Moreover, differences in the bioavailability of various exposure routes can be directly compared.

Approaches for assessing bioavailability

Previous research indicates that observed toxicity to aquatic organisms correlates to the freely dissolved rather than total chemical concentrations. As a result, analytical methods or predictive models that enable freely dissolved concentrations to be estimated often have been used as a first step in accounting for contaminant bioavailability. Three commonly used techniques to determine freely dissolved concentrations are gas-sparging, equilibrium dialysis with semipermeable membrane devices (SPMDs), and filtration in conjunction with solid-phase extraction (Gustafson and Dickhut 1997). Equilibrium partitioning models have been widely used to account for the differences in bioavailability of sediment and soil contaminants (DiToro et al. 1991; van Leeuwen et al. 1992; Belfroid 1996; Ankley et al. 1996). While this approach is pragmatic, such models may not accurately represent bioavailability in the field (Ronday et al. 1997). Numerous studies examining contaminant desorption from sediments/soils indicate a biphasic behavior that consists of an easily desorbable and more resistant fraction (Pignatello and Xing 1995). The relationship between contaminant sequestration in soils/sediments and bioavailability has enormous implications for regulatory and remediation decisions and is consequently a major focus of current research (McGroddy et al. 1996; Loehr and Webster 1996; Kelsey and Alexander 1997; Gustafsson et al. 1997; Tomson and Pignatello 1997). In addition to sequestration mechanisms, other factors may limit the applicability of simple equilibrium partitioning theory (Belfroid et al. 1996; Peijenburg et al. 1997).

To provide a more accurate characterization of contaminant bioavailability, a variety of experimental approaches have been proposed. One commonly used technique is the use of a passive sampling device (SPMD containing lipid, solid-phase sorbent) that acts as a surrogate for organism lipid (Huckins et al. 1993; Corrol et al. 1994; Verhaar et al. 1994; Parkerton and Stone 1996; Lake et al. 1996; van Loon et al. 1997). To better account for organism-specific factors that may influence contaminant bioavailability, bioaccumulation and/or toxicity tests may be used. Such tests have been used for determining bioavailability of both organics (Harkey et al. 1995; McFarland 1995; Boesa et al. 1997; Meier et al. 1997; White et al. 1997; Kane-

Driscoll and Landrum 1997) and metals (Ankley et al. 1996; Ruby et al. 1996; van Gestel and Van Diepen 1997). Bioassays may be more realistic than biomimetic extraction approaches, but have the drawback of being significantly more cost- and time-intensive.

Effect of animal life history and behavior on chemical exposure

Allometric relationships controlling relative rates of diffusive and dietary uptake

The 2 main routes of exposure are diffusive uptake across gill and skin surfaces and dietary exposure via active feeding. While the latter is driven by contaminant levels in prey and bioenergetic-based food consumption rates, the former depends largely upon the diffusional gradient between the chemical in the organism's circulatory system and in the surrounding fluid (dissolved in water for aquatic organisms and in the gas phase for terrestrial organisms). Provided that chemical delivery to the surface is not limiting delivery of chemical via diffusive uptake scales to the concentration gradient and to the interfacial surface area, but the resulting concentration of chemical within the organism is determined by normalizing the delivered load to the mass of organism. Thus, the organism's surface-area-to-volume ratio is a direct index of the potential importance of diffusive uptake as an exposure pathway.

The surface-area-to-volume ratio of organisms that maintain reasonably constant shape during growth (i.e., finfish) decreases nearly exponentially, with dramatic decreases occurring in the period between egg hatching and growth to juvenile. Coupled with the lack of feeding during the early stages of larval development, this simple allometric relationship implies that diffusive uptake is likely to be the dominant exposure pathway to organisms with small embryonic and larval stages. Therefore, exposure of dissolved chemical via diffusive uptake may be an important route of exposure to sensitive life stages, even for those species whose adults are exposed primarily through their diets. This analysis cannot be extrapolated to those organisms, such as birds, who experience their exponential growth within the confines of an egg.

Changes in diet as animal matures

Animals may experience variations in dietary exposure of contaminants during their lifetimes because of shifts in their diets. Most fish begin with primarily planktonic or detrital diets and progress to larger prey items as they grow. As both contaminant levels and nutritional value of these various foods differ, exposure via diet varies. For example, concentrations of PCBs and polycyclic aromatic hydrocarbons (PAHs) are more than 10-fold enriched in macro-zooplankton relative to those of smaller plankton in the Chesapeake Bay (Ko and Baker 1995). Changes in diet with development can alter xenobiotic exposure. For example, white perch shift from a primarily phytoplanktonic diet to a mixed diet of zooplankton and benthos after 2 years of age. This shift has been shown to alter exposure pathways of Kepone to white perch age classes in the James River (Connolly and Tonnelli 1985). Kepone levels in phytoplankton reflect dissolved Kepone levels in the surface waters, which declined rapidly after discharge of Kepone to the river stopped. Dietary exposure of Kepone to young herbivorous white perch declined as well. In contrast, older white perch eating a mixed diet of zooplankton and benthos (primarily the polychaete *Nereis*) maintained higher levels. Presumably Kepone was concentrated in zooplankton and mobilized from the sediments through ingested benthos.

Varying exposure due to migration

Mobile animals may experience varying exposure as they move among areas with varying contaminant levels. Movements include both local meanderings on daily time scales and long-distance migrations. Recurring seasonal migrations, such as those of migratory waterfowl and anadromous fishes and single migrations to spawn by species such as salmon, may drive temporally varying exposure. Assessing exposure and the resulting internal disposition of the chemicals is complicated by the fact that migrating organisms are also generally undergoing major physiological stress during migrations, resulting in changes in lipid reservoirs and bioenergetics.

As an example of seasonal migration as a control on exposure variability in an anadromous fish population, Zlokovitz and Secor (1999) have recently employed otolith microchemistry techniques to relate PCB levels in Hudson River striped bass (*Morone saxatilis*) to their migration patterns. Using energy-dispersive, x-ray spectroscopy coupled with scanning electron microscopy, the strontium/calcium ratios of individual otolith annuli can be measured (Secor 1992). Based on laboratory-derived calibration curves between salinity and Sr/Ca rations, the lifetime habitat salinity and migration patterns of each individual can be recreated. Figure 2-2 shows the migration history of a 15-year-old female striped bass collected from the Hudson River. After a 4- to 5-year period of irregular movement in the mesohaline reaches of the river, this individual female undertook regular annual migrations from the coastal ocean into less saline water. Other individual striped bass undergo dramatic shifts in their selected habitat, quickly moving from the river to the coastal waters or vice versa. The 7-year-old male striped bass in Figure 2-3 spent the first half of its life in high salinity water before moving into the upper reaches of the tidal Hudson. Although this fish contains fairly high levels of PCBs in its tissue (10.4 mg/g wet weight), it seems likely that this level reflects exposure during the second half of its life. In contrast, the other male striped bass shown in Figure 2-3 lived in low salinity waters under elevated PCB exposure before moving into saline waters after 2 years. Although levels of PCBs in its tissue at the time of capture were relatively low (0.4 mg/g wet weight), the reconstructed salinity history suggest that this fish was exposed to higher levels of PCBs during its early life. Movement into and out of contaminated areas by parental stock or during early developmental stages may very well determine the success or apparent impact of xenobiotics upon local populations.

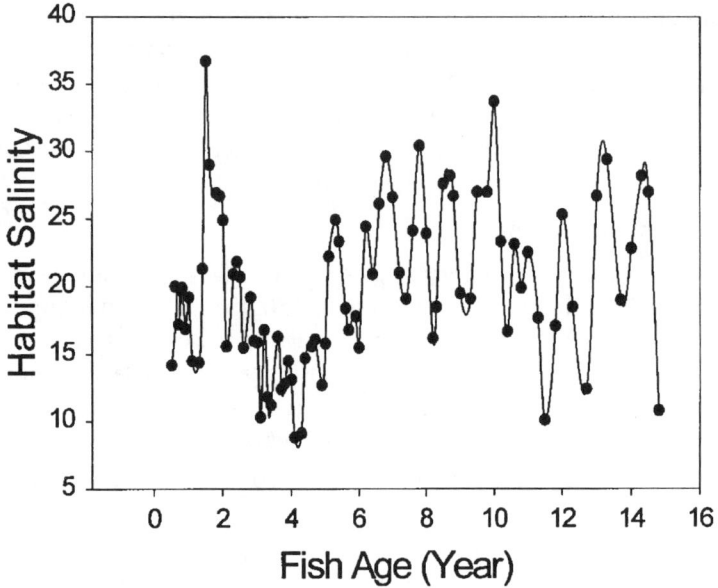

Figure 2-2 Reconstructed migration history of 15-year-old female Hudson River striped bass (Zlokovitz and Secor 1999)

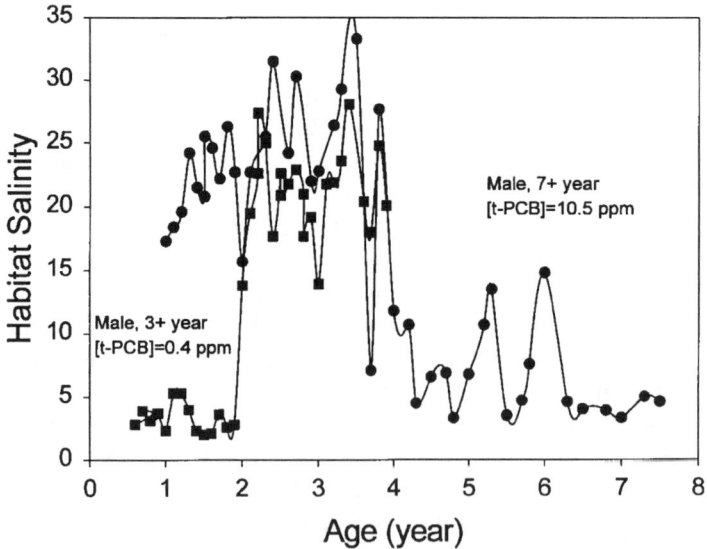

Figure 2-3 Contrasting migration behaviors of 2 male striped bass collected in the Hudson River. Total PCB concentrations reflect elevated exposure in freshwater regions of the river (Zlokovitz and Secor 1999).

Food-web structure and trophic transfer

Productivity and bioenergetics

Several recent studies have demonstrated that organisms living in more productive (e.g., more eutrophic) systems are exposed to lower levels of contaminants. In a study of 61 lakes in southern Scandinavia by Larsson et al. (1992), it was determined that levels of persistent pollutants (e.g., PCBs and DDT) in northern pike decreased as lake productivity increased, despite similar external pollutant loadings. They attributed this relationship either to the higher growth rate of the fish or to the decreased exposure resulting from enhanced sedimentation and complexation of the chemicals in the more eutrophic lakes. Using a model planktonic ecosystem, Millard et al. (1993) demonstrated that a larger fraction of added PCBs was bound by settling particles and colloids in more productive mesocosms. This model also showed that exposure to added PCBs decreased more rapidly in highly mixed, low-productivity treatments because of enhanced volatilization. In laboratory culture studies, Sijm et al. (1995) found that the bioconcentration of hydrophobic organic chemicals (HOCs) by 2 species of algae was inversely dependent upon algal density. They demonstrated an algal density-dependent production of organic exudates that bind HOCs and lower exposure levels. Paradoxically, evidence now suggests that relatively remote, oligotrophic water bodies with low external contaminant loadings (i.e., northern Great Lakes, Arctic Ocean) support elevated levels of contaminants in higher trophic levels (Tanabe et al. 1983; Muir et al. 1988; Norstrom et al. 1988; Swackhamer and Hites 1988; Hesselberg et al. 1990; Evans et al. 1991). These elevated tissue levels in water bodies with relatively low contaminant levels imply extremely efficient trophic transfer of particle-reactive chemicals in these oligotrophic systems.

Growth of aquatic organisms dilutes chemical concentrations within tissues even without net chemical exchange with the surrounding water (Thomann 1981). Swackhamer and Skoglund (1993) suggest that rapidly growing plankton fail to reach sorptive equilibrium with dissolved HOCs in the surrounding water due to the continual production of sorptive phase (i.e., biomass). This sustained disequilibrium maintains diffusional gradients and likely "pumps" dissolved contaminants into the base of the pelagic food web at a rate proportional to the algal growth rate. In these studies, algal growth rates were determined by temperature, both in the laboratory (Swackhamer and Skoglund 1993) and in the field (Swackhamer and Skoglund 1991). Contaminant exposure to secondary producers and predators, both through diffusive exposure across gill surfaces and through dietary exposure, likely scales to the organism's metabolic activity (Thomann 1981). Bioenergetic models have been used to explore the functional relationship between energetic requirements (i.e., prey consumption and gill ventilation rates), feeding strategy (i.e., food preference), and contaminant exposures (e.g., Thomann et al. 1992; Madenjian et al. 1993). These relationships are commonly studied in isolation on individual fishes in short-term laboratory studies (e.g., Gobas and Mackay 1987), which lack the feedback between exposure levels, predator-prey interactions, and physiological functioning. Mesocosm-level studies, with adequate control and characterization of these feedbacks, will greatly improve our understanding of the interplay between bioenergetics and contaminant exposure.

Food-web structure

Recent studies have suggested that changes in trophic complexity (i.e., length of food chain) strongly impact dietary exposures of persistent chemicals to higher trophic levels. In a survey of lakes in Ontario, Rasmussen et al. (1990) found that concentrations of PCBs in the tissue of top predators increased significantly as the length of the food chain increased. Lake trout collected in lakes containing *Mysis* and pelagic forage fish (smelt, ciscoes, alewife, whitefish) contained significantly higher lipid-normalized levels of PCBs than those taken from lakes lacking these intermediate trophic levels. Cabana and Rasmussen (1994) report a strong positive correlation between mercury levels and trophic position (as determined by $d^{15}N$) in biota from Ontario lakes.

Benthic-pelagic coupling may play an important role in enhancing exposure of aquatic organisms to particle-reactive chemicals. Epibenthic fishes common to coastal areas are exposed to contaminants both from water column sources and from transfer of sediment-associated contaminants through the benthic food web. Kucklick and Baker (1998) and Thomann et al. (1992) suggest that sediment-associated contaminants are responsible for the relatively slow response times of contaminant levels in higher trophic levels in the Great Lakes and elsewhere. While sediments are recognized as the "geochemical memory" of aquatic systems, the mechanisms and extent of bioavailability of historically contaminated sediments are open questions (e.g., Landrum et al. 1992; McGroddy et al. 1996).

Chemical Disposition in Oviparous Vertebrates as Related to Reproduction and Development

Chemical uptake and disposition in early life stages of fish

Overview

In the 1940s the lake trout (*Salvelinus namaycush*) in the Great Lakes demonstrated large population declines due to predation by the sea lamprey, overfishing, and loss of prime spawning habitat. Resolution of such issues and extensive restocking failed to foster natural reproduction. Subsequent studies established that the lack of reproductive success was related to the failure to produce viable offspring (Jude et al. 1981; Nester and Poe 1984; Marsden et al. 1988). Numerous studies documented the bioaccumulation of halogenated aromatic hydrocarbons in the Great Lakes lake trout (Stalling et al. 1983; Schmitt et al. 1985; Huckins et al. 1988; Niimi and Oliver 1989; DeVault et al. 1989) as additional investigations lead to correlations between environmental contamination and mortality of early life stages (Willford et al. 1981; Mac et al.1985, 1988). Burdick et al. (1964), Mac et al. (1985), Spitsbergen et al. (1991), and Walker, Spitsbergen et al. (1991) provided evidence that concentrations of organochlorine compounds in lake trout eggs were inversely related to hatching success or fry survival. Xenobiotics transferred from the parental stock to the gametes were suggested as a reason for the declines noted. Contaminant concentrations found in eggs now have been correlated with reduced hatchability or survival in a number of fish species, including Atlantic salmon (*Salmo salar*) (Jensen et al. 1970), rainbow trout (*Oncorhychus mykiss*) (Hogan and Brauhn 1975), dace fry (*Phoxinus phoxinus*) (Bengtsson 1980), Arctic char (*Salvelinus alpinus*) (Monod 1985), white perch (*Morone americana*) (Monosson 1992), and chinook salmon (*Oncorhynchus tshawytscha*) (Ankley et al. 1991).

Numerous studies have demonstrated the toxicity of contaminants to the early life stages of fish. As compared to adults, developmental stages often appear to be more sensitive to the toxicity of xenobiotics (McKim et al. 1975; McKim 1977; Macek and Sleight 1977; Eaton et al. 1978; Sauter et al. 1976; Pickering and Thatcher 1970; Pickering and Gast 1972). Contaminant uptake, accumulation, and disposition are important modulators of toxicity in early life stages, as in the adult. Uptake of xenobiotic chemicals may occur at any stage of the fish life cycle. Early life stages may be exposed to xenobiotics as a result of maternal transfer to the egg prior to parturition or by direct environmental exposure post-parturition. Varying maternal contributions to the early life stage are presented throughout development as xenobiotics stored in the yolk and associated structures are mobilized and depleted. Direct exposure for the egg occurs primarily via water/chorion exchange. Following hatch, the yolk-sac, larval stage may be directly exposed through dermal and branchial routes. With consumption of the yolk, post-yolk-sac larvae present a new dosing paradigm, as, in addition to dermal and branchial exposure, the animal is now feeding. Similar to absorption, possible elimination routes vary with the stage of development. Xenobiotic elimination from the egg or embryo is a multiphasic event including loss from the embryo proper into the surrounding structures and fluids of the egg and from the totality of the egg across the chorion (or comparable membrane) to the environment. Development from the egg to the yolk-sac larvae (or equivalent stage) alters the complement of elimination routes with the advent of a functional circulatory system in conjunction with newly functional kidneys and gills. Branchial and renal routes of elimination in combination with dermal processes are likely to play an important role. Past the yolk-sac stage (swim-up), the biliary and gastrointestinal tracts become functional, again altering the complement of routes of xenobiotic elimination. Interposed upon the development of these elimination routes is the ontogeny of biotransformation enzymes and their ensuing activity.

This section reviews chemical uptake and disposition in early life stages of fish. A variety of excellent reviews are available regarding these issues in adult fishes for the reader to peruse and contrast. In the current discussion, the adult is included only for those topics in which they are an integral part of the process, such as with maternal transfer. This discussion follows the order of exposure and perceived importance: maternal then direct exposure. The stage will be set by discussing the scope of the contaminant problem in early life stages and the relationship with adult body burdens. This will be followed by a more in depth discussion of maternal transfer of xenobiotics. Factors influencing maternal transfer and the processes involved will be addressed in this section. Direct uptake and disposition of chemicals in early life stages will follow. Routes and patterns of absorption, elimination, and biotransformation in early life stages will be emphasized.

Xenobiotic residues in gametes of fish

A variety of contaminants have been found in the eggs of wild fish with the implication of potential effects upon reproduction and development. Most of the studies examining contaminant concentrations in eggs have focused on persistent organochlorine compounds, although some work has been carried out on metals or metaloids such as selenium, mercury, lead, copper, and cadmium. A fairly diverse group of fish species has exhibited xenobiotic residues in eggs; however, the predominance of data resides with salmonids. Studies as early as the 1960s have identified DDT and PCB residues in eggs of salmonid species (Willford et al. 1969). Studies by Miller (1993) and Miller and Amrhein (1995) have provided a more comprehensive examination of a wider variety of organochlorine residues in eggs from chinook salmon (*Oncorhynchus tshawytscha*), lake trout (*Salvelinus namaycush*) and siscowet (*Salvelinus namaycush siscowet*). PCBs, toxaphene, p,p' DDE, o,p' DDE, p,p'DDT, and p,p'DDD have been identified in eggs from Lake Michigan chinook salmon (Miller 1993). Similarly, Lake Michigan lake trout eggs were shown to contain PCBs, toxaphene, p,p'DDE, p,p'DDT, and p,p'DDD. Lake Superior lake trout (siscowet), while demonstrating a profile similar to their Lake Michigan counterparts, exhibited overall lower xenobiotic concentrations. Egg concentrations of these compounds (collected in 1982) ranged from 0.3 to 8.3, 1.0 to 2.6, 0.15 to 2.3, 0.08 to 0.14, and below detectable limits to 0.15 ppm for PCBs, toxaphene p,p'DDE, p,p'DDT, and p,p'DDD, respectively. Eggs collected in 1991 from Lake Superior lake trout, also have demonstrated detectable concentrations of PCBs (0.45 ppm), trans-nonchlordane (0.03 ppm), p,p'DDE (0.09 ppm), and dieldrin (0.02 ppm) (Miller and Amrhein 1995). With the exception of PCBs, concentrations of like compounds in siscowet were generally lower than their 1982 lake trout counterparts (Miller 1993; Miller and Amrhein 1995). Eggs from Lake Ontario rainbow trout contained PCB, p,p'DDE, ΣDDT, α chlordane, Σchlordane, mirex, heptachlor epoxide, dieldrin, hexachlorobenzene (HCB), and mercury at concentrations of 2050 ± 700, 293 ± 97, 458 ± 149, 61 ± 18, 76 ± 25, 71 ± 29, 2 ± 1, 20 ± 9, 14 ± 7, and 11 ± 8 mg/kg, respectively (Niimi 1983). Heavy metals and organochlorine compounds also have been detected in coho salmon (*Oncorhynchus kisutch walbaum*) eggs collected from Lakes Michigan, Erie, and Ontario (Morrison et al. 1985). PCB, p,p'TDE, o,p'DDT, dieldrin, p,p'DDE, chlordane, zinc, lead, mercury, copper, and cadmium were found in the eggs from all 3 sources. Additional studies have expanded the list of detected compounds to include polychlorinated dibenzo-p-dioxins (PCDDs) and dibenzofurans (PCDFs), which were identified in eggs from lake trout (Walker et al. 1992).

Species of fish other than salmonids have also been shown to exhibit gonadal accumulation of contaminants. Eggs collected from winter flounder (*Pseudopleuronectes americanus*) (New Bedford Harbor, MA) demonstrated significantly higher levels of PCBs (39.6 mg/g dry weight) than at a control site (Fox Island: 1.08 mg PCB/g dry weight) (Black et al. 1988). Studies with Baltic flounder (*Platichthys flesus*) demonstrated significant ovarian tissue residues of a variety of contaminants including PCBs, hexachlorocyclohexane (Lindane -αHCH), γHCH, DDD, dieldrin, HCB, heptachlorepoxide, and zinc (Von Westernhagen et al. 1981). Residue levels in the ovaries from these fish ranged from 5.0 to 317.0, 0.7 to 6.0, 0.4 to 5.6, 3.0 to 30.0, 0.1 to 49.0, 0.6 to 2.0, 0.08 to 3.0 ng/g wet weight, and 3.7 to 31.7 mg/g for the foregoing contaminants, respectively. Eggs collected from Hudson River striped bass (*Morone saxatilis*) also exhibited PCB concentrations ranging from 1.1 to 8.1 mg/g wet weight (Westin et al. 1983). An interesting feature of these data was that the highest PCB concentrations in the eggs came from the smallest female, while the lowest concentration was associated with the largest animal. Paddelfish (*Polyodon spathula*) have also been shown to accumulate significant levels of PCBs in both ovaries and testis (Gundersen and Pearson 1992). Aroclor 1260 concentrations ranged from 0.05 to 18.70 (mean = 7.3) and 5.63 to 23.00 (mean = 16.2) mg/g for ovaries and testes, respectively. Interestingly, PCB levels in mature roe of paddlefish were lower than that of immature ovaries. The authors attributed this finding to a change in lipid content. Mature ovaries, while much larger, had considerably lower lipid content than their immature counterparts. Niimi (1983) examined the deposition of a variety of contaminants in the eggs of various fish species collected from Lakes Ontario and Erie, including white sucker (*Catostomus commersoni*), white bass (*Morone chrysops*), smallmouth bass (*Micropterus dolomieui*), yellow perch (*Perca flavescens*), and rainbow trout (*Salmo gairdneri*). Residues of PCBs, p,p'-DDE, DDT, α chlordane, Σchlordane, heptachlor, dieldrin, HCB, and mercury were found in the eggs of all species. Smallmouth bass eggs maintained the highest residue levels on a mg/kg basis for 5 of the 9 chemicals analyzed in all species. Conversely, yellow perch contained the lowest levels for 7 of the 9 chemicals. Not surprisingly, given the nature of the compounds, this general trend correlated to the percentage of lipids in the eggs of these 2 species.

A considerable amount of work has focused upon the reproductive and developmental toxicity of selenium. Hence, a number of studies have focused upon selenium content in the gametes of fish from impacted waters. Investigations by Gillespie and Baumann (1986) focused upon selenium concentrations and reproductive and developmental performance of bluegills collected from Hyco Reservoir (Roxboro,

NC), a selenium-contaminated environment, and Roxboro City Lake as a reference site. Selenium concentrations in the ovaries and testes mirrored carcass and environmental exposure. For the control site, mean selenium concentrations were 0.37 and 0.50 mg/kg for carcass and testis, respectively for males, and 0.37 and 0.66 mg/kg for the carcass and ovaries of females. There was a similar comparison for Hyco reservoir animals: males had concentrations of 7.81 and 4.37 mg/kg for carcass and testis, respectively, and females had concentrations of 5.91 and 6.96 mg/kg for carcass and ovaries. These results suggest that, while perhaps not preferentially accumulating in gametes, selenium does distribute to gametes as a reflection of exposure. A similar finding was evident for selenium in eggs and milt of the razorback sucker (*Xyrauchen texanus*) (Hamilton and Waddell 1994).

Clearly, the existing literature identifies a wide range of fish species that contain contaminants in their eggs. It is apparent that for numerous fish species, persistent lipophilic organochlorine compounds can be passed along from the parent to their gametes. Although not as extensively documented, it has been demonstrated that a number of metals or metaloids are forwarded to gametes in natural settings. Studies with less persistent compounds have not been reported with regard to parental transport to the gametes.

Correlation of maternal nonreproductive tissue residues to egg or ovarian contaminant levels

Correlations have been made between egg or ovarian contaminant levels and maternal nonreproductive tissue residues, lending insight into xenobiotic partitioning and providing a potential tool for predictive assessments. Total concentrations of PCBs, p,p'DDE, and dieldrin in lake trout eggs and total concentrations of PCBs and p,p' DDE concentrations in chinook salmon eggs have been significantly correlated with concentrations of these chemicals in the muscle tissue of gravid fish (Miller 1993). While this is true, the relationship between muscle and egg organochlorine concentrations differs between the two species. Higher contaminant concentrations were found in chinook salmon eggs relative to their muscle tissue, whereas in lake trout higher contaminant levels were found in the muscle tissue relative to the eggs (Miller 1993). Correlations between organochlorine concentrations in the muscle tissue and eggs were greatest for compounds found at higher concentrations. In addition, the correlation coefficients for the relationships between muscle and egg organochlorine concentrations were higher for the lake trout than for chinook salmon (R = 0.62 to 0.67 for chinook salmon versus 0.94 to 0.96 for lake trout). The author suggested that reproductive staging of collected individuals may have a strong effect on observed residues in both the eggs and muscle of chinook salmon as both loss of appetite prior to spawning and somatic lipid mobilization for egg development are pronounced components of the chinook life history. In a similar study, with siscowet lake trout, concentrations of PCBs and p,p'DDE in eggs were also positively correlated with muscle concentrations (Miller and Amrhein 1995). In contrast, cis-chlordane, cis-nonchlordane, and p,p' DDT evident in muscle at concentrations ranging from 0.06 to 0.14 mg/kg were undetected in eggs (0.05mg/kg detection limits). For those compounds detectable in both muscle tissue and eggs, concentrations in muscle ranged from 5 to 8 times higher than concentrations in eggs. Again, like the Lake Michigan lake trout, organochlorine concentrations were significantly higher in the muscle than in the eggs of the siscowet. TCDD concentrations in muscle of maternal lake trout also appear to correlate to TCDD concentrations in their eggs (Walker et al. 1994). The TCDD concentration in eggs were 42 and 43% of skeletal muscle levels when expressed on a lipid and wet-weight basis. This compares to values for total PCB concentrations in lake trout eggs of 22 to 38% of the total PCB in maternal skeletal muscle on a wet weight basis and 66 to 78% on a lipid-normalized basis (Miller 1993). Similar residue comparisons have been made among muscle, liver, and ovaries of Northwest Atlantic cod (Hellou et al. 1993). Generally, concentrations of 23 specific organochlorine contaminants were undetectable in cod muscle tissue, while the ovaries presented detectable concentrations that were 10 times lower than liver tissue on a wet-weight basis (Hellou et al. 1993). A number of compounds including αHCH, β-HCH , γHCH, and oxychlordane were shown to have substantial negative correlations between concentrations in the ovaries and the liver. Cis/trans chlordane, trans-nonachlor, cis-nonachlor, *p,p*'DDE, *p,p*'DDT, Aroclor1254 (weak), *p,p*' DDD, and *o,p*' DDD generally exhibited positive, or at least non-negative, correlations between concentrations in liver and ovaries in cod. In studies with paddlefish, PCB concentrations were much higher in reproductive tissues than in corresponding muscle tissues (Gundersen and Pearson 1992). Average ovarian PCB concentrations were 1.7 to 18.2-fold higher than those of muscle, dependent on whether the comparison was made with white muscle (1.7) or red muscle (18.2). Male gonadal tissue concentrations were 23-fold higher than were white muscle concentrations. Other studies examining PCB concentrations in winter flounder from New Bedford Harbor, MA, showed that the content of PCB congeners in the liver is influenced little by sex or reproductive condition (Elskus et al. 1994). Compositely, these studies suggest that the relationship of somatic or visceral xenobiotic residue levels to that deposited in the ovaries is highly species dependent.

Studies by Kammann et al. (1993) and Knickmeyer and Steinhart (1989) suggest that organ residues may vary with the spawning season. For example, at the beginning of the dab (*Limanda limanda*) spawning period, during high egg production, PCB and HCB concentrations are low in female livers, while the levels of these compounds increased during the same interval in the ovaries (Kammann et al. 1993). Maximal PCB and HCB burdens in dab ovaries were observed at the end of the spawning period in April. Studies with wild-caught dab suggest that PCB patterns in liver and ovaries were dominated by penta- and hexachlorobiphenyls with a large contribution from PCB 138 and 153 (Kammann et al. 1993). The PCB pattern in testes, on the other hand, was dominated by tri- and tetrachlorobiphenyls. PCB patterns of female dab livers collected in February and May were similar, with the exception of PCB 138. This congener was shown to be preferentially transferred from dab liver to ovarian tissues during ovarian development, resulting in significant alterations in liver patterns (Knickmeyer and Steinhart 1989). PCB 28, 60, 66, and 99 were also present in higher concentrations in ovary than in liver at select samplings (February). Significant differences in congener profile occurred in female dab at the May sampling with greater representation by tri- and tetrachlorobiphenyls. The authors concluded that a change in lipids for ovary anabolism may have occurred. In winter months, when nutrition is poor, lipids transferred from the liver may be the main source for anabolism in ovaries leading to similar PCB patterns in these two organs. Increased feeding in spring resulted in the incorporation of the less-chlorinated compounds present in prey with ovarian maturation (Knust 1986). These conclusions will require further verification with additional studies.

Von Westernhagen et al. (1995) examined age- and length- dependent concentrations of chlorinated hydrocarbons in ovary and muscle of herring, flounder, dab, whiting, and horse mackerel. From these data, no generally acceptable pattern was recognizable between the development of tissue residues and age in sexually mature female fish (Von Westernhagen et al. 1995). The authors suggested that this may be because of gonadal maturation, whereby there is a transfer of lipids and associated xenobiotics from the liver to the gonads and a subsequent loss of gonadal materials upon spawning. It is thought that the seasonal release of eggs thus acts to remove excessive accumulation of contaminants.

Xenobiotic residues in gametes: laboratory studies

A number of experimental studies have demonstrated and quantitatively characterized transfer of chemicals to fish gametes. Hall and Oris (1991) demonstrated significant bioconcentration of anthracene in the eggs of fathead minnows (*Pimephales promelas*) exposed at 12 mg/L in the water. The fathead minnow BCFs noted for the male carcass was 1126, male testes 769, female carcass 3581, ovaries 1452, and eggs 759. Female yellow perch dosed with 2,5,2',5'-tetrachloro[14C] biphenyl in the water demonstrated that 2 weeks after exposure, 30% of the initial biphenyl body burden was distributed to the eggs, whereas just prior to spawning, about 50% was present in this tissue (Vodicnik and Peterson 1985). In these studies eggs contained anywhere from 23.5 to 41.8 mg/kg of 2,5,2',5'-tetrachloro[14C] biphenyl for the 18 weeks prior to spawning. Mean total ng PCB concentrations were much lower in perch fillet and viscera than in carcass and eggs during ovarian maturation. Total carcass PCB levels were high early in egg development (469 to 502 ng), but dropped precipitously (57 to 228 ng) as the egg PCB burden raised from 270 to 498 ng before spawning. This change could be accounted for in part by a large increase in egg mass. Additional laboratory studies with 2,5,2',5',-tetrachloro[14C] biphenyl in rainbow trout demonstrate accumulation in maturing eggs and sperm following redistribution of residues from within the fish's body (Guiney et al. 1979). Following a 36-hour static, waterborne exposure to PCB, concentrations of radiolabelled PCB dropped in the whole fish, skin, carcass,

skeletal muscle, and liver. During this same interval, the percent of initial ^{14}C
residue increased from 0.1 to 5% and from 0.1 to nearly 2% for eggs and sperm,
respectively. Similarly, several experimental studies have demonstrated transfer of
maternally derived TCDD to the eggs of lake trout (Walker et al. 1994). Female lake
trout administered TCDD in the diet (total exposures 22.8 to 62.2 mg) for 11 weeks
demonstrated widely varying TCDD concentrations for individual fish even within a
treatment. Egg TCDD concentrations, which were generally lower than liver and
muscle levels and higher than blood levels, ranged from 71 to 311 pg/g wet weight.
Studies by Ungerer and Thomas (1996) examined the accumulation of *o,p'*-DDT and
Aroclor 1254 in gonadal and hepatic tissues of the Atlantic croaker (*Micropogonias
undulatus*) following dietary administration. For both compounds and sexes, the
liver contained greater concentrations of the contaminants. For females the go-
nadal/liver contaminant ratios ranged from 0.645 to 0.794, while for males the
ratios ranged from 0.009 to 0.037 for both compounds. Female gonadal tissue
contained 118 and 38.6-fold higher levels than that of males on a mg/g basis for
*o,p'*DDT and Aroclor 1254, respectively.

Factors influencing contaminant deposition in gametes
A number of studies have investigated the role of parental body composition upon
contaminant disposition in the ovary. Niimi (1983) examined the relationship of
lipid content of the fish, mean egg weight as a percent of the total fish weight,
percent of total maternal lipid deposited in eggs, and percent of whole body
contaminants in fish transferred to eggs. When the data were compiled for 5 fish
species (rainbow trout, yellow perch, smallmouth bass, white bass, and white
sucker), several interesting features emerged as related to a variety of organic
contaminants, lipid content, and transfer efficiency. On a percentage basis, the
yellow perch, although the leanest (5.1% lipid) species examined, transferred the
greatest percentage of whole-body contaminants (25.5%). Rainbow trout, while
having the highest lipid content (11.4%), had the lowest percentage of contaminants
transferred to the eggs (5.5%). Interplaying on these 2 factors are mean egg weight as
percent of total weight and percent of total lipid deposited in eggs. Contributing to
the high percent of contaminants transferred to eggs in yellow perch was a relatively
high mean egg weight as percent of total weight (yellow perch: 22.3 versus 13.6% for
rainbow trout) and the high percent of total lipid transferred to eggs (yellow perch
27.1 versus 10.3% for rainbow trout). When all species are examined relative to total
organic contaminant load in eggs on a mg/kg basis and also as related to the egg as
percent lipid, the following comparison becomes available, given in order of highest
to lowest composite contaminant concentration on a per kg basis (lipid percent in
parentheses): smallmouth bass, 4080 mg/kg (13.4% lipid in eggs); rainbow trout,
2985 mg/kg (8.7%); white bass, 2771 mg/kg (9.7%); white sucker, 2501 mg/kg
(5.7%); and yellow perch, 1078 mg/kg (5.8%). While it is clear that the composite
concentration of these compounds in the eggs appear to be generally correlated to
their lipid content, similarities in lipid content and disparities in residue levels

(perch and sucker) and similarities in residue levels and disparities in lipids (suckers and white bass) also appear. These results suggest that other factors may be influential modulators of residue loading in eggs on a mg/kg basis. It is worthy to point out that differences in egg concentrations of contaminants may be related to a variety of factors including species differences, contaminant levels in the environment, or even differences in gonadal development relative to contaminant deposition at the time of egg collection.

Comparative differences in the residues imparted to eggs are also evident between the lake trout and the chinook salmon (Miller 1993). In this case the reproductive life history also appears to play a determinant role in this process. Chinook salmon, semelparous (once-bearing) in nature, transfer most of their somatic lipid stores to developing gonads. The lake trout, an iteroparous (multiple-bearing) species, transfers a much lower proportion of their somatic lipids to the developing gonads (Miller 1993). This fact translates to estimates that gravid lake trout lose 3 to 5% of their total PCB body burden during spawning, whereas chinook salmon lose 28 to 39%.

There are numerous indications that different species have different life strategies for resource mobilization for ovarian development. Nassour and Leger (1989) suggested that carcass and visceral lipid reserves are mobilized for rainbow trout ovarian development. For the same purpose, freshwater catfish (*Clarias batrachus*) utilize abdominal fat (Lal and Singh 1987) and Atlantic salmon (*Salmo salar*) use muscle lipid and protein (Aksnes et al. 1986). In contrast, dietary lipid sources appear to be primary for ovarian development of the gilthead bream (*Sparus aurata*) (Harel et al. 1994) and the northern pike (*Esox lucius*) (Medford and MacKay 1978; Diana and MacKay 1979). Another feature that may come into play is the concept of differential distribution of contaminants in the parental stock. There are a couple indications in the literature beyond the foregoing discussion that differential distribution may play a role in contaminant deposition in gametes. It has been demonstrated that the disposition and elimination of organochlorine compounds may differ between fish species. Guiney and Peterson (1980) demonstrated that in rainbow trout, the major distribution sites for 2,5,2',5' [^{14}C]-tetrachlorobiphenyl were the skeletal muscle and carcass, accounting for 60% of the dose, while in yellow perch, 70% of the administered dose was contained in the viscera and carcass. This finding, as the authors suggest, may be related to species differences in the lipid content of the various tissues. Similar findings have been noted in other studies. Zitko et al. (1974) found that Atlantic herring (*Clupea harengus harengus*) had both a muscle PCB and a lipid content 5 to 10 times higher than those of yellow perch. While the direct source and the transfer dynamics of contaminants are unclear, such findings again suggest that life history and physiology may play a determinate role in the relative importance of maternal transport of lipophilic contaminants in a given species.

The influence of dietary energetics on the process of parental transfer of contaminants is largely unknown. Miller (1993) suggested that variations in chinook organochlorine concentrations in both muscle tissue and eggs may be related to energetic constraints resulting from forage limitations. Reduced size, lipid content, and growth rate have been noted for chinook salmon with such limitations. As compared to lake trout, chinook appear to have greater variability in somatic lipid stores which in turn may induce variability in lipids and contaminants transferred to the eggs. It has been shown that starvation-induced depletion of lipid stores in rainbow trout and coho salmon (*Oncorhynchus kisutch*) have failed to increase elimination of PCBs from the whole fish (Lieb et al. 1974; Gruger et al. 1975). It is plausible that if contaminants are not co-eliminated with lipid loss under conditions of nutrient limitations, then lipid mobilization to the ovaries under such conditions may increase the exposure potential to the eggs. Again, further investigations are necessary to examine the relative importance of these issues.

In totality, this information suggests that the deposition of contaminants in the egg mass may be directly linked with species-specific characteristics. Lipids and lipid dynamics are critical to lipophilic contaminant transfer to eggs. This transfer may be altered by increased or decreased lipid mobilization in response to the maternal lipid content and/or the reproductive life history of the animal. The toxicological significance of this transfer may center not only on the initial maternal concentration of contaminants, but also on the source and amount of the lipids transferred. High contaminant concentrations in maternal lipid coupled with low lipid transfer, such as with rainbow trout, may approach similar contaminant transfer conditions as those of low contaminant concentrations, such as in perch. The similarity in contaminant transfer with differences in contaminant load and lipid transfer may mean that tissue (muscle) concentrations of contaminants and contaminants in eggs may be related by some formulation on only a species-to-species basis.

Transport of contaminants to ovaries

Vitellogenesis and oogenesis in oviparous fishes

The onset of oogenesis initiates a host of physiological changes, including alterations in serum protein profiles, that may play an important role in the maternal transfer of endogenous and exogenous hormones, and xenobiotics may be transferred to oocytes. This alteration in serum lipoprotein (LP) content, a part of vitellogenesis, results from the dramatic, estrogen-induced synthesis of vitellogenin (VTG) and other lipoproteins. In oviparous fishes, the synthesis and sequestration of large quantities of LPs, particularly VTG, are critical elements of oocyte development. While catalytically processed and sequestered LPs serve primarily as a nutrient source for embryogenesis and early-life-stage development, the oocytic accumulation of some trace metals, micronutrients, and hormones also may be associated, in part, with oocytic accumulation of lipoprotein-associated ligands (Specker and Sullivan 1994). The oocytic sequestration of serum LPs, however, also may serve as a vector of maternal transfer of xenobiotic compounds, facilitating embryonic and early-life-stage exposures.

Vitellogenesis and oogenesis in oviparous fishes, like all oviparous vertebrates, are endocrine-orchestrated responses to environmental stimuli such as changes in temperature, photoperiod, lunar cycle, and/or diet. During vitellogenesis, increases in serum estradiol (E2) levels lead to the production of a variety of lipoproteins, including the phospholipoglycoprotein VTG and very low-density lipoproteins (VLDLs). These proteins are taken into maturing oocytes by selective, receptor-mediated endocytosis. On a dry-weight basis, VTG is the major serum lipoprotein in vitellogenic females and forms the basis of yolk protein sequestered in maturing oocytes. In addition, while VLDLs typically contain greater molar quantities of lipid, the majority of lipid found in the oocyte of oviparous fishes is derived from proteolytic cleavage of VTG, not VLDL (Wallace 1985).

A great deal is known about oogenesis and the formation of oocytes in oviparous vertebrates, including the uptake of vitellogenin and other serum proteins. Detailed information regarding these processes is available in a number of seminal reviews, including those contributed by Chapman (1980), Wallace (1985), Byrne et al. (1989), Maller (1985), Wallace and Selman (1990), and Nagahama (1994). Likewise, a great deal of recent work continues to be done examining the effects of xenobiotic chemicals on early life stages of a number of oviparous vertebrate species including fish (e.g., Guiney et al. 1997; Olivieri and Cooper 1997; Fahraeus-Van Ree and Payne 1997), birds (e.g., Hoffman et al. 1987; Nosek et al. 1993; Stanley et al. 1994), amphibians (e.g., Bernardini et al. 1996; Dawson et al. 1996), and reptiles (Crews et al. 1996). Less is known, however, about the precise mechanisms of pollutant uptake by oocytes and the relative role LP incorporation plays in maternal transfer of xenobiotics.

While substantial similarities exist between oogenesis in different phyla of oviparous vertebrates, significant differences in protein biochemistry and life histories both within and between these phyla complicates overreaching generalizations. Critical questions addressed below, and in similar subsection for birds, focus on the role played by LPs in the transfer of xenobiotics from the female piscine to her maturing oocytes.

Elements of lipoprotein synthesis and transport

Vitellogenesis and oogenesis have been the focus of a great deal of research for over 40 years. Of central importance in the current discussion is the fact that most oviparous species, including fish, produce a variety of lipoproteins, typically differentiated in terms of their lipid density: VLDL, low-density lipoproteins (LDL), high-density lipoproteins (HDL), and the VTG-containing very-high-density lipoproteins (VHDL). Of these, VTG and VLDL are the predominant LPs incorporated by maturing oocytes.

Once transported from the circulation into the oocyte, VTG is processed into smaller yolk proteins consisting of lipovitellins, phosvitins, phosvettes, and other cleavage products that represent the major source of nutrition during embryonic development and early life stages (Wallace and Selman 1990; LaFleur et al. 1995). Analysis of VTG-cleavage products from oocytes suggests that VTGs of oviparous fishes are frequently more lipidated and contain more widely divergent (though typically smaller) quantities of protein-phosphorous (Jared and Wallace 1968; de Vlaming et al. 1980; Craik 1982) than VTGs of other oviparous vertebrates.

Vitellogenin is perhaps the best studied oviparous lipoprotein and, while multiple forms of circulating VTG have been identified within specific species of fishes (Ding et al. 1989; Chan et al. 1991; Kishida and Specker 1993), birds (Wang and Williams 1980, 1983; Evans et al. 1988), and amphibians (Wiley and Wallace 1981, Wahli and Ryffel 1985), differences in VTG composition also exist between species within each of these phyla. Except where noted, however, vitellogenin and VLDL will be used as general terms inclusive of all forms of the proteins in a specific phyla (i.e., VTGs of all oviparous fishes will be considered to possess similar physicochemical composition and properties). During the discussion that follows, it should be kept in mind that intra- and interspecies differences in lipoprotein composition and oocytic uptake may result in differences in rates and magnitudes of maternal transfer. Future studies that improve our understanding of the structure and binding dynamics of lipoproteins from a greater diversity of oviparous fishes will greatly improve our ability to elucidate the role lipoprotein sequestration plays in maternal transfer of ligands to maturing oocytes.

VTG- and VLDL-associated xenobiotic transfer

Vitellogenin and VLDL are the primary lipoproteins sequestered by maturing oocytes of oviparous fish. While VLDL may plausibly be assumed to associate with greater molar amounts of lipophilic xenobiotics, VTG is typically produced and sequestered in far greater quantities than VLDL. Examination of LP-associated maternal transfer, therefore, must acknowledge that binding and transport are dependent upon a multiplicity of factors, including, but not limited to, the physicochemical properties of ligand and lipoprotein, as well as the serum chemistry and synthesis and sequestration kinetics of vitellogenesis.

Central to the concept of VTG as a transporter of xenobiotics to maturing oocytes is the hypothesis that the lipid and ionic regions of VTG can bind xenobiotic compounds in significant quantities and with adequate affinity to allow for oocytic accumulation. The coexistence of lipid and ionic regions on VTG makes it a potential carrier molecule for a wide array of endogenous (e.g., hormones, vitamins, minerals, etc.) and exogenous compounds possessing diverse physicochemical properties. While VTG transport of many of these compounds has not been proven, Specker and Sullivan (1994) and Sullivan et al. (1989) presented data supporting the role of VTG as a transporter of hormones in plasma samples from coho salmon

(*Oncorhynchus kisutch*). Further, Babin (1992) identified lipoproteins (potentially VTG) as the main plasma vector for transport of thyroid hormones, and Cyr and Eales (1989) reported binding of thyroxine to a presumed lipoprotein. Tagawa and Specker (1993), however, have shown that cortisol is taken up by the oocytes in the absence of VTG and in molar ratios similar to serum levels, implicating simple diffusion as the mode of uptake.

Although many ubiquitous aquatic pollutants possess structural and chemical similarities to steroid and thyroid hormones (Specker and Sullivan 1994), data implicating the role of VTG as a carrier protein for exogenous compounds is limited. Babin (1992) and Cyr and Eales (1992) have conducted studies showing that plasma lipoproteins (including VTG) are the principle binding sites for thyroid hormones in fish plasma. In recent studies, the thyroid hormones thyroxine (T_4) and 3,5,3'-triiodothyronine (T_3) have been shown to associate with VTG and to accumulate in oocytes of gravid *Fundulus heteroclitus* (Monteverdi 1999). In addition, the highly charged phosphate groups of the phosvitin region of VTG are presumed to be responsible for the significant binding capacity of the molecule for ions such as calcium, magnesium, iron, copper, and zinc (Specker and Sullivan 1994; Ghosh and Thomas 1995). However, little is known about the role this binding may play in the transport of nutrient ions or pollutant metals to maturing oocytes.

Fish vitellogenins and other serum lipoproteins have been associated with a number of exogenous ligands including persistent organochlorine compounds (e.g., PCBs; Mohammed et al. 1990) and a number of pesticides, including DDT (Ungerer and Thomas 1996; Mohammed et al. 1990). Plack et al. 1979), toxaphene (Mohammed et al. 1990), and dieldrin (Skalsky and Gutherie 1977). A number of studies, including those by Monteverdi (1999), Ankley et al. (1989), and Westin et al. (1983), have illustrated an apparent correlation between the onset of vitellogenesis and the delivery of common aquatic pollutants to the ovaries of gravid (i.e., egg-producing) fish. Further, a study by Ungerer and Thomas (1996) suggested that serum lipoproteins are intimately involved with the ovarian accumulation of *o,p'*-DDT in the Atlantic croaker (*Micropogonias undulatus*).

While conclusive, mechanistic descriptions of maternal transfer of xenobiotics by VTG and other lipoproteins have yet to be provided, a growing body of data suggests that such transfer is likely and serves a dual purpose as an important mechanism of xenobiotic depuration from maternal stores. For example, Guiney et al. (1979), Vodicnik and Peterson (1985), and Ankley et al. (1989) have all shown that spawning can enhance the elimination of maternal stores of PCBs.

Factors left unaddressed here that may prove important in the study of maternal xenobiotic transfer include the oocytic accumulation of xenobiotics by simple diffusion, in association with uptake of non-lipidated proteins, or (in the case of many marine teleosts) during oocyte hydration. Further, it has not been established whether lipophilic xenobiotics are preferentially sequestered by oocytes or whether their transfer is determined by simple lipid-partitioning kinetics. Finally, an issue that overlies the entire discussion of lipoprotein-associated maternal transfer is the effect species- and environment-specific factors may play in the composition, transport, and sequestration of VTG and VLDL. For example, because the major source of lipid (i.e., fat stores versus diet) used in lipoprotein synthesis can vary as a result of genetic or environmental factors (e.g., availability of food), rates of maternal transfer can reflect variability that complicates our ability to predict xenobiotic sequestration in oocytes. Further work will be necessary to delineate the specific roles lipoproteins play in xenobiotic transport and sequestration by maturing oocytes.

Direct xenobiotic uptake and disposition in early life stages

For lipophilic compounds, current information suggests that maternal routes of xenobiotic exposure may be relatively more important for early developmental stages than is direct uptake. However, exposure data and the indirect, but more comprehensive, toxicity database suggest that eggs are capable of direct uptake of xenobiotics from the water. It is likely that environmental contaminant levels, contaminant and egg location, contaminant characteristics, structural and chemical characteristics of the eggs' surface, and the length of time the embryos develop in the impacted area are important determinants of direct exposure. Many of the comparisons between maternal and direct exposure have yet to be made quantitatively in natural systems under relevantly complex exposure conditions.

Xenobiotic uptake into eggs by direct exposure is highly variable depending on the characteristics of the chemical. Trans-chorionic permeability values for the eggs of medaka (*Oryzias latipes*), for example, have been shown to range from 10.5 to 82.9% for diethylnitrosamine, dipropylnitrosamine, dibutyltin dichloride, tributyltin chloride, lindane, pentachlorophenol, and aldrin (Helmstetter and Alden 1995). Log K_{ow}s of the compounds, when regressed against permeability factors, exhibited an R^2 value of 0.96. The authors suggested that the amount of xenobiotic entering the egg from external exposure was controlled in large measure by the lipid solubility of the chemical even in the presence of a membrane permeable carrier. Lake trout eggs exposed to TCDD at 0, 10, 20, 40, 62, or 100 ppt (ng[^{3}H] TCDD/L water) in an acetone carrier for 48 hours attained TCDD concentrations in a dose-related fashion with levels approximately 3 times greater than the nominal exposure concentrations (Walker, Spitsbergen et al. 1991). These concentrations were maintained through the egg and sac fry stages. In a stepwise fashion for each concentration, coho salmon embryos exposed under daily static renewal conditions to 0, 6, 13, 29, 62, and 139 mg/L of methyl mercury hydroxide rapidly attained near steady state levels of mercury by 15 days (Devlin and Mottet 1992). The mercury concentrations attained were approximately 16 to 24 times greater than the mercury levels in the exposure water. Investigations with inorganic mercury (mercuric chloride) have demonstrated even greater uptake rates for the ricefish embryo (Heisinger and Green 1975). Studies examining selenium accumulation in fathead minnow eggs reported that 24-hour-old eggs from adults reared for 1 year in artificial streams containing 10 mg selenium/L (waterborne and dietary) accumulated 16 mg/g selenium whereas control eggs amassed 1.2 mg/g selenium (Schultz and Hermanutz 1990). When control eggs were exposed to selenium via the water under like experimental dosage conditions for 24 hours, the eggs took up 0.56 mg/g selenium. The authors suggested from this data that selenium in eggs comes primarily from parental exposure rather than from waterborne exposure of the new eggs.

A few studies have examined xenobiotic uptake, accumulation, and elimination throughout early development. These investigations highlight a number of interesting differences between each of the life stages. Studies with TBT and the minnow *Phoxinus phoxinus* show that TBT uptake rate was considerably lower in embryos than in yolk-sac larvae (Fent 1991). A 91-hour exposure of embryos to TBT resulted in residues of 0.85 mg/g, while an additional 64 hours of exposure to larvae resulted in residues of 4.65mg/g. The author suggested that the difference in TBT accumulation rates between embryos and yolk-sac larvae were related to the fact that larvae absorb TBT via skin and gills while the chorion of the egg modulates absorption. Likewise, embryos lack blood circulation, limiting distribution of the TBT. The authors also suggested that the uptake of TBT does not follow a simple hydrophobicity model, probably as a result of such morphological and physiological determinants.

Bioconcentration experiments with lindane in early life stages of rainbow trout indicate significantly higher BCFs at the hatching and yolk sac stages as compared to early juveniles (Vigano et al. 1992). Similar findings were demonstrated for *p*-dichlorobenzene (Galassi et al. 1982), 1,2,3 trichlorobenzene and 1,2,4 trichlorobenzene (Galassi and Calamari 1983). Interestingly, both the uptake and depuration coefficients of lindane in rainbow trout were consistently higher for young juveniles than in the other early stages. These findings suggest that the later developmental stages have a greater inward and outward flux of chemicals than do eggs or yolk sac larvae. Such results are not confined to lindane, as the same phenomenon has been noted with dieldrin (Van Leeuwen et al. 1985) and the elimination phase of trichlorobenzenes (Galassi and Calamari 1983). The authors of the lindane study suggested that the higher BCF for hatching and yolk sac stages relative to juveniles may be related to decreased lipid content or increased gill function for older life stages. They demonstrated that total lipid content in early life stages dropped from 8.0% in the eyed egg to a low of 1.95% in the early juvenile. An attempt to correlate this to an age dependent decline in BCFs was only partially successful. It was suggested that perhaps changes in the actual lipid composition during larval development may be responsible. They supported this premise with findings by Galassi et al. (1982), which determined that neutral lipids were 49.9% of the total lipids in eyed eggs, whereas at hatching and in the early fry stage neutral lipids comprised 61.8 and 73.9% of the total. Other features, such as biotransformation were not examined.

Solbakken et al. (1984) examined the uptake, accumulation and elimination of ^{14}C phenanthrene, naphthalene, benzo(a)pyrene (BaP), and 2,4,5,2',4',5' hexachlorobiphenyl by eggs and newly hatched larvae of cod (*Gadus morhua*). For the 24-hour exposure, the maximal accumulation in eggs occurred for phenanthrene, while the lowest uptake was noted for PCB. Examination of the yolk and chorion from cod eggs exposed to ^{14}C-labeled phenanthrene revealed that most of the compound was associated with the yolk. For each of the 4 compounds, most of the radioactivity that accumulated in the eggs was transferred to the larvae upon hatching. Larvae similarly exposed for 24 hours also exhibited the greatest uptake with phenanthrene, as seen with the eggs. However, the order of accumulation was different for the other compounds; the PCB showed the second highest accumulation, followed by BaP, and lastly by naphthalene. In general, larvae accumulated much more of each compound (except naphthalene) than did the eggs for a corresponding treatment. The authors suggested that the low accumulation of PCB in eggs may be due to the low lipid content of cod eggs and the high molecular weight of the PCB. The lower accumulation ranking of BaP in the larvae than in the eggs was attributed to increased metabolism in the larval stages, whereas the low accumulation and rapid elimination of naphthalene in both life stages was explained by a low lipid/water partition coefficient and high water solubility. Other studies have examined some of these same compounds in other species. Kuhnhold and Busch (1978) found that the relative accumulation of naphthalene and BaP in salmon eggs was dependent upon the duration of exposure. For short exposures naphthalene accumulated to a greater extent than did BaP; however, after 7 days the accumulation factor was 60% higher for BaP than for naphthalene. Both naphthalene and BaP were found to be associated with the yolk and vitelline fluid of the salmon egg (Kuhnhold and Busch 1978). Similar findings regarding distribution to the yolk have been reported for BaP in the sand sole egg (Hose et al. 1982). Investigations by Sharp et al. (1979) also have reported that the uptake of chrysene in killifish eggs was only 10% of that of naphthalene following a 2-hour exposure.

A variety of studies have examined the elimination of xenobiotics from early life stages of fish. For many compounds and fish species, xenobiotic elimination is a biphasic process through early development. Guiney and Peterson (1980) examined the elimination of the PCB 2,2',5,5' tetrachlorobiphenyl (TCB) during early development of rainbow trout. The half-life of 2,2',5,5' TCB was approximately 231 days in eggs and sac fry, while the compound was rapidly eliminated from fry (t_2 = 15 days). Other studies with PCBs have demonstrated this general theme. PCB losses from striped bass eggs and during stages of yolk absorption were variable; however, the loss of PCB appeared to increase rapidly once the yolk-sac was absorbed and the larvae began feeding (Westin et al. 1983). Even with continued dietary PCBs, larvae demonstrated a consistent reduction of parentally imparted PCB burdens (Westin et al. 1983). Striped bass larvae lost 78 to 97% of the initial egg PCB concentrations after 20 days of feeding. The PCB losses were most notable on a mg/g basis and were far less dramatic on a per larvae basis, suggesting that at least part of the apparent loss was due to growth dilution. A similar reduction in PCB body burden and in concentration was observed for post-yolk-sac lake trout (Mac and Seelye 1981).

Elimination studies with other compounds in early life stages of other species parallel the studies with PCB. The elimination of TCDD from lake trout eggs and sac fry was very slow, with the absolute amounts of TCDD remaining relatively constant. Once reaching the fry stage (no yolk sac), TCDD was rapidly eliminated (t_2 = 35 to 37 days) in terms of absolute amounts (Walker, Spitsbergen et al. 1991). Studies with TBT and the minnow *Phoxinus phoxinus* show that elimination of TBT was very slow in all stages up through the yolk-sac stage, at which point elimination increased (Fent 1991). Rainbow trout eggs statically exposed to either 10 mg/L of 2,2'-dichlorobiphenyl (DCB) or 200 mg/L of 2,4,6-trichlorophenol (TCP) demonstrated biphasic elimination curves which in large measure correlated to life stage and the feeding process (Freitag et al. 1991). During the yolk-sac stage, elimination of both compounds was slow, while in the feeding larval stage, a rapid increase in depuration rate was noted. The elimination rate was higher on a wet-weight basis than on a per-animal basis. This again suggests that at least part of the apparent elimination was due to growth dilution. Dilution due to growth has been suggested in other studies for larval lake trout (Berlin et al. 1981) and fathead minnows (Defoe et al. 1978).

While biphasic elimination of xenobiotics is a prominent feature in early life stages of fishes, other processes are also operative. Trout embryos exposed at 1, 15, and 25 days post-fertilization to a 24-hour waterborne dose of [^{14}C] benzo(a)pyrene (BaP) demonstrated differential kinetics dependent on stage of exposure (Kocan and Landolt 1984). As long as the egg to volume ratio remained constant, the amount of BaP taken up by the eggs was consistent among all exposure-time treatments. In contrast, the elimination of BaP from the developing egg was highly dependent upon the stage at which the embryo was exposed. Embryos exposed prior to blastogenesis (27 to 30 days for depuration) lost only 2 to 3% of the initial concentration in the egg, while 15 to 20% was lost during a 5-to7-day period for the late exposed animals. Those embryos exposed prior to blastula formation, upon hatching, retained 80% of the initial BaP dose in the sac fry. Those dosed later retained approximately 60%. Such studies suggest that perhaps the time of exposure may influence the accessibility of the contaminant for elimination. Likewise, as has been shown for chemical uptake, other work suggests that compound character may be strongly influential in the early-life-stage elimination process. Solbakken et al. (1984) examined the elimination of ^{14}C phenanthrene, naphthalene, benzo(a)pyrene, and 2,4,5,2',4',5' hexachlorobiphenyl by eggs and newly hatched larvae of cod (*Gadus morhua*). Naphthalene was rapidly eliminated from eggs, while phenanthrene was slowly eliminated BaP and PCB showed no apparent elimination from the eggs after 12 hours in freshwater.

The foregoing studies provide some recurring themes regarding chemical uptake and disposition in early developmental stages. It is clear from the existing data that early life stages may obtain a contaminant burden from both maternal sources and by direct exposure of the life stage to the contaminant in the environment. It is also evident that a variety of compounds can bioaccumulate in embryos and larval stages above ambient exposure levels. The compounds that show the greatest propensity to bioaccumulate, as in adults, appear to be those that are more lipophilic. Studies that have examined chemical distribution in early life stages indicate that contaminants were largely associated with the yolk. Several studies suggest that a greater chemical flux occurs with later development. It is unknown if this is due to development of uptake and depuration pathways, loss of contaminant storage areas in the form of yolk loss, developing competency of metabolic systems, or by some other factors. It is recognized that elimination rates from embryos and yolk-sac larvae are generally slow as compared to rapid increases in the feeding post-yolk-sac stage. There are numerous indications that a substantial portion of these apparent losses are due to growth dilution rather than actual elimination.

Biotransformation: P450 induction and in vitro activities with prototypic substrates

Much of the work with biotransformation in early life stages has focused upon P450 activity and induction as measured by prototypic substrates. Early ontogenic studies with killifish and brook trout demonstrated P450 mediated biotransformational activities (aryl hydrocarbon hydroxylase [AHH]) and their induction in both embryos (eggs pre-hatch) and in larvae (post-hatch) following exposure of embryos to oil or to PCBs (Binder and Stegeman 1980, 1983). Further investigations with hatchery and PCB-contaminated lake trout embryos and swim-up fry correlated contaminant burdens with P450 activity (Binder and Lech 1984). In these investigations, AHH activity was approximately 5-fold higher in lake trout embryos from Lake Michigan and Green Bay parental stock when compared to hatchery controls. Likewise, AHH activity was 4-fold higher in swim-up fry from contaminated sites than from hatchery derived stock. The total PCB burdens for swim-up stages were 0.175, 4.30, and 2.19 mg/g for hatchery, Lake Michigan and Green Bay swim-up fry, respectively. The authors suggested that burdens of contaminants imparted to progeny during maternal development resulted in induction of monooxygenase activities in these early life stages. When juvenile lake trout were allowed to depurate contaminants out to 210 days post-hatch in a control-laboratory situation, the PCB levels in contaminated Lake Michigan animals dropped to near-control levels, as did AHH activities. P450 activities as measured by hepatic AHH activity in control swim-up fry of lake trout were 3-fold higher than those of control embryos. Similarly, Binder et al. (1985) demonstrated that in killifish, embryos have a very low basal cytochrome P450-dependent enzyme activity. An interesting feature of these latter studies is that induction of activity, while present in all life stages, occurred at lower inducer concentrations for the eleutheroembryos (yolk-sac larvae) than for the embryos (Binder et al. 1985). The maximal activities attained, however, were much the same when exposed at higher PCB dosages.

Monod et al. (1996) expanded upon earlier studies by adding a diversity of biotrans-formation reactions and fish species. P450-mediated 7-ethoxyresorufin-O-deethylase (EROD) activity along with NADPH-cytochrome-c-reductase and glutathione-S-transferase (GST) activity was examined in arctic charr (*Salvelinus alpinus*), whitefish (*Coregonus lavaretus*), and grayling (*Thymallus thymallus*). In these studies, enzyme activities increased during embryolarval development for all 3 species. The rate of increase of EROD activities in whitefish, grayling and arctic charr was 3-fold higher following hatching, while NADPH-cytochrome-c-reductase activity appeared to reach maximal rates of increase prior to hatching. GST activity appeared to mirror that of EROD in the grayling and whitefish, while the charr exhibited an increase in catalytic rate prior to hatching. EROD, NADPH-cytochrome-c-reductase, and GST activities in subcellular fractions prepared from arctic charr embryos were 0.00124, 2.03, and 50 nmol/min/mg protein, respectively. In eleuteroembryos the EROD, NADPH-cytochrome-c-reductase, and GST activities were higher, with values of 0.0047, 2.53, and 90 nmol/min/mg protein, respectively. Exposure of arctic charr embryos and eleuteroembryos for 72 hours to 0.12 ppm of the inducer β-naphthoflavone (BNF) in the water resulted in a 6-fold induction of EROD for both life stages and no change for either NADPH-cytochrome-c-reductase or GST.

EROD activity has been measured in larval, juvenile, and adult turbot (*Scophthalmus maximus* L.) following exposure to contaminants and model inducers (Peters and Livingstone 1995). In the turbot, basal EROD activity was not measurable in embryos, but it increased for 3-day larvae through adults. The activities were 0.45, 0.57, 10.8, and 12.3 pmol/min/mg protein for 3-day larvae, 11-day larvae, 90-day juveniles, and adults, respectively. Exposure of 4-day turbot larvae for 24 hours to 5 ppb BaP increased whole-body EROD activity 3-fold, while hepatic EROD activity following exposure to 25 ppb BaP for 48 hours was elevated 2-fold in 90-day juveniles and 13-fold (BNF injection) in adults. Seven-day turbot larvae exposed to lindane for 48 hours demonstrated a 6-fold increase in whole-body EROD activity.

Induction studies have been performed with cod larvae and juveniles following differing exposures to the water soluble fraction (WSF) of North Sea Crude Oil (Goksøyr et al. 1991). Measurement of P4501A1 by an ELISA assay indicated that exposure to levels of WSF as low as 40 mg/L elevated P4501A1 levels in larvae. Even though the exposure was started during the early egg stage, in all of the larval experiments induction was delayed until after hatching. WSF which contains predominately compounds such as benzene, toluene and xylene resulted in P4501A1 induction in juvenile cod as evidenced by EROD, western blot and ELISA assays. When exposed juveniles were transferred to clean water P4501A1 levels dropped to 85% and EROD activities dropped to 64% of exposed animals within 2 days.

Investigations examining waterborne 3,3',4,4' tetrachlorobiphenyl (TCB) accumula-tion in gonads, liver, and muscle of adult fathead minnows, as well as immunohis-tochemical localization of P4501A in the adult and F1 larval fishes, suggests that

TCB transfer to larval fishes may result in P450 induction in those early life stages (Lindstrom-Seppa et al. 1994). The major sites of induction in post-hatched larvae were in the endothelium of yolk-sac vessels and branchial vessels. This may be a result of localized dosimetry as PCB was mobilized from the lipid storage depots in the yolk sac. In studies with rainbow trout eggs, where an inducer (BNF) was injected directly into the egg, CYP1A was strongly induced, particularly in cellular sites that would have been induced in the adult (Lauren et al. 1990). Lake trout eggs injected with TCDD also demonstrate strong induction of P450-dependent activities in multiple cell types in embryos (Guiney et al. 1992).

Taken together, the results of the foregoing studies suggest a number of important features regarding P450 induction and Phase I biotransformation in early life stages. The first being that Phase I xenobiotic metabolism may occur in both the embryo and the larval stages. Basal levels of P450 activity are generally higher in larval stages when compared to the low levels present in embryos. Eggs and larvae may be induced environmentally, through maternal routes, and by experimental routes such as injection. Although CYP1A is inducible before and after hatching in fish, there appears to be an increased sensitivity in the induction response after hatching. P450 activity in eggs and larvae, while inducible, is transitory at induced levels when the source of the inducer is removed. Levels of P450 activity are considerably lower in embryo, yolk-sac, and early swim-up stages than in adults following similar treatments. Potential etiologies for differences in biotransformation and induction response between life stages could be multifold. The fact that eggs respond at higher concentrations and the strong response in eggs upon injection suggest that dosimetry considerations relative to the inducer may play an important role in these findings. Obvious issues such as permeability of the egg chorion, the high lipid reserves in eggs, lack of a defined circulation in the egg, and changes in diffusive surface area upon hatch may be operative factors. Furthermore, it has been suggested that variation in the responsiveness of the Ah receptor may occur among the various life stages (Stegeman and Hahn 1994). Very little is known about both the dosimetry and molecular aspects of P450 induction in early life stages. As is evident from this discussion, even less is known regarding ontogeny, induction, and activity of Phase II reactions in developmental stages.

In vivo biotransformation by early life stages
A limited number of studies have examined in vivo xenobiotic metabolism in early life stages of fish. Eggs or embryos appear to exhibit, by the methods employed, limited in vivo biotransformation. Studies examining metabolism of TBT in the embryos of the minnow *Phoxinus phoxinus* indicated that metabolism, as indicated by the appearance of metabolites in the water, was very low or absent (Fent 1991). Likewise, Sharp et al. (1979) reported that killifish embryos did not metabolize either naphthalene or chrysene when the water was examined. Extracts of lake trout eggs following exposure to TCDD also contained only parent compound with no metabolites being detected (Walker, Spitsbergen et al. 1991). In contrast to the

foregoing studies, trout embryos exposed to a 24-hour, waterborne dose of ^{14}C benzo(a)pyrene excreted water soluble products into water (Kocan and Landolt 1984). Approximately 30 to 45% of the metabolites were extractable with ethyl acetate (phenols and diols) and 60% were conjugated products. However, there was no quantitative indication of how much of the initial BaP dose was metabolized.

Evidence for in vivo xenobiotic metabolism is also inconsistent for the latter stages of early fish development. The in vivo metabolism of aniline and pentachlorophenol (PCP) by Arctic charr eleutheroembryos at the end of yolk-sac resorption were examined by Cravedi et al. (1995). Following a 48-hour static exposure to PCP, 18.3, 24.2, and 49.4% of the radioactivity in water was found as PCP, PCP-glucuronide, and PCP-sulfate, respectively. This compositely amounted to 13.7% of the administered radioactivity. A total of 51.9% of the administered radioactivity was recovered from water following aniline exposure. Parent compound accounted for 13.8%, while acetanilide accounted for 76.4% and p-aminophenol 2.1%. The authors presented this data as evidence that early life stages of salmonids were capable of biotransformation reactions such as N-acetylation, sulfation, and glucuronidation.

Biotransformation of dietary hexachlorobenzene (HCB) was examined in steelhead trout fry (Frankovic et al. 1995). In 1 gram fry, there was evidence of reductive dechlorination, as traces of 2,5-dichlorophenol, 2,3,6-trichlorophenol, and 2,4,5 trichlorophenol were evident in fry extracts late in the depuration phase. However, no metabolites including pentachlorophenol or its conjugates were evident earlier in the time course. HCB in 10g fry was metabolized to low levels of pentachlorophenol and its glucuronic acid conjugate. Biotransformation of tributyltin has been demonstrated to be very slow or absent in yolk-sac larvae (Fent 1991). Additional studies suggest that TBT may in fact inhibit its own metabolism (Fent and Stegeman 1991). As in the embryo, metabolites of TCDD were not detected in sac-fry stages of lake trout (Walker, Spitsbergen et al. 1991).

The in vivo studies with early life stages appear less consistent in demonstrating biotransformation than in vitro experiments using prototypic substrates. What is an uncertainty in the existing knowledge base is what role analytic sensitivity and differential metabolite solubility play on the detectability of inherently low levels of metabolites. The small body size, whole body, and water dilution of metabolites, as well as apparently low-basal, uninduced metabolic rates, complicate assessment of biotransformation in vivo. Biotransformation by eggs or larvae with their low metabolic rates (P450) may well go undetected because of these factors. In eggs, transchorionic movement of both parent and metabolites and xenobiotic partitioning into lipid stores are added processes that may play a modulating role in vivo, but not in vitro. In general, more work will have to be performed on kinetic and biotransformational issues before the determinant factors and relative importance of real-world early-life-stage biotransformation can be ascertained.

Chemical disposition in birds

Overview

Accumulating evidence suggests that a variety of compounds have adversely affected reproduction and development in wild birds. Included among these compounds are DDT and related metabolic products (Ratcliffe 1967; Hickey and Anderson 1968; Cooke 1973), PAHs (Kubiak et al. 1989; Gilbertson et al. 1991; Harris et al. 1993; Hoffman et al. 1993; Larson et al. 1996), methylmercury (Barr 1986), selenium (Ohlendorf et al. 1986), and crude oil mixtures (Ainley et al. 1981; Trivelpiece et al. 1981; Fry et al. 1986). Case studies detailing some of these effects and the mechanisms thought to underlie them are described in Chapter 4 of this volume. For additional information, the reader is referred to several excellent reviews (Giesy et al. 1994; Fry 1995; Barron et al. 1995), as well as a text on interpretation of contaminant residues in wildlife (Beyer et al. 1996).

The response of the scientific community to these observations has been impressive. A survey of literature published in the last 5 years shows that residue levels have been characterized in birds and eggs from numerous contaminated sites, revealing both temporal and geographic trends (e.g., Ormerod and Tyler 1994; Hebert et al. 1994; Hothem et al. 1995). Embryotoxic effects have been described for eggs collected in the field (Harris et al. 1993; Hoffman et al. 1993; Sanderson et al. 1994; Larson et al. 1996) and after maternal dosing in the laboratory (Heinz and Hoffman 1996; MacLellan et al. 1996; Summer et al. 1996; Sanderson et al. 1997). A variety of compounds have been injected directly into eggs to study metabolic biotransformation as well as toxicity (e.g., Nosek et al. 1993; Van Den Berg et al. 1994; Sanderson and Bellward 1995; Janz and Bellward 1996a, 1996b; Sanderson et al. 1997; Zhao et al. 1997), and kinetic studies have been performed to characterize the transfer of accumulated residues from adults to eggs and from eggs to chicks (Nichols et al. 1995; Custer and Custer 1995).

Despite this progress, however, relationships among environmental exposure, absorbed dose, tissue concentration time-course, and toxic effect are seldom known. Traditionally, contaminant levels in wildlife have been related to residues which, on the basis of laboratory or other data, are thought to be toxic (Beyer et al. 1996). These comparisons are often complicated by differences in dosing route, dose level, and exposure duration. Environmental exposures are in general more complex than those carried out in the laboratory, due to fluctuating concentrations and natural history considerations (e.g., migration and changes in dietary preference). Residue levels in field-collected specimens must therefore be viewed as the integrated result of numerous factors, many of which may be poorly understood. Quantitative models embodying a variety of approaches have been used extensively in environmental toxicology to describe contaminant uptake and disposition (reviewed by Landrum et al. 1992; Newman 1995). Unfortunately, efforts to develop such models for birds have lagged considerably behind similar efforts with fish and mammals. It is reasonable to expect that basic toxicokinetic principals established in studies with other taxa would apply also to birds. However, because so little information exists, any discussion of chemical kinetics in birds must be labeled as informed speculation.

The goal of this section is to review factors that could have a large impact on chemical uptake and disposition in birds. Consistent with the goals of this volume, special emphasis is placed on attributes of reproducing adults and their developing offspring. The following topics are discussed in the sections below.

- A general discussion of routes of exposure and other factors relevant to adult birds;
- Metabolic biotransformation, with an emphasis on the metabolic capabilities of developing embryos;
- The reproductive biology of birds, in relationship to maternal transfer of contaminants and the disposition of compounds in eggs and juveniles;
- Two kinetic models for birds, and;
- Suggestions for future research, including the proposed development of linked bioenergetics and physiologically based kinetic models.

Biological attributes of adults that may impact exposure assessment

The diet is the most important route by which adult birds are exposed to toxic compounds. Direct chemical application as a result of, for example, airborne spraying of a pesticide, could lead to a significant dermal exposure, although even in this instance the oral route may predominate because of preening behaviors. The consumption of contaminated water could also contribute to the total applied dose, particularly in exposures to metals and metalloids. Inhalation is unlikely to represent an important route of uptake except in exposures to highly contaminated atmospheres, occurring when animals live within the exhaust plume of an industrial discharge.

Historically, demonstrated impacts on raptorial and piscivorous birds have focused attention on the importance of chemical bioaccumulation and biomagnification. Chemical attributes that contribute to bioaccumulation include environmental persistence and hydrophobicity . Bioaccumulation in aquatic biota is commonly expressed using a bioaccumulation factor (BAF), which is defined as the concentration of a substance in the animal, accumulated by all possible routes, divided by that in water. BAFs exceeding 10,000 are common for hydrophobic compounds that do not readily undergo metabolic biotransformation. Bioaccumulation can also be referenced to a chemical concentration in aquatic sediment. In general, biota-sediment accumulation factors (BSAFs) are much lower than BAFs, owing to the substantial chemical capacity of most sediments. Chemical bioaccumulation also occurs in terrestrial systems. In such cases, residues are generally referenced to chemical concentrations in soil or in dietary constituents. Biomagnification is defined as a progressive increase in chemical concentration at successively higher trophic levels, and it is always associated with a predominantly oral route of exposure. Biomagnification occurs in both aquatic and terrestrial food webs. By definition, biomagnification is associated with an increase in the BAF at each trophic level. Together, bioaccumulation and biomagnification can result in substantial delivered doses to animals that feed at the top of either aquatic or terrestrial webs.

Attention has also been given to birds that feed at intermediate trophic levels. American robins (*Turdus migratorius*) were shown to accumulate substantial residues of DDT and other pesticides after consuming contaminated earthworms (Johnson et al. 1976; Beyer and Gish 1980). More recently, attention has been focused on species that consume emergent aquatic insects, including both tree swallows (*Tachycineta bicolor*) and red-winged blackbirds (*Agelaius phoeniceus*) (DeWeese et al. 1985; Ankley et al. 1993; Nichols et al. 1995; Bishop et al. 1995). Birds also can consume toxic levels of some compounds without a requirement for food web effects or bioaccumulation. One notable example is the ingestion of granulated pesticides by granivorous species (USEPA 1992). In general, however, granivorous birds are less likely to accumulate high body burdens of persistent environmental contaminants than insectivores, piscivores, or carnivores (Enderson et al. 1982; Elliot et al. 1994).

Field surveys suggest that the extent of chemical biomagnification in piscivorous birds usually exceeds that of predatory fish. For example, biomagnification factors (BMFs) ranging from 2 to 10 were reported for a variety of persistent organic compounds in large salmonids from Lake Ontario (Connolly and Pedersen 1988; Oliver and Niimi 1988). Herring gulls (*Larus argentatus*) feeding on the same prey base exhibited BMFs ranging from about 10 to 200 (Clark T et al. 1988; Braune and Norstrom 1989). The reason for this difference is not entirely clear, although it can be speculated that the absence in birds of branchial elimination is a contributing factor. Methylmercury may represent an exception to this general rule. BMFs of approximately 7 were observed in common loons (*Gavia immer*) and herring gulls feeding on rainbow smelt (*Osmerus mordax*) and blunt nosed minnows (*Pimephales notatus*) (Wren et al. 1983). Common mergansers (*Mergus merganser*) feeding on yellow perch exhibited a BMF of 2.5 (Vermeer et al. 1973). BMFs of 2 to 10 are commonly reported for large predatory fish (e.g., Wren et al. 1983; Cope et al. 1990; Lindqvist 1991; Mason and Sullivan 1997). Unlike persistent organic compounds, methylmercury can be eliminated from birds by incorporation into feathers (Braune and Gaskin 1987). Cadmium and selenium have also been shown to deposit in feathers, leading to the use of feathers as non-invasive bioindicators of exposure (Bowerman et al. 1994; reviewed by Scheuhammer 1991; Burger 1993).

Limited data suggest that the oral bioavailability of lipophilic compounds in birds is highly variable and depends upon both the dosing vehicle and the species. For example, the bioavailability of 2,3,7,8-TCDD ranged from 30% (earthworm suspension) to 58% (cricket suspension) when fed to hen pheasants (*Phasianus colchicus*)

(Nosek et al. 1992). Oral bioavailabilities of 90% were observed when Arochlor 1254 was loaded into a gelatin capsule and fed to pigeons (*Columbia livia*) (de Freitas and Norstrom 1974) or spiked into grain and fed to pheasants (Dahlgren et al. 1972). Using an in vitro perfusion technique, Serafin (1984) found that intestinal absorption of 2,2',4,4',5,5'-hexachlorobiphenyl, dieldrin and mercury (as $HgCl_2$) varied greatly among 5 bird species and suggested that these differences could result in dissimilar uptake in intact animals. Unfortunately, the oral bioavailability of most chemical classes in adult birds is essentially unknown, and no data are available for young birds of any species.

Energy metabolism in birds is reviewed by Whittow (1986). Birds are largely homeothermic, although many smaller species allow core temperatures to drop for short periods of time as a means of conserving energy. Standard metabolic rates in non-passerine birds are similar to or somewhat higher than those of placental mammals of similar size, while those of passerine birds may be a factor of 2 higher (McNab 1988; Daan et al. 1990). Metabolic rates across species tend to scale to a fractional exponent of body weight (about 0.7 to 0.8 in adults), resulting in marked differences in weight-adjusted metabolism and consumption. A potential consequence of this fact is that small birds ingesting large quantities of moderately contaminated prey can, within a similar period of time, consume the same dose (on a weight-adjusted basis) as a larger bird eating relatively smaller quantities of more highly contaminated prey.

Migration, courtship, breeding, and parental caregiving behaviors require high expenditures of energy and are often accompanied by periods of starvation. Birds respond to these situations by mobilizing stored lipid. In such cases, toxicity may occur as a result of the release of lipophilic contaminants that, prior to that time, had been sequestered in body-fat stores. Mirex concentrations in brain tissues of 4 species of passerine birds increased during a period of food depletion, resulting in the attainment of lethal residues (Stickle et al. 1973). Similar findings were reported for cowbirds (*Molothrus ater*) and American kestrels (*Falco sparverius*) exposed to DDT and DDE, respectively (VanVelsen et al. 1972; Porter and Wiemeyer 1972). DDE concentrations in brain tissues from sparrowhawks (*Accipiter nisus*) collected in the field were found to be inversely related to whole-body lipid status (Bogan and Newton 1977). Whole-body concentrations of DDE and total PCBs increased in herring gulls during seasonal periods of lipid depletion (Anderson and Hickey 1976). A net movement of PCBs from adipose fat to muscle occurred in pigeons during starvation (de Freitas and Norstrom 1974); upon subsequent feeding, PCBs moved back into fat.

Metabolic biotransformation

Adults

A bird's ability to eliminate a compound once it has been absorbed depends on the attributes of the compound and of the bird itself. In general, relatively polar, organic compounds can be efficiently eliminated in urine or feces. Relatively non-polar compounds tend instead to partition to somatic lipid stores and are eliminated very slowly unless they can be transformed to more polar products. Enzyme systems capable of metabolically transforming xenobiotic compounds exist in many tissues, although enzyme activities tend to be highest in organs of elimination such as the liver and kidney. Because it facilitates elimination, metabolic biotransformation often leads to detoxification of the substance in question. It is important, however, to recognize that in some instances these biotransformation reactions create products that are more toxic than the parent compounds. This outcome, referred to as bioactivation, has attracted a great deal of attention because of the potential for unanticipated toxicity, including effects (e.g., carcinogenesis) that occur long after the initial exposure.

A comprehensive review of metabolic biotransformation in birds is beyond the scope of this chapter. Nevertheless, several generalizations may be attempted regarding metabolism in adult birds. Constitutive levels of metabolic enzymes and the activity of these enzymes toward a variety of substrates have been determined in tissues from several species (reviewed by Pan and Fouts 1978; Ronis and Walker 1989; Walker and Ronis 1989; Walker, Brealey et al. 1991). In general, mixed function oxidase (MFO) activities are highest in the liver. Substantial activity also may be present in the kidney, possibly due to the existence of a renal portal shunt that carries blood from the gastrointestinal tract to the kidney (Pan and Fouts 1978). The cytochrome P450 content of hepatic microsomes varies greatly among species but is, on average, about one-quarter that of most mammals. Differences among species may in some cases be related to feeding ecology. The content and activity of oxidative enzymes are in general higher in omnivorous birds than in piscivorous species. Indeed, enzyme activities in several species of seabirds appear to be similar to those of the fish that they consume (Walker and Knight 1981).

The activities of specific P450 isoforms are generally lower than those of mammals. Gender-related differences in enzyme activity have been noted (Knight and Walker 1982; Walker and Ronis 1989), although in other studies there were no differences between sexes (Knight and Walker 1982; Husain et al. 1984; Peakall et al. 1986). Biotransforming capabilities can in some cases be determined from chemical residue patterns in field-collected birds. For example, the metabolism of specific PCB congeners has been inferred by comparing congener patterns in tissues to those of defined Aroclor mixtures (Borlakoglu et al. 1990a, 1990b, 1990c, 1991). Alternatively, PCB congener levels can be expressed in relationship to the concentration of a congener which, on the basis of other information, is thought to be poorly metabolized (Borlakogu et al. 1990c). Both approaches have been used to establish structural "rules" for metabolism of PCBs by birds, based on the number and position of chlorine substituent groups (Borlakoglu et al. 1990c). Controlled laboratory studies have, for the most part, provided support for these generalizations (de Freitas and Norstrom 1974; Rozemeijer et al. 1995).

In birds, as in mammals, a large number of chemicals have been shown to induce metabolic enzyme activity, leading to the proposal that induction can be used as a biomarker for exposure (Payne et al. 1987; Rattner et al. 1989; Ronis and Walker 1989). Interestingly, variation in the inducibility of specific enzymes tends to be greater among adults than among embryos and juveniles of the same species (Rattner et al. 1989). Additional studies have explored the role of metabolism in determining species sensitivity, either by affecting the kinetics of a toxic parent compound or by creating toxic levels of a reactive metabolite (Walker 1978; Knight et al. 1981; Knight and Walker 1982; Walker et al. 1987; Walker, Brealey et al. 1991).

Embryonic metabolism

In contrast to mammals, birds exhibit high levels of MFO activity during the embryonic and neonatal periods. In chicken embryos (e.g., white leghorn, *Gallus domesticus*), AHH activity 5 days after fertilization (the earliest point at which an embryo can be sampled) is comparable to adult levels (Hamilton et al. 1983). Similar findings have been reported in herring gulls (Boersma et al. 1986; Peakall et al. 1986) and black-crowned night herons (*Nycticorax nycticorax*) (Hoffman et al. 1986). High levels of MFO activity are maintained and may even increase somewhat during the first few days after hatching, declining slowly thereafter (Powis et al. 1976; Haug et al. 1980; Brunstrom 1986).

AHH activity in the chicken embryo is inducible with 3,3',4,4'-tetrachlorobiphenyl (IUPAC no. 77) as early as 5 days after fertilization (Hamilton et al. 1983). A partial listing of compounds that have been shown to induce enzyme activity in ovo includes 3,3',4,4',5-pentachlorobiphenyl (IUPAC no. 126), 3,3',4,4',5,5'-hexachlorobiphenyl (IUPAC no. 169), 3-methylcholanthrene, phenobarbital, allylisopropylacetamide, β-naphthoflavone, and TCDD (Mitani et al. 1971; Poland and Glover 1977; Rifkind et al. 1982, 1985; Brunstrom 1986, 1991; Brunstrom and Andersson 1988; Sanderson and Bellward 1995; Janz and Bellward 1996b; Sanderson et al. 1997). The extent of enzyme induction has been correlated with contaminant residues from field-collected eggs (Hoffman et al. 1987; Bellward et al. 1990; Sanderson et al. 1994, Van Den Berg et al. 1994). Correlations have also been reported between AHH induction and hatching success (Hoffman et al. 1987). A comparison of data from 4 bird species suggested that the extent of EROD induction in ovo by TCDD was related to differences in Ah receptor affinity (Sanderson and Bellward 1995). The potential toxicological significance of embryonic metabolism was demonstrated by injecting a mixture of PAHs into eggs of the chicken on day 4 of incubation. By day 18, greater than 95% of all PAHs taken up by the embryo had been metabolized (Naf et al. 1992). Hydroxylation of 3,3',4,4'-tetrachlorobiphenyl by the chicken embryo yield products that, in comparison to the parent compound, are much less toxic and exhibit lower affinity for the Ah receptor (Wehler et al. 1990).

Aspects of reproductive biology that may affect exposure assessment
Maternal transfer

The process of vitellogenesis and protein deposition in bird eggs is similar to that described previously for fish (Verrinder Gibbins and Robinson 1982; Wallace 1985). Environmental cues give rise to endocrine-modulated synthesis of VTG and other lipoproteins in the liver. These substances are then transported by the circulatory system to the ovary and selectively sequestered as critical components of oocyte maturation. In ovo proteolytic cleavage of these lipoproteins provides for oocyte growth and is the primary nutrient source for the embryonic bird. Unlike the early-life stages of many fish species, however, yolk ceases to be a nutrient source soon after hatching.

From the standpoint of lipoproteins functioning as transporters of xenobiotic compounds, the most notable difference between birds and non-avian oviparous vertebrates lies in the source of lipids for the maturing oocyte. While VTG is the source of the vast majority of yolk lipid in fishes, amphibians, and reptiles, 80 to 90% of the protein-associated lipid in avian oocytes comes from VLDL (Hillyard et al. 1956; Wallace 1985). Furthermore, on a dry-weight basis, VLDL is the major yolk-related protein in birds (Schjeide 1954; McIndoe 1959).

By virtue of its greater lipidation, it can be speculated that VLDL has a greater capacity to transport lipophilic xenobiotics than VTG. However, Borlakoglu et al. (1989) reported that partitioning of various PCB congeners into serum lipoprotein fractions (VLDL, LDL, and HDL) cannot be predicted based solely on solubility in the lipid components of these fractions. The possibility also exists that compounds may bind to VTG and VLDL in a specific manner, facilitating their deposition in the egg. Studies on the maternal transfer of calcium (Packard and Clark 1996), estradiol (Adkins-Regan et al. 1995), and immunoglobulins (Lung et al. 1996) suggest the importance of this mechanism in determining both egg composition and successful embryogenesis. In general, however, the extent that lipoprotein participates in the oocytic sequestration of both endogenous and xenobiotic compounds in avian species has yet to be clearly defined.

Chemical residues in eggs and tissues from the same maternal parent have been shown to be statistically correlated (Mineau 1982). Similar correlations exist when eggs from the same nest are compared to eggs from different nests (Thompson et al. 1977; Custer et al. 1990). When normalized for lipid content, DDE levels in eggs and carcasses of black ducks (*Anas rubripes*) were essentially identical (Longcore and Stendell 1977). A similar result was reported for DDE in sparrowhawks (Bogan and Newton 1977). Collectively, these observations suggest that a chemical equilibrium is established between a female and its eggs before the eggs are laid.

In other cases, an internal equilibrium does not appear to have been achieved. Lipid-normalized PCB residues in eggs and liver from herring gulls were about half the levels measured in adipose tissue (Norstrom, Clark, Jeffrey et al. 1986). It was suggested by these authors that diet, rather than adipose tissue, was the main source of lipid deposited in eggs. Residue levels in eggs thus came to resemble those of the liver, which is the organ within which yolk lipids are synthesized and packaged. Low chemical concentrations in the liver (lipid-normalized) were thought to be due to dilution of existing lipid stores with "new" lipid absorbed directly from the digestive tract or synthesized from dietary protein.

Work with several bird species supports the suggestion that lipids required for egg formation derive primarily from dietary sources (Roudybush et al. 1979). However, the extent to which this observation can be generalized to all birds is unclear. Moreover, the relative contribution of stored and dietary lipid may change with body condition and food availability, and perhaps even within a single clutch of eggs. Thus, Mineau (1982) found that organochlorine contaminant levels increased in sequentially laid herring gull eggs, consistent with an increase in lipid content.

There is also a lack of consensus on whether egg laying represents an important route of chemical elimination for female birds. Nosek et al. (1992) found that female pheasant transfer about 1% of their TCDD body burden to each of 15 sequentially laid eggs. Based on the observation that wild pheasant lay as many as 30 eggs a year, it was concluded that egg laying contributes substantially to chemical elimination. Sparrowhawks eliminated as much as 50% of their DDE burden in a single clutch of eggs (Bogan and Newton 1977), while Arctic terns (*Sterna paradisaea*) and herring gulls transferred 45% and 24%, respectively, of their maternal PCB load to eggs (Lemmetyinen et al. 1982).

It may not be possible, however, to determine the importance of egg laying as an elimination route from a simple comparison of residue masses (Σ all eggs/total in the parent). For example, Norstrom, Clark, Jeffrey et al. (1986) found that although 16% of [^{14}C] DDE was transferred from female gulls to eggs, the combined weight of the eggs comprised 28% of the total body weight. The concentration of DDE in the eggs was therefore considerably lower than that of the gulls. Under such circumstances, it may be speculated that egg laying results in an *increase* in whole-body residue concentration. Norstrom, Clark, Jeffrey et al. (1986) concluded that while important, egg laying accounted for only 15% of the total elimination of DDE. Taken together, these data suggest that the importance of egg laying as a route of chemical elimination is species-specific and depends on a number of factors including egg and clutch size, maternal fat reserves, and the extent to which these reserves are mobilized to provide for yolk deposition.

Data pertaining to maternal transfer of compounds other than lipophilic organics are extremely limited. Heavy metals, including lead and cadmium, have been detected in eggs of several bird species (Burger and Gochfeld 1991, 1993, 1995). However, relationships between metal concentrations in parents and their eggs are not commonly known. In a study of common terns, lead levels in eggs were positively correlated with levels in feathers from the maternal parent (Burger and

Gochfeld 1991), but were much lower overall. Cadmium levels in eggs were also much lower than those in feathers and did not exhibit a statistical correlation with maternal concentrations. The significance of these observations is difficult to determine. As indicated previously, several metals are eliminated from birds by incorporation into feathers. Comparing the metal concentration in an egg to that of a feather does not, therefore, provide a straightforward basis for comparisons with internal tissue levels. In the same study, lead levels were higher in feathers from female terns than in feathers from males. However, because of possible sex-related differences in dietary loading, this finding does not by itself indicate that eggs are unimportant as an elimination route for lead.

Interestingly, concentrations of some heavy metals may be higher in the shell portion of the egg than in the egg contents (Burger 1994). This is important, since many researchers analyze only the egg contents when they are reporting on maternal transfer of metals. The distribution of metal between the shell and the egg contents may also have toxicological implications for the developing embryo.

Selenium concentrations in eggs taken from waterfowl at a contaminated site were comparable to those in the livers of adult birds (Ohlendorf et al. 1986). Subsequent laboratory studies with mallard ducks (*Anas platyrhynchos*) confirmed that selenium levels in eggs and livers of breeding females are similar and highly correlated (Heinz et al. 1989). Correlations among selenium levels in other tissues have also been reported (reviewed by Heinz 1996). However, these levels can change rapidly as selenium levels in the diet change. In contrast to selenium, mercury levels in eggs from common loons were much lower than those in the livers or feathers of adult birds (Belant and Anderson 1990).

The interpretation of these findings is complicated by the fact that selenium and mercury do not partition with tissue lipid, but instead are incorporated into protein. The uptake and retention of both compounds are enhanced by prior transformation to organometallic forms (principally selenomethionine and methylmecury). As indicated previously, deposition into feathers represents a route by which birds eliminate mercury and selenium (Braune and Gaskin 1987; Bowerman et al. 1994). Hepatic demethylation may also play a role in the elimination of mercury. Evidence for demethylation is provided by the observation that mercury found in the liver often exists as Hg^{2+}, while in other tissues the predominant form is methylmercury. Studies suggest that the mechanism of demethylation depends upon the presence of selenium (Palmisano et al. 1995; Cavalli and Cardellicchio 1995), possibly explaining the protective effect of dietary selenium against methylmercury toxicity to birds (Ganther et al. 1972). Limited evidence also suggests that the extent to which this pathway is developed depends upon a bird's feeding habits. Among adult ducks, Fimreite (1974) found that fish-eating mergansers contained the lowest levels of methylmercury in liver (12% of total), while in goldeneyes, mallards, and pintails, methylmercury constituted 32, 38, and 52% of the total, respectively. Work by Fimreite (1974) also suggests that this detoxifying ability appears early in life.

Thus, methylmercury in livers taken from ducklings constituted 27, 49, 53, and 58% of the total in mergansers, mallards, goldeneyes and pintails, respectively. Methylmercury levels in breast muscle from all 4 species were essentially identical, averaging about 60% of total.

Embryonic development

Once an egg is laid, its protective shell limits chemical interactions between the embryo and its surroundings. Toxic substances applied directly to eggs may in some instances cause embryotoxicity and reduced hatchability (reviewed by Hoffman 1990; Hoffman and Albers 1984), but this occurs only rarely in the wild. In a majority of cases, chemical dose to the developing embryo is determined by maternal inheritance and subsequent internal events. Chemical residues in eggs are usually measured and expressed on a whole-egg basis. Limited data suggest that very early in development, hydrophobic organic compounds partition almost exclusively to yolk, due to its high lipid content (Nosek et al. 1992). Later, as embryos grow, these chemicals diffuse across the yolk membrane and redistribute to developing tissues and organs (Naf et al. 1992; Sanderson and Bellward 1995). It is unclear, however, whether this redistribution results in an equilibrium distribution among tissues. Because the egg is a closed system, this question may be of particular importance. As noted previously, contaminant concentrations in eggs may approach maternal levels when normalized for lipid content. During development, most of the yolk originally deposited in an egg is consumed to support the growth tissues which are, by comparison, relatively lean. Assuming an absence of metabolic biotransformation, chemical concentration within the egg must remain the same. The potential exists, therefore, for chemical concentrations in "lean" embryonic tissues to exceed those of the same tissues of the maternal parent.

Juvenile development

Avian physiologists frequently distinguish among species based on the extent of development of the young at hatching. Birds that are in an advanced state of development are called precocial, while those in an early stage of development are termed altricial. An examination of these two breeding strategies suggests several differences that could have toxicological and toxicokinetic consequences. Relative to the size of the adult bird, the eggs of altricial species are smaller than those of precocial species and contain less yolk. Altricial species do not thermoregulate upon hatching and can therefore utilize a higher percentage of ingested energy for growth than precocial birds. On the other hand, demands on the adults of altricial species may be greater due to relatively narrower limits for brooding behavior. Anecdotal evidence suggests that in several cases, toxicity to young birds was exacerbated by reduced quality of parental care (Fox et al. 1978; Kubiak et al. 1989).

Limited data are available on chemical kinetics in very young birds. Nosek et al. (1992) estimated the $t_{1/2}$ for whole-body elimination of [³H]TCDD (total TCDD-derived radioactivity) in pheasant chicks to be 13 d, while that of adult hens was determined to be 378 d. Pheasant chicks deplete inherited yolk reserves during the first few days after hatching, causing whole-body lipid content to decline. Based on this observation, it was suggested that high rates of TCDD elimination in chicks occur primarily via partitioning from tissues into contents of the gastrointestinal tract.

Modeling Exposure and Disposition of Contaminants in Oviparous Vertebrates

On an absorbed-dose basis, early life stages of oviparous vertebrates often exhibit greater sensitivity to chemical contaminants than do adult life stages. For example, lake trout fry exposed as eggs to TCDD exhibit a suite of toxic effects including pericardial and yolk sac edema, subcutaneous hemorrhaging, cranio-facial alterations, and arrested development. The concentration of TCDD in eggs required to elicit this response is between 40 and 80 pg/g, regardless of whether the compound is injected directly, taken up from contaminated water, or inherited via maternal transfer (Walker et al. 1994). In contrast, adult lake trout live apparently "normal" lives even when whole-body concentrations of TCDD approach 100 pg/g. These and similar observations underscore the importance of accurately assessing the exposure of early life stages.

One means of performing exposure assessments is through the development and use of mathematical models. These models formalize and simplify complex phenomena and can be used to extrapolate limited information. Research is conducted in support of model development and as a means of evaluating model performance. Descriptive research is frequently conducted in advance of more mechanistic work to define the system under study and to collect an empirical dataset, which then becomes the basis for developing mechanistic hypotheses. To date, modeling efforts with oviparous vertebrates have focused primarily on descriptions of chemical uptake and disposition in adult fish. Several different modeling approaches have been used, depending upon the specific application. These models vary considerably with respect to underlying assumptions and the level of mechanistic detail. All, however, are based on established principals of chemical mass-balance. Readers interested in more information on modeling efforts with adult fish are referred to several excellent reviews (e.g., Barron et al. 1990; McKim and Nichols 1994; Gobas and Morrison 1999).

Material presented earlier in this chapter indicates that many environmental contaminants are transferred from female fish or birds to their eggs. Several authors have suggested that this transfer provides a means by which adult females eliminate accumulated chemical residues. By comparison, the disposition of chemical within the developing embryo has received less attention. In the following sections, a novel approach is described for modeling in ovo exposures to lipophilic organic compounds. Based on the principal of chemical fugacity, this approach is deliberately presented as a simple and generic method that can be used to derive a first-order assessment of chemical exposure. In subsequent sections, an effort is made to describe 2 kinetic models for contaminant bioaccumulation in birds. The first is a compartmental model for organic chemical accumulation in the herring gull; the second is a bioenergetics-based model for PCB uptake by nestling tree swallows. Together these 2 efforts represent the totality of exposure modeling efforts with birds published to date. A concluding section provides suggestions for future research, with an emphasis on the development of advanced kinetic models for birds.

Fugacity approach to describing in ovo exposure

The goal of an exposure assessment is to relate chemical concentrations in a range of environmental media (e.g., water, sediment, diet) to the effective concentration at the site of toxic action. It is well recognized that physicochemical characteristics of both environmental media and biological material (i.e., tissues) have a profound effect on the relationship between an external chemical concentration and that which exists within the organism. For example, the long-term exposure of a fish to 1 ng/L of pentachlorobenzene in water results in a whole-body concentration similar to that which is achieved when the fish is exposed to a much greater concentration (10,000 ng/kg) in the diet. Contaminant concentrations also can vary among different tissues as a result of differences in their chemical makeup.

The role of physicochemical factors in controlling exposure is very important for toxicological assessments. Environmental monitoring studies provide an indication of chemical exposure in terms of whole organism or tissue concentrations. The relevance of these concentration data for toxicological assessments may be unclear, however, because the chemical makeup of tissues that comprise the suspected target organ is different from that of tissues that were sampled. To avoid this problem, it is important to express concentrations in the ambient environment and inside the organisms in terms of "effective" concentrations. These effective concentrations are equivalent to the "chemical potential" or "chemical activity" used by chemists and engineers to express the thermodynamic status of the chemical. There are various ways to express, quantify and measure chemical activity in the ambient environment and the organism. One approach that has already been discussed is to express chemical residues in tissues as lipid-normalized values. As indicated previously, differences in concentrations of persistent organic chemicals among tissues often disappear when they are expressed on a lipid-normalized basis, indicating that a "common" concentration applies to these tissues (Geyer et al. 1985). In these cases, concentrations measured in one tissue can be interpreted in terms of the effective concentration in another tissue or organ.

A second technique is to use the principal of chemical fugacity to characterize the thermodynamic status of a system. Applications of the fugacity concept to environmental modeling have been discussed by Mackay (1991). The second law of thermo-dynamics states that when a chemical is allowed to exchange between 2 adjoining media, it will tend to approach a situation wherein chemical fugacities in both media are the same. In an environmental setting, which contains media of differing chemical composition, this means that the diffusing chemical substance will tend toward equal fugacities but not equal concentrations. The main advantage of the fugacity approach is that physicochemical factors affecting relationships between external and internal chemical concentrations can be isolated from other factors because all concentrations are expressed on a common basis. The advantage to toxicologists is that fugacity can be directly related to chemical concentration at a target organ regardless of its composition. Developments in gas sparging (Yin and Hassett 1986; Sproule et al. 1991; Horstmann and McLachlan 1992) and solid phase extraction techniques (Zhang and Pawliszyn 1990; Zhang et al. 1994; Parkerton and Stone 1996) provide a very promising future for the measurement of chemical fugacities in environmental media and tissues.

A fugacity-based model for maternal deposition of hydrophobic organic contaminants into eggs

The primary assumption of the egg-deposition model for hydrophobic organic chemicals is that chemical transport from maternal tissues to developing eggs follows a set of passive (non-energy consuming) transport processes resulting in a thermodynamic equilibrium. This assumption is based on the observations that

- the internal distribution of hydrophobic organic chemicals within organisms tends to be relatively fast, often resulting in a tissue distribution that is homogenous when expressed on an appropriate basis (e.g., fugacity or lipid based concentration) (Clark et al. 1987; Nichols et al. 1990);
- hydrophobic organic chemicals generally exhibit high rates of permeability in biological membranes (Stein 1981);
- egg formation involves the transfer of lipoproteins from maternal tissues to the eggs; and
- the metabolic biotransformation of contaminants in eggs is often negligible because the necessary enzymes are not yet active.

Adopting this assumption, a chemical equilibrium can be formulated in terms of the chemical fugacities (Pa) in the maternal tissues (f_M) and the eggs (f_E): $f_E = f_M$. Chemical fugacity is equal to the ratio of the chemical concentration C (in mol/m^3) in a tissue and its fugacity capacity Z (mol/m^3 Pa): $f = C/Z$. It follows that chemical concentrations in the eggs (C_E) and the maternal tissues (C_M) reflect the ratio of the fugacity capacities of the eggs (Z_E) and the maternal tissues (Z_M), resulting in an egg-to-mother concentration factor (EMF) of:

$$EMF = C_E/C_M = Z_E/Z_M$$

Equation 2-1.

Because a large number of studies have demonstrated that hydrophobic organic chemicals in organisms largely reside in lipids (Geyer et al. 1985) and that the solubility of contaminants does not differ substantially between different types of lipids (Dobbs and Williams 1983), it is reasonable to assume that Z_E and Z_L are approximately equal to $L_E Z_L$ and $L_M Z_M$ respectively, where L_E is the lipid content of the eggs, L_M is the maternal tissues, and Z_L is the fugacity capacity of both maternal and egg lipids. Substituting these relationships in Equation 2-1 gives us Equation 2-2:

$$EMF = C_E/C_M = L_E/L_M \qquad \text{Equation 2-2.}$$

Equation 2-2 suggests that the relationship between chemical concentrations in the egg and the maternal tissue simply reflects relative differences in lipid content. It follows, therefore, that if chemical concentrations are expressed on a lipid-weight basis as C_{EL} and C_{ML}, the relationship between the lipid based chemical concentrations in the eggs and the maternal tissues, which will be referred to as EMFL, is

$$EMFL = C_{EL}/C_{ML} = 1.0 \qquad \text{Equation 2-3.}$$

Equation 2-3 states that on a lipid-weight basis, chemical concentrations in eggs and maternal tissues will tend to approach equality. An important implication of this result is that the concentration of a chemical contaminant in an egg may be predicted from the concentration of the chemical in maternal tissues, provided that the lipid contents of the egg and maternal tissues are known.

It should be stressed that Equations 2-2 and 2-3 can be used to represent a chemical equilibrium only if 1) the fugacity capacity of different types of lipids is approximately the same and 2) tissue components other than the lipids do not contribute significantly to the total fugacity capacity of the eggs or the maternal tissues. If the lipid content is very low, the EMFL can be expected to differ from 1.0 even if a chemical equilibrium between eggs and maternal tissues exists. Also, the EMFL can be expected to differ from 1.0 while an equilibrium exists if there are significant differences in lipid composition between maternal tissues and eggs and in solubilities among different types of lipid in the eggs and maternal tissues. The error associated with the assumption that lipids are the sole component of egg and muscle tissue providing fugacity capacity for chemicals and that the fugacity capacity of different kinds of lipids are similar may contribute to the variability in observed EMFLs around the predicted model value of 1.0.

Model versus data

A considerable amount of research has been done on the transfer of chemical contaminants from maternal tissues into eggs of both birds and fish. For example, Braune and Norstrom (1989) published ratios of egg/female concentrations in herring gulls and suggested a mechanism for maternal transfer; Niimi (1983) provided data on the transfer of several contaminants into rainbow trout eggs; Miller (1993) examined differences in maternal transfer of organochlorines into

eggs of lake trout and chinook salmon; and Miller and Amrhien (1995) characterized the maternal transfer of PCBs and several chlorinated pesticides in Lake
Superior siscowet, a subspecies of lake trout noted for its high lipid content.

A comprehensive collection of maternal transfer data was recently published by
Russell et al. (1999). These investigators combined existing data with the results of
field studies on Lake Erie to determine maternal transfer and in ovo bioaccumulation of 44 hydrophobic organic chemicals in 9 species of fish, one species of bird
(herring gull), and 1 reptile species (snapping turtle, *Chelydra serpentina*) (Figure
2-4). An examination of these data suggests that there are very large differences
among fish species in the egg/female lipid content ratio, ranging from approximately 0.78 for carp (*Cyprinus carpio*) to 41.9 for the freshwater drum (*Aplodinotus
grunniens*). The egg/female concentration ratios (EMF) for test chemicals also varied
considerably among species, ranging from 0.69 to 51.5. In terms of wet-weight-
based concentrations, therefore, the eggs of some species appeared to receive a
greater dose than others. However, when chemical concentrations in the eggs and

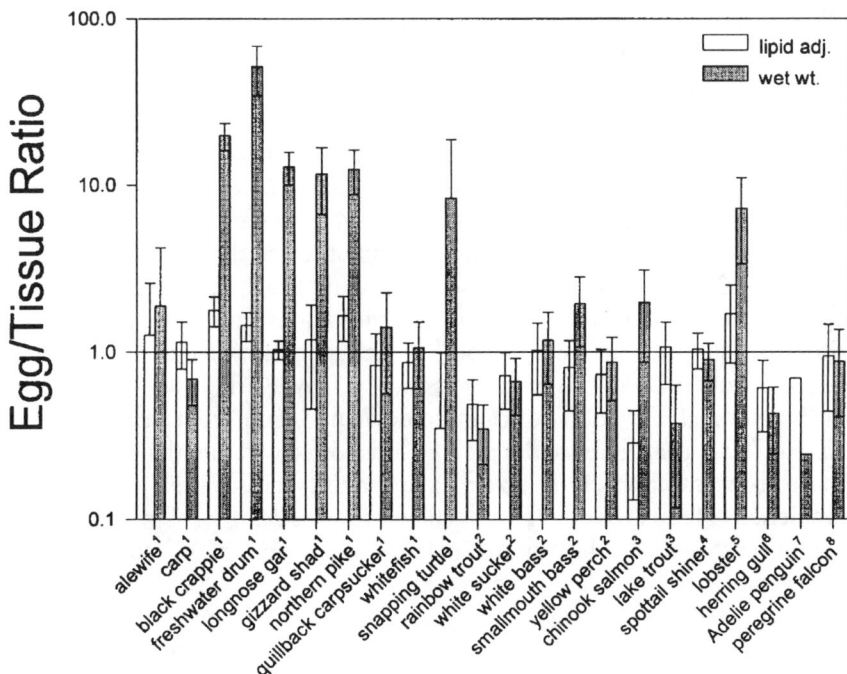

Figure 2-4 Observed relationships between chemical concentration in eggs and maternal tissues, as
expressed by lipid-weight-based (gray bars) and wet-weight-based (white bars) egg/maternal tissue
concentration ratios, in several classes of oviparous organisms. The solid horizontal line represents
the chemical equilibrium model prediction for lipid-based concentration ratios. Data are from
[1]Russell et al. (1999), [2]Niimi (1983), [3]Miller(1993), [4]Noguchi and Hesselberg (1991), [5]Guarino et al.
(1974), [6]Braune and Norstrom (1989), [7]Tanabe et al. (1986), and [8]Cade et al. (1968).

the females were adjusted for lipid content, the egg/female concentration ratios (EMFL) were normally distributed with a mean of 1.22, a 2.5 percentile of 0.56, and a 97.5 percentile of 2.51. Importantly, EMF and EMFL values did not vary with chemical log K_{ow}, suggesting that maternal transfer in fish is essentially independent of chemical hydrophobicity. The mean EMFL for 24 chemicals in snapping turtles was 0.35. A statistical analysis suggested that this value was significantly less than 1.0 ($\alpha = 0.05$). The mean EMFL for 47 chemicals in herring gulls, as reported by Braune and Norstrom (1986), was 0.61. This value was also significantly lower than 1.0.

On average, therefore, eggs from snapping turtles and herring gulls did not contain as much chemical as the fugacity model would have predicted from chemical residues in female parents; at the same time, fish eggs contained chemical residues that were slightly higher than expected. Nevertheless, it is remarkable that concentrations in eggs and maternal tissues were so similar across such a large range of species and chemicals. This suggests that at the time of egg deposition, contaminant concentrations in the eggs and maternal tissues of fish, turtles, and birds are close to chemical equilibrium.

When all data were combined, the variability in the mean EMFL, expressed as 95% probability intervals, was approximately a factor of 2. This means that 95% of the EMFL values were found to be between 0.56 and 2.51. Part of this variability may originate from differences in lipid composition (not lipid content) between the egg and the mother. Differences in lipid composition may provide a somewhat different fugacity capacity to the egg and the mother if chemicals exhibit different solubilities and fugacity capacities in different types of lipids (e.g., phospholipids versus nonpolar triglycerides). In addition, tissue components other than lipids may have a small effect on the fugacity capacity of the egg contents. The lipid contents, as determined by the extraction techniques used in the various studies, may not account for the role of non-lipid components of the eggs and the female fish on the fugacity capacity of the chemicals in the eggs and the female fish, hence causing an apparent (but not real) deviation from a thermodynamic equilibrium. This apparent deviation may "disappear" if fugacities are measured directly rather than indirectly as lipid-based concentrations. If the observed uncertainty in the EMFL is applied to model predictions of egg exposure for new chemicals or species, it follows that lipid-based egg concentrations can be expected to fall within a range of 0.56 to 2.51 times the lipid-based concentration in the mother. Because all of the current observations regarding maternal transfer of organic chemicals involve persistent hydrophobic organic chemicals, it is unclear whether this simple partitioning model applies to less hydrophobic chemicals or to chemicals that undergo metabolic biotransformation. Nevertheless, the information for persistent chemicals allows for some extrapolation. If the chemical is metabolized, it is possible that the model does not apply. For example, if the chemical is preferentially metabolized in the eggs, the fugacity in the eggs can be expected to be lower than that of maternal tissues.

Preferential metabolism by the female may have little effect on model predictions if concentrations of the parent compound remain relatively constant. This is because the egg would "inherit" parent compound in the same manner as an unmetabolized compound; that is, chemical activity in the female and the egg would be reduced by metabolism to a similar extent.

Chemical exchange between deposited eggs and the environment

Maternal transfer is not the only process that affects the concentration and fugacity of chemicals in the eggs. Eggs deposited by fish may be exposed to contaminants in water and benthic sediments, while for reptiles there may be exposure to contaminants in soil and groundwater. Bird eggs may also experience some exposure to environmental contaminants, although the amount of chemical exchange that occurs may be limited by a relatively impermeable shell. Because of the multiple media involved in egg exposure, it is advantageous to characterize the potential for chemical transfer in terms of the differences in chemical fugacity in the eggs and those in the ambient environment. If the chemical fugacity in water and/or sediment exceeds the fugacity in the eggs, there is a potential for the concentration and fugacity of the chemical in the eggs to increase. This may occur, for example, when fish return from relatively clean sites (e.g., ocean or main river stem) to deposit their eggs in areas that are subject to contamination (e.g., gravel beds or sloughs). In these situations, the fugacity in the eggs will tend toward the prevailing fugacity in the environment. The extent to which ambient fugacity is achieved is dependent on the rate of diffusion of the chemical into the eggs and the duration over which the exposure occurs. These diffusion rates are largely unknown but can be expected to depend on hydrodynamic conditions (water flow over eggs), the thickness and composition of the chorion, the surface area and size of the eggs, and molecular characteristics related to diffusion (e.g., molecular volume). Although the relevant data do not exist, it is not inconceivable that chemical concentrations and fugacities in eggs can be substantially increased (perhaps approaching the ambient fugacity) under conditions when relatively uncontaminated eggs are exposed to high concentrations of chemicals at contaminated sites.

If the chemical fugacity in deposited eggs is less than that in the ambient environment, there is a potential for the chemical fugacity in the eggs to decrease. The rate of depuration of contaminants from the eggs is affected by the same diffusional and temporal factors that control the rate of uptake. The limited amount of data on chemical elimination from fish eggs was discussed previously. This elimination is expected to occur more slowly than that from yolk-sac fry and may be negligible for high log K_{ow} compounds even in a relatively clean environment. If significant depuration occurs, however, as can be expected for some lower K_{ow} chemicals, then the eggs may loose some of their chemical burden and approach the chemical fugacity in the environment.

Finally, it is well known that some very hydrophobic organic chemicals (log $K_{ow} > 6$) have a tendency to accumulate in aquatic food chains resulting in chemical fugacities that increase with each trophic level. Under these circumstances organisms at the highest trophic levels can be expected to deposit eggs that possess a chemical fugacity greater than that in surrounding water or sediment. Unless the eggs are being deposited in a location of greater contamination than that in which the female resided, there is a good possibility that chemical fugacity in the eggs is greater than that of the environment. Under these conditions, in ovo concentrations and fugacities of these chemicals could be expected to fall.

Chemical exposure during embryo development

During embryonic development, material that was transferred to the egg for nutritional purposes is consumed, changing the environment within which the embryo resides. With respect to the distribution of hydrophobic organic chemicals, the most important change is a potential decline in lipid content. For example, Peakall and Gilman (1979) reported a 2-fold decline in the lipid content of herring gull eggs during incubation. This means that although there may be no change in chemical mass or concentration in the egg (assuming, for simplicity, no exchange with the environment), the lipid-based concentration can be expected to increase. In terms of the fugacity theory, this is equivalent to a drop in the fugacity capacity of the egg. Assuming that the chemical concentration does not change, this will result in an increase in chemical fugacity, because fugacity is equal to the ratio of concentration (C) and the fugacity capacity (Z): $f = C/Z$. This increase in fugacity would then be associated with an increase in the chemical's effective concentration, as illustrated in Figure 2-5. Because it is a "closed" system, the potential for consumption of stored yolk to increase fugacity may be greatest in bird eggs. In fish eggs an increase in fugacity can potentially be offset by elimination to the environment (assuming that the fugacity of the environment is similar to or less than that of the egg when it was deposited). The final concentration, therefore, would represent the net result of these 2 competing processes.

Conclusions

There is a considerable body of evidence indicating that concentrations of hydrophobic organic chemicals in eggs closely reflect the concentration in maternal tissues as long as the concentrations are expressed on a lipid-weight basis. The available data are reasonably consistent with a simple model that assumes that chemical concentrations in eggs and maternal tissues achieve a chemical equilibrium. The data and the model imply that chemical transfer into eggs is a passive process that follows fugacity gradients. The result is that the chemical fugacity in the egg is close to that in the maternal tissues. This simple equilibrium model can be used to provide first-order estimates of in ovo chemical exposure at the time of egg deposition based upon knowledge of maternal tissue residues, provided that the lipid contents of the maternal tissues and egg are known.

	Adult	Egg	Embryo Development (1)	Embryo Development (2)	Embryo Development (3)	Fry
f (Pa):	1	1	1.1	1.3	2	4
Z (mol/m^{3}Pa):	10	20	18	15	10	5
Z_L (mol/m^{3}Pa):	100	100	100	100	100	100
C (mol/m^3):	10	20	20	20	20	20
C_L (mol/m^3):	100	100	110	130	200	400

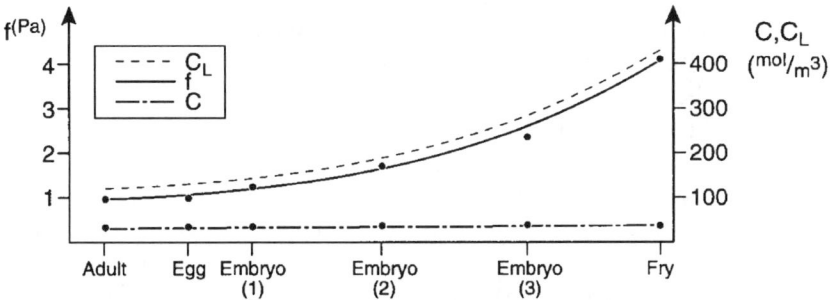

Figure 2-5 Illustrative example of the expected effect of in ovo embryonic development on the wet-weight-based chemical concentration in the egg (C), the lipid-based chemical concentration in the egg (C$_L$), the chemical fugacity capacity of the egg (Z) and the chemical fugacity (f) in the egg.

During embryo development, the lipid-based chemical concentration has the potential to increase, particularly within bird eggs, resulting in an effective concentration that is greater than that of the mother. If the embryo exhibits the same toxicological sensitivity to the contaminant as the mother, the expected increase in effective chemical concentration in the egg will cause a degree of toxicological risk to the embryo that is greater than that to the mother. An increase in the sensitivity of the embryo relative to that of the mother would tend to make this situation even worse.

Chemical uptake and accumulation models for birds

The foregoing discussion highlights the fact that in many instances it is possible to describe chemical uptake and bioaccumulation in oviparous vertebrates using an equilibrium partitioning approach (i.e., by assuming that an organism is in thermodynamic equilibrium with all or part of its environment). Alternatively, a dynamic equilibrium may occur as the result of a balance between uptake and elimination, including metabolic biotransformation. In other cases, however, chemicals do not attain chemical equilibrium, either within an organism or between an organism and its environment. Factors that tend to promote this condition include fluctuating exposures, rapid growth, rapid mobilization of lipid stores, and changing metabolic capabilities, particularly when these factors are combined with chemical attributes (e.g., hydrophobicity, low bioavailability) that result in low rates of uptake and elimination. An examination of field residue data suggests that as a generalization, the potential for chemical disequilibrium between an organism and its environment increases at progressively higher trophic levels (Clark T et al. 1988). Thus, while an equilibrium may exist between a female and its eggs, the female may not be in equilibrium with its environment.

Kinetic processes can be included in fugacity-based models, which then combine information on partitioning, transport, and biotransformation (Mackay 1991). The advantage of these models is that they distinguish between kinetic and physico-chemical factors controlling chemical distribution, which improves insights into the chemical distribution process. A more "classical," compartmental kinetic modeling approach may also yield insights, particularly when fitted rate constants are interpreted in the context of relevant physiological information (e.g., glomerular filtration rate for a compound eliminated in urine). Physiologically based toxicokinetic (PBTK) models are based on anatomical, physiological, and biochemical information, and they do not require a priori collection of data to fit the value of kinetic rate constants. In practice, however, they incorporate many of the same principals employed in fugacity-based modeling, insofar as mass-balance equations for individual tissues require the specification of chemical capacity relative to that of blood. For many compounds this chemical capacity is determined largely by tissue lipid content.

The following sections describe 2 models that have been developed to describe contaminant uptake and accumulation by birds. The first is a 2-compartment kinetic model for herring gulls published by Clark et al. (1987). Although originally formulated with forward and reverse rate constants for internal chemical distribution, implementation of the model was subsequently simplified by assuming that the 2 compartments were in chemical equilibrium. The second model is a bioenergetics-based model for tree swallows given by Nichols et al. (1995). This model is notable because it yields concentration predictions based upon continuously changing rates of chemical uptake and organism growth.

Two-compartment, open model for herring gulls

The herring gull is the only year-round, resident piscivorous bird species in the Great Lakes region. In the 1960s it was found that gull populations in Lakes Ontario and Erie were declining and that embryo mortality was the principal cause (Keith 1966; Ludwig and Tomoff 1966). Efforts to identify the compounds responsible for this mortality and the mechanisms by which they act were complicated by the fact that gulls are simultaneously exposed to many bioaccumulative contaminants, including pesticides, PCBs, dioxins, and methylmercury. Herring gulls are relatively resistant to eggshell thinning by DDE (Gilbertson 1974; Gilman et al. 1977). Perhaps the strongest case to date has been made for compounds with a TCDD-like mode of action, resulting in a suite of toxic effects that include porphyria, wasting syndrome, and edema (reviewed by Gilbertson 1991; Giesy et al. 1994; Chapter 4 of this volume).

Efforts were initiated in the late 1960s to monitor contaminant levels in herring gull eggs (Mineau et al. 1984). These efforts continue to this day and have provided detailed descriptions of geographical and temporal trends in organochlorine contaminants in eggs collected from each of the Great Lakes (Weseloh et al. 1990, 1994; Ewins et al. 1992; Hebert et al. 1994). Organochlorine levels in eggs declined substantially in the 1970s, coinciding with improvements in herring gull reproduction. Efforts to understand these changes in relationship to contaminant levels in environmental media and adult birds were complicated, however, by the long-lived nature of gulls and by seasonal changes in locality, dietary choice, and lipid status. Studies were therefore initiated to better characterize contaminant uptake, distribution, and clearance in both captive and free-living gulls, maternal transfer of chemical residues to eggs (Anderson and Hickey 1976; Norstrom, Clark, Jeffrey et al. 1986), and basic aspects of gull biology and bioenergetics (Norstrom, Clark, Kearney et al. 1986).

These studies eventually led to the development of a 2-compartment open model for gulls (Figure 2-6; Clark et al. 1987), with elimination from a central "plasma" compartment and chemical exchange between plasma and a peripheral "fat" compartment. Employing symbols used by the authors, the contaminant concentration in plasma is C_p, contaminant burden in fat is X_f, and the weight of the fat pool is W_f. The fat compartment was allowed to change in size as a means of describing seasonal changes in whole-body lipid content. First-order rate constants for chemical transport into and out of fat are k_{pf} and k_{fp}, respectively. The first-order rate constant for clearance by all mechanisms is k_{pc}. A simplification of this model was also achieved by assuming that the distribution of chemical between fat and plasma occurs rapidly such that the ratio of distribution rate constants can be represented as a "partition coefficient" K_{pf}. With this simplification, the final equation for calculation of plasma concentration becomes: $C_p = X_f/W_f) C K_{pf}$.

Data collected from dosing studies with 10 different compounds were used to estimate the values of all relevant model parameters. The model was then evaluated by comparing simulated chemical residues with measured values from a second group of dosed animals (Figure 2-7). Model performance was very good, as might have been expected, since a similar protocol was used for both calibration and evaluation. Nevertheless, the model performed better for some compounds than others, suggesting in several cases that basic modeling assumptions had been violated (principally the assumption of instantaneous uptake of the administered dose). The value of the gull model as an aid to interpreting chemical residues was

FAT

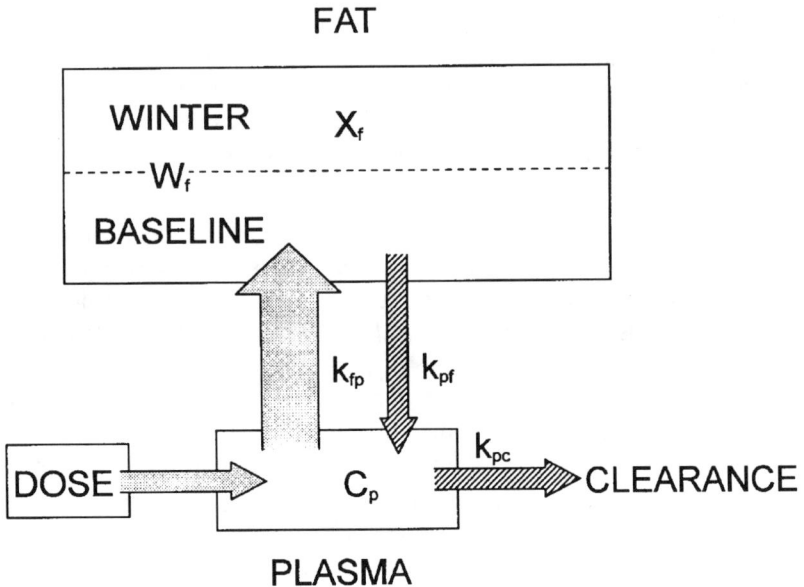

Figure 2-6 2-compartment, open model for contaminant toxicokinetics in the herring gull (adapted from Clark et al. 1987)

demonstrated by simulating tissue-residue concentrations under a variety of exposure scenarios. In this manner, it was possible to show that the time-constant for reestablishing an annualized steady-state concentration after a hypothetical change in food concentration was as much as 3 years for slow-clearing compounds like mirex (Clark T et al. 1988).

Bioenergetics-based model for tree swallows

As with herring gulls, a model for swallows was developed to aid in the interpretation of measured tissue residues (Nichols et al. 1995). Tree swallows have attracted special attention because they feed extensively on emergent aquatic insects. Persistent organic compounds that are present in sediments tend to accumulate in larval forms of these insects, often approaching an equilibrium distribution, as indicated by BSAFs (lipid and carbon normalized) of approximately 1.0 (Ankley et al. 1992). Upon emergence as adults, these insects provide a vector for translocation of sediment contaminants to insectivorous birds (Larsson 1984; Clements and Kawatski 1984; Gobas et al. 1989; Kovats and Ciborowski 1989; Fairchild et al. 1992). The tree swallow is a widely distributed passerine species. Throughout their breeding range, population densities are limited by the availability of nesting cavities. By placing nest boxes in suitable locations it is possible to create a study population of swallows, often within the first year. In contrast to adults, which may accumulate chemical residues while overwintering, nestling swallows are exposed primarily to compounds present around the nest site, combined with those inher-

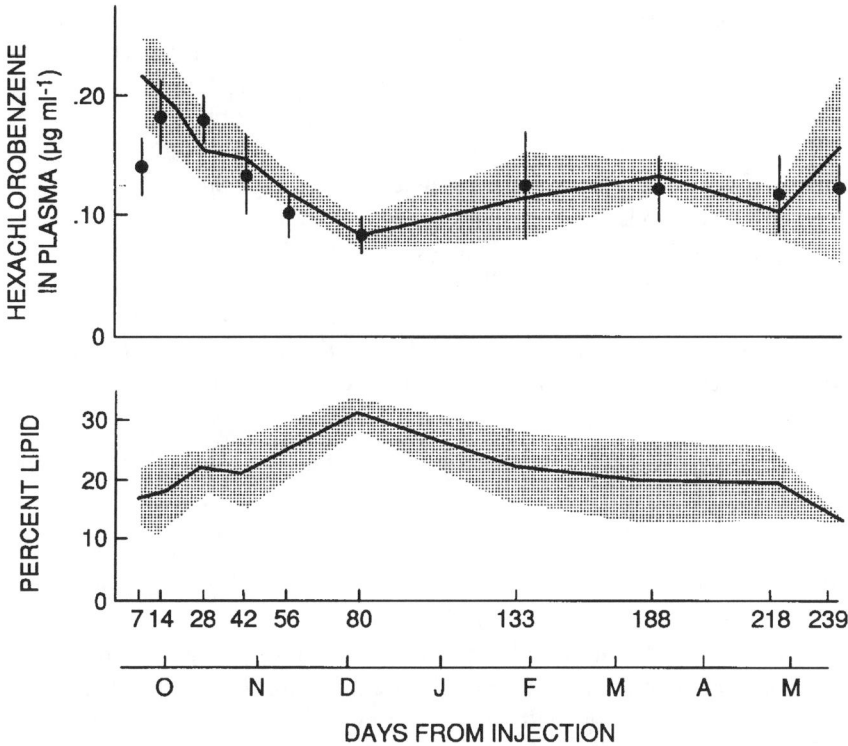

Figure 2-7 Hexachlorobenzene concentrations in gull plasma following a single i.p. injection (adapted from Clark et al. 1987). The model simulation is shown as a solid line, measured values are given as individual points.

ited as egg residues. For this reason, nestling tree swallows have been proposed for use as sentinels of local sediment contamination.

A bioenergetics-based modeling approach was used to simulate chemical accumulation by swallows because of the need to explicitly describe both dietary uptake of contaminant and changing growth rate during the nestling development period. Contaminant concentrations in nestling tree swallows have been observed to decline for a time after hatching. Termed "growth dilution," this result can occur only if the efficiency with which ingested energy is converted to tissue mass exceeds the assimilation efficiency for the toxicant. Later, as nestlings become more active and start to thermoregulate, the proportion of ingested energy allocated to growth declines and contaminant levels increase. Although varying in detail, this pattern appears to be a general one and has been reported for nestlings of both precocial and altricial species (Robinson et al. 1967; Persson 1971; Anderson and Hickey 1976; Harris et al. 1993; Ankley et al. 1993).

A bioenergetics-based model defines a system in terms of energy fluxes. A common form of such a model is to define the growth of an individual organism as the net result of an energy-mass balance. This is accomplished by using conversion factors to transform biomass into energy equivalents, utilizing the calorie scale of measurement. Thus, the "energy density" of a whole organism, individual tissues, or prey item can be defined in calories per gram, as determined using bomb calorimetry or other techniques. Methods also have been developed to estimate energy expenditures for physiological "housekeeping" functions, such as basal metabolism, digestion, and waste processing, as well as energy requiring activities such as prey capture, migration, thermoregulation, and mating. Excellent reviews are available on the application of this modeling approach to fish (Kitchell 1983; Adams and Breck 1990; Hansen et al. 1993; Lucas 1996) and birds (Kendeigh et al. 1977; Whittow 1986).

Typically, the energy-mass balance on an individual organism is solved for the value of the production term:

$$P = C - (F + U + R) \qquad \text{Equation 2-4,}$$

where P = tissues formed, C = food consumed, F = fecal losses, U = urinary losses, and R = respiration. All terms in the equation appear as rates. Loss terms are commonly summed and subtracted from consumption. Integrating the equation transforms production into mass as total calorie equivalents. Mass expressed as grams of animal is obtained by multiplying calories by the appropriate energy density function.

Simulations for tree swallows were obtained using PCB concentrations in insects and eggs as inputs to the model. The model reproduced observed rates of growth and the reported pattern of PCB growth dilution. The predicted contribution of maternally derived residues to total residues at fledging ranged from 10 to 25% and varied inversely with the extent of site contamination. PCB concentrations in birds from a relatively uncontaminated reference site were well described, but residues from an area of known sediment contamination were overestimated (Figure 2-8). An examination of the levels of individual congeners suggested that this result was not due to differences in metabolic biotransformation. Instead, it was suggested that there may have been differences in dietary composition among sites or a general underestimation of prey consumption rate.

Future research needs

Data needs for model development

To further explore the mechanism and the applicability of fugacity-based models of in ovo exposure of embryos to contaminants, a number of suggestions regarding research and data needs can be made. One of the key assumptions in the fugacity-based, maternal transfer model presented in this chapter is that the fugacity capacity of eggs and maternal tissues for hydrophobic organic chemicals is exclu-

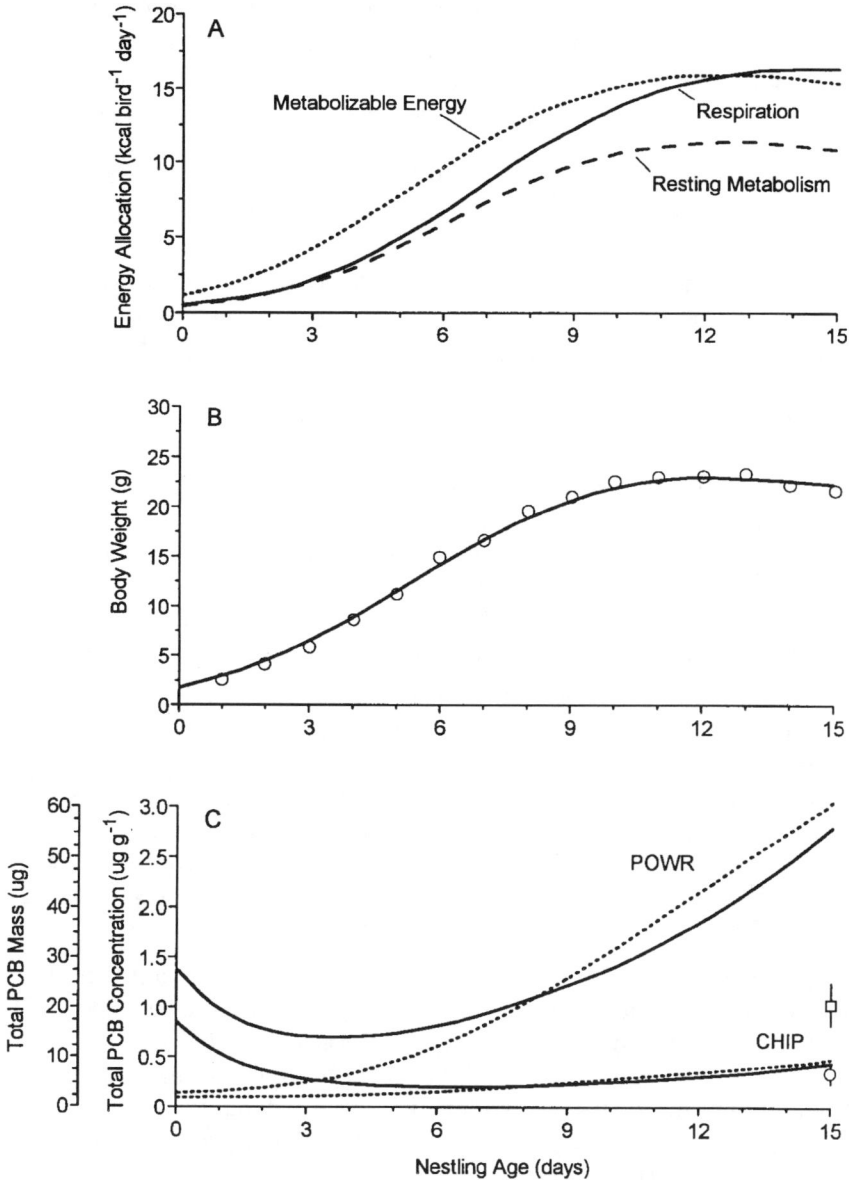

Figure 2-8 A) Energy budget predicted by a tree swallow bioenergetics model. Simulations correspond to model predictions of metabolizable energy (dotted line), total respiration (solid line), and resting metabolic rate (dashed line). B) Growth of nestling tree swallows from the Saginaw River watershed, Michigan. Measured values are shown as individual points. C) Predicted and observed total PCB residues in nesting tree swallows from an unpolluted upstream site (CHIP) and a polluted downstream site (POWR) on the Saginaw River, Michigan. The mean concentration of total PCBs measured in 15-d old nestlings is shown as an open circle (CHIP) or open square (POWR). Model simulations are shown as solid (PCB concentration) and dashed (PCB mass) lines. Reprinted with permission from Nichols et al. 1995. Copyright 1995 American Chemical Society.

sively the result of the amount of lipids in the eggs and maternal tissues. This ignores the contribution of non-lipid components of the eggs and tissues (e.g., proteins, glycogen) to the fugacity capacity. To further test the assumption of equilibrium partitioning as a mechanism of maternal transfer into eggs, it is important to first develop methods for determining fugacities and the fugacity capacity of contaminants in eggs and maternal tissues and then to apply these methods to investigate fugacity gradients between maternal tissues and eggs.

While this chapter focuses on maternal transfer and initial in ovo exposure of eggs to contaminants, it is important to realize that there may be toxicologically significant changes in the effective concentration of a compound during in ovo embryo development simply because of changes in physicochemical composition of the embryo. Fugacity analysis of contaminants during embryo development is probably one of the better methods to investigate changes in these effective concentrations. Understanding the effective exposure during embryo development also plays a role in determining the sensitivity of early life stages relative to adult life stages. Finally, while maternal transfer is an important mechanism controlling in ovo exposure of embryos to contaminants, it is not the only one. Chemical exchange between the embryo and its ambient environment can play an important role as well. Currently, it is difficult to assess the exposure of fish embryos under actual environmental conditions because trans-chorionic diffusion rates are not commonly known.

The scientific information needed to develop kinetic models for birds consists broadly of mechanistic, or "process," information and high-quality kinetic data. Among the many factors that have potential to impact chemical uptake and disposition, there is a particular need to understand the biological and physico-chemical determinants of oral bioavailability. Modeling efforts with fish have shown that differences in dietary uptake efficiency can contribute substantially to differences in bioaccumulation rate among chemicals, species, and life stage. Limited modeling efforts with birds have led to the same conclusion (Nichols et al. 1995).

An improved understanding of metabolic biotransformation is also critical. Although it is unreasonable to expect that detailed information can be collected for all species of interest, it may be possible to generalize activity and specificity information to taxonomic subgroups. The consideration of compounds that undergo metabolic biotransformation also requires improved methods for metabolite identification and quantitation. Most of the metabolism information collected to date has been obtained using in vitro systems. A need exists for improved methods to monitor the kinetics of labile metabolites in vivo. Such information could be used to investigate metabolite disposition within exposed animals and would make possible direct in vitro/in vivo comparisons.

High-quality kinetic data is needed to relate the chemical time-course in whole-animal dosing studies or environmental exposures to the dosages given directly to eggs in the laboratory. This will require additional information on the biochemistry and physiology of yolk deposition in eggs, with particular emphasis on the sources and fate of lipid and lipoprotein. One approach that could be used to address these questions would be to conduct a chemometric analysis of chemical residues in eggs, maternal parents (liver and fat), and dietary sources, thereby quantifying the "resemblance" of chemical residue profiles.

Finally, there is a general need to obtain detailed kinetic information for specific tissues and organs that can be linked in turn to mechanistic studies of effect. PBTK models are particularly well suited for this purpose. Information that would be required to develop a physiological model for birds is described below. Additionally, it is proposed that physiological models for birds could be combined with bioenergetics and natural history information to obtain chemical time-course predictions for individual tissues in realistic environmental exposures.

Development of physiologically based toxicokinetic models for birds

Physiologically based toxicokinetic models are developed from the physiology, biochemistry, and anatomy of the exposed organism and, as a result, can be used to predict chemical uptake and disposition without the prior need for dosing information. In a typical model, the animal is divided up into several compartments representing tissues of kinetic as well as toxicological significance. Chemical distribution among tissues is assumed to occur only via circulation. Thus, for each tissue group, the following information is required: 1) tissue volume, 2) arterial blood flow as a percentage of cardiac output, 3) equilibrium chemical partitioning between the tissues and blood, and 4) biotransformation rate and capacity parameters. In addition, mathematical descriptions are required for chemical flux across relevant exchange surfaces. Mass-balance expressions are written for each compartment and are solved simultaneously by numerical integration to obtain a solution set for each time point. A schematic illustration of a PBTK model is given in Figure 2-9. The symbols used in this figure refer to the following modeled quantities: Q_{alv} – ventiliation volume; C_{inh} – chemical concentration in inspired air; Q_c – cardiac output; C_{ven} – chemical concentration in mixed venous blood; C_{art} – chemical concentration in arterial blood; Q_f, Q_m, Q_r, Q_l, Q_s – blood flows to the fat, muscle, richly perfused, liver, and skin tissue groups, respectively; C_{vf}, C_{vm}, C_{vr}, C_{vl}, C_{vs} - chemical concentrations in blood exiting the fat, muscle, richly profused, liver, and skin tissue, respectively; K_m, V_{max} – rate and capacity parameters for saturable metabolic biotransformation. Modeling assumptions and the structural details that follow from them have been extensively reviewed (Gerlowski and Jain 1983; Rowland 1985; McKim and Nichols 1994).

The principal advantage of a PBTK model is that it yields predictions of the chemical time-course in specific tissues of interest. Moreover, because these models have biological and physicochemical integrity, they can be used to extrapolate predictions well beyond the range of experimental conditions. In practice, the development of such models follows an iterative process of simulation and data collection.

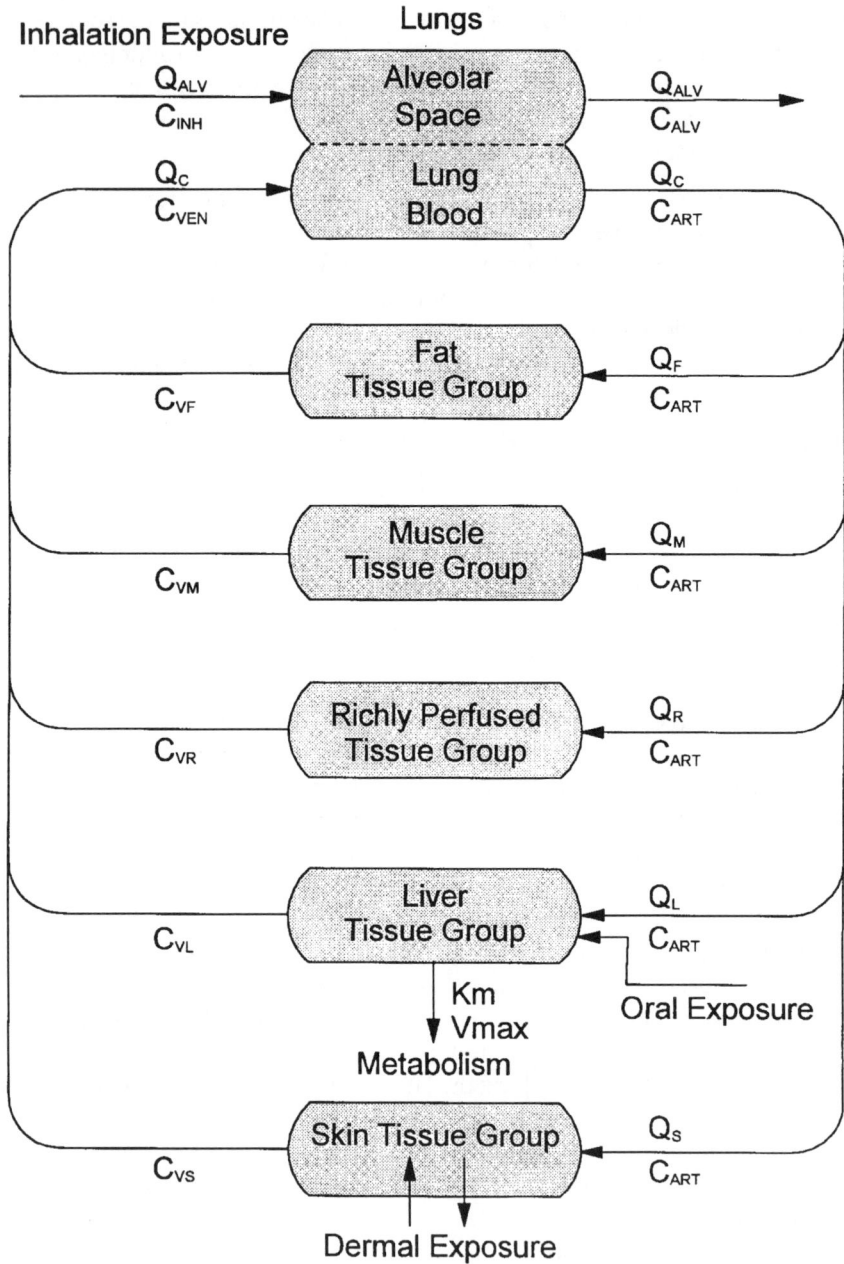

Figure 2-9 Schematic representation of a physiologically based toxicokinetic model for an air-breathing vertebrate

As experience is gained, sensitivity analyses are performed to assess the relative contribution of individual parameters to model performance. This information can then be considered along with the modeler's confidence in the various parameters to determine whether and how the model should be amended and which of the parameters requires further investigation.

PBTK models have been developed for more than 100 chemicals and for a dozen or more mammalian species, including man, and presently are being used in human-health risk assessment (Rietz et al. 1996). PBTK models have also been developed for fish, and have been employed to extrapolate kinetic information among species (McKim and Nichols 1994), evaluate metabolic rate and capacity parameters (Law et al. 1991), and investigate maternal transfer of hydrophobic organic compounds (Nichols et al. 1998). To date, there have been no attempts to develop PBTK models for birds.

Use of bioenergetics-based models to simulate the combined effects of chemical and non-chemical stressors on reproduction

It is important to recognize that avian reproduction and development can be impacted by non-chemical as well as chemical stressors. Perhaps the best example is that of habitat alteration, with associated impacts on food-web dynamics and the availability of suitable nesting sites. A potential consequence of these non-chemical impacts is a reduction in energy available for production of young and the growth of neonates. Numerous studies have shown that clutch size, egg quality, number of nesting attempts, and survival of young vary with the amount and quality of available food (Hepp et al. 1987; Esler and Grand 1994). These impacts have the potential to complicate the assessment of toxic effects. For example, the combined effect of DDE and a small reduction in ration on the reproductive performance of turtle doves (*Streptopelia risoria*) was much greater than the effect of either treatment by itself (Keith and Mitchell 1993). As demonstrated previously, bioenergetics-based models can be used to simulate dietary uptake of chemical contaminants on a whole-organism basis. In theory, this approach could also be used to describe chemical and non-chemical impacts on growth through their effects on organism energetics.

Bioenergetics models that explicitly describe reproduction can be developed by expanding the balanced energy equation as appropriate for the adult or offspring. For the adult (male or female), the respiration term is expanded to include energy expenditures for breeding behavior and parental care. The production term for the female is also expanded to include growth of both the parent and gametes. Summing these energy expenditures gives the total reproductive effort. For the offspring, respiratory expenditures may change dramatically due to increases in activity associated with fledging and the onset of thermoregulation. A period of weight loss frequently occurs as the energy demands of the young come to exceed the foraging capability of the parents. The data needs of this approach include both bioenergetics and natural history information. Fortunately, there is a great deal of this information in the literature, due to long-standing interest in avian biology.

Linked bioenergetics and physiologically based toxicokinetic models

Finally, it should be possible to link both bioenergetics and PBTK models. Bioenergetics-based models describe organism growth, while PBTK models are generally formulated to calculate tissue volumes as fractions of total body weight. Individual tissues therefore change in size along with the organism. In some cases, such as the growth of gametes, tissue volume as a percent of body size changes in time. The growth of these tissues must be modeled explicitly and appropriate adjustments must be made to other tissue volumes. Physiological functions in both modeling approaches are usually scaled to body weight using allometric relationships.

The critical point of contact between these modeling approaches is provided by the dietary route of exposure. Bioenergetics-based models, combined with appropriate natural history information, describe what and how much an organism consumes. Knowledge of chemical residues in prey provides the delivered dose. PBTK models then translate this delivered dose into a tissue-level dose that can then be interpreted in the context of toxic effects. As indicated previously, respiration and growth terms in a bioenergetics-based model can be subdivided to account for energy expenditures associated with reproductive effort. Output from both types of models can be structured as input to population-level models. Thus, the production and growth of gametes and young can be expressed as measures of survival, fecundity, and recruitment.

Summary

Exposure of early life stages to xenobiotics is dependent upon a complexity of issues. Chemical release in areas of potential impact is an obvious prerequisite for exposure. Equally as critical are parental, early life stage, and environmental features that modulate exposure. Bioaccessibility (organismic exposure as modified by external environmental determinants) and bioavailability (systemic organismic exposure as defined by target organism determinants) categorically scale the toxicological importance of xenobiotics in the environment. Clearly, oviparous vertebrates are a special case for the early-life-stage exposure paradigm. There is the potential for not only direct environmental exposure with the accompanying modulating factors, but also for maternal transfer with its own array of determinants. With maternal transfer, historical residues of parental exposure are presented to early life stages as a concentrated source of contaminants within transferred nutrients. The existing literature, biased by the limited number of classes of chemicals examined, suggests that the compounds that show the greatest propensity to be transferred to early life stages are those that are more lipophilic, correlating in part to the characteristics of the transferred nutrients. Although conclusive evidence for the role of the yolk lipoprotein vitellogenin in xenobiotic transfer is lacking for oviparous vertebrates, a number of studies have identified the presence of contaminants in the yolk and correlated the onset of vitellogenesis with the delivery of pollutants to the ovary. Lipids and lipid transfer dynamics from maternal sources to the developing ovaries do appear to be the critical determinants in the maternal transfer of lipophilic contaminants. Issues such as maternal concentrations of contaminants, the percent of total maternal lipid transferred to the eggs, egg weight as a percentage of the total maternal weight, and total maternal lipid burden appear to relate to the percent of whole-body contaminants that are transferred to eggs. Initial indications suggest reproductive life histories of the animals, different life strategies for resource mobilization for ovarian development, and dietary energetics may play an important role in determining the source and amount of the lipids and thus contaminants transferred by maternal transfer to early life stages. In birds, unique features such as homoeothermy and, in some cases, energetically expensive migration, courtship, breeding, and parental caregiving behaviors provide for unusual xenobiotic uptake and mobilization paradigms within this contextual framework. In totality, this information suggests that the deposition of contaminants in the egg mass by lipid association may be linked directly with species specific characteristics, maternal nutrient status, and the maternal resident xenobiotic body burden.

Available systemically upon mobilization of nutrient stores, maternally derived contaminants are present during critical phases of early development. Embryonic contaminant stores are mobilized at a time when there is an isolation of early life stages from parental assistance with regard to biotransformation and elimination. Limitations of like systems during early ontogeny put developing oviparous organisms at risk. For all practical purposes, the same mechanisms that provide (e.g., maternal nutrients) and protect (e.g., chorion, shell) during early development put early life stages of oviparous vertebrates in jeopardy when they are challenged by a chemical insult, especially those originating internally. For fish, elimination rates from embryos and yolk-sac larvae are generally much lower than the feeding post-yolk-sac stage which in turn is slower than more advanced stages. Changes in xenobiotic disposition with development, while of unknown etiology, are probably caused by development of uptake and depuration pathways, loss of contaminant storage areas in the form of yolk loss, and increased competency of developing metabolic systems. Generally, it is accepted that a substantial portion of apparent losses with development are also a result of growth dilution rather than of actual elimination. Phase I xenobiotic biotransformation, when measured under in vitro conditions, may occur in both the embryo and the larval stages of fish. P450 activities are generally higher in larval stages than in embryos, and activities for both are much lower than in adults. Fish egg and larval P450 activities may be induced by environmental routes, by experimental routes (e.g., by injection), and by maternal routes. While CYP1A is inducible pre- and post-hatch in fish, the induction response is greater after hatching. The reasons for differences in biotransformation and the induction response between life stages are unknown; however, they could be related to dosimetry considerations or variation in the responsiveness of the Ah receptor. A number of conjugation reactions have been identified for early life stages of fish. Although this is true, relatively little is known with regard to Phase II activities, induction, and ontogeny. Compositely, in vivo studies with early life stages have been inconsistent in demonstrating biotransformation. Small body size of fish in early life stages, whole-body and water dilution of metabolites, analytic sensitivity, differential metabolite solubility, and apparently low basal-uninduced metabolic rates complicate assessment of biotransformation. In general, more work will have to be performed on kinetic and biotransformational issues before the determinant factors and relative importance of real-world early-life-stage biotransformation in fish can be ascertained.

The importance of egg laying as a route of elimination for adult birds and as a route of exposure for early life stages appears to be species-specific and dependent on factors including egg and clutch size, maternal fat reserves, and the extent to which the reserves are mobilized. In terms of biotransformation, birds present a somewhat different picture than their piscine counterparts. P450 biotransformation varies greatly between adult birds of varying species, largely dependent on feeding ecology. P450 activities in birds are lower relative to mammals but higher than generally exhibited by other oviparous species. In contrast to mammals and other oviparous species, high levels of MFO activity are present during the embryonic and neonatal periods, suggesting that developing birds may be more independent in a biotransformational sense. The activity of these enzymes appears to be inducible by a wide variety of compounds, and the extent of this induction has been correlated with contaminant residues in field-collected eggs. Chemical kinetic data in neonates and juvenile birds are extremely limited. It is difficult, therefore, to assess the relative importance of growth dilution, metabolic biotransformation, and other routes of elimination in determining chemical residues. The oral bioavailability of most chemical classes in birds of any age remains essentially unknown.

Exposure and fate are critical determinants of chemical toxicity in early life stages of oviparous vertebrates. Clearly, the processes involved in direct exposure, maternal transfer, xenobiotic mobilization, biotransformation and excretory ontogeny are complex and interrelated with regard to their role in early-life-stage toxicity. The development and use of mathematical models provides a means to simplify complex phenomena and make predictive assessments. Thus far, most of the chemical modeling efforts with oviparous vertebrates are oriented toward the adult. Simplistic generic modeling approaches such as chemical fugacity show promise in the first tier assessment of chemical exposure for early life stages. Further understanding and better predictability are likely with model development incorporating physiological and bioenergetic-linked approaches.

References

Adams SM, Breck JE. 1990. Bioenergetics. In: Schreck CB, Moyle PB, editors. Methods of fish biology. Bethesda MD: American Fisheries Society. p 389–415.

Adkins-Regan E, Ottinger MA, Park J. 1995. Maternal transfer of oestradiol to egg yolks alters sexual differentiation of avian offspring. *J Exp Zool* 271:466–470.

Ahel M, McEvoy J, Giger W. 1993. Bioaccumulation of the lipophilic metabolites of nonionic surfactants in freshwater organisms. *Environ Pollut* 79:243–248.

Ahel M, Giger W, Schaffner C. 1994. Behaviour of alkylphenol polyethoxylate surfactants in the aquatic environment. 2. Occurrence and transformation in rivers. *Water Res* 28:1143–1152.

Ahel M, Schaffner C, Giger W. 1996. Behaviour of alkylphenol polyethoxylate surfactants in the aquatic environment. 2. Occurrence and elimination of their persistent metabolites during infiltration of river water to groundwater. *Water Res* 30:37–46.

Ainley DG, Grau CR, Roudybush TE, Morrell SH, Utts JM. 1981. Petroleum ingestion reduces reproduction in Cassin's auklets. *Mar Pollut Bull* 12:314–317.

Aksnes A, Gjerde B, Roald SO. 1986. Biological, chemical and organoleptic changes during maturation of farmed Atlantic salmon, *Salmo salar*. *Aquaculture* 53:7–20.

Alexander M. 1995. Critical review: how toxic are toxic chemicals in soil? *Environ Sci Technol* 29:2713–2717.

Anderson DW, Hickey JJ. 1976. Dynamics of storage of organochlorine pollutants in herring gulls. *Environ Pollut* 10:183–200.

Ankley GT, Cook PM, Carlson AR, Call DJ, Swenson JA, Corcoran HF, Hoke RA. 1992. Bioaccumulation of PCBs from sediments by oligochaetes and fishes: comparison of laboratory and field studies. *Can J Fish Aquat Sci* 49:2080–2085.

Ankley GT, Niemi GJ, Lodge KB, Harris HJ, Beaver DL, Tillitt DE, Schwartz TR, Giesy JP, Jones PD, Hagley C. 1993. Uptake of planar polychlorinated biphenyls and 2,3,7,8-substituted polychlorinated dibenzofurans and dibenzo-*p*-dioxins by birds nesting in the Lower Fox River and Green Bay, Wisconsin, USA. *Arch Environ Contam Toxicol* 24:332–344.

Ankley GT, Tillitt DE, Geisy JP, Jones PD, Verbrugge DA. 1991. Bioassay-derived 2,3,7,8-tetrachloro-*p*-dioxin equivalents in PCB-containing extracts from the flesh and eggs of Lake Michigan chinook salmon (*Oncorhynchus tshawytscha*) and possible implications for reproduction. *Can J Fish Aquatic Sci* 48:1685–1690.

Ankley GT, Tillitt DE, Geisy JP. 1989. Maternal transfer of bioactive PAHs in spawning chinook salmon (*O. tschawytsch*). *Mar Environ Res* 28:231–234.

Ankley GT, DiToro DM, Hansen DJ, Berry WJ. 1996. Technical basis and proposal for deriving sediment quality criteria for metals. *Environ Toxicol Chem* 15:2056–2066.

Augustijn-Beckers PWM, Hornsby AG, Wauchope RD. 1994. The SCS/ARS/CES pesticide properties database for environmental decision-making. II. Additional compounds. *Rev Environ Contam Toxicol* 137: 1–82.

Babin PJ. 1992. Binding of thyroxine and 3,5,3'-triiodothyronine to trout plasma lipoproteins. *Am J Physiol* 262:E712–720.

Barr JF. 1986. Population dynamics of the common loon (*Gavia immer*) associated with mercury-contaminated waters in northwestern Ontario. Ottawa, Ontario: Canadian Wildlife Service. Occasional Paper No. 56

Barron MG, Galbraith H, Beltman D. 1995. Comparative reproductive and developmental toxicology of PCBs in birds. *Comp Biochem Physiol* 112C:1–14.

Barron, MG, Stehly, GR, Hayton WL. 1990. Pharmacokinetic modeling in aquatic animals. I. Models and concepts. *Aquat Toxicol* 17:187–211.

Belant JL, Anderson RK. 1990. Environmental contaminants in common loons from northern Wisconsin. *Pass Pigeon* 52:306–310.

Belfroid AC. 1996. Toxicokinetics of hydrophobic chemicals in earthworm: validation of the equilibrium partitioning theory [Ph.D. Dissertation]. Utrecht, The Netherlands: Research Institute of Toxicology, University of Utrecht. 96 p.

Belfroid AC, Sijm DTHM, and van Gestel CAM. 1996. Bioavailability and toxicokinetics of hydrophobic aromatic hydrocarbons in benthic and terrestrial invertebrates. *Environ Rev* 4:276–299.

Bellward GD, Norstrom RJ, Whitehead PE, Elliott JE, Bandiera SM, Dworschak C, Chang T, Forbes S, Cadario B, Hart LE, Cheng KM. 1990. Comparison of polychlorinated dibenzodioxin levels with hepatic mixed-function oxidase induction in great blue herons. *J Toxicol Environ Health* 30:33–52.

Bengtsson BE. 1980. Long-term effects of PCB (Clophen A50) on growth, reproduction and swimming performance in the minnow *Phoxinus phoxinus*. *Water Res* 14:681–687.

Bernardini G, Spinelli O, Presutti C, Vismara C, Bolzacchini E, Orlandi M, Settimi R. 1996. Evaluation of the developmental toxicity of the pesticide MCPA and its contaminants phenol and chlorocresol. *Environ Toxicol Chem* 15:754–760.

Berlin WH, Hesselberg RJ, Mac MJ. 1981. Growth and mortality of fry of Lake Michigan trout during chronic exposure to PCBs and DDE. In: Chlorinated hydrocarbons as a factor in the reproduction and survival of lake trout (*Salvelinus namaycush*) in Lake Michigan. Washington DC: U.S. Fish and Wildlife Service. U.S. Fish and Wildlife Serv Tech Paper 105. p 11–22.

Beyer WN, Gish CD. 1980. Persistence in earthworms and potential hazards to birds of soil applied DDT, dieldren and heptachlor. *J Appl Ecol* 17:295–307.

Beyer WN, Heinz GH, Redmon-Norwood AW. 1996. Environmental contaminants in wildlife. Interpreting tissue concentrations. Boca Raton FL: Lewis. 494 p.

Binder RL, Lech JJ. 1984. Xenobiotics in gametes of Lake Michigan lake trout (*Salvelinus namaycush*) induce hepatic monooxygenase activity in their offspring. *Fundament Appl Toxicol* 4:1042–1054.

Binder RL, Stegeman JJ. 1980. Induction of aryl hydrocarbon hydroxylase activity in embryos of an estuarine fish. *Biochem Pharmacol* 29:949–951.

Binder RL, Stegeman JJ. 1983. Basal levels and induction of hepatic aryl hydrocarbon hydroxylase activity during the embryonic period of development in brook trout. *Biochem Pharmacol* 32:1324–1327.

Binder RL, Stegeman JJ, Lech JJ. 1985. Induction of cytochrome P-450-dependent monooxygenase systems in embryos and eleutheroembryos of the killifish *Fundulus heteroclitus. Chem Biol Interact* 55:185–202.

Bintein S, Devillers J, and Karcher W. 1993. Nonlinear dependence of fish bioconcentration on n-octanol/water partition coefficient. *SAR QSAR Env Res* 1:29–39.

Bishop CA, Koster MD, Chek AA, Hussell DJT, Jock K. 1995. Chlorinated hydrocarbons and mercury in sediments, red-winged blackbirds (*Agelaius phoeniceus*) and tree swallows (*Tachycineta bicolor*) from wetlands in the Great Lakes-St. Lawrence River basin. *Environ Toxicol Chem* 14:491–501.

Black DE, Phelps DK, Lapan RL. 1988. The effect of inherited contamination on egg and larval winter flounder, *Pseudopleuronectes americanus. Mar Environ Res* 25:45–62.

Black MC, McCarthy JF. 1988. Dissolved organic macromolecules reduce the uptake of hydrophobic organic contaminants by the gills of rainbow trout (*Salmo gairdneri*). *Environ Toxicol Chem* 7:593–600.

Boersma DC, Ellenton JA, Vagminas A. 1986. Investigations on the hepatic mixed-function oxidase system in herring gulls in relation to environmental contaminants. *Environ Toxicol Chem* 5:309–318.

Boesa BL, Lee H, Echols ES. 1997. Evaluation of a first-order model for the prediction of the bioaccumulation of PCBs and DDT from sediment into the marine deposit feeding clam *Macoma nasuta. Environ Toxicol Chem* 7:1545–1553.

Boethling RS, Howard PH, Beauman JA, Larosche ME. 1995. Factors for intermedia extrapolation in biodegradability assessment. *Chemosphere* 30:741–752.

Boethling RS, Howard PH, Meylan WM, Stiteler WM, Beauman JA, Tirado N. 1994. Group contribution method for predicting probability and rate of aerobic biodegradation. *Environ Sci Technol* 28:459–465.

Bogan JA, Newton I. 1977. Redistribution of DDE in sparrowhawks during starvation. *Bull Environ Contam Toxicol* 18:317–321.

Borlakoglu JT, Welch VA, Walker CH, Dils RR. 1989. Role of lipoproteins in the transport of polychlorinated biphenyls in birds. *Mar Environ Res* 28:235–239.

Borlakoglu JT, Wilkins JPG, Walker CH, Dils RR. 1990a. Polychlorinated biphenyls (PCBs) in fish-eating sea birds. I. Molecular features of PCB isomers and congeners in adipose tissue of male and female razorbills (*Alca tarda*) of British and Irish coastal waters. *Comp Biochem Physiol* 97(C):151–160.

Borlakoglu JT, Wilkins JPG, Walker CH, Dils RR. 1990b. Polychlorinated biphenyls (PCBs) in fish-eating sea birds. II. Molecular features of PCB isomers and congeners in adipose tissue of male and female puffins (*Fratercula arctica*), guillemots (*Uria aalga*), cormorants (*Phalocrocorax carbo*) of British and Irish coastal waters. *Comp Biochem Physiol* 97(C):161–171.

Borlakoglu JT, Wilkins JPG, Walker CH, Dils RR. 1990c. Polychlorinated biphenyls (PCBs) in fish-eating sea birds. III. Molecular features and metabolic interpretations of PCB isomers and congeners in adipose tissues. *Comp Biochem Physiol* 97(C):173–177.

Borlakoglu JT, Wilkins JPG, Dils RR. 1991. Distribution and elimination in vivo of polychlorinated biphenyl (PCB) isomers and congeners in the pigeon. *Xenobiotica* 21:433–445.

Bowerman WW IV, Evans ED, Giesy JP, Postupalsky S. 1994. Using feathers to assess risk of mercury and selenium to bald eagle reproduction in the Great Lakes region. *Arch Environ Contam Toxicol* 27:294–298.

Braune BM, Gaskin DE. 1987. Mercury levels in Bonaparte's gulls (*Larus philadelphia*) during autumn molt in the Quoddy region, New Brunswick, Canada. *Arch Environ Contam Toxicol* 16:539–549.

Braune BM, Norstrom RJ. 1989. Dynamics of organochlorine compounds in herring gulls: III. Tissue distribution and bioaccumulation in Lake Ontario gulls. *Environ Toxicol Chem* 8:957–968.

Brown Jr JF, Bedard DL. 1987. Polychlorinated biphenyl dechlorination in aquatic sediments. *Science* 236:709–712.

Brunstrom B. 1986. Activities in chick embryos of 7-ethoxycoumarin O-deethylase and aryl hydrocarbon (benzo[a]pyrene) hydrolase and their induction by 3,3'4,4'-tetrachlorobiphenyl in early embryos. *Xenobiotica* 16:865–872.

Brunstrom B. 1991. Toxicity and EROD-inducing potency of polychlorinated biphenyls (PCBs) and polycyclic aromatic hydrocarbons (PAHs) in avian embryos. *Comp Biochem Physiol* 100(C):241–243.

Brunstom B, Andersson L. 1988. Toxicity and 7-ethoxyresorufin O-deethylase-inducing potency of coplanar polychlorinated biphenyls (PCBs) in chick embryos. *Arch Toxicol* 62:263–266.

Budavari S, Smith A, Heckelman P, Kinneary J, O'Neill MJ, editors. 1996.The Merck index: an encyclopedia of drugs, chemicals, and biologicals. 12th edition. Rahway NJ: Merck and Co., Inc. 1741 p.

Burdick GE, Harris EJ, Dean HJ, Walker TM, Skea J, Colby D. 1964. The accumulation of DDT in lake trout and the effects on reproduction. *Trans Am Fish Soc* 93:127–136.

Burger J. 1993. Metals in avian feathers: bioindicators of environmental pollution. *Rev Environ Toxicol* 5:203–311.

Burger J. 1994. Heavy metals in avian eggshells: another excretion method. *J Toxicol Environ Health* 41:207–220.

Burger J, Gochfeld M. 1991. Cadmium and lead in common terns (Aves: *Sterna hirundo*): relationship between levels in parents and eggs. *Environ Monit Assess* 16:253–258.

Burger J, Gochfeld M. 1993. Lead and cadmium accumulation in eggs and fledgling seabirds in the New York Bight. *Environ Toxicol Chem* 12:261–267.

Burger J, Gochfeld M. 1995. Heavy metal and selenium concentrations in eggs of herring gulls (*Larus argentatus*): temporal differences from 1989 to 1994. *Arch Environ Contam Toxicol* 29:192–197.

Buser HR. 1988. Rapid photolytic decomposition of brominated and brominated/chlorinated dibenzodioxins and dibenzofurans. *Chemosphere* 17:889–903.

Byrne BM, Gruber M, Ab G.1989. The evolution of egg yolk proteins. *Prog Biophys Mol Biol* 53:33–69.

Cabana G, Rasmussen JB. 1994. Modeling food chain structure and contaminant bioaccumulation using stable nitrogen isotopes. *Nature* 372:255–257.

Cavalli S, Cardellicchio N. 1995. Direct determination of seleno-amino acids in biological tissues by anion-exchange separation and electrochemical detection. *J Chromatog* 706(A):429–436.

Chan SL, Tan CH, Pang MK, Lam TJ. 1991. Vitellogenin purification and development of assay for vitellogenin receptor in oocyte membranes of the tilapia (*Oreochromis niloticus*, Linnaeus 1766). *J Exper Zool* 257:96–106.

Chapman MJ. 1980. Animal lipoproteins: chemistry, structure and comparative aspects. *J Lipid Res* 21:789–853.

Clark EA, Steritt RM, Lester JN. 1988. The fate of tributyltin in the aquatic environment. *Environ Sci Technol* 22:600–603.

Clark T, Clark K, Paterson S, Mackay D, Norstrom RJ. 1988. Wildlife monitoring, modeling, and fugacity. *Environ Sci Technol* 22:120–127.

Clark TP, Norstrom RJ, Fox GA, Won HT. 1987. Dynamics of organochlorine compounds in herring gulls (*Larus argentatus*): II. A two-compartment model and data for ten compounds. *Environ Toxicol Chem* 6:547–559.

Clements JR, Kawatski JA. 1984. Occurrence of polychlorinated biphenyls (PCBs) in adult mayflies (*Hexagenia bilineata*) of the upper Mississippi River. *J Fresh Ecol* 2:611–614.

Connolly JP, Pedersen CJ. 1988. A thermodynamic-based evaluation of organic chemical accumulation in aquatic organisms. *Environ Sci Technol* 22:99–103.

Connolly JP, Tonnelli R. 1985. Modeling kepone in the striped bass food chain of the James River estuary. *Est Coast Shelf Sci* 20:349–366.

Cooke AS. 1973. Shell thinning in avian eggs by environmental pollutants. *Environ Pollut* 4:85–152.

Cope WG, Wiener JG, Rada RG. 1990. Mercury accumulation in yellow perch in Wisconsin seepage lakes: Relation to lake characteristics. *Environ Toxicol Chem* 9:931–940.

Corrol KM, Harkness MR, Bracco AA, Balcarcel RR. 1994. Application of a permeant/ polymer diffusional model to the desorption of polychlorinated biphenyls from Hudson river sediments. *Environ Toxicol Chem* 28:253–258.

Cowan CE, Mackay D, Feijtel TCJ, Van de Meent D, DiGuardo A, Davies J, Mackay N. 1995. The multi-media fate model: a vital tool for predicting the fate of chemicals. Pensacola FL: SETAC. 78 p.

Craik JCA. 1982. Levels of phosphoprotein in the eggs and ovaries of some fish species. *Comp Biochem Physiol* 72B:507–510.

Cravedi JP, Gillet C, Monod G. 1995. In vivo metabolism of pentachlorophenol and aniline in arctic charr (*Salvelinus alpinus L.*) larvae. *Bull Environ Contam Toxicol* 54:711–716.

Crews D, Bergeron JM, McLachlan JA. 1996. The role of estrogen in turtle sex determination and the effect of PCBs. *Environ Health Perspect* 103:73–77.

Custer TW, Custer CM. 1995. Transfer and accumulation of organochlorines from black-crowned night-heron eggs to chicks. *Environ Toxicol Chem* 14:533–536.

Custer TW, Pendleton G, Ohlendorf HM. 1990. Within and among-clutch variation of organochlorine residues in eggs of black-crowned night-herons. *Environ Monit Assess* 15:83–89.

Cyr DG, Eales JG. 1989. Effects of short-term 17β-estradiol treatment on the properties of T$_3$-binding proteins in the plasma of immature rainbow trout (*Salmo gairdneri*). *J Exp Zool* 252:245–251.

Cyr DG, Eales JG. 1992. Effects of short-term 17β-estradiol treatment on the properties of T$_4$-binding proteins in the plasma of immature rainbow trout, *Onchorhynchus mykiss*. *J Exper Zool* 262:414–419.

Daan S, Masman D, Groenewold A. 1990. Avian basal metabolic rates: their association with body composition and energy expenditure in nature. *Am J Physiol* 259:R333–R340.

Dahlgren RB, Linder RL, Carlson CW. 1972. Polychlorinated biphenyls: their effects on penned pheasants. *Environ Health Perspec* 1:89–101.

Dawson DA, Schultz TW, Hunter RS. 1996. Developmental toxicity of carboxylic acids to Xenopus embryos: A quantitative structure-activity relationship and computer-automated structure evaluation. *Teratogen Carcin Mutagen* 16:109–124.

de Boer J, Van Der Valk F. Kerkhoff MAT, Hagel P. 1994. 8-year study on the elimination of PCBs and other organochlorine compounds from eel (anguilla) under natural conditions. *Environ Sci Technol* 28:2242–2248.

Defoe DL, Veith GD, Carlson RW. 1978. Effects of Arochlor 1248 and 1260 on the fathead minnow (*Pimephales promelas*). *J Fish Res Board Can* 35:997–1002.

de Freitas AS, Norstrom RJ. 1974. Turnover and metabolism of polychlorinated biphenyls in relation to their chemical structure and the movement of lipids in the pigeon. *Can J Physiol Pharmacol* 52:1080–1094.

Degnen P, Muller M, Nendza M, Klein W. 1993. Structure activity relationships for biodegradation. Paris, France: Organization for Economic Cooperation and Development. OECD Environmental Monographs No. 68, OECD/GD (9) 126.

Delorme PD, Muir DCG, Lockhart WL, Mills KH, Ward FJ. 1993. Depuration of toxaphene in lake trout and white suckers in a natural ecosystem following a single i.p. dose. *Chemosphere* 27:1965–1973.

de Vlaming VL, Wiley HS, Delahunty G. Wallace RA. 1980. Goldfish (*Carassius auratus*) vitellogenin: induction, isolation, properties and relationship to yolk proteins. *Comp Biochem Physiol* 67B:613–623.

DeVault D, Dunn W, Bergqvist P, Wiberg K, Rappe C. 1989. Polychlorinated dibenzofurans and polychlorinated dibenzo-*p*-dioxins in Great Lakes fish: a baseline and interlake comparison. *Environ Toxicol Chem* 8:1013–1022.

Devlin EW, Mottet NK. 1992. Embryotoxic action of methyl mercury on coho salmon embryos *Bull Environ Contam Toxicol* 49:449–454.

DeWeese LR, Cohen RR, Stafford CJ. 1985. Organochlorine residues and eggshell measurements for tree swallows *Tachycineta bicolor* in Colorado. *Bull Environ Contam Toxicol* 35:767–775.

Diana JS, Mackay WC. 1979. Timing and magnitude of energy deposition and loss in body, liver and gonads of northern pike *Esox lucius. J Fish Res. Board of Canada* 36:481–487.

Dickson KL, Geisy JP, Parrish R, Wolfe L. 1994. Bioavailability: physical, chemical and biological interactions. Society of Environmental Toxicology and Chemistry 15th Annual Meeting; 30 Oct–3 Nov 1994; Denver CO. Pensacola, FL: SETAC. p. 221–230.

Ding JL, Hee PL, Lam TJ. 1989. Two forms of vitellogenin in the plasma and gonads of male *Oreochromis aureus*. *Comp Biochem Physiol* 93:363–370.

DiToro DM, Zabra CS, Hansen DJ, Berry WJ, Swartz RC, Cowan CE, Pavlou SP, Allen HE, Thomas NA, Pagquin PR. 1991. Technical basis for establishing sediment quality criteria for nonionic organic chemicals using equilibrium partitioning. *Environ Toxicol Chem* 10:1541–1583.

Dobbs AJ, Williams N. (1983). Fat solubility—a property of environmental relevance? *Chemosphere* 12:97–104.

Dowson PH, Bubb JM, Lester JN. 1993. Temporal distribution of organotins in the aquatic environment. *Mar Pollut Bull* 26:487–484.

Dung MH, O'Keefe PW. 1994, Comparative rates of photolysis of polychlorinated dibenzofurans in organic solvents and in aqueous solutions. *Environ Sci Technol* 28:549–554.

Eaton JG, McKim JM, Holcombe GW. 1978. Metal toxicity to embryos and larvae of seven freshwater fish species I: Cadmium. *Bull Environ Contam Toxicol* 19:95–103.

Elliot JE, Martin PA, Arnold TW, Sinclair PH. 1994. Organochlorines and reproductive success of birds in orchard and non-orchard areas of central British Columbia, Canada 1990–91. *Arch Environ Contam Toxicol* 26:435–443.

Elskus AA, Stegeman JJ, Gooch JW, Black DE, Pruell RJ. 1994. Polychlorinated biphenyl congener distributions in winter flounder as related to gender, spawning site and congener metabolism. *Environ Sci Technol* 28:401–407.

Enderson JH, Craig GR, Burnham WA, Berger DD. 1982. Eggshell thinning and organochlorine residues in Rocky Mountain peregrines, *Falco pereginus*, and their prey. *Can Field-Nat* 96:255–264.

Esler D, Grand JB. 1994. The role of nutrient reserves for clutch formation by northern pintails in Alaska. *The Condor* 96:422–432.

Evans MS, Noguchi GE, Rice CP. 1991. The biomagnification of polychlorinated biphenyls, toxaphene, and DDT compounds in a Lake Michigan offshore food web. *Arch Environ Contam Toxicol* 20:87–93.

Evans MI, Silva R, Burch JB. 1988. Isolation of chicken vitellogenin I and III cDNAs and the developmental regulation of five estrogen-responsive genes in the embryonic liver. *Genes Dev* 2:116–124.

Ewins PJ, Weseloh DV, Mineau P. 1992. Geographical distribution of contaminants and productivity measures of herring gulls in the Great Lakes: Lake Huron 1980. *J Great Lakes Res* 18:316–330.

Fahraeus-Van Ree GE, Payne JF. 1997. Effect of toxaphene on reproduction of fish. *Chemosphere* 34:855–867.

Fairchild WL, Muir DCG, Currie RS, Yarechewski AL. 1992. Emerging insects as a biotic pathway for movement of 2,3,7,8-tetrachlorodibenzofuran from lake sediments. *Environ Toxicol Chem* 11:867–872.

Federle TW, Gasior SD, Nuck BA. 1997. Extrapolating mineralization rates from the ready CO_2 screening test to activated sludge, river water and soil. *Environ Toxicol Chem* 16:127–134.

Fent K. 1991. Bioconcentration and elimination of tributyltin chloride by embryos and larvae of minnows *Phoxinus phoxinus*. *Aquat Toxicol* 20:147–158.

Fent K, Stegeman JJ. 1991. Effects of tributyltin chloride in vitro on the hepatic microsomal monooxygenase system in the fish *Stenotomus chrysops*. *Aquat Toxicol* 20:159–168.

Field JA, Reed RL. 1996. Nonylphenol polyethoxycarboxylate metabolites of nonionic surfactants in U.S. paper mill effluents, municipal sewage treatment plant effluents and river waters. *Environ Sci Technol* 30:3544–3550.

Fimreite N. 1974. Mercury contamination of aquatic birds in northwestern Ontario. *J Wildl Manage* 38:120–131.

Finizio A, Mackay D, Bidleman T, Harner T. 1997. Octanol-air partition coefficient as a predictor of partitioning of semi-volatile organic chemicals to aerosols. *Atmos Environ* 31:2289–2296.

Fordham CL, Regan DP. 1991. Pathways analysis method for estimating water and sediment criteria at hazardous waste sites. *Environ Toxicol Chem* 10:949–960.

Fox GA, Gilman AP, Peakall DB, Anderka FW. 1978. Behavioral abnormalities of nesting Lake Ontario herring gulls. *J Wildl Manage* 42:477–483.

Frankovic L, Khan MAQ, Ghais SMA. 1995. Metabolism of hexachlorobenzene in the fry of steelhead trout, *Salmo gairdneri* (*Oncorhynchus mykiss*). *Arch Environ Contam Toxicol* 28:209–214.

Freitag D, Haas-Jobelius M, Yiwei Y. 1991. Influence of 2,2'-dichlorobiphenyl and 2,4,6-trichlorophenol on development of rainbow trout eggs and larvae. *Toxicolog Environ Chem* 31–32:401–408.

Fry DM, Swenson J, Addiego LA, Grau CR, Kang A. 1986. Reduced reproduction of wedge-tailed shearwaters exposed to a single dose of 2 ml of weathered Santa Barbara crude oil. *Arch Environ Contam Toxicol* 15:453–463.

Fry DM. 1995. Reproductive effects in birds exposed to pesticides and industrial compounds. *Environ Health Perspec* 103(Suppl 7):165–171.

Galassi S, Calamari D. 1983. Toxicokinetics of 1,2,3 and 1,2,4 trichlorobenzenes in early life stages of *Salmo gairdneri*. *Chemosphere* 12:1599–1603.

Galassi S, Calamari D, Setti F. 1982. Uptake and release of p-dichlorobenzene in early life stages of *Salmo gairdneri*. *Ecotoxicol Environ Saf* 6:439–447.

Ganther HE, Goudie C, Sunde ML, Kipecky MJ, Wagner P, Oh SH, Hoekstra WG. 1972. Selenium relation to decreased toxicity of methyl mercury added to diets containing tuna. *Science* 175:1122–1124.

Garten CT, Trabalka JR. 1983. Evaluation of models for predicting terrestrial food chain behavior of xenobiotics. *Environ Sci Technol* 17:590–595.

[GEA] German Environmental Agency. 1996. Endocrinically active chemicals in the environment Expert Round. Umweltbundesamt; 1995 Mar 9–10; Berlin, Germany: Nr. 96/03.

Gerlowski LE, Jain RK. 1983. Physiologically based pharmacokinetic modeling: principles and applications. *J Pharm Sci* 72:1103–1128.

Geyer H, Scheunert I, Korte F. 1985. Relationship between the lipid content of fish and their bioconcentration potential of 1,2,4-trichlorobenzene. *Chemosphere* 14:545–551.

Ghosh P, Thomas P. 1995. Binding of metals to red drum vitellogenin and incorporation into oocytes. *Mar Environ Res* 39:165–168.

Giesy JP, Ludwig JP, Tillitt DE. 1994. Deformities in birds of the Great Lakes region. Assigning causality. *Environ Sci Technol* 28:128–134.

Gilbertson ME. 1974. Pollutants in breeding herring gulls in the Lower Great Lakes. *Can Field-Nat.* 88:2273–2280.

Gilbertson M, Kubiak T, Ludwig J, Fox GJ. 1991. Great Lakes embryo mortality, edema, and deformitites syndrome (GLEMEDS) in colonial fish-eating birds: similarity to chick-edema disease. *Toxicol Environ Health* 33:455–520.

Gillespie RB, Baumann PC. 1986. Effects of high tissue concentrations of selenium on reproduction by bluegills. *Trans Am Fish Soc* 115:208–213.

Gilman AP, Fox GA, Peakall DB, Teeple SM, Carroll TR, Haynes GT. 1977. Reproductive parameters and egg contaminant levels of Great Lakes herring gulls. *J Wildl Manage* 41:458–468.

Glassmeyer ST, De Vault DS, Myers TR, Hites RA. 1997. Toxaphene in Great Lakes fish: a temporal, spatial and trophic study. *Environ Sci Technol* 31:84–88.

Glotfelty DE, Taylor AW. 1984. Atrazine and Simazine movement to Wye River estuary. *J Environ Qual* 13:115–121.

Gobas FAPC. 1993. A model for predicting the bioaccumulation of hydrophobic organic chemicals in aquatic food-webs: Application to Lake Ontario. *Ecolog Model* 69:1–17.

Gobas FAPC, Bedard DC, Ciborowski JJH, Haffner GD. 1989. Bioaccumulation of chlorinated hydrocarbons by the mayfly (*Hexagenia limbata*) in Lake St. Clair. *J Great Lakes Res* 15:581–588.

Gobas FAPC, Mackay D. 1987. Dynamics of hydrophobic organic chemical bioconcentration in fish. *Environ Toxicol Chem* 6:495–504.

Gobas FAPC, Morrison HA. 1999. Bioconcentration and bioaccumulation in the aquatic environment. In: Boethling R, Mackay D, editors. Handbook for environmental properties. Boca Raton FL: Lewis. In press.

Goksoyr A, Solberg TS, Serigstad B. 1991. Immunochemical detection of cytochrome P450IA1 induction in cod larvae and juveniles exposed to a water soluble fraction of North sea crude oil. *Marine Pollution Bull* 22:122–127.

Gruger EH, Karrick NL, Davidson AI, Hruby T. 1975. Accumulation of 3,4,3',4'-tetrachlorobiphenyl and 2,4,5,2',4',5' and 2,4,6, 2',4',6' hexachlorobiphenyl in juvenile coho salmon. *Environ Sci Technol* 9:121–127.

Guarino AM, Pritchard JB, Anderson JB, Rall DP. 1974. Tissue distributions of [14C]DDT in the lobster after administration via intravascular or oral routes or after exposure form ambient sea water. *Toxicol Appl Pharmacol* 29:277–288.

Guiney PD, Peterson RE. 1980. Distribution and elimination of a polychlorinated biphenyl after acute dietary exposure in yellow perch and rainbow trout. *Arch Environ Contam Toxicol* 9:667–674.

Guiney PD, Melancon MJ, Lech JJ, Peterson, RE. 1979. Effects of egg and sperm maturation and spawning on the distribution and elimination of a polychlorinated biphenyl in rainbow trout (*Salmo gairdneri*). *Toxicol Appl Pharmacol* 47:261–272.

Guiney PD, Stegeman JJ, Smolowitz RM, Walker MK, Peterson RE.1992. Localization of 2,3,7,8-tetrachlorodibenzo-*p*-dioxin (TCDD)-induced cytochrome P4501A1 in vascular endothelium of early life stages of lake trout. Society of Environmental Toxicology and Chemistry 13th Annual Meeting; 8–12 Nov 1992; Cincinnatti OH. Pensacola FL: SETAC. p 45.

Guiney PD, Smolowitz RM, Peterson RE, Stegeman JJ. 1997. Correlation of 2,3,7,8-tetrachlorodibenzo-*p*-dioxin induction of cytochrome P4501A in vasculature epithelium with toxicity in early life stages of lake trout. *Toxicol Appl Pharmac* 143:256–273.

Gundersen DT, Pearson WD. 1992. Partitioning of PCBs in the muscle and reproductive tissues of the paddlefish, *Polyodon spathula*, at the falls of the Ohio River. *Bull Environ Contam Toxicol* 49:455–462.

Gustafson KE, Dickhut RM. 1997. Distribution of polycyclic aromatic hydrocarbons in southern Chesapeake Bay surface water: evaluation of three methods for determining freely dissolved water concentrations. *Environ Toxicol Chem* 16:452–461.

Gustafsson O, Haghseta F, Chan C, McFarlane J, Gschwend PM. 1997. Quantitation of the dilute sedimentary soot phase: implications for PAH speciation and bioavailability. *Environ Sci Technol* 31:203–209.

Hales SG, Feijtel T, King H, Fox K, Verstraete W. 1996. Biodegradation kinetics: generation and use of data for regulatory decision-making. Brussels, Belgium: SETAC Europe. 165 p.

Hall AT, Oris JT. 1991. Anthracene reduces reproductive potential and is maternally transferred during long-term exposure in fathead minnows. *Aquat Toxicol* 19:249–264.

Hamilton JW, Denison MS, Bloom SE. 1983. Development of basal and induced aryl hydrocarbon (benzo[a]pyrene) hydroxylase activity in the chicken embryo in ovo. *Proc Natl Acad Sci* 80:3372–3376.

Hamilton SJ, Waddell B. 1994. Selenium in eggs and milt of razorback sucker (*Xyrauchen texanus*) in the middle Green River, Utah. *Arch Environ Contam Toxicol* 27:195–201.

Hansen MJ, Boisclair D, Brandt SB, Hewett SW, Kitchell JF, Lucas MC, Ney JJ. 1993. Applications of bioenergetics models to fish ecology and management: where do we go from here? *Trans Am Fish Soc* 122:1019–1030.

Hansch C, Hoekman D, Leo A, Zhang LT, Li P. 1995. The expanding role of quantitative structure-activity relationships (QSAR) in toxicology. *Toxicol Letters* 79:45–53.

Harel M, Tandler A, Kissil GW, Applebaum SW. 1994. The kinetics of nutrient incorporation into body tissues of gilthead sea bream ((*Sparus aurata*) females and the subsequent effects on egg composition and egg quality. *Br J Nutr* 72:45–58.

Harkey GA, Van Hoff PL, Landrum PF. 1995. Bioavailability of polycyclic aromatic hydrocarbons from a historically contaminated sediment core. *Environ Toxicol Chem* 14:1551–1560.

Harris HJ, Erdman TC, Ankley GT, Lodge KB. 1993. Measures of reproductive success and polychlorinated biphenyl residues in eggs and chicks of Forster's terns on Green Bay, Lake Michigan, Wisconsin—1988. *Environ Contam Toxicol* 25:304–314.

Haug LT, Dybing E, Thorgeirsson SS. 1980. Developmental aspects of 2-acetamidofluorene metabolism and mutagenic activation in the chick. *Xenobiotica* 10:863–872.

Hebert CE, Norstrom RJ, Simon M, Braune BM, Weseloh DV, Macdonald CR. 1994. Temporal trends and sources of PCDDs and PCDFs in the Great Lakes: herring gull egg monitoring 1981–1991. *Environ Sci Technol* 28:1266–1277.

Heinz GH. 1996. Selenium in birds. In: Beyer WN, Heinz GH, Redmon-Norwood AW, editors. Environmental contaminants in wildlife. Interpreting tissue concentrations. Boca Raton FL: Lewis. p 447–458.

Heinz GH, Hoffman DJ. 1996. Comparison of the effects of seleno-L-methionine, seleno-DL-methionine, and selenized yeast on reproduction of mallards. *Environ Pollut* 91:169–175.

Heinz GH, Hoffman DJ, Gold LG. 1989. Impaired reproduction of mallards fed an organic form of selenium. *J Wildl Manage* 53:418–428.

Heisinger JF, Green W. 1975. Mercuric chloride uptake by eggs of the ricefish and resulting teratogenic effects. *Bull Environ Contam Toxicol* 14:665–673.

Hellou J, Warren WG, Payne JF. 1993. Organochlorines including polychlorinated biphenyls in muscle, liver and ovaries of cod, *Gadus morhua*. *Arch Environ Contam Toxicol* 25:497–505.

Helmstetter MF, Alden RW. 1995. Passive trans-chorionic transport of toxicants in topically treated Japanese medaka (*Oryzias latipes*) eggs. *Aquat Toxicol* 32:1–13.

Hepp GR, Stangohr DJ, Baker LS, Kennamer RA. 1987. Factors affecting variation in the egg and duckling components of wood ducks. *The Auk* 104:435–443.

Hesselberg RJ, Hickey JP, Nortrup DA, Willford WA. 1990. Contaminant residues in the bloater (*Coregonus hoyi*) of Lake Michigan 1969–1986. *J Great Lakes Res* 16:121–129.

Hickey JJ, Anderson DW. 1968. Chlorinated hydrocarbons and egg-shell changes in raptorial and fish-eating birds. *Science* 162:271–273.

Hillyard LA, Entenman C, Chaikoff IL. 1956. Concentration and composition of serum lipoproteins of cholesterol-fed and stilbestrol-injected birds. *J Biol Chem* 223:359–368.

Hites RA. 1990. Environmental behavior of chlorinated dioxins and furans. *Acc Chem Res* 23:194–201.

Hoffman DJ. 1990. Embryotoxicity and teratogenicity of environmental contaminants to bird eggs. *Rev Environ Contam Toxicol* 115:39–89.

Hoffman DJ, Albers PH. 1984. Evaluation of potential embryotoxicity and teratogenicity of 42 herbicides, insecticides, and petroleum contaminants to mallard eggs. *Arch Environ Contam Toxicol* 13:15–27.

Hoffman DJ, Rattner BA, Bunck CM, Krynitsky A, Ohlendorf HM, Lowe RW. 1986. Association between PCBs and lower embryonic weight in black-crowned night herons in San Francisco Bay. *J Toxicol Environ Health* 19:383–391.

Hoffman DJ, Rattner BA, Sileo L, Docherty D, Kubiak TJ. 1987. Embryotoxicity, teratogenicity, and aryl hydrocarbon hydroxylase activity in Forster's terns on Green Bay, Lake Michigan. *Environ Res* 24:176–184.

Hoffman DJ, Smith GJ, Rattner BA. 1993. Biomarkers of contaminant exposure in common terns and black-crowned night herons in the Great Lakes. *Environ Toxicol Chem* 12:1095–1103.

Hogan JW, Brauhn JL. 1975. Abnormal rainbow trout fry from eggs containing high residues of PCB (Aroclor 1242). *Prog Fish-Cult* 37:229–230.

Horstmann M, McLachlan MS. 1992. Initial development of a solid-phase fugacity meter for semivolatile organic compounds. *Environ Sci Technol* 26:1643–1649.

Hose JE, Hannah JB, DiJulio D, Landolt ML, Miller BS, Iwaoka WT, Felton SP. 1982. Effects of benzo(a)pyrene on early development of flatfish. *Arch Environ Contam Toxicol* 11:167–171.

Hothem RL, Roster DL, King KA, Keldsen TJ, Marois DC, Wainwright SE. 1995. Spatial and temporal trends of contaminants in eggs of wading birds from San Francisco Bay, California. *Environ Toxicol Chem* 14:1319–1331.

Howard PH. 1991. Handbook of environmental fate and exposure data for organic chemicals. II. Pesticides. Chelsea MI: Lewis.

Howard PH, Boethling RS, Jarvis WF, Meyland WM, Michalenko EM. 1991. Handbook of environmental degradation rates. Chelsea MI: Lewis. 725 p.

Huckins JN, Manuweera GK, Petty J, Mackay D, Lebo JA. 1993. Lipid containing semipermeable membrane devices for monitoring organic contaminants in water. *Environ Sci Technol* 27:2489–2496.

Huckins JN, Schwartz TR, Petty JD, Smith LM. 1988. Determination, fate and potential significance of PCBs in fish and sediment samples with emphasis on selected AHH-inducing congeners. *Chemosphere* 17:1995–2016.

Husain MM, Kumar A, Mukhtar H. 1984. Hepatic mixed-function oxidase activities of wild pigeons. *Xenobiotica* 14:761–766.

Ingersoll CG, Dillon T, Biddinger GR. 1997. Ecological risk assessment of contaminated sediments. Pensacola FL: SETAC. 390 p.

Janz DM, Bellward GD. 1996a. In ovo 2,3,7,8-tetrachlorodibenzo-*p*-dioxin exposure in three avian species. 1. Effects on thyroid hormones and growth during the perinatal period. *Toxicol Appl Pharmacol* 139:281–291.

Janz DM, Bellward GD. 1996b. In ovo 2,3,7,8-tetrachlorodibenzo-*p*-dioxin exposure in three avian species. 2. Effects on estrogen receptor and plasma sex steroid hormones during the perinatal period. *Toxicol Appl Pharmacol* 139:292–300.

Jared DW, Wallace RA. 1968. Comparative chromatography of the yolk proteins of teleosts. *Comp Biochem Physiol* 24:437–443.

Jensen S, Johansson N, Olsson M. 1970. PCB-indicators of effects on salmon. Stockholm, Sweden: Rep. LFI MEDD of the Swedish Research Institute. 42 p.

Johnson EV, Mack GL, Thompson DQ. 1976. The effects of pesticide applications on breeding robins. *Wilson Bull* 88:16–35.

Jude DJ, Klinger SA, Enk MD. 1981. Evidence of natural reproduction by planted lake trout in Lake Michigan. *J Great Lakes Res* 7:57–61.

Kammann U, Landgraff O, Steinhart H. 1993. Distribution of aromatic organochlorines in liver and reproductive organs of male and female dabs from the German Bight. *Marine Pollution Bulletin* 26:629–635.

Kane-Driscoll S, Landrum PF. 1997. A comparison of equilibrium partitioning and critical body residue approaches for predicting toxicity of sediment-associated fluoranthene to freshwater amphipods. *Environ Toxicol Chem* 16:2179–2186.

Karickhoff SW. 1981. Semi-empirical estimate of sorption of hydrophobic pollutants on natural sediments and soils. *Chemosphere* 10:833–846.

Karickhoff SW, McDaniel VK, Melton C, Vellino AN, Nute DE, Carreira LA. 1991. Predicting chemical reactivity by computer. *Environ Toxicol Chem* 10:1405–1416.

Keith JA. 1966. Reproduction in a population of herring gulls (*Larus argentatus*) contaminated by DDT. *J Appl Ecol* 3:57–70.

Keith JO, Mitchell CA. 1993. Effects of DDE and food stress on reproduction and body condition of ringed turtle doves. *Arch Environ Contam Toxicol* 25:192–203.

Kelsey JW, Alexander M. 1997. Declining bioavailability and inappropriate estimation of risk of persistent compounds. *Environ Toxicol Chem* 16:582–585.

Kendeigh SC, Dol'nik VR, Gavrilov VM. 1977. Avian energetics. In: Pinowski J, Kendeigh SC, editors. Granivorous birds in ecosystems. London, UK: Cambridge University Press. p 127–204.

Kishida M, Specker JL. 1993. Vitellogenin in tilapia (*Oreochromis mossambicus*): induction of two forms by estradiol, quantification in the plasma and characterization in oocyte extract. *Fish Physiol Biochem* 12:171–182.

Kitchell JF. 1983. Energetics. In: Webb PW, Weihs D, editors. Fish biomechanics. New York: Praeger. p 313–338.

Klopman G, Zhang Z, Balthasar DM, Rosenkranz HS. 1995. Computer-automated predictions of aerobic biodegradation of chemicals. *Environ Toxicol Chem* 14:395–403.

Knickmeyer R, Steinhart H. 1989. On the distribution of polychlorinated biphenyl congeners and hexachlorobenzene in different tissues of dab (*Limanda limanda*) from the North Sea. *Chemosphere* 19:1309–1320.

Knight GC, Walker CH. 1982. A study of the hepatic microsomal epoxide hydrolase in sea birds. *Comp Biochem Physiol* 73(C):211–221.

Knight GC, Walker CH, Cabot DC, Harris MP. 1981. The activity of two hepatic microsomal enzymes in sea birds. *Comp Biochem Physiol* 68(C):127–132.

Knust R. 1986. Food selection of the dab (*Limanda limanda* L.): diet and seasonal changes. ICES C. M. 1986/g:63.

Ko F-C, Baker JE. 1995. Partitioning of hydrophobic organic contaminants to resuspended sediments and plankton in the mesohaline Chesapeake Bay. *Mar Chem* 49:171–188.

Kocan RM, Landolt ML. 1984. Uptake and excretion of benzo(a)pyrene by trout embryos and sac fry. *Marine Environ Res* 14:433–436.

Kovats ZE, Ciborowski JJH. 1989. Aquatic insect adults as indicators of organochlorine contamination. *J Great Lakes Res* 15:623–634.

Kraaij H, Connell DW. 1997. Bioconcentration and uptake kinetics of chlorobenzenes in soybean roots. *Chemosphere* 34:2607–2620.

Kubiak TJ, Harris HJ, Smith LM, Schwartz TR, Stalling DL, Trick JA, Sileo L, Docherty DE, Erdman TC. 1989. Microcontaminants and reproductive impairment of the Forster's tern in Green Bay, Lake Michigan—1983. *Environ Contam Toxicol* 18:706–727.

Kucklick JR, Baker JE. 1998. Organochlorines in Lake Superior's food web. *Environ Sci Technol* 32:1192–1198.

Kuehl DW, Butterworth BC, McBride A, Kroner S, Bahnick D. 1989. Contamination of fish by 2,3,7,8-tetrachlorodibenzo-p-dioxin: a survey of fish from major watersheds in the United States. *Chemosphere* 18:1997–2014.

Kuhnhold WW, Busch F. 1978. On the uptake of the three different types of hydrocarbons by salmon eggs (*Salmo salar* L). *Meeresforsch* 26:50–59.

Kvestak R, Ahel M. 1995. Biotransformation of nonylphenol polyethoxylate surfactants by estuarine mixed bacrerial cultures. *Arch Environ Contam Toxicol* 29:551–556.

Kwok ES, Atkinson R. 1995. Estimation of hydroxyl radical reaction rate constants for gas-phase organic compounds using a structure-reactivity relationship: an update. *Atmos Environ* 14:1685–1695.

LaFleur Jr GJ, Byrne BM, Kanungo J, Nelson LD, Greenberg RM, Wallace RA. 1995. Fundulus heteroclitus vitellogenin: the deduced primary structure of a piscine precursor to noncrystalline, liquid-phase yolk protein. *J Mol Evol* 41:505–521.

Lake JL, McKinney R, Osterman FA, Lake CA. 1996. C-18-Coated silica particles as a surrogate for benthic uptake of hydrophobic compounds from bedded sediments. *Environ Toxicol Chem* 15:2284–2289.

Lal B, Singh TP. 1987. Changes in tissue lipid levels in the freshwater catfish *Calarias batrachus* associated with the reproductive cycle. *Fish Physiol Biochem* 3:191–201.

Landrum PF, Eadie BJ, Fayst WR. 1992. Variation in the bioavailability of polycyclic aromatic hydrocarbons to the amphipod *Diporeia* (spp.) With sediment aging. *Environ Toxicol Chem* 11:1197–1208.

Landrum PF, Lee II H, Lydy MJ. 1992. Toxicokinetics in aquatic systems: Model comparisons and use in hazard assessment. *Environ Toxicol Chem* 11:1709–1725.

Landrum PF, Reinhold MD, Nihart SR, Eadie BJ. 1985. Predicting the bioavailability of organic xenobiotics to *Pontoporeia hoyi* in the presence of humic and fulvic materials and natural dissolved organic matter. *Environ Toxicol Chem* 4:459–467.

Langston WJ, Pope ND. 1995. Determinants of TBT adsorption and desorption in estuarine sediments. In: Trace metals in the aquatic environment. Proceedings of the 3rd International Conference; 16–20 May 1994; Aarhus, Denmark. *Mar Pollut Bull* 31:32–43.

Larson JM, Karasov WH, Sileo L, Stromborg DL, Hanbidge BA, Giesy JP, Jones PD, Tillitt DE, Berbrugge DA. 1996. Reproductive success, developmental anomalies, and environmental contaminants in double-crested cormorants (*Phalacrocorax auritus*). *Environ Toxicol Chem* 15:553–559.

Larsson P. 1984. Transport of PCBs from aquatic to terrestrial environments by emerging chironomids. *Environ Pollut* 34:283–289.

Larsson P, Collvin L, Okla L, Meyer G. 1992. Lake productivity and water chemistry as governors of the uptake of persistent pollutants in fish. *Environ Sci Technol* 26:346–352.

Lauren DJ, Okihiro MS, Hinton DE, Stegeman JJ. 1990. Localization of cytochrome P450 IA1 induced by β-naphthoflavone (BNF) in rainbow trout (*Oncorhynchus mykiss*) embryos. *FASEB J* 4:A739.

Law FCP, Abedini S, Kennedy CJ. 1991. A biologically based toxicokinetic model for pyrene in rainbow trout. *Toxicol Appl Pharmacol* 110:390–402.

Lawrence GS, Gobas FAPC. 1997. A pharmacokinetic analysis of interspecies extrapolation in dioxin risk assessment. *Chemosphere* 35:427–452.

Lemmetyinen R, Rantamaki P, Karlin A. 1982. Levels of DDT and PCBs in different stages of the life cycle of the Arctic tern *Sterna Paradisaea* and the herring gull *Larus argentatus*. *Chemosphere* 11:1059–1068.

Lewis SK, Lech JJ. 1996. Uptake, disposition and persistence of nonyl-phenol from water in rainbow trout (*Oncorhynchus mykiss*). *Xenobiotica* 26:813–819.

Lieb AJ, Bills DD, Sinnhuber RO. 1974. Accumulation of dietary polychlorinated biphenyls (Arochlor 1254) by rainbow trout (*Salmo gairdneri*). *J Agr Food Chem* 22:638–642.

Lindqvist O. 1991. Mercury in forest lake ecosystems : Bioavailability, bioaccumulation and biomagnification. *Water Air Soil Pollut* 55:131–157.

Lindstrom-Seppa P, Korytko PJ, Hahn ME, Stegeman JJ. 1994. Uptake of waterborne 3,3',4,4'-tetrachlorobiphenyl and organ and cell-specific induction of cytochrome P4501A in adult and larval fathead minnow *Pimephales promelas*. *Aquat Toxicol* 28:147–167.

Loehr RC, Webster MT. 1996. Behavior of fresh versus aged chemicals in soil. *J Soil Contam* 5(4):361–383.

Longcore JR, Stendell RC. 1977. Shell thinning and reproductive impairment in black ducks after cessation of DDE dosage. *Arch Environ Contam Toxicol* 6:293–304.

Lucas A. 1996. Bioenergetics of aquatic animals. Bristol PA: Taylor and Francis. 169 p.

Ludwig JP, Tomoff CS. 1966. Reproductive success and insecticide residues in Lake Michigan herring gulls. *The Jack-Pine Warbler* 44:77–85.

Lung NP, Thompson JP, Kollias GVJ, Olsen JH, Zdziarshi JM, Klein PA. 1996. Maternal immunoglobulin G antibody transfer and development of immunoglobulin G antibody responses in blue and gold macaw (*Ara ararauna*) chicks. *Am J Vet Res* 57:1162–1167.

Lyman CE, Lakis RE, Stenger Jr HG 1995. Composition-size distribution diagrams for alloy catalysts. Conference on Microscopy and Microanalysis; Kansas City, MO; 13–17 Aug 1995.

Mac MJ, Edsall CC, Seelye JG. 1985. Survival of lake trout eggs and fry reared in water from the upper Great Lakes. *J Great Lakes Res* 11:520–529.

Mac MJ, Schwartz TR, Edsall CC. 1988. Correlating PCB effects on fish reproduction using dioxin equivalents. Society of Environmental Toxicology and Chemistry Ninth Annual Meeting; 13–17 Nov 1988; Arlington VA. Pensacola FL: SETAC. p 116.

Mac MJ, Seeley JG. 1981. Patterns of PCB accumulation by fry of lake trout. *Bull Environ Contam Toxicol* 27:368–375.

Macek KJ, Sleight BH. 1977. Utility of toxicity tests with embryos and fry of fish in evaluating hazards associated with the chronic toxicity of chemicals to fishes. In: Mayer FL,

Hamelink JL, editors. Aquatic toxicology and hazard evaluation. Philadelphia PA: ASTM. ASTM STP 634. p 137–146.

Mackay D. 1982. Correlation of bioconcentration factors. *Environ Sci Technol* 16:274–276.

Mackay D. 1991. Multimedia environmental fate models: the fugacity approach. Chelsea MI: Lewis Publishers. 272 p.

Mackay D, Shiu WY, Ma KC. 1992a. Illustrated handbook of physical-chemical properties and environmental fate for organic chemicals. Volume I: Monoaromatic hydrocarbons, chlorobenzenes, and PCBs. Boca Raton FL: Lewis. 697 p.

Mackay D, Shiu WY, Ma KC. 1992b. Illustrated handbook of physical-chemical properties and environmental fate for organic chemicals. Volume II: Polynuclear aromatic hydrocarbons, polychlorinated dioxins, and dibenzofurans. Boca Raton FL: Lewis. 597 p.

MacLellan KNM, Bird DM, Fry DM, Cowles JL. 1996. Reproductive and morphological effects of *o,p'*-dicofol on two generations of captive American kestrels. *Arch Environ Contam Toxicol* 30:364–372.

Madenjian CP, Carpenter SR, Eck GW, Miller MA. 1993. Accumulation of PCBs by lake trout (*Savelinus namaycush*): an individual-based model approach. *Can J Fish Aquat Sci* 50:97–109.

Maller JL. 1985. Oocyte maturation in amphibians. *Dev Biol* 1:289–311.

Marsden JE, Krueger CC, Schneider CP. 1988. Evidence of natural reproduction by stocked lake trout in Lake Ontario. *J Great Lakes Res* 14:3–8.

Mason RP, Sullivan KA. 1997. Mercury in Lake Michigan. *Environ Sci Technol* 31:942–947.

Mayer FL, Mehrle PM, Dwyer WP. 1977. Toxaphene: chronic toxicity to fathead minnows and channel catfish. Washington DC: U.S. Environmental Protection Agency. EPA 600/3-77-069.

McCarthy JF, Jimenez BD. 1985. Reduction in bioavailability to bluegills of polycyclic aromatic hydrocarbons bound to dissolved humic material. *Environ Toxicol Chem* 4:511–521.

McFarland VA. 1995. Evaluation of field generated accumulation factors for predicting the bioaccumulation potential of sediment associated pah compounds. Washington DC: U.S. Army Corps of Engineers. Technical Report D-95-2. 139 p.

McGroddy SE, Farrington JW, Gschwend PM. 1996. Comparison of the in situ and desorption sediment-water partitioning of polycyclic aromatic hydrocarbons and polychlorinated biphenyls. *Environ Sci Technol* 30:172–177.

McIndoe WM. 1959. A lipophosphoprotein complex in hen plasma associated with yolk production. *Biochem J* 72:153–159.

McKone TE. 1993. The precision of QSAR methods for estimating intermedia transfer factors in exposure assessments. *SAR and QSAR in Environmental Research* 1:41–51.

McKim JM. 1977. Evaluation of tests with early life stages of fish for predicting long-term toxicity. *J Fish Res Board Can* 34:1148–1154.

McKim JM, Arthur JW, Thorslund TW. 1975. Toxicity of a linear alkylate sulfonate detergent to larvae of four species of freshwater fish. *Bull Environ Contam Toxicol* 14:1–7.

McKim JM, Nichols JW. 1994. Use of physiologically based toxicokinetic models in a mechanistic approach to aquatic toxicology. In: Malins DC, Ostrander GK, editors. Aquatic toxicology, molecular, biochemical and cellular perspectives.Boca Raton FL: Lewis. p 469–519.

McLeese DW, Zitko V, Sergeant DB, Burridge L, Metcalfe CD. 1981. Lethality and accumulation of alkylphenols in aquatic fauna. *Chemosphere* 10:723–730.

McNab BK. 1988. Food habits and the basal rate of metabolism in birds. *Oecologia* 77:343–349.

Medford BA, Mackay WC. 1978. Protein and lipid content of gonads, liver and muscle of northern pike (*Esox lucius*) in relation to gonad growth. *J Fish Res Board of Canada* 35:213–219.

Meier JR, Chang LW, Jacobs S, Torsella J, Meckes MC, Smith MK. 1997. Use of plant and earthworm bioassays to evaluate remediation of soil from a site contaminated with polychlorinated biphenyls. *Environ Toxicol Chem* 16:928–938.

Meylan WM, Howard PH. 1995. Users guide for EPIWIN. Syracuse NY: Syracuse Research Corporation.

Millard ES, Halfon E, Minns CK, Charlton CC. 1993. Effect of primary productivity and vertical mixing of PCB dynamics in planktonic model ecosystems. *Environ Toxicol Chem* 12:931–946.

Miller MA 1993. Maternal transfer of organochlorine compounds in Salmonines to their eggs. *Can J Fish Aquat Sci* 50:1405–1413.

Miller MA, Amrhein JF. 1995. Maternal transfer of organochlorine compounds in Lake Superior Siscowet (*Salvelinus namaycush sisowet*) to their eggs. *Bull Environ Contam Toxicol* 55:96–103.

Mineau P. 1982. Levels of major organochlorine contaminants in sequentially laid herring gull eggs. *Chemosphere* 11:679–685.

Mineau P, Fox GA, Norstrom RJ, Weseloh DV, Hallett DJ, Ellenton JA. 1984. Using the herring gull to monitor levels and effects of organochlorine contamination in the Canadian Great Lakes. In: Nriagu JO, Simmons MS, editors. Toxic contaminants in the Great Lakes. New York: Wiley. p 425–452.

Miskimmin BM, Muir DCG, Schindler DW, Stern GA, Grift NP. 1995. Chlorobornanes in sediments and fish thirty years after toxaphene treatment of lakes. *Environ Sci Technol* 29:2490–2495.

Mitani F, Alvares AP, Sassa S, Kappas A. 1971. Preparation and properties of a solubilized form of chytochrome P-450 from chick embryo liver microsomes. *Mol Pharmacol* 7:280–292.

Mohammed A, Eklund A, Ostlund-Lindqvist AM, Slanina P. 1990. Distribution of toxaphene, DDT, and PCB among lipoprotein fractions in rat and human plasma. *Arch Toxicol* 64:567–571.

Monad G. 1985. Egg mortality of Lake Geneva charr (*Salvelinus alpinus L.*) contaminated by PCB and DDT derivatives. *Bull Environ Contam Toxicol* 35:531–536.

Monod G, Boudry MA, Gillet C. 1996. Biotransformation enzymes and their induction by β-naphthoflavone during embryolarval development in salmonid species. *Comp Biochem Physiol* 114C:45–50.

Monosson E. 1992. Effects of 3,3',4,4'-tetrachlorobiphenyl on reproductive processes in white perch. Society of Environmental Toxicology and Chemistry 13th Annual Meeting; 8–12 Nov 1992; Cincinnati OH. Pensacola FL: SETAC. p 259.

Monteverdi GH. 1999. Vitellogenin-mediated chemical uptake by oocytes of the estuarine killifish (*Fundulus heteroclitus*). [Ph.D. dissertation]. Durham NC: Duke University.

Morrison PF, Leatherland JF, Sonstegard RA. 1985. Proximate composition and organochlorine and heavy metal contamination of eggs from Lake Ontario, Lake Erie and Lake Michigan coho salmon (*Oncorhynchus kisutch walbaum*) in relation to egg survival. *Aquat Toxicol* 6:73–86.

Muir DCG. 1984. Phosphate esters. In: Hutzinger O, editor. Handbook of environmental chemistry. Heidelberg, Germany: Springer-Verlag. p 41–66.

Muir DCG, Norstrom RJ, Simon M. 1988. Organochlorine contaminants in arctic marine food chains: Accumulation of specific PCB and chordane-related compounds. *Environ Sci Technol* 22:1071–1079.

Naf C, Broman D, Brunstrom B. 1992. Distribution and metabolism of polycyclic aromatic hydrocarbons (PAHs) injected into eggs of chicken (*Gallus domesticus*) and common eider duck (*Somateria mollissima*). *Environ Toxicol Chem* 11:1653–1660.

Nagahama Y. 1994. Endocrine regulation of gametogenesis in fish. *Int J Dev Biol* 38:217–229.

Nassour I, Leger CL. 1989. Deposition and mobilization of body fat during sexual maturation in female trout *Salmo gairdneri R. Aquat Living Resour* 2:153–159.

Naylor CG, Mieure JP, Adams WJ, Weeks JA, Castaldi FJ, Ogle LD, Ramano RR. 1992. *J Am Oil Chem Soc* 69:695–703.

Nester RT, Poe TP. 1984. First evidence of successful natural reproduction of planted lake trout in Lake Huron. *N Am J Fish Manage* 4:126–128.

Newman MC. 1995. Quantitative methods in aquatic ecotoxicology. Boca Raton FL: Lewis. 426 p.

Nichols JW, Jensen KM, Tietge JE, Johnson RD. 1998. Physiologically based toxicokinetic model for maternal transfer of 2,3,7,8-tetrachlorodibenzo-*p*-dioxin in brook trout (*Salvelinus fontinalis*). *Environ Toxicol Chem* 17:2422–2434.

Nichols JW, Larsen CP, McDonald ME, Niemi GJ, Ankley GT. 1995. Bioenergetics-based model for accumulation of polychlorinated biphenyls by nestling tree swallows, *Tachycineta bicolor. Environ Sci Technol* 29:604–612.

Nichols JW, McKim JM, Andersen ME, Gargas ML, Clewell III HJ, Erickson RJ. 1990. A physiology based toxicokinetic model for the uptake and disposition of waterborne organic chemicals in fish. *Toxicol Appl Pharmacol* 106:433–447.

Niimi AJ. 1983. Biological and toxicological effects of environmental contaminants in fish and their eggs. *Can J Fish Aquat Sci* 40:306–312.

Niimi AJ. 1987. Biological half-lives of chemicals in fishes. *Rev Environ Contam Toxicol* 99:1–46

Niimi AJ, Oliver BG. 1989. Assessment of relative toxicity of chlorinated dibenzo-*p*-dioxins, dibenzofurans, and biphenyls in Lake Ontario salmonids to mammalian systems using toxic equivalent factors (TEF). *Chemosphere* 18:1413–1423.

Noguchi GE, Hesselberg RJ. 1991. Parental transfer of organic contaminants to young-of-the-year spottail shiners, *Notropis hudsonius*. *Bull Environ Contam Toxicol* 46:745–750.

Norstrom RJ, Clark TP, Jeffrey DA, Won HT, Gilman AP. 1986. Dynamics of organochlorine compounds in herring gulls (*Larus argentatus*) I. Distribution and clearance of [^{14}C] DDE in free-living herring gulls (*Larus argentatus*). *Environ Toxicol Chem* 5:41–48.

Norstrom RJ, Clark TP, Kearney JP, Gilman AP. 1986. Herring gull energy requirements and body constituents in the Great Lakes. *Ardea* 72:1–23.

Norstrom RJ, Simon M, Muir DCG, Schweinsburg RE. 1988. Organochlorine contaminants in Arctic marine food chains: identification, geographical distribution, and temporal trends in polar bears. *Environ Sci Technol* 22:1063–1071.

Nosek JA, Craven SR, Sullivan JR, Olson JR, Peterson RE. 1992. Metabolism and disposition of 2,3,7,8-tetrachlorodibenzo-*p*-dioxin in ring-necked pheasant hens, chicks, and eggs. *J Toxicol Environ Health* 35:153–164.

Nosek JA, Sullivan JR, Craven SR, Gendron-Fitzpatrick A, Peterson RE. 1993. Embryotoxicity of 2,3,7,8-tetrachlorodibenzo-*p*-dioxin in the ring-necked pheasant. *Environ Toxicol Chem* 12:1215–1222.

[OECD] Organization for Economic Cooperation and Development. 1992. OECD guideline for testing of chemicals: ready biodegradability 301 A-F. Paris, France: OECD. 62 p.

Ohlendorf HM, Hoffman DJ, Saiki MK, Aldrich TW. 1986. Embryonic mortality and abnormalities of aquatic birds: apparent impacts of selenium from irrigation drainwater. *Sci Tot Environ* 52:49–63.

Oliver BG, Niimi AJ. 1988. Trophodynamic analysis of polychlorinated biphenyl congeners and other chlorinated hydrocarbons in the Lake Ontario ecosystem. *Environ Sci Technol* 22:388–397.

Olivieri CE, Cooper KR. 1997. Toxicity of 2,3,7,8-tetrachlorodibenzo-*p*-dioxin (TCDD) in embryos and larvae of the fathead minnow (*Pimephales promelas*). *Chemosphere* 34:1139–1150.

Opperhuizen A, Sijm DTHM. 1990. Bioaccumulation and biotransformation of polychlorinated dibenzo-*p*-dioxins and dibenzofurans in fish. *Environ Toxicol Chem* 9:175–186.

Ormerod SJ, Tyler SJ. 1994. Inter- and intra-annual variation in the occurrence of organochlorine pesticides, polychlorinated biphenyl congeners, and mercury in the eggs of a river passerine. *Arch Environ Contam Toxicol* 26:7–12.

Packard MJ, Clark NB. 1996. Aspects of calcium regulation in embryonic lepidosaurians and chelonians and a review of calcium regulation in embryonic archosaurians. *Physiol Zool* 69:435–466.

Palmisano F, Cardellicchio N, Zambonin PG. 1995. Speciation of mercury in dolphin liver: a two-stage mechanism for the demethylation process and role of selenium. *Mar Environ Res* 40:109–121.

Pan HP, Fouts JR. 1978. Drug metabolism in birds. *Drug Metab Rev* 7:1–253.

Parkerton TF. 1993. Estimating toxicokinetic parameters for modeling the bioaccumulation of non-ionic organic chemicals in aquatic organisms [Ph.D. dissertation]. New Brunswick NJ: Rutgers University. 336 p.

Parkerton TF, Stone MA. 1996. Ecotoxicity on a stick: a novel analytical method for assessing the ecotoxicity of hydrocarbon contaminated samples. Exxon Biomedical Sciences Draft Publication. East Millstone NJ: Exxon.

Payne JF, Fancey LL, Rahimtula AD, Porter EL. 1987. Review and perspective on the use of mixed-function oxygenase enzymes in biological monitoring. *Comp Biochem Physiol* 86(C):233–245.

Peakall DB, Gilman AP. 1979. Limitations of expressing organochlorine levels in eggs on a lipid weight basis. *Bull Environ Contam Toxicol* 23:287–290.

Peakall DB, Norstrom RJ, Rahimtula AD, Butler RD. 1986. Characterization of mixed-function oxidase systems of the nestling herring gull and its implications for bioeffects monitoring. *Environ Toxicol Chem* 5:379–385.

Peijenburg WJGM, Posthuma L, Eijsackers HJP, Allen HE. 1997. A concepted framework for implementation of bioavailability of metals for environmental management. *Ecotox Environ Safety* 37:163–172.

Persson B. 1971. Chlorinated hydrocarbons and reproduction of a south Swedish population of whitethroats Synvia communis. *Oikos* 22:248–255.

Peters LD, Livingstone DR. 1995. Studies on cytochrome P4501A in early and adult life stages of turbot (*Scophthalmus maximus L.*). *Marine Environ Res* 39:5–9.

Pickering QH, Gast MH. 1972. Acute and chronic toxicity of cadmium to the fathead minnow *Pimephales promelas. J Fish Res Board Can* 29:1099–1106.

Pickering QH, Thatcher TO. 1970. The chronic toxicity of linear alkylate sulfonate (LAS) to *Pimephales promelas* Rafinesque. *J Water Pollut Control Fed* 42:243–254.

Pignatello JJ, Xing B. 1995. Mechanisms of slow sorption of organic chemicals to natural particles. *Environ Sci Technol* 30:1–11.

Plack PA, Skinner ER, Rogie A, Mitchell AI 1979. Distribution of DDT between lipoproteins of trout serum. *Comp Biochem Physiol* 62C:119–125.

Poland A, Glover E. 1977. Chlorinated biphenyl induction of aryl hydrocarbon hydroxylase activity: a study of the structure-activity relationship. *Mol Pharmacol* 13:924–938.

Porter RD, Wiemeyer SN. 1972. DDE at low dietary levels kills captive American kestrels. *Bull Environ Contam Toxicol* 8:193–199.

Powis G, Drummond AH, MacIntyre DE, Jondorf WR. 1976. Development of microsomal oxidations in the chick. *Xenobiotica* 6:69–81.

Rasmussen JB, Rowan DJ, Lean DRS, Carey JH. 1990. Food chain structure in Ontario lakes determines PCB levels in lake trout (*Salvelinus namaycush*) and other pelagic fish. *Can J Fish Aquat Sci* 47:2030–2038.

Ratcliffe DA. 1967. Decrease in eggshell weight in certain birds of prey. *Nature* 215:208–210.

Rattner BA, Hoffman DJ, Marn CM. 1989. Use of mixed-function oxygenases to monitor contaminant exposures in wildlife. *Environ Toxicol Chem* 8:1093–1102.

Reitz RH, Gargas ME, Andersen ME, Provan WM, Green TL. 1996. Predicting cancer risk from vinyl chloride exposure with a physiologically based pharmacokinetic model. *Toxicol Appl Pharmacol* 137:253–267.

Rhee G, Sokol RC. 1993. A long-term study of anaerobic dechlorination of PCB congeners by sediment microorganisms: pathways and mass balance. *Environ Toxicol Chem* 12:1829–1834.

Rifkind AB, Toreger M, Muschick H. 1982. Kinetic evidence for heterogeneous responsiveness of mixed function oxidase isozymes to inhibition and induction by allylisopropylacetamide in chick embryo liver. *J Biol Chem* 257:11717–11727.

Rifkind AR, Sassa S, Reyes J, Muschick H. 1985. Polychlorinated aromatic hydrocarbon lethality, mixed-function oxidase induction, and uroporphyrinogen decarboxylase inhibition in the chick embryo: dissociation of dose-response relationships. *Toxicol Appl Pharmacol* 78:268–279.

Robinson J, Richarson A, Crabtree AN, Coulson JC, Potts JR. 1967. Organochlorine residues in marine organisms. *Nature* 214:1307–1311.

Ronday R, van Kammen-Polman AMM, Dekker A, Houx NWW, Leistra M. 1997. Persistence and toxicological effects of pesticides in topsoil: use of equilibrium partitioning theory. *Environ Toxicol Chem* 16:601–607.

Ronis MJJ, Walker CH. 1989. The microsomal monooxygenases of birds. In: Hodgson E, Philpot R, Bend J, editors. Reviews in biochemical toxicology. Volume 2. New York: Elsevier. p 301–384.

Roudybush TE, Grau CR, Petersen MR, Ainley DG, Hirse KU, Gillman AP, Patten SM. 1979. Yolk formation in some charadriform birds. *Condor* 81:293–298.

Rowland M. 1985. Physiologic pharmacokinetic models and interanimal species scaling. *Pharmac Ther* 29:49–68.

Rozemeijer MJC, Boon JP, Swennen C, Brouwer A, Murk AJ. 1995. Dioxin type and mixed type induction of the cytochrome P-450 system of common eider ducklings (*Somateria mollissima*) by PCBs: with indications for biotransformation. *Aquat Toxicol* 32:93–113.

Ruby MV, Davis A, Schoof R, Eberle S, Sellstone CM. 1996. Estimation of lead and arsenic bioavailability using a physiologically based extraction test. *Environ Sci Technol* 30:422–430.

Russell RW, Gobas FAPC, Haffner GD. 1999. Maternal transfer and in-ovo exposure of organochlorines in oviparous organisms: a model and field verification. *Environ Sci Technol* 33:416–420.

Sanderson JT, Norstrom RJ, Elliott JE, Hart LE, Cheng KM, Bellward GD. 1994. Biological effects of polychlorinated dibenzo-*p*-dioxins, dibenzofurans, and biphenyls in double-crested cormorant chicks (*Phalacrocorax auritus*). *J Toxicol Environ Health* 41:247–265.

Sanderson JT, Bellward GD. 1995. Hepatic microsomal ethoxyresorufin o-deethylase-inducing potency in ovo and cytosolic Ah receptor binding affinity of 2,3,7,8-tetrachorodibenzo-*p*-dioxin: comparison of four avian species. *Toxicol Appl Pharmacol* 132:131–145.

Sanderson JT, Janz DM, Bellward GD, Giesy JP. 1997. Effects of embryonic and adult exposure to 2,3,7,8-tetrachlorodibenzo-*p*-dioxin on hepatic microsomal testosterone hydroxylase activities in great blue herons (*Ardea herodias*). *Environ Toxicol Chem* 6:1304–1310.

Sauter S, Buxton KS, Macek KJ, Petrocelli SR. 1976. Effects of exposure to heavy metals on selected freshwater fish. Washington DC: USEPA. EPA-600/3-76-105.

Scheuhammer AM. 1991. The chronic toxicity of aluminum, cadmium, mercury, and lead in birds: a review. *Environ Pollut* 46:263–295.

Schjeide OA. 1954. Studies of New Hampshire chicken embryo III. Nitrogen and lipid analyses of ultracentrifugal fractions of plasma. *J Biol Chem* 211:355–362.

Schmitt CJ, Zajicek JL, Ribick MA. 1985. National pesticide monitoring program: residues of organochlorine chemicals in freshwater fish 1980–81. *Arch Environ Contam Toxicol* 14:225–260.

Schultz R, Hermanutz R. 1990. Transfer of toxic concentrations of selenium from parent to progeny in the fathead minnow (*Pimephales promelas*). *Bull Environ Contam Toxicol* 45:568–573.

Secor DH. 1992. Application of otolith microchemistry analysis to investigate anadromy in Chesapeake Bay striped bass, *Morone saxatilis*. *Fish Bull U.S.* 90:798–806.

Serafin JA. 1984. Avian species differences in the intestinal absorption of xenobiotics (PCB, dieldren, Hg^{2+}). *Comp Biochem Physiol* 78C:491–496.

Sharp JR, Fucik KW, Neff JM. 1979. Physiological basis of differential sensitivity of fish embryonic stages to oil pollution. In: Vernberg FJ, Vernberg WB, Calabrese A. editors. Marine pollution: functional responses. New York: Academic Press. p 85–108.

Sijm D, de Bruijn J, de Voogt P, de Wolfe W. 1997. Biotransformation in environmental risk assessment. Brussels, Belgium: SETAC Europe.

Sijm DTHM, Middelkoop J, Vrisekoop K. 1995. Algal density dependent bioconcentration factors of hydrophobic chemicals. *Chemosphere* 31:4001–4012.

Skalsky HL, Gutherie FE. 1977. Affinities of parathion, DDT, dieldrin and carbaryl for macromolecules in the blood of the rat and American cockroach and the competitive interactions of steroids. *Pest Biochem Physiol* 7:289–296.

Solbakken JE, Tilseth S, Palmork KH. 1984. Uptake and elimination of aromatic hydrocarbons and a chlorinated biphenyl in eggs and larvae of cod *Gadus morhua*. *Marine Ecology Progress Series* 16:297–301.

Soto AM, Sonnenschein C, Chung KL, Fernandez MF, Olea N, Serrano FO. 1995. The E-Screen assay as a tool to identify estrogens: an update on estrogenic environmental pollutants. *Environ Health Perspect* 103(7):113–122.

Specker J, Sullivan CV. 1994. Vitellogenesis in fishes: status and perspectives. In: Davey KG, Peter RG, Tobe SS, editors. Perspectives in comparative endocrinology. Toronto, ON, Canada: NRC Canada. p 304–315.

Spitsbergen JM, Walker MK, Olson JR, Petersen RE. 1991. Pathologic alterations in early life stages of lake trout (*Salvelinus namaycush*), exposed to 2,3,7,8-tetrachlorobenzo-*p*-dioxin as fertilized eggs. *Aquat Toxicol* 19:41–72.

Sproule JW, Shiu WY, Mackay D, Schroeder WH, Russell RW, Gobas FAPC. 1991. In situ measurement of the truly dissolved concentration of hydrophobic chemicals in natural waters. *Environ Toxicol Chem* 10:9–20.

Stalling DL, Smith LM, Petty JD, Hogan JW, Johnson JL, Rappe C, Buser HR. 1983. Residues of polychlorinated dibenzo-*p*-dioxins and dibenzofurans in Laurentian Great Lakes fish. In: Tucher RE, Young AL, Gray AP, editors. Human and environmental risks of chlorinated dioxins and related compounds. New York: Plenum Press. p 221–240.

Stanley TRJ, Spann JW, Smith GJ, Rosscoe R. 1994. Main and interactive effects of arsenic and selenium on mallard reproduction and duckling growth and survival. *Arch Environ Contam Toxicol* 26:444–451.

Stegeman JJ, Hahn, ME. 1994. Biochemistry and molecular biology of monooxygenases: current perspectives on forms, functions, and regulation of cytochrome P450 in aquatic species. In: Malins DC, Ostrander GK, editors. Aquatic toxicology: molecular, biochemical and cellular perspectives. Boca Raton FL: Lewis Publishers. p 152.

Stein WD. 1981. Permeability for lipophilic molecules. In: Bonting SL, de Pont JJHHM, editors. Membrane transport. Amsterdam, The Netherlands: Elsevier p 1–28.

Stewart C, de Mora SJ. 1990. A review of the degradation of tri(n-butyl)tin in the marine environment. *Environ Technol* 11:565–570.

Stickle WH, Galyen JA, Dyrland RA, Hughes DL. 1973. Toxicity and persistence of mixex in birds. In: Deichmann WB, editor. Pesticides and the environment: a continuing controversy. Eighth Inter-American Conference on Toxicology and Occupational Medicine. New York: Intercontinental Medical Book Corp. p 437–467.

Sullivan CV, Bernard MG, Hara A, Dickoff WW. 1989. Thyroid hormones in trout reproduction: enhancement of gonandotropin-releasing hormone analogue and partially purified salmon gonadotropin induced ovarian maturationin vivo and in vitro. *J Exper Zool* 250:188–195.

Summer CL, Giesy JP, Bursian SJ, Render JA, Kubiak TJ, Jones PD, Verbrugge DA, Aulerich RJ. 1996. Effects induced by feeding organochlorine-contaminated carp from Saginaw Bay, Lake Huron, to laying white leghorn hens. I. Effects on health of adult hens, egg production, and fertility. *J Toxicol Environ Health* 49:389–407.

Swackhamer DL, Hites RA. 1988. Occurrence and bioaccumulation of organochlorine compounds in fishes from Siskiwit Lake, Isle Royale, Lake Superior. *Environ Sci Technol* 22:543–548.

Swackhammer DL, Skoglund RS. 1991. The role of phytoplankton in the partitioning of hydrophobic organic contaminants in water. Baker RA, editor. Organic substances and sediments in water. Washington DC: American Chemical Society. p 91–106.

Swackhamer DL, Skoglund RS. 1993. Bioaccumulation of PCBs by algae: kinetics versus equilibrium. *Environ Toxicol Chem* 12:831–838.

Tagawa M, Specker JL. 1993. Incorporation and metabolism of cortisol in tilapia oocyte. In: Abstracts of the XII International Congress on Comparative Endocrinology. Toronto Canada. p A-155.

Tanabe S, Hidaka H, Tatsukawa R. 1983. PCBs and chlorinated hydrocarbon pesticides in Antarctic atmosphere and hydrosphere. *Chemosphere* 12(2):277–288.

Tanabe S, Subramanian AN, Hidaka H, Tatsukawa R. 1986. Transfer rates and patterns of PCB isomers and congeners and *p,p'*-DDE from mother to egg in Adelie penguin (*Pygoscelis adeliae*) *Chemosphere* 15:343–351.

Tell JG, Parkerton TF. 1997. A comparison of multimedia models used in regulatory decision-making. *SAR and QSAR Env Res* 6:29–45.

Thomann RV. 1981. Equilibrium model of fate of microcontaminants in diverse aquatic food chains. *Can J Fish Aquat Sci* 38:280–296.

Thomann RV. 1989. Bioaccumulation model of organic chemical distribution in aquatic food chains. *Environ Sci Technol* 23:699–707.

Thomann RB, Connolly J, Parkerton TF. 1992. An equilibrium model of organic chemical accumulation in aquatic food webs with sediment interaction. *Environ Toxicol Chem* 11:615–629.

Tomlin C, editor. 1994. The pesticide manual. 10th edition. London, UK: British Crop Protection Council and Royal Society of Chemistry. 1341 p.

Thomson WT. 1992. Agricultural chemicals. Book I: insecticides. Fresno CA: Thomson Publications.

Thompson NP, Rankin PW, Cowan PE, Williams Jr LE, Nesbitt SA. 1977. Chlorinated hydrocarbon residues in the diet and eggs of the Florida brown pelican. *Bull Environ Contam Toxicol* 18:331–339.

Tolls J, Melachlan MS. 1994. Partitioning of semivolatile organic compounds between air and *Lolium multiflorum* (Welsh Ray Grass). *Environ Sci Technol* 28:159–166.

Tomson MB, Pignatello JJ. 1997. Mechanisms and effects of resistant sorption processes of organic compounds in natural particles. Preprints of Papers Presented at the American Chemical Society Meeting; Las Vegas, NV. 37(2):130.

Travis CC, Arms AD. 1988. Bioconcnetration of organics in beef, milk, and vegetation. *Environ Sci Technol* 22(3):271–274.

Trivelpiece W, Butler RG, Miller DS, Peakall DB. 1981. Reduced growth and survival of chicks of oil-dosed adult Leach's storm-petrals. *Condor* 86:81–82.

Ungerer J, Thomas P. 1996. Transport and accumulation of organochlorines in the ovaries of Atlantic croaker. *Mar Environ Res* 42:167–171.

[USEPA] U.S. Environmental Protection Agency. 1991. Bioaccumulation of selected pollutants in fish. A national study. Washington DC: Office of Water Regulations and Standards. EPA 506/6-90/001a.

[USEPA] U.S. Environmental Protection Agency. 1992. Comparative analysis of acute avian risk from granular pesticides. Washington DC: USEPA. E371.

[USEPA] U.S. Environmental Protection Agency. 1993. Wildlife exposure factors, Volumes I and II. Washington DC: Office of Research & Development. EPA/600/R-93/187a,b.

[USEPA] U.S. Environmental Protection Agency. 1995. Final water quality guidance for the Great Lakes system: final rule. *Federal Register* 60(56):15366–15425.

[USEPA] U.S. Environmental Protection Agency. 1996. Proposed guidelines for ecological risk assessment (review draft). EPA/630/R-95/002B.

[USEPA] U.S. Environmental Protection Agency. 1997. Special report on Environmental Endocrine disruption: An Effects Assessment and Analysis. Washington DC: USEPA Risk Assessment Form. 111 p. EPA/630/R-96/012.

Van Den Berg M, Craane BLHJ, Sinnige T, Van Mourik S, Dirksen S, Boudewijn T, Van Der Gaag M, Lutke-Schipholt IJ, Spenkelink B, Brouwer A. 1994. Biochemical and toxic effects of polychlorinated biphenyls (PCBs), dibenzo-*p*-dioxins (PCDDs) and dibenzofurans (PCDFs) in the cormorant (*Phalacrocorax carbo*) after in ovo exposure. *Environ Toxicol Chem* 13:803–816.

Van der Poel P, Van Leeuween CJ, Hermens JLM, Ros JPM. 1995. Emissions of chemicals in risk assessment of chemicals, an introduction. Dordrecht, The Netherlands: Kluwer Academic. p 19–36.

Van Gestel CAM, van Diepen AMF. 1997. The influence of soil moisture content or the bioavailability and toxicity of cadmium for *F. Lsomia candida* Willem (Collembula: Isotomidae). *Ecotox Environ Safety* 36:123–132.

Van Leeuwen CJ. Griffioen PS, Vergouw WHA, Maas-Diepeveen JL. 1985. Differences in susceptibility of early life stages of rainbow trout (*Salmo gairdneri*) to environmental pollutants. *Aquat Toxicol* 7:59–78.

Van Leeuwen CJ, Van Der Zandt PTJ, Aldenberg T, Verhaar HJM, Hermens JLM. 1992. Application of QSARs, extrapolation and equilibrium partitioning. In: Aquatic effects assessment. I. Narcotic industrial pollutants. *Environ Toxicol Chem* 11:267–282.

Van Loon WMGM, Verwoerd ME, Wijnker FG, van Leeuwen CJ, van Duyn P, van de Gutche C, Hermens JLM. 1997. Estimating total body residues and baseline toxicity of complex organic mixtures in effluents and surface waters. *Environ Toxicol Chem* 16:1358–1365.

Van Velsen AC, Stiles WB, Stickle LF. 1972. Lethal mobilization of DDT by cowbirds. *J Wildl Manage* 36:733–739.

Verhaar HJM, Busser FJM, Hermens JLM. 1994. A surrogate parameter for the baseline toxicity content of contaminated water: simulating bioconcentration and counting molecules. *Environ Sci Technol* 29:726–734.

Vermeer K, Armstrong FAJ, Hatch DRM. 1973. Mercury in aquatic birds at Clay Lake, Western Ontario. *J Wildl Manage* 37:58–61.

Verrinder Gibbins AM, Robinson GA. 1982. Comparative study of the physiology of vitellogenesis in Japanese quail. *Comp Biochem Physiol* 72(A):149–155.

Vetter RD. 1983. The uptake of hydrophobic toxicants and biochemical measurement of stress in marine fishes. [Ph.D. Dissertation] Athens GA: University of Georgia.

Vigano L, Galassi S, Gatto M. 1992. Factors affecting the bioconcentration of hexachlorocyclohexanes in early life stages of *Oncorhynchus mykiss*. *Environ Toxicol Chem* 11:535–540.

Vodicnik MJ, Peterson RE. 1985. The enhancing effect of spawning on elimination of a persistent polychlorinated biphenyl from yellow perch. *Fund Appl Toxicol* 5:770–776.

Voldner EC, Ellenton G. 1987. Production, usage and atmospheric emission of 14 priority toxic chemicals with emphasis on North America. Downsview, ON, Canada: Atmospheric Environment Service, Environment Canada. Report ARD-88-4.

Voldner EC, Li YF. 1993. Global usage of toxaphene. *Chemosphere* 27:2073–2078.

Von Westernhagen H, Cameron P, Janssen D, Kerstan M. 1995. Age and size dependent chlorinated hydrocarbon concentrations in marine teleosts. *Mar Pollut Bulletin* 30:655–65.

Von Westernhagen H, Rosenthal H, Dethlefsen V, Ernst W, Harms U, Hansen PD. 1981. Bioaccumulating substances and reproductive success in Baltic flounder *Platichthys flesus. Aquat Toxicol* 1:85–99.

Wahli W, Ryffel GU. 1985. Xenopus vitellogenin genes. *Oxf Surv Eukaryot Genes* 2:96–120.

Walker CH. 1978. Species differences in microsomal monooxygenase activity and their relationship to biological half-lives. *Drug Metab Rev* 7:295–323.

Walker CH, Brealey CJ, Mackness MI, Johnston G. 1991. Toxicity of pesticides to birds: the enzymatic factor. *Biochem Soc Transact* 19:741–745.

Walker CH, Knight GC. 1981. The hepatic microsomal enzymes of sea birds and their interactions with liposoluble pollutants. *Aquat Toxicol* 1:343–354.

Walker CH, Newton I, Hallam SD, Ronis MJJ. 1987. Activities and toxicological significance of hepatic microsomal enzymes of the kestrel (*Falco tinnunculus*) and sparrowhawk (*Accipiter nisus*). *Comp Biochem Physiol* 86(C):379–382.

Walker CH, Ronis MJJ. 1989. The monooxygenases of birds, reptiles, and amphibians. *Xenobiotica* 19:1111–1121.

Walker MK, Cook PM, Batterman AR, Butterworth BC, Berini C, Libal JJ, Hufnagle LC, Peterson RE. 1994. Translocation of 2,3,7,8- tetrachlorodibenzo-*p*-dioxin from adult female lake trout (*Salvelinus namaycush*) to oocytes: effects on early life stage development and sac fry survival. *Can J Fish Aquatic Sci* 51:1410–1419.

Walker MK, Cook PM, Marquis P, Zabel E, Peterson RE. 1992. Risk posed by polychlorinated dibenzo-*p*-dioxins (PCDDs) dibenzofurans (PCDFs) and biphenyls (PCBs) to lake trout early life stage survival in the Great Lakes. Society of Environmental Toxicology and Chemistry 13th Annual Meeting; 8–12 Nov 1992; Cincinnati, OH. Pensacola FL: SETAC. p 45.

Walker MK, Spitsbergen JM, Olson JR, Peterson RE. 1991. 2,3,7,8-tetrachlorodibenzo-*p*-dioxin (TCDD) toxicity during early life stage development of lake trout (*Salvelinus namaycush*). *Can J Fish Aquat Sci* 48:875–883.

Wallace RA. 1985. Vitellogenesis and oocyte growth in nonmammalian vertebrates. In: Browder LW, editor. Developmental biology: a comprehensive synthesis. Volume 1: Oogenesis. New York: Plenum Press. p 127–166.

Wallace RA, Selman K. 1990. Ultrastructural aspects of oogenesis and oocyte growth in fish and amphibians. *J Electron Microsc Tech* 16:175–201.

Wang SY, Williams DL. 1980. Identification, purification, and characterization of two distinct avian vitellogenins. *Biochem* 19:1557–1563.

Wang SY, Williams DL. 1983. Differential responsiveness of avian vitellogenin I and vitellogenin II during primary and secondary stimulation with estrogen. *Biochem Biophys Res Commun* 112:1049–1055.

Wauchope RD, Buttler TM, Hornsby AG, Augustijn-Beckers PWM, Burt JP. 1992. The SCS/ARS/CES pesticide properties database for environmental decision-making. *Rev Environ Contam Toxicol* 123:1–155.

Wehler EK, Brunstrom B, Rannug U, Bergman A. 1990. 3,3',4,4'-tetrachlorobiphenyl: metabolism by the chick embryo in ovo and toxicity of hydroxylated metabolites. *Chem-Biol Interactions* 73:121–132.

Weseloh DVC, Ewins PJ, Struger J, Mineau P, Norstrom RJ. 1994. Geographical distribution of organochlorine contaminants and reproductive parameters in herring gulls on Lake Superior in 1983. *Environ Monit Assess* 29:229–251.

Weseloh DV, Mineau P, Struger J. 1990. Geographical distribution of contaminants and productivity measures of herring gulls in the Great Lakes: Lake Erie and connecting channels 1978/1979. *Sci Tot Environ* 91:141–159.

Westin DT, Olney CE, Rogers BA. 1983. Effects of parental and dietary PCBs on the survival, growth and body burdens of larval striped bass. *Bull Environ Contam Toxicol* 30:50–57.

White JC, Kelsey JW, Hatzinger PB, Alexander M. 1997. Factors affecting sequestration and bioavailability of pheranthrene in soils. *Environ Toxicol Chem* 10:2040–2045.

Whittow GC. 1986. Energy Metabolism. In: Sturkie PD, editor. Avian physiology. New York: Springer-Verlag. p 253–268.

Wiley HS, Wallace RA. 1981. The structure of vitellogenin. Multiple vitellogenins in *Xenopus laevis* give rise to multiple forms of the yolk proteins. *J Biol Chem* 256:8626–8634.

Willford WA, Bergstedt RA, Berlin WH, Foster NR, Hesselberg RJ, Mac MJ, Passino DR, Reinert RE, Rottiers. 1981. Introduction and summary. In: Chlorinated hydrocarbons as a factor in the reproduction and survival of lake trout (*Salvelinus namaycush*) in Lake Michigan. *US Fish Wildl Serv Fish Dis Leafl* 15:(4):1–7.

Willford WA, Sills JB, Whealdon. 1969. Chlorinated hydrocarbons in the young of Lake Michigan coho salmon. *Prog Fish Cult* 31:220.

Wren CD, MacCrimmon HR, Loescher BR. 1983. Examination of bioaccumulation and biomagnification of metals in a precambrian shield lake. *Water Air Soil Pollut* 19:277–291.

Yin C, Hassett JP. 1986. Gas-partitioning approach for laboratory and field studies of mirex fugacity in water. *Environ Sci Technol* 20:1213–1217.

Zhang Z, Pawliszyn. J. 1990. Head-space solid phase microextraction. *Anal Chem* 65:1843–1852.

Zhang ZM, Yang J, Pawilszyn J. 1994. Solid-phase microextraction. *Anal Chem* 66:844A–853A.

Zhao F, Mayura K, Kocurek N, Edwards JF, Kubena LF, Safe SH, Phillips TD. 1997. Inhibition of 3,3',4,4',5-pentachlorobiphenyl-induced chicken embryotoxicity by 2,2',4,4',5,5'-hexachlorobiphenyl. *Fund Appl Toxicol* 35:1–8.

Zitko V, Choi PMK, Wildish DJ, Monaghan CF, Lister NA. 1974. Distribution of PCB and *p,p'* DDE residues in Atlantic herring (*Clupea harengus harengus*) and yellow perch *(Perca flavescens)* in eastern Canada—1972. *Pestic Mont J* 8:105–109.

Zlokovitz ER, Secor DH. 1999. Effects of habitat use on PCB body burden in fall-collected, Hudson River striped bass (*Morone saxatilis*). *Can J Fish Aquat Sci* 56: Suppl 1 (in press).

CHAPTER 3

Basic Physiology

Anne McNabb, Carl Schreck, Charles Tyler, Peter Thomas, Vince Kramer,
Jennifer Specker, Monte Mayes, Kyle Selcer

The aim of this chapter is to determine what basic developmental and reproductive physiological processes of oviparous vertebrates are vulnerable to chemical perturbation. We define physiology very broadly to include the fields called general physiology, developmental physiology, cellular physiology, physiological toxicology, behavioral physiology, and environmental physiology. We consider physiological processes to include all functional events and processes in the organism and to include cellular biochemical events, functional histology and anatomy, and endocrine and nervous control and regulation of other functions. Critical to understanding the effects of chemical pollutants on physiology is research that elucidates the mechanisms whereby pollutants disrupt normal physiological functions.

This chapter first provides a basic treatment of the control of physiological function by the endocrine system to set the stage for the remainder of the chapter. We then move to a consideration of the fundamental events in development (life-cycle stages) and reproduction in oviparous vertebrates that are potentially vulnerable to disruption by chemical pollutants. We consider all of the oviparous vertebrate classes—fish, amphibians, reptiles, and birds—while recognizing that the literature is richest for fish and birds. Thus, phylogeny adds a third dimension to our consideration of development and reproduction.

During development, vertebrates pass through a number of transitional stages in which new functions emerge. These periods of transition often are critical "windows" during which relatively brief, key developmental events occur or a sequence of developmental events is initiated. At these times, organisms are particularly vulnerable to disruption that interferes with the expression of the genetic potential for development in the organism. The expression of this potential is directed by the

Reproductive and Developmental Effects of Contaminants in Oviparous Vertebrates. Richard T. Di Giulio and Donald E. Tillitt, editors.
©1999 Society of Environmental Toxicology and Chemistry (SETAC). ISBN 1-880611-37-6

physiological control systems (nervous and endocrine) that regulate developmental processes. It should be noted that nervous and endocrine control are integrated at the level of neuroendocrine control substances ("hormones" produced by neurons) and hypothalamic–pituitary–other-endocrine-gland axes. A general description of biological control, with emphasis on the endocrine system, is presented to provide a framework for discussion of development and reproduction.

In ovo and posthatching development (i.e., differentiation/maturation and growth) and reproduction may be viewed as phases in a continuum in the life of the animal. "Normal" development and survival of the young must underlie all of the functional processes leading to and involved in reproduction, which is necessary for the survival of the species. We have chosen to describe development before reproduction, although beginning with reproducing adults and their production of developing young would have been an equally credible approach.

Phenotypic Variation

Physiological operations and the capacity of functions are regulated ultimately by the organism's genotype and proximally by the environment (Schreck 1981). The maintenance of the variation in physiological traits in individuals is critical to maintaining the evolutionary capacity of species. This is implicit in the "evolutionarily significant unit" definition used for distinct population segments of vertebrates that must be considered under the endangered Species Act (Waples 1991). In other words, the variation in genotypes, and hence phenotypes, allows for evolution on the one hand and buffers against genetic canalization on the other (Endler 1986; Meffe and Carroll 1994). It is our contention that assessments of the effects of chemical pollutants must consider the increased vulnerability that occurs with losses of variation in a population or species, as well as a change in the mean status of the phenotype under consideration. In addition, the natural variation extant within and between populations must be considered in the selection of sentinel and surrogate species.

Genetics

There clearly are large genetic differences between the major taxa of vertebrates; between-species differences in tolerance of toxicants and natural stressors is obvious. However, it is equally important to recognize the relatively large amount of heterozygosity that exists within and between many, if not most, populations of oviparous vertebrates in the wild (Harrison 1993; Meffe and Carroll 1994). While these different groups within a species are readily apparent through taxonomic analyses that typically employ classifications based on allozyme or DNA patterns, it is now evident that distinct genetic groups also may differ in physiological traits that have survival value. For example, Currens (1997) found that 2 forms of rainbow trout (*Oncorhynchus mykiss*) recognizable by taxonomic analysis of allozyme patterns also differed in their ability to resist the bacterial pathogen *Ceratomyxa shasta*. Even though both of these genotypes are present in the same river basin, one form is totally resistant to the pathogen, while the other has little tolerance. Similar information exists for interpopulational differences in tolerance of other stressors, although perhaps on a larger geographic scale (Suzumoto et al. 1977; Winter et al. 1979; Pottinger et al. 1992). The existence of genetically discrete populations of the same species, with different physiological tolerance limits that can be distributed within the same watershed, is important to recognize from a toxicological assessment perspective. One cannot assume a priori that all subpopulations of a population, all populations of a metapopulation, or all metapopulations of a species have the same physiological responses to environmental stressors.

Environment

Physiological control, tolerance, and directing factors are affected by the environment (Fry 1947). In general, environments are predictable, and the biological rhythms that underlie the timing of many physiological events in animals provide evidence of that predictability (Wingfield 1994). Stressors or environmental insults to animals are superimposed on the predictability of the environment and require physiological responses by animals for survival. The maximum capability of the animal to tolerate environmental insults is determined genetically, but the expression of that genetic capability in physiological performance can vary depending on the habitat in which the organism resides (Schreck 1981; Schreck and Li 1991). Numerous habitat "quality" variables are important in establishing the individual organism's physiological tolerance limits and the nature of the variation among individuals of a population, and this must be considered in the design or interpretation of environmental assessment.

Temperature, for example, is a key factor affecting the performance of ectothermic organisms. The Q10 concept is well established that for biological systems the rates of physiological processes tend to double or more for every 10 °C increase in temperature within normal limits for the species (Warren 1971). This concept implies that individuals of a species living at different temperatures have different physiological phenotypes; they may achieve the same function at different rates or may go about achieving that function through different operations (mechanisms). However, the Q10 concept does not apply to birds that are homeothermic, except in altricial nestlings that are poikilothermic during much of nestling life.

Animals frequently are exposed to more than one stressor or chemical pollutant, either concurrently or sequentially at different points in the life cycle. This exposure can contribute significantly to the overall variation in the physiology of organisms.

Thus, physiological indicators of malfunction can differ among members of a species at different times in the life cycle and because of the animal's prior history of environmental stress.

Biological Signaling and Control Mechanisms

Control and communication

Control of homeostasis and development

Control systems are instrumental in homeostasis and in regulation of developmental and reproductive processes. Thus, perturbation of control systems can have major detrimental effects on the health, survival and reproduction of an organism. Environmental contaminants are among the factors that can disrupt biological control systems.

A hierarchy of importance among homeostatic processes exists. The most important physiological processes will typically be regulated at the expense of less vital ones. Also, the most critical functions are generally regulated by multiple control mechanisms, providing a level of protection if one of the mechanisms is rendered inoperative. Furthermore, there may be multiple effectors capable of initiating a change in a particular physiological variable. This redundancy allows the organism to compensate for problems that may arise with a particular effector.

The presence of compensatory or acclimatory mechanisms and redundancy in physiological functions allows organisms to adapt to a certain amount of perturbation. However, compensatory mechanisms have limits, and when these limits are reached, physiological variables may go out of control rapidly (time may be critical to assessing such effects). This means that for a particular variable, environmental perturbation may have no obvious effect up to the point where the compensatory mechanisms are overloaded, at which time the effects may be dramatic.

It can be difficult to assess dose-responsiveness to contaminants when the variable in question is regulated by compensatory mechanisms. Dose-response relationships are commonly assessed using in vitro assays, where the magnitude of the response is related to the dose of the compound. However, in vitro assays are not reflective of the situation in whole animals because they lack the compensatory mechanisms involved in maintenance of organismal homeostasis. In contrast, in whole-animal assays, the magnitude of the response may not simply be linked to dose; rather, compensatory (adaptive) mechanisms may lead to the appearance of no effect followed by an apparent all-or-none response under conditions where the compensatory mechanisms fail. In such instances, the dose-response relationship must be based on the percentage of responders and non-responders at each dose rather than on the magnitude of the response.

Types of control systems

Multicellular organisms rely on communication by the nervous and endocrine systems for the control of body functions. Because of their importance in the control of physiological processes, disruption of either of these systems by environmental contaminants can have a serious impact on the health, development, and reproduction of an organism. Perturbations by chemical toxicants are known to interfere with the chemical and electrical signaling mechanisms of both of these control systems (Timbrell 1991).

There is a high degree of integration between the nervous and endocrine systems; however, there are some important distinctions between the signaling mechanisms they employ. Both systems use chemical signals, but the nervous system also uses electrical signals. The speed and duration of the signal also varies. Nervous signals tend to be rapid, typically reaching the target sites in milliseconds, and the signal effects are typically of short duration. Endocrine signals are slower, taking minutes to hours to reach the target site, but the duration of the signal and the effects may be quite long, from hours to even years in some cases. There are implications concerning the duration of a response and the vulnerability of the signaling pathway to contaminants. If the length of the response to a signal is short, then a brief interference by a toxicant may have only a transient, reversible effect on the physiology of the organism. However, if the duration of the response is long or permanent, a brief exposure to an environmental agent could result in irreversible alterations in the physiology of the organism. This is especially true if the exposure occurs during development. For example, there are critical windows during sex determination and differentiation when the pattern of hormones present determines whether the reproductive structures and the brain will develop in the male or female direction (Witschi 1971; Arnold and Gorski 1984; Adkins-Regan 1987). Interference with these hormonal signals can result in permanent alterations in the sexual development of the organism.

There are known toxicants that affect both endocrine and nervous function. Toxicants that interfere with nervous communication may interfere with the electrical or neurochemical signals. Some of these effects can be analogous to cutting cables within the anatomical communication network of the nervous system. Many effects on the nervous system occur by the same mechanisms as in the endocrine system (e.g., effects on some neurotransmitters and neurohormones), perhaps reflecting the inherent overlap in function of these physiological control systems. However, there are examples of disruption that are peculiar to the nervous system, e.g., effects on monoamine synthesis. Overall, there is evidence that the neuroendocrine system is particularly vulnerable to disruption by chemical pollutants (Thomas and Khan 1997; Khan and Thomas 1997). Effects on the endocrine system by pollutants acting at the level of receptors in target organs (either receptor blocking by antagonists or "false-positive signals" by agonists that mimic hormones) have received a great deal of attention. However, other mechanisms of hormone disruption by pollutants are known, for example alterations in the hormone dynamics (e.g., for thyroid hormones) or disruption of post-receptor signaling pathways (e.g., cadmium effects on fish gonadotropin secretion by the pituitary and disruption of steroidogenesis in the ovary, both of which appear to be mediated by alterations in calcium homeostasis) (Thomas and Khan 1997).

Reflex arcs and feedback loops

The reflex arc, composed of a stimulus, a receptor, an afferent pathway, an integrating center, an efferent pathway, and an effector, is a common communication network design for maintaining homeostasis in both the nervous and endocrine systems. Homeostatic control systems rely on feedback mechanisms to keep variables within appropriate physiological ranges. In negative feedback, which is most common, a change in a variable (in this case, a metabolic stimulus) results in actions that move the variable in the opposite direction (e.g., increasing blood glucose leads to increased insulin release, which results in lowering of blood glucose). Negative feedback leads to stability of a particular variable; this type of function is common in physiological systems. In contrast, in instances such as the acquisition of new functions in development, positive feedback is used. Positive feedback involves responding to a change in a variable with actions that augment changes in the original variable in the same direction as the stimulus. Positive feedback can be destabilizing, driving the variable in one direction. Thus, this mechanism is typically used under conditions that are self-limiting, such as blood clotting, pregnancy, or irreversible developmental changes.

Negative feedback loops are especially important in regulating hormone levels in the blood. Virtually all hormones are under feedback control (Wilson and Foster 1992). Sometimes the feedback loop is direct, such as blood calcium levels and parathyroid hormone release or blood glucose and insulin release. Often, there are several levels to the feedback mechanism. This is exemplified by the hypothalamic–anterior pituitary–gonadal axis (Catt and Dufau 1991). In this axis, gonadotropin-releasing hormone stimulates the release of luteinizing hormone and follicle stimulating hormone, which in turn stimulate gonadal steroid production. The gonadal steroids then negatively feed back on the hypothalamus (long-loop feedback) to decrease gonadotropin-releasing hormone and on the anterior pituitary (short-loop feedback) to decrease production of luteinizing hormone and follicle stimulating hormone. Ultra short-loop feedback also may be present, and in this case the hormone feeds back directly on the tissue producing that hormone. For example, in the hypothalamic–anterior pituitary–adrenal axis, increasing concentrations of cortisol can inhibit cortisol production by adrenal cortical tissue (e.g., in fish) (Bradford et al. 1992).

Signaling molecules

Nomenclature

Chemical signals are widely used in communication and control systems. Some signals must travel a long distance to reach the target site, while others need to travel only to the adjacent cell. A few signals even leave the body to impact another organism. An extensive terminology has been created to refer to these various types of signals (Bolander 1994). A *hormone* in the classical sense is a signal secreted by one type of cell that travels through the blood to act on target cells. A *neurohormone* is a signal secreted by a nerve cell that travels in the blood to act on target cells. There are several types of signals that act locally, rather than traveling in the blood. A *neurotransmitter* is a signal secreted by one neuron that travels through the interstitial fluid to act on another neuron or on an effector cell, usually at a specialized junction called a synapse. A *paracrine* is a signal secreted by one type of cell that travels through the interstitial fluid to act on nearby target cells. An *autocrine* is similar to a paracrine except that the secreted signal acts on the same cell that produced it. A *juxtacrine* is a signal produced by a cell as a membrane-bound precursor that may function as a signal while attached to the cell by a spacer arm or that may be cleaved to form a soluble protein signal. An *intracrine* is a signal that is produced or modified intracellularly and acts inside the same cell. *Ectohormone* is a collective term for signals that are secreted outside one organism and travel through the air or water to act on another organism. These include *pheromones*, which are chemical signals that communicate between different individuals of the same species (this often involves sexual attraction); *allochemics*, which involve communication between individuals of different species (e.g., repellant odors of potential prey species or *allomones*, which are interspecies attractants); and *gamones*, which are inducers of sexual development.

Major classes of signaling molecules

Chemical signals usually travel through the blood or interstitial fluid. As a result, they may come into contact with essentially all body cells, but they have effects only on those target cells that have highly specific receptors for binding that signal. Together the endocrine glands, nervous system, and other types of cells of the body are able to send and receive a large number of different messages because of the great amount of structural diversity of the chemical signals. While there are probably well over a hundred different chemical messengers used by a given organism, most signals fall into a few major classes: peptides, lipids, and amino acid derivatives (Bolander 1994; Wilson and Foster 1992). Figure 3-1 shows the structure of a number of the lipid signaling and amino-acid-derived signaling molecules. Peptide and protein hormones are not shown because many of them are extremely large and the total range of diversity in structure is greater.

A large number of chemical signals are peptides, which are chains of amino acids linked by peptide bonds. Familiar examples include the hypophysiotropic hormones

Figure 3-1 Lipid- and amino-acid-derived signaling chemicals

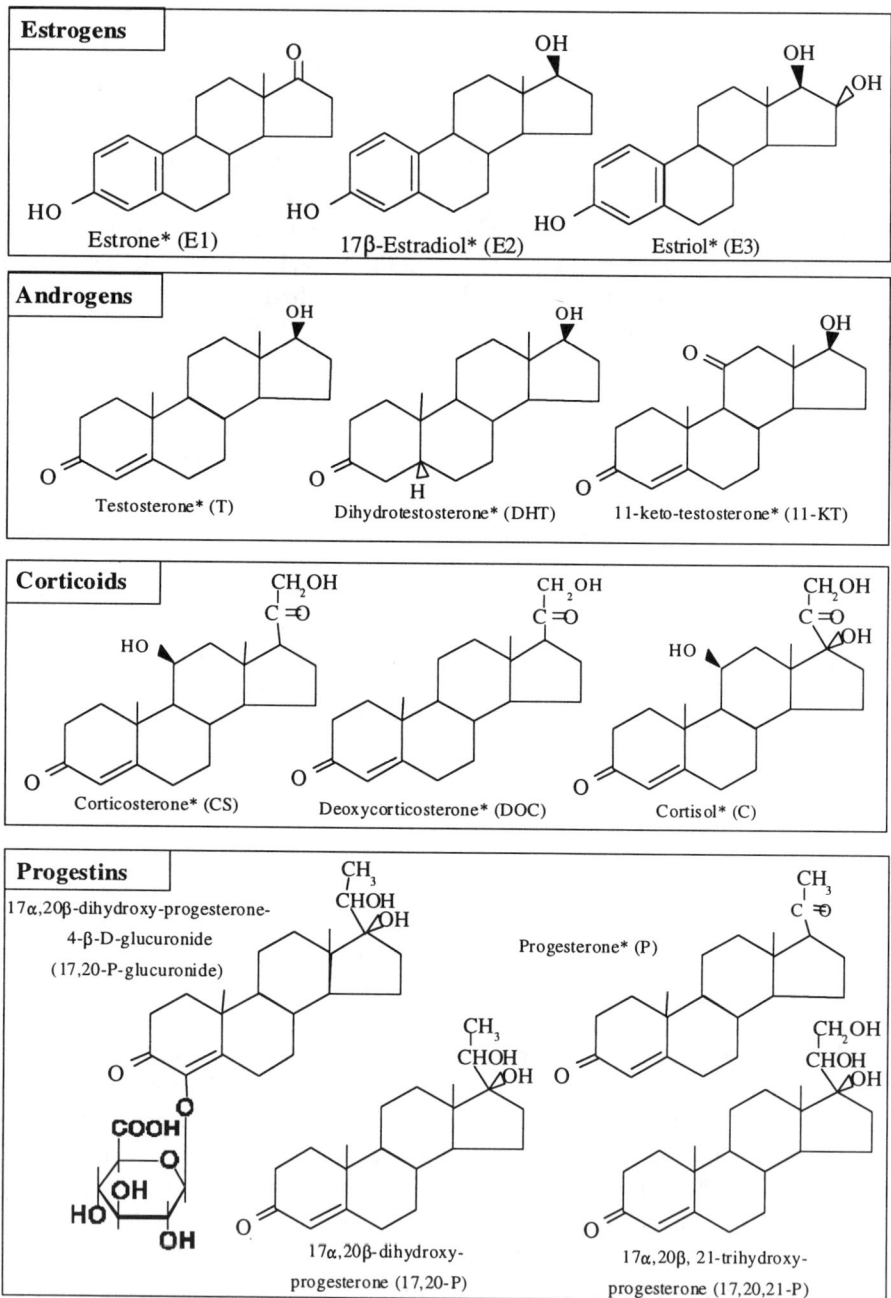

Estrogens

Estrone* (E1)   17β-Estradiol* (E2)   Estriol* (E3)

Androgens

Testosterone* (T)   Dihydrotestosterone* (DHT)   11-keto-testosterone* (11-KT)

Corticoids

Corticosterone* (CS)   Deoxycorticosterone* (DOC)   Cortisol* (C)

Progestins

17α,20β-dihydroxy-progesterone-
4-β-D-glucuronide
(17,20-P-glucuronide)

Progesterone* (P)

17α,20β-dihydroxy-
progesterone (17,20-P)

17α,20β, 21-trihydroxy-
progesterone (17,20,21-P)

* designates those that bind to nuclear receptors; others bind to plasma membrane receptors except for the second messengers which are involved in the transduction of signals from membrane receptors into the cell

Figure 3-1 continued

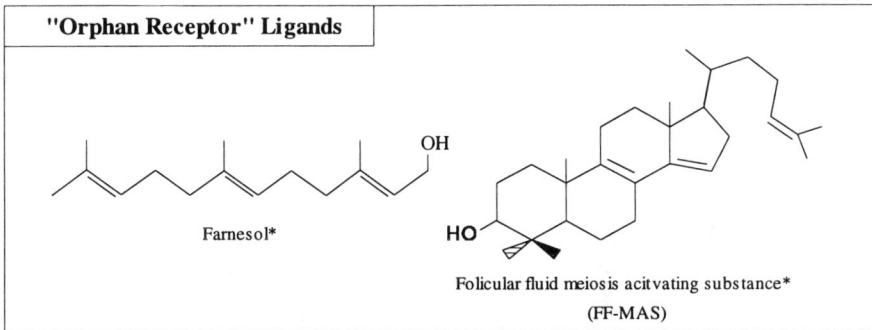

"Orphan Receptor" Ligands

Farnesol*

Folicular fluid meiosis acitvating substance*
(FF-MAS)

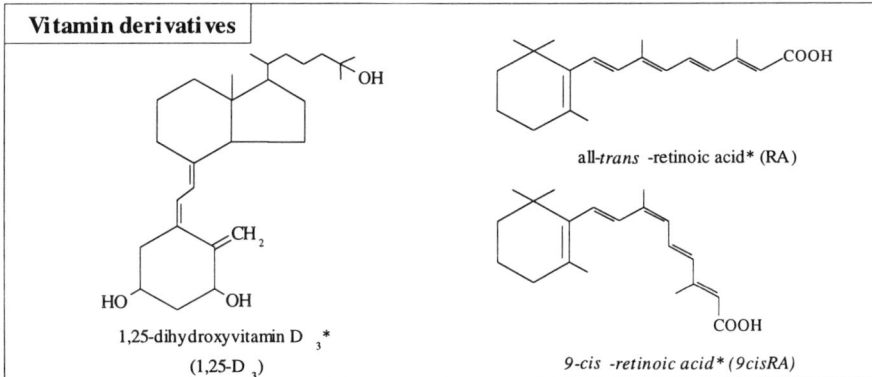

Vitamin derivatives

all-*trans* -retinoic acid* (RA)

1,25-dihydroxyvitamin D$_3$*
(1,25-D$_3$)

9-*cis* -retinoic acid* (9cisRA)

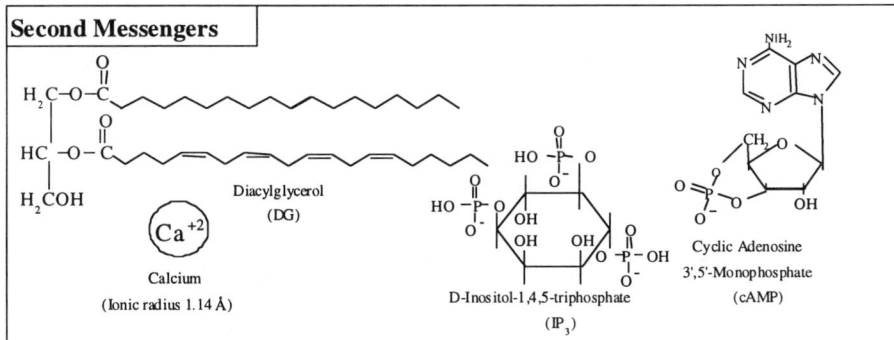

Second Messengers

Diacylglycerol
(DG)

Calcium
(Ionic radius 1.14 Å)

D-Inositol-1,4,5-triphosphate
(IP$_3$)

Cyclic Adenosine
3',5'-Monophosphate
(cAMP)

Figure 3-1 continued

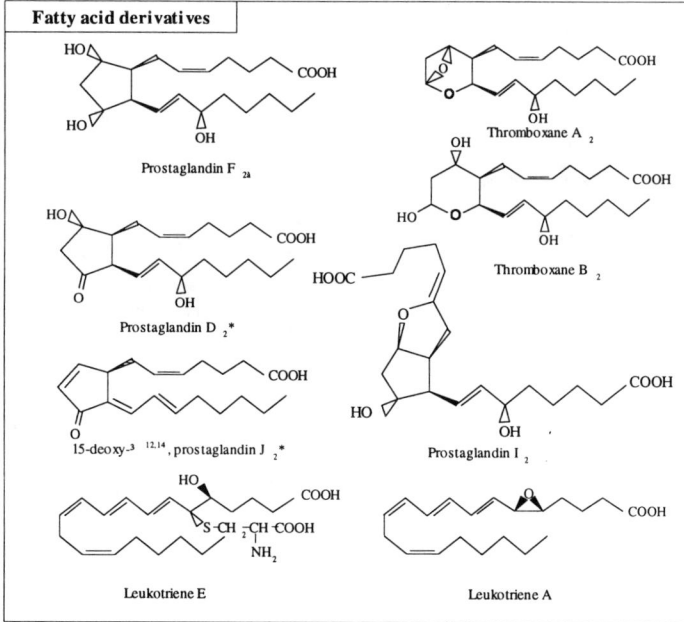

Fatty acid derivatives

Prostaglandin F $_{2\alpha}$

Thromboxane A $_2$

Thromboxane B $_2$

Prostaglandin D $_2$*

15-deoxy-$^{\Delta}$ ^{12,14}, prostaglandin J $_2$*

Prostaglandin I $_2$

Leukotriene E

Leukotriene A

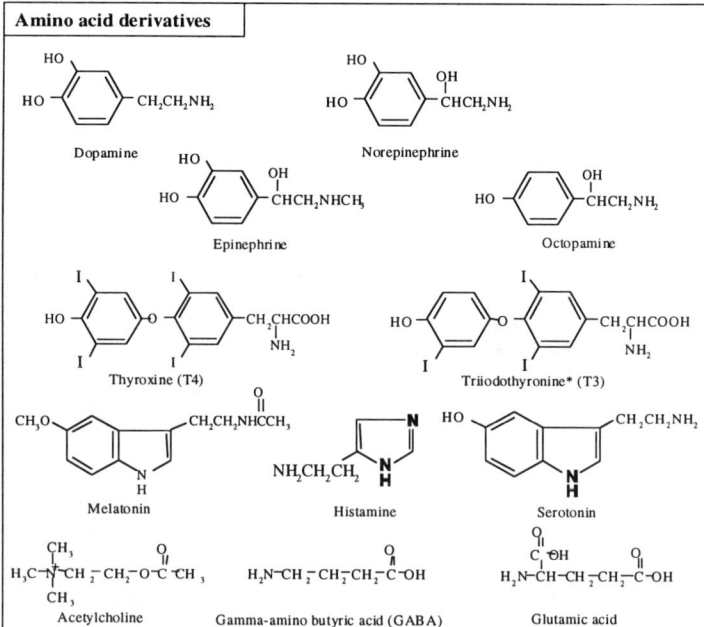

Amino acid derivatives

Dopamine

Norepinephrine

Epinephrine

Octopamine

Thyroxine (T4)

Triiodothyronine* (T3)

Melatonin

Histamine

Serotonin

Acetylcholine

Gamma-amino butyric acid (GABA)

Glutamic acid

from the hypothalamus (e.g., thyrotropin releasing hormone [TRH], corticotropin releasing hormone [CRH], gonadotropin-releasing hormone [GnRH], and growth hormone releasing hormone [GHRH]), the anterior pituitary hormones (e.g., adrenocorticotropic hormone [ACTH], follicle stimulating hormone [FSH], luteinizing hormone [LH], thyroid stimulating hormone [TSH], prolactin, and growth hormone), the posterior pituitary hormones (e.g., oxytocin and vasopressin), and the pancreatic hormones (e.g., insulin and glugagon). Peptide signaling molecules may be as short as 3 amino acids, as for thyrotropin releasing hormone, or they may extend to hundreds of amino acid residues, as for prolactin and growth hormone.

Most peptide hormones fall into one of a limited number of hormone families that share homologous sequences of amino acids but differ in their target organ effects (Bolander 1994). Most peptide hormone families are thought to have evolved from a single precursor amino acid sequence. Examples include the anterior pituitary family of glycoprotein hormones (e.g, FSH, LH, chorionic gonadotropin), the anterior pituitary family of growth hormone and prolactin, the posterior pituitary neuropeptides (e.g., oxytocin and vasopressin), and several different families of growth factors that are produced in a number of tissues (e.g., insulin-like growth factors [IGFs], epidermal growth factors [EGFs]). Some members of hormone families are different parts of a single precursor protein. For example, a group of peptides from the anterior pituitary (including ACTH, α-melanocyte stimulating hormone [α-MSH or α-melanotropin], β-MSH, β-lipotropin [β-LPH], β-endorphin) are all derived from different parts of the precursor protein proopiomelanocortin.

Derivatives of amino acids often are used as signals (Figure 3-1). The amino acid tyrosine gives rise to the catecholamines, a group of neurotransmitters and neurohormones that includes dopamine, epinephrine, and norepineprine. Tyrosine also serves as the precursor to the iodine-containing thyroid hormones (thyroxine [T_4] and triiodothyronine [T_3]), which are regulators of many aspects of development and metabolism in all the vertebrate classes. Although amino-acid-derived, thyroid hormones are very similar to steroid hormones in their receptor binding and mechanism of action. Melatonin (N-acetyl-5-methoxytryptamine), a pineal hormone involved in circadian rhythms, and the neurotransmitter serotonin (5-hydroxytryptamine) are both derived from tryptophan. Nitric oxide (NO) is a chemical signal synthesized in the vascular endothelium by oxidation of arginine. This molecule is known to be involved in relaxation of vascular smooth muscle but may have other functions as well (Marletta 1989). Some amino acids serve directly as signals. For example, glutamate and its derivative gamma aminobutyric acid, as well as glycine, can serve as neurotransmitters in some parts of the nervous system.

Lipids or lipid-soluble molecules also are used extensively as signaling molecules. Steroids constitute a major class of lipid-based hormones that are important in regulation of many areas of development, reproduction, and metabolism. Steroid hormones are derived from cholesterol and are structurally related through a common nucleus of 4 fused rings (cyclopentanoperhydrophen-anthrene nucleus).

The particular properties of the various steroid hormones are conferred by the specific functional groups attached to the common nucleus. Eicosanoids are another group of lipid signaling molecules (Figure 3-1). These compounds, derived from the membrane lipid arachidonic acid, typically operate as local (paracrine) signals. Eicosanoids include prostaglandins, thromboxanes, and leukotrienes. Their functions generally fall into the areas of inflammation, blood clotting, and smooth muscle contraction. Some hormones are derived from lipid-soluble vitamins, e.g., the retinoids from Vitamin A and the Vitamin D derivatives, and are grouped with the lipid hormones (Hadley 1996; Norris 1997).

Receptors for signaling molecules

A receptor is a molecule that interacts with a signal or ligand in a highly specific manner such that a characteristic response or group of responses is initiated by the cell. The signal must chemically interact with the receptor, which is invariably a protein, although the interaction may involve carbohydrate moieties on the protein. Whether or not a signal is recognized by a receptor depends on the 3-dimensional shapes of the signal and the receptor. The match is often compared to a lock and key, but the relationship is actually more complex because the shape of the signal (the "key") or that of the receptor (the "lock") may be changed by the interaction. The binding of the signal or the receptor is the result of bonds developed between specific regions of both the ligand and the receptor protein. Binding of the ligand often alters the conformation of the receptor (Brzozowski et al. 1997). The interaction is typically through weak bonds rather than covalent bonds. Therefore, a receptor will have the highest affinity for a ligand that "fits" most precisely with the shape and charge of the receptor's binding site. The binding site usually consists of a cavity formed by a specific arrangement of amino acids on the receptor protein's surface and usually represents only a minor portion of the receptor protein. The rest of the protein may have a number of functions, including maintaining the proper configuration of the binding site, anchoring the receptor to the membrane, serving as binding sites for other regulatory molecules, and performing any enzymatic functions of the receptor. Receptors are generally allosteric proteins that can reversibly change their shape. Such changes in conformation can alter receptor function. For example, nuclear steroid receptors do not usually bind to DNA in the absence of ligand. However, when bound to ligand, the steroid receptor changes conformation, resulting in a high affinity for binding to specific DNA sequences and resultant effects on gene expression.

While there are many types of receptors, they all share certain general characteristics. Receptors generally have a high binding affinity for hormones, often in the 10^{-12} to 10^{-8} mol/L range. They are also usually quite specific in their binding, recognizing only one signal or class of signals. Receptors also occur in low concentrations, ranging from hundreds to thousands per cell. Finally, receptors are capable of initiating a characteristic response when bound to a signal. These features separate receptors from the transport proteins that also can bind to signals (Bolander 1994; Goodman 1994).

As indicated above, endogenous or exogenous chemicals other than the "native" ligands can bind to receptors if they "fit" the receptor binding sites. Such ligands are called agonists if their receptor binding results in a cellular response, or antagonists, if they block or inhibit the cellular response.

Receptor regulation

Receptor concentrations (i.e., numbers of receptors per cell), in part, determine the responsiveness of a cell to a particular ligand. The control of receptor concentrations is an important aspect of the overall control of chemical communication. As with all proteins, the concentration of a receptor at any time is the balance between processes of synthesis and degradation. The factors that influence these fall into 2 categories: homologous controls, where the ligand is involved in the control of its own receptor, and heterologous controls, where other factors control the receptor levels (Bolander 1994). The regulation of nuclear steroid hormone receptors provides examples of both these types of control. Homologous control is illustrated by estradiol receptor up-regulation (an increase in the concentration or numbers of receptors) by the native ligand, estradiol (Selcer and Leavitt 1991a). Heterologous control is illustrated by the up-regulation of the progesterone receptor by estrogen in certain tissues (Nardulli et al. 1988) and by progesterone down-regulation of estrogen receptors in the uterus of mammals and in the oviduct of birds and reptiles (Selcer and Leavitt 1988; Selcer and Leavitt 1991b). Heterologous control also can be achieved by non-hormonal chemical signals. For example, binding of dioxin to the aryl hydrocarbon receptor (AhR), a member of a family of transcription factors that are important in many developmental events (Burbach et al. 1992), modulates expression of various xenobiotic-metabolizing cytochrome P450 enzymes (Bannister and Safe 1987; Landers and Bunce 1991; Whitlock et al. 1996) and causes down-regulation of estrogen receptor in vitro (Zacharewski et al. 1991) and in vivo (DeVito et al. 1992).

Receptor location

In general, the location of the receptor on the target cells reflects the properties of the signal with respect to cell entry. Small, nonpolar, or lipophilic molecules can cross the plasma membrane by diffusion; consequently, the receptor can be located inside the cell. Thus the receptors for steroid hormones, thyroid hormones, and retinoic acid and its derivatives are intracellular. Large or polar molecules (peptide hormones and catecholamines) cannot cross the plasma membrane by diffusion; therefore, the receptors for these signals must be located on the outer surface of the plasma membrane. Several of the steroid hormones have both membrane receptors and intracellular receptors, so steroid hormones may act through at least 2 different types of receptors. Examples include the progestin and cortisol membrane receptors involved in oocyte maturation in frogs and fishes (Sadler and Maller 1982; Patiño and Thomas 1990; Nagahama et al. 1994; Bandyopadhyay et al. 1998) and in hyperactivation and motility of fish sperm (Thomas et al. 1997).

Receptors and environmental pollutants

Environmental chemical pollutants can interfere with biological communication systems at the level of receptors by acting as agonists or antagonists of the natural ligand or by altering the regulation of receptor concentrations. Agonists may elicit a physiological response at an inappropriate time or alter the magnitude of the response. Antagonists can prevent or delay a normal biological response. Both can be disruptive to control systems. Important considerations in evaluating the potential physiological impact of an agonist (or antagonist) are both its concentration and the affinity with which it binds to the receptor.

Receptors that have strict specificity for their ligand are less subject to perturbation by contaminants than those that are less specific in their binding requirements. For example, some environmental endocrine disruptors operate by binding to the estrogen receptor, which initially appeared to bind to a rather broad range of chemical structures (McLachlan 1985, 1993). However, recent studies show that nearly all of these chemicals have the same or similar functional units that interact with the estrogen receptor (Waller et al. 1996; Ankley et al. 1997). In considering the effects of chemical pollutants on endocrine systems, it should be noted that binding of a chemical to a receptor does not categorically disrupt endocrine function and that endocrine disruption can occur by mechanisms other than effects on receptor binding.

Signal transduction

Signal transduction is the process through which binding of a chemical-control signal to a receptor is translated into a biological response. This process differs dramatically between types of receptors and even can vary between tissues for the same receptor. There are many types of signal transduction pathways, but certain biochemical processes are common to many of them. Two of the most prominent are transcriptional regulation (mediated by nuclear receptors) and phosphorylation (cascades of phosphorylations initiated by plasma membrane receptors).

Transcriptional regulation

There are estimated to be tens of thousands of genes that code for proteins. Most cells possess the complete complement of these protein coding genes and are thus potentially capable of synthesizing all of the proteins. However, at any given time only several thousand, or fewer, proteins are produced in a particular cell. Clearly, the control of which proteins are produced, the timing of their production, and the quantity of protein produced are all tightly regulated. This control can be at the level of gene transcription or protein translation. Transcriptional regulation determines which of the many genes in the nucleus are copied into mRNAs and which genes are not. Translational regulation controls the rate at which mRNAs are used by the protein synthetic machinery. These controls are supplemented by post-transcriptional controls, such as mRNA processing and transport of mRNAs from the nucleus to the cytoplasm, and by post-translational controls that regulate the rate of protein maturation and degradation (Bolander 1994). Thus, protein regulation is not only tightly regulated but also is dependent on cell and tissue types.

Many proteins in the cell serve to regulate transcription of specific genes. These proteins, termed transcription factors, are important regulators of gene expression and, as such, are instrumental in the processes of cell differentiation and development. They have the capacity to recognize individual genes and to turn their transcriptional activity on or off (Maniatis et al. 1987; Mitchell and Tijan 1989; Latchman 1990; Harrison 1991; Suzuki 1993). Transcription factors function by binding to short DNA control sequences of genes and attracting or repelling RNA polymerase, which is responsible for the initiation of RNA transcription (Ptashne 1988; Pabo and Sauer 1992). Transcription factors often have one or more structural features in common, such as zinc fingers, leucine zippers, or helix-turn-helix motifs. These elements aid in DNA recognition and binding. Transcription factors also have domains that serve to enhance or suppress transcription. It should be noted that control of a given gene usually involves the cooperation of several different transcription factors, each of which may be separately regulated. Because of the importance of transcription factors in cellular differentiation and development, any perturbation of these factors, or of the pathways controlling them, could have significant effects on the organism, including developmental defects and cancers.

Signal transduction pathways have profound influences on transcriptional regulation, often by altering intracellular levels and activities of specific transcription factors. A well known example is the pathway for nuclear receptors. Nuclear receptors are actually transcription factors that are activated by binding of the hormone. Membrane receptor pathways also influence transcriptional regulation, usually by activating or deactivating transcription factors through phosphorylation or dephosphorylation.

Protein phosphorylation/dephosphorylation
Chemical-control signals that bind to receptors on the plasma membrane trigger cascades of intracellular events involving primarily enzymatic proteins. The function of many such proteins is regulated by the covalent addition of side groups, such as methyl groups, acetyl groups, and phosphate groups. Of these, phosphorylation is the most common and most important regulating mechanism for biological activity of enzymes and other proteins. Addition of a phosphate group often results in allosteric changes in the protein. Phosphorylation can result in turning on, or turning off, the normal catalytic activity of a particular protein. Likewise, dephosphorylation can result in deactivation or activation of a protein's function. The site of protein phosphorylation may be important in metabolic regulation; phosphate groups are usually added to serine, threonine, or tyrosine residues of proteins of

eukaryotic cells (Blackshear et al. 1988; Cohen 1988). The enzymatic proteins that add these groups to other proteins are called kinases. Removal of phosphate groups occurs through enzymes termed phosphoprotein phosphatases. Nearly 100 protein kinases have been discovered, but this number represents only a fraction of the actual number (Hunter 1987). Important examples include protein kinase A, protein kinase C, phosphorylase kinase, and mitogen activating protein (MAP) kinase. Many of the kinases are associated with signal transduction pathways. Especially prevalent are kinases associated with regulation of cell growth. Also important are a variety of phosphoprotein phosphatases that counterbalance the activity of the kinases (Hunter 1989; Walton and Dixon 1993).

Phosphorylation that involves covalent modification of plasma membrane receptors also plays an important role in the regulation of the function of these receptors (Sibley et al. 1987; Inglese et al. 1993). Examples include desensitization of β-adrenergic receptors, deactivation of rhodopsin receptors, desensitization or down-regulation of G protein-coupled receptors, enhanced activity of insulin receptors, and regulation of epidermal growth factor (EGF) receptors.

Nuclear receptor superfamily

The nuclear receptor superfamily consists of intracellular transcription factors that bind to and are activated by a wide variety of non-peptide ligands (Table 3-1; Figure 3-1). These factors are found in all invertebrate and vertebrate classes of animals. In vertebrates, ligands of the nuclear receptor superfamily include sex steroid hormones, adrenocortical steroids, Vitamin D derivatives, thyroid hormones, and retinoic acid derivatives. In addition, there are intracellular receptors that share amino acid sequence homology with the nuclear receptor superfamily but for which the precise chemical structures of the "native" ligands are not known, thus earning them the name "orphan receptors." New receptors and ligands that bind to such orphan receptors continue to be identified. For example, in mammals, proposed ligands for the orphan receptor liver X-receptor a (LXRa) (Willy et al. 1995) include a C29 sterol called follicular fluid meiosis activating substance (FF-MAS) (Byskov et al. 1995) as well as other hydroxylated cholesterol metabolites (Janowski et al. 1996). The function and structure of the proteins comprising the nuclear receptor superfamily has been reviewed by Tsai and O'Malley (1994). In addition to their nuclear actions, a number of steroid hormones also have plasma membrane receptors that trigger different cellular actions (Revelli et al. 1998).

Functional domains

The receptors of the nuclear receptor superfamily possess 6 functional domains, some with overlapping functions. The first 2 domains at the N terminus of the receptor protein, generally referred to as the A/B region, possess transactivation (gene transcription activation function 1 [AF-1]) and transcription factor binding functions. Proceeding towards the carboxy terminus, the C region contains 2 "zinc finger" motifs that are responsible for DNA binding and specific DNA sequence

Table 3-1 Biological control pathways potentially vulnerable to perturbation by chemical substances

Signaling molecule	Target receptor	Target tissue	Vulnerable functions
Estrogens 17β-Estradiol Estrone Estriol	Nuclear estrogen receptor	Liver	Induce vitellogenesis Inhibit some Cyt P450 activity
		Brain	Male courtship, singing behavior, female receptivity HPG axis feedback control of steroidogenesis
		Embryonic gonads	Gonadal differentiation
	Membrane estrogen receptor	Neuro-endocrine tissue	HPG feedback
	Steroid binding proteins	Blood	Control free hormone concentrations, possibly plasma membrane transport
Androgens Testosterone 11-Keto-testosterone Dihydrotestosterone	Nuclear receptor (Androgen receptor)	Epidermis	Secondary sex characteristics (comb/wattle; tubercles; coloration; kype; dewlap)
		Various	Accessory sex organ development (gonopodium; intromittant organs; claspers; fighting structures, cloacal glands)
		Embryonic gonads	Wolffian duct development
		Brain	Mating behavior
	Steroid binding proteins	Blood, Testis	Control free hormone concentrations, plasma membrane transport

Table 3-1 continued

Signaling molecule	Target receptor	Target tissue	Vulnerable functions
Progestins			
Progesterone (P)	Nuclear receptor (Progesterone receptor, PR)	Ovary	Ovulation, gestation*, shell deposition
		Brain	Incubation behavior, female receptivity
	Steroid binding proteins	Blood	Control free hormone concentrations, plasma membrane transport
17α, 20β-dihydroxy-progesterone (17α, 20β-P)	Membrane receptor	Oocyte	Germinal vesicle breakdown (fish)
17α, 20β, 21-trihydroxy-progesterone (17α, 20β, 21-P)	Membrane receptor	Oocyte	Oocyte maturation (fish)
Progesterone glucuronides	Pheromone membrane receptor	Olfactory epithelium	Coordinates oogenesis (fish)
Thyroid hormones			
Triiodothyronine (T3)	Nuclear receptor (Thyroxine receptor, TR)	Various	Development, growth, metabolism
Thyroxine (T4)		Larvae	Metamorphosis
	Transthyretin (carrier protein) Thyroxine binding globulin	Blood	T3, T4, retinoic acid transport
Retinoids			
all trans-Retinoic acid	Nuclear receptor (Retinoic acid receptor, RAR)	Embryo	Development, visual systems
9-cis-Retinoic acid	(Retinoid X receptor, RXR)	Embryo	Development, limb bud differentiation
Calcium	Protein kinases Second messenger systems	Oocyte, sperm	Oocyte maturation, gonadotropin signal transduction

*Ovoviviparous and viviparous fish, amphibians, and reptiles

Table 3-1 continued

Signaling molecule	Target receptor	Target tissue	Vulnerable functions
Prostaglandins	Membrane receptor Nuclear receptor (Peroxisome proliferator activated receptor, PPAR)	Shell gland, ovary brain	Calcium deposition in shell, Induce ovulation, Inhibit female receptivity
Leukotrienes	Membrane receptor	Lymphocytes	Immune function
Corticosteroids Cortisol	Nuclear receptor (Glucocorticoid receptor, GR)	Ovary Liver	Oocyte maturation (fish), ovulation Potentiate vitellogenesis (fish)
Aldosterone	(Mineralocorticoid receptor, MR)	Kidney	Ion homeostasis
Catecholamines Epinephrine Norepinephrine	Membrane receptor	Brain	Inhibits oviposition (reptiles)
Gases Nitric oxide (NO) Carbon monoxide (CO)	Intracellular enzymes, guanylyl cyclase	Vascular smooth muscle, brain	NO known to mediate erection of intromittent organs in mammals CO role unknown
Peptide growth factors Epidermal growth factor (EGF) Platelet derived growth factor (PDGF) Transforming growth factor α (TGFα) Transforming growth factor β (TGFβ) Insulin-like growth factor (IGF) Hepatocyte growth factor (HGF)	Membrane receptors	Epithelial and mesodermal cells, liver, other various tissues	Growth, development

recognition. The D region is a variable-length section devoted to nuclear localization that occurs after ligand binding and is referred to as the "hinge" region because it is thought to confer flexibility to the receptor related to transactivation. The E region is the largest domain, typically 250 amino acids, with the combined functions of ligand binding, dimerization, heat shock protein binding, transactivation (gene transcription activation function 2 [AF-2]) and gene repression, and some nuclear localization and transcription factor binding. The intramolecular interactions of the separate regions of nuclear receptors are very complex and are incompletely understood (Gandini et al. 1997). Significant progress has been made in under-standing the role of the ligand binding domain (Kumar and Chambon 1988; Mattick et al. 1997; Tsai and O'Malley 1994), which is a potential site of vulnerability to perturbation through binding of xenobiotic substances that are structural agonists or antagonists of native hormones (Anstead et al. 1997).

Receptor action in the cell

The primary function of nuclear receptors is the intracellular transduction of a chemical signal received by a cell to induce or repress the transcription of genes. In addition to this important role as transcription factors, other functions of nuclear receptors include stabilization of mRNA transcripts and modulation of the availabil-ity of other types of transcription factors (Tsai and O'Malley 1994). The major steps in the signal-transduction pathway (Figure 3-2) mediated by members of the nuclear receptor superfamily are

- ligand binding,
- receptor activation,
- nuclear localization,
- receptor dimerization,
- binding to hormone response elements (HREs) on DNA,
- transcription factor recruitment, and
- transcription enhancement or repression (Torchia et al. 1997).

Ligand binding is an essential step in activation of the receptor to a form that enhances transcription of target genes mediated by the AF-2 region of the receptor (Fawell et al. 1990). There are a few known examples of receptor activity in the absence of ligand, e.g., through DNA binding of monomeric thyroid receptor, and these may be related to the activity of the AF-1 region of the receptor (Murdoch et al. 1990; Tsai and O'Malley 1994). Receptor tyrosine phosphorylation is required in some cases to facilitate high affinity binding of ligand to receptor (Migliaccio et al. 1986, 1989). Serine phosphorylation also is required for transcriptional activation by estrogen receptors (Kato et al. 1995). Receptor dimerization can involve homodimerization (e.g., estrogen [ER], androgen [AR], progesterone [PR], gluco-corticoid [GR], and mineralocorticoid receptors [MR]). Heterodimerization involves combinations of retinoid X-receptor (RXR), thyroid receptor (TR), retinoic acid receptor (RAR), vitamin D3 (VDR), and peroxisome proliferator-activated

Target Cell

Cell growth
Differentiation
Proteins

Enhanced or
Repressed
Transcription

DNA

Plasma
Binding
Proteins

DNA Binding with
Transcription Factors

Response
Element

Hydrophobic
binding pocket

Further
Phosphorylation
(P)

DNA-binding
"zinc fingers"

mRNA
Stabilization

Diffusion

Phosphorylation

Ligand Binding
& Conformation
Change

Ligand-stabilized
Receptor
Dimerization

Homodimer or
Heterodimer

Nuclear Receptor
Superfamily
Ligand

Heat Shock Proteins
Dissociate

Nuclear
Translocation

Nucleus

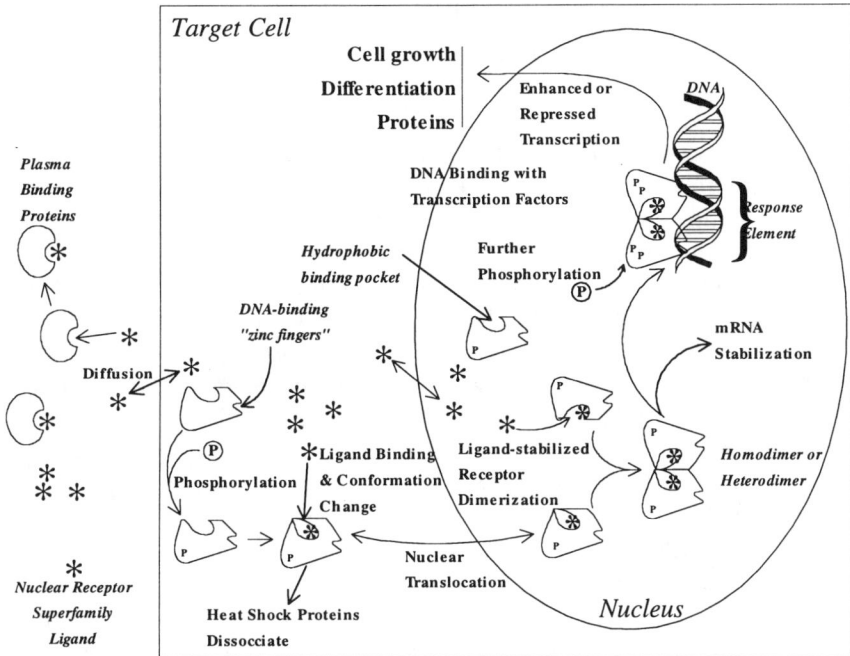

Figure 3-2 Nuclear receptor function. Shown are circulatory transport of lipophilic hormone by binding proteins, interaction of hormone and unoccupied receptor in the cytoplasm or in the nucleus (varies with different receptors) and interactions of the hormone-receptor complexes with DNA in the nucleus. Diagrammatic presentation of the changes in the receptor as it interacts with the hormone are based on studies of the estrogen receptor by Brzozowski et al. (1997).

receptors (PPAR) and imparts differential binding affinity to HREs depending upon the receptors involved (Glass 1994).

Biochemical interactions at the level of receptors are referred to as receptor cross-talk, and study of these interactions is emerging as an important field. For example, nuclear receptors can be activated or inhibited by the action of membrane-bound receptors by the binding of other transcription factors such as the aryl hydrocarbon receptor (AhR) or by necrosis factor-kB (NF-kB) (Tsai and O'Malley 1994; Khara and Saatcioglu 1996; Kurebayashi et al. 1997). Activation of transcription by nuclear receptors is strongly affected by the binding of transcription factors to the ligand-bound receptor (Halachmi et al. 1994; Anzick et al. 1997; Torchia et al. 1997). A recently discovered second form of the estrogen receptor, ERβ (Kuiper et al. 1997), paradoxically, is activated by the anti-estrogen tamoxifen and inactivated by the natural hormone 17β-estradiol when ERβ binds to AP-1 promoter sites with the co-activators Fos and Jun (Paech et al. 1997).

Vulnerabilities of nuclear receptors to chemical perturbation

The highly complex, interrelated pathways of biological action of nuclear receptors provide multiple sites for potential perturbation by xenobiotics. The most obvious and often studied site of action is ligand binding of nuclear receptors (e.g., Kelce et al. 1995; Kramer et al. 1997). Ligand binding is thought to induce a conformational change in the structure of the nuclear receptor, causing dissociation of heat shock protein, promoting the binding of transcription factors and bringing the N-terminal AF-1 region into a conformation that promotes transcriptional activation (Gandini et al. 1997). Chemical pollutants that act as nuclear receptor agonists activate the receptor in a similar manner by binding to the ligand-binding region and inducing a "productive" conformational change that promotes transcription. In some cases the binding of such receptor agonists may not be at the ligand-binding site but may nevertheless cause "productive" conformational changes (Tsai and O'Malley 1994). Conversely, nuclear receptor antagonists may inactivate the receptor by binding to and changing the conformation of the ligand binding domain to an "unproductive" form. Some evidence suggests that binding sites for antagonists differ from those of agonists (McDonnell et al. 1995) and that the conformation changes induced by antagonists are qualitatively different from those produced by agonists (Klinge et al. 1992). The complexities of the ligand binding domain responses to binding by pollutants, in particular the binding of xenobiotics to sites different than the native ligand-binding site, may limit the usefulness of structure-activity relationships for making predictions of the potential effects that will be triggered by these pollutants (Waller et al. 1996).

Nuclear receptors are subject to the influence of membrane-bound receptors and other intracellular transcription factors. For example, activated AhR interferes with estrogen-receptor binding to DNA HRES (Khara and Saatcioglu 1996). Estrogen-receptor-mediated gene transcription can be induced by IGF-1 (Lee et al. 1997), EGF, and transforming growth factor-α (TGF-α) (Ignar-Trowbridge et al. 1996) in the absence of 17β-estradiol. RXR/PPAR heterodimer can bind to estrogen-response elements, leading to enhanced transcription of genes in the absence of estrogen receptor (Nuñez et al. 1997). RXR/RAR heterodimer inhibits ER-mediated gene transcription in mammalian cells but not in a yeast assay system (Joyeux et al. 1996). Receptor phosphorylation, and therefore transcription activation and hormone binding (Migliaccio et al. 1992), can be achieved through common cellular pathways of phosphorylation, such as the MAP kinase cascade (Kato et al. 1995).

Membrane receptors

Large or highly polar chemical signals cannot readily enter the cell and therefore typically have their receptors located on the external surface of the plasma membrane. Membrane receptors are used by the many types of peptide hormones and by the catecholamine hormones. Even some nonpolar molecules elicit responses through membrane receptors. For example, several steroid hormones bind to specific membrane receptors in addition to the nuclear receptors.

There are at least as many different membrane receptors as there are signals. However, they can be classified into several groups based on common structure and functions. Some of the more prominent groups are ion channel receptors, catalytic receptors, and G-protein-linked receptors. After a signal binds to one of these membrane receptors, some type of signal transduction mechanism must be initiated to translate the signal into a cellular response. The type of mechanism used varies among the different classes of membrane receptors and between receptors in the same class. Furthermore, the same chemical signal may operate through different receptors and through different signal transduction mechanisms in different cells.

Ion channel receptors

These receptors open or close ion channels in the plasma membrane upon binding of a chemical communication signal. They are often referred to as ligand-gated channels to distinguish them from ion channels that respond to changes in voltage (Unwin 1993). The receptor molecule itself is thought to form the walls of the channel, and specific amino acids serve to select ions on the basis of size and charge. Binding of the ligand to the receptor results in a conformational change that alters the position of amino acids in the channel wall, allowing certain ions to pass (Galzi et al. 1991). The movement of ions is then associated with changes in the target cell membrane potential or changes in cell function. Ion channel receptors are usually associated with neurotransmitters. Examples include the nicotinic acetylcholine receptor and the glutamate receptor.

Catalytic receptors

Many chemical signals that regulate growth and the cell cycle utilize catalytic receptors (Figure 3-3) (e.g., epidermal growth factor, insulin and insulin-like growth factors, and platelet derived growth factor) (Ellis et al. 1991; Schlessinger and Ullrich 1992). Binding of signal to this type of receptor results in activation of the endogenous enzyme activity of the receptor. The enzyme activity is usually tyrosine kinase activity; however, the transforming growth factor b receptor family has serine kinase activity (Barnard et al. 1990; Lin et al. 1992). Catalytic receptors have at least 3 important functional domains: an extracellular domain that binds to the signal, a transmembrane domain that anchors the receptor within the membrane, and an intracellular domain that has catalytic activity when activated. Binding of a signal to a catalytic receptor triggers oligomerization between receptors or receptor subtypes and results in autophosphorylation of tyrosine residues of the receptors.

A number of different mechanisms may be activated by the phosphorylated receptors. First, the phosphotyrosine residues may serve as binding sites for domains located within certain proteins, most notably the SH2 domains. Proteins possessing SH2 domains are attracted to the phosphotyrosines and bind to the receptor

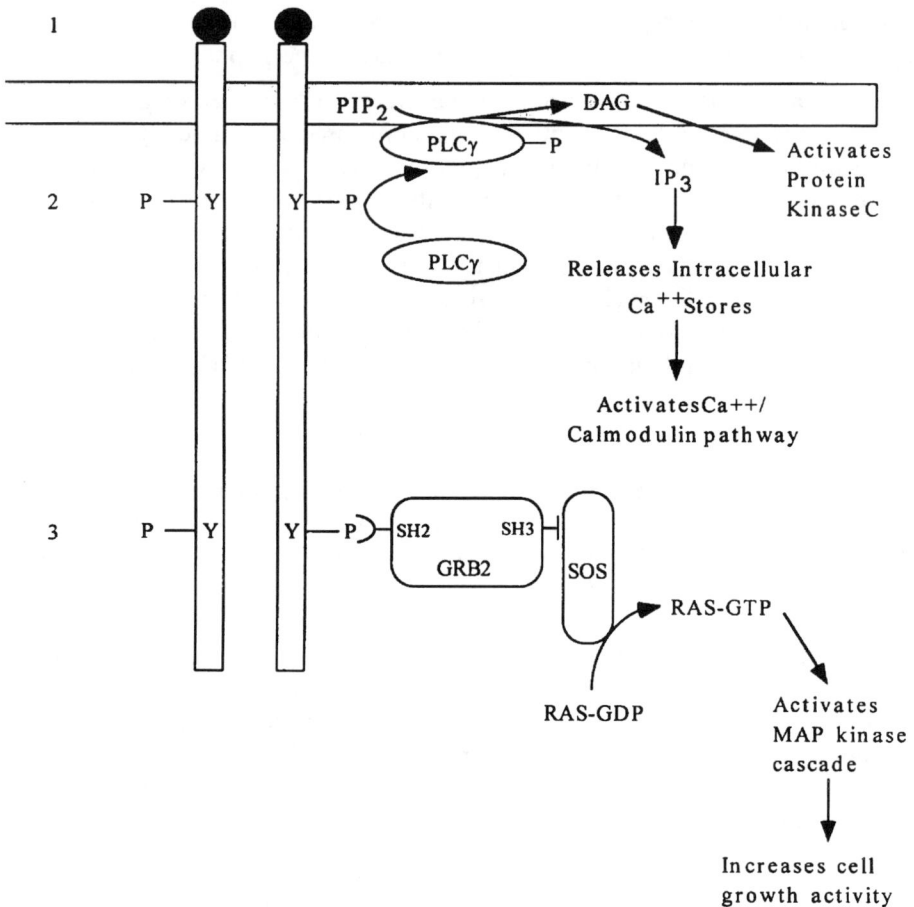

Figure 3-3 Plasma membrane receptor function: Signal transduction by a catalytic receptor (receptor tyrosine kinase). The receptor is a transmembrane protein with an extracellular ligand binding domain, a transmembrane domain, and an intracellular catalytic (tyrosine kinase) domain.

(Pawson and Schlessinger 1993; Margolis 1992). This can have several results, including phosphorylation of the SH2-containing protein or bringing together proteins with their substrates. The type of proteins that are attracted and activated by a particular catalytic receptor will determine the biological responses that result. One example is phospholipase Cc (Margolis et al. 1990; Fantl et al. 1992; Vega et al. 1992). This enzyme binds to the activated catalytic receptor and is phosphorylated on tyrosine residues. This activates the phospholipase Cc, which then catalyzes the hydrolysis of phosphatidyl inositol 4,5, bisphosphate to inositol trisphosphate and diacylglycerol, which have their own actions.

Figure 3-3 shows the following events for catalytic receptors:

1) Upon ligand binding, the receptors aggregate (dimerize) and autophosphorylate on specific tyrosine (Y) residues. The phosphotyrosines may then phosphorylate other proteins on tyrosine residues or they may serve as recognition sites for the binding of proteins containing SH2 domains, as illustrated in 2 and 3.

2) In this example, the enzyme phospholipase Cg (PLCg) is phosphorylated on a specific tyrosine residue by the catalytic receptor. This activates the enzymatic activity of PLCg, which is to stimulate the conversion of phosphotidylinositol bisphosphate (PIP_2) in the plasma membrane into inositol trisphosphate (IP_3) and diacylglycerol (DAG). DAG can go on to activate protein kinase C, which changes cellular function by phosphorylating specific proteins on serine and threonine residues. IP_3 travels to intracellular membranes and causes the release of stored Ca^{++}, which then activates calcium-dependent proteins, including calmodulin.

3) In this example, the phosphotyrosine on the catalytic receptor serves as a docking site for GRB2, a linker protein. GRB2 binds to phosphotyrosine by its SH2 domain and also binds to SOS by a protein-binding domain called SH3. SOS is a nucleotide exchange protein which, when complexed to GRB2 and the catalytic receptor, stimulates the exchange of GDP for GTP on RAS. RAS-GTP is an important intermediate in activation of the MAP kinase cascade, which regulates cellular growth.

G protein-linked receptors

Probably the most common type of membrane-receptor action (Figure 3-4) is through a G-protein-mediated mechanism (Bolander 1994; Kahn et al. 1992; Taylor 1990). Examples include rhodopsin, the light sensitive protein of the retina; odorant receptors; several neurotransmitter receptors (e.g., β-adrenergic, muscarinic acetylcholine); and many of the receptors for peptide hormones (e.g., ACTH, FSH, LH, TSH). These receptors do not possess intrinsic enzyme activity. Rather, binding of signal to the receptor acts indirectly through a G protein located in the plasma membrane. The G protein transduces the message by activating a membrane-bound enzyme, called an effector, which then amplifies the message by catalyzing the activation of one or more intracellular signals called second messengers. The second messengers then elicit changes in cellular function.

G-protein-linked receptors are often called serpentine receptors because of the way they "snake through" the plasma membrane. All have 7 transmembrane regions that anchor the proteins in the membrane, as well as an external ligand-binding domain and a G-protein-binding domain. Binding of the ligand causes a conformational change in the receptor that makes the G-protein-binding domain accessible, resulting in interaction of the receptor with the G protein.

Ligand Absent

Ligand Present

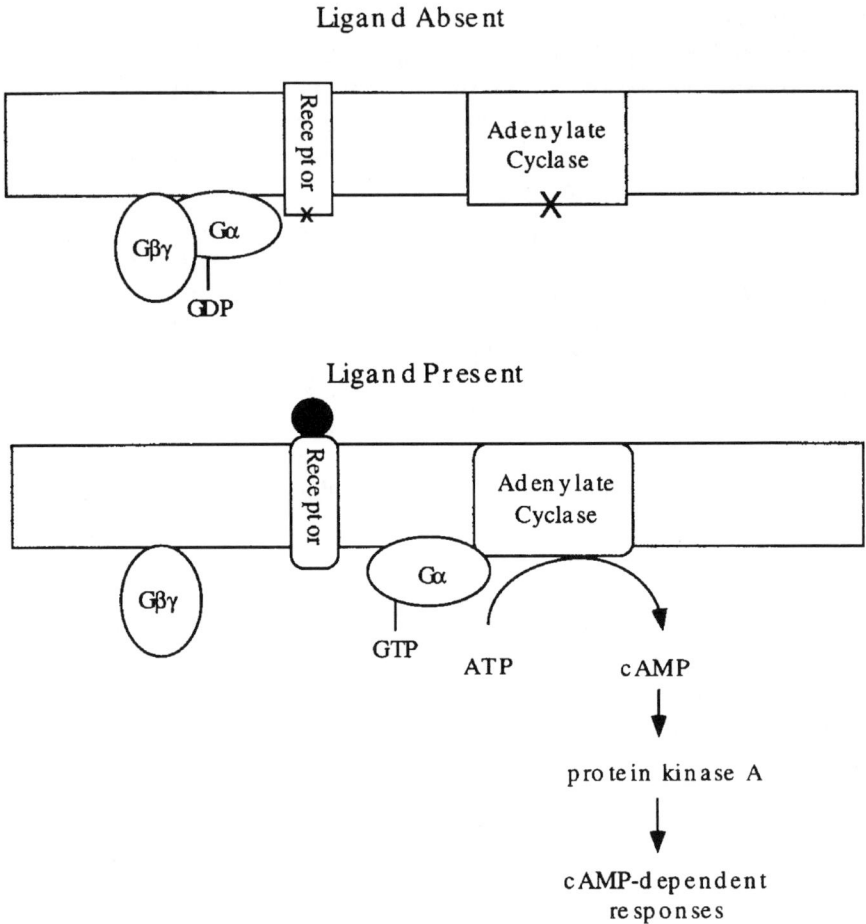

Figure 3-4 Plasma membrane receptor function: Signal transduction by a G protein-linked receptor

The G protein is a guanine nucleotide binding protein. Binding of the receptor to the G protein allows the G protein to exchange a bound molecule of guanosine diphosphate (GDP) for a guanosine triphosphate (GTP) molecule. In the presence of GTP, the G protein splits into subunits. The subunit bound to GTP (Ga) then activates an effector enzyme located on the inner surface of the plasma membrane. The effector remains activated until the Ga spontaneously hydrolyzes the GTP to GDP. At this time, the G protein reassembles and the effector is turned off. Thus, the G protein resembles a switch that is turned on by the activated (ligand-bound) receptor and turned off by its own rate of spontaneous GTP hydrolysis.

Figure 3-4 illustrates the details of these events for a G protein-linked receptor. The receptor is a serpentine protein that is looped back and forth through the membrane 7 times. It contains an extracellular domain that binds ligand (hormone) and an intracellular domain that interacts with the G-protein. The G-protein is a heterotrimeric complex containing Ga, Gb, and Gg subunits. In this example, the Ga subunit is of the Ga_s type, which is stimulatory for adenylate cyclase. In the absence of ligand (top), the receptor is inactive, the G-protein subunits are complexed together and Ga is bound to a molecule of GDP. Upon ligand binding (bottom) the receptor undergoes a conformational change, allowing it to interact with the G-protein. This causes the Ga subunit to exchange the GDP for a GTP and, subsequently, the G-protein complex to separate into Ga and Gbg subunits. The now active Ga then interacts with adenylate cyclase and activates it, resulting in the conversion of ATP to the second messenger, cAMP. cAMP activates protein kinase A, which alters cell function by phosphorylating specific proteins on serine and threonine residues. The process is turned off by the spontaneous hydrolysis of GTP to GDP on the Ga subunit and the reformation of the G-protein complex.

There are several different types of second messengers that can be activated, or inactivated, by G proteins. The two best known are adenylate cyclase and phospholipase Cb. Both catalyze the conversion of highly phosphorylated precursor molecules into second messengers. Adenylate cyclase converts ATP into cAMP. Phospholipase Cb converts phosphatidylinositol 4,5 bisphosphate into inositol trisphosphate and diacylglycerol.

The G-protein mechanism is important in amplifying the signal. One signal molecule can result in the production of hundreds or thousands of second messenger molecules. The actual changes in cellular function are accomplished by the second messengers. The second messengers differ considerably in their mechanism of changing cell function.

Cyclic AMP can have direct effects such as activating ion channels in the plasma membrane. However, most of the effects of cAMP are indirect ones mediated by activation of protein kinase A (cAMP-dependent protein kinase). Protein kinase A is a tetramer of 2 regulatory and 2 catalytic subunits. Binding of cAMP results in separation and activation of the catalytic subunits, which are serine-threonine protein kinases. The catalytic subunits then phosphorylate a wide variety of substrate proteins, altering cell function.

The products of phospholipase C (both Cβ and Cχ) activity have several mechanisms of action. First, inositol trisphosphate, which is soluble, binds to specific receptors on intracellular membranes and causes the opening of calcium channels, releasing stored calcium (Berridge 1993; Putney and Bird 1993). The calcium then acts as a messenger itself (Sanderson et al. 1994), primarily by activating the calcium-binding protein calmodulin. Calmodulin in turn activates specific protein kinases, which alter cell function. Diacylglycerol remains in the membrane but also has functions in signaling. Diacylglycerol activates protein kinase C (Stabel and Parker 1991; Nishizuka 1992), which is also a serine-threonine protein kinase that has numerous substrates within the cell. Diacylglycerol can also be converted to arachidonic acid, which is used as substrate for eicosanoid production.

Vulnerabilities of membrane receptors to chemical perturbation

Interest in the effects of pollutant agonists and antagonists to reproductive steroid hormones has focused a great deal of attention on nuclear receptors as mechanistic sites of action for disruptive developmental and reproductive effects (Table 3-1). However, membrane receptors for both steroid and non-steroid hormones also are important in some mechanisms of physiological disruption by chemical pollutants. For example, some estrogenic chemicals display relatively high binding affinities for progestin membrane receptors of teleost fish (Thomas et al., in press) and thereby block the steroid hormone induction that triggers final maturation induction of both male and female gametes (Ghosh and Thomas 1995; Thomas et al., in press). Formamidines, which are used in pest control, can block the action of norepinephrine by binding to μ-noradrenergic receptors (Costa and Murphy 1987; Costa et al. 1988), and the blocking of the GnRH-triggered LH surge in male rats by these chemicals is thought to be a result of this membrane receptor binding (Goldman et al. 1991; Goldman and Cooper 1993). This disruption by formamidines can alter the LH surge and ovulatory patterns and thereby affect the number of implantations in a subsequent pregnancy (Cooper et al. 1994).

Life history of signaling molecules

Most signaling molecules go through a distinct series of stages, each of which is subject to physiological regulation. The level of signal found at any given time is the result of the differential activities of the processes regulating these stages. The stages are synthesis, secretion, transport, metabolism, and degradation.

Synthesis

Peptide signals are synthesized through the transciptional/translational machinery of the cell. Since peptide signals are destined for secretion, the mRNAs for these proteins usually are processed by the ribosomes of the rough endoplasmic reticulum. Then the nascent peptide is processed for secretion in the rough endoplasmic reticulum and the Golgi apparatus. The peptide eventually resides in a secretory vesicle awaiting release. Many peptide signals are produced as inactive precursor molecules that are converted into active signals during processing or after secretion.

The lipid signaling molecules, primarily steroids and eicosanoids, are synthesized in the cell by enzymatic pathways. Control of lipid signaling is primarily achieved by controlling levels of the synthetic enzymes. The particular lipid signals that are produced by a cell are the result of the specific complement of synthetic enzymes present.

Steroids are synthesized in the adrenal glands and the gonads (Kime 1987). Cholesterol is the substrate for formation of the steroid hormones. Cholesterol is converted to pregnenolone by removal of a 6-carbon side chain. This process is called side-chain cleavage and is catalyzed by the cytochrome P450 side-chain cleavage enzyme complex. Side-chain cleavage is the rate-limiting enzymatic step in steroid hormone synthesis. Pregnenolone is then converted to the various other steroid hormones through one of several enzymatic pathways. Typical transformations include addition of ketone or aldehyde side groups and addition or removal of double bonds. The enzymes responsible for these transformations differ between steroidogenic tissues. The specific enzymes present in a given cell determine which steroid hormones can be produced by that cell. Many of the steroid hormones can be further metabolized in peripheral (nonsteroidogenic) tissues and the particular steroids produced may be critical to the tissue effects at that site. For example, in mammals, 5-hydroxytestosterone (5DHT), a metabolite of testosterone, triggers the differentiation of some genital structures while testosterone triggers the differentiation of others (Hadley 1996). Enzymatic conversions of thyroid hormones also differ with tissue type. For example, in birds and mammals, central nervous tissues, which are critically dependent on thyroid hormones, protect their own supply of the most active thyroid hormone (T_3) by increased enzymatic conversion of T_4 to T_3 and by altering their hormone uptake and loss if the circulating hormone supply is decreased (Rudas and Bartha 1993; Rudas et al. 1993, 1994).

The eicosanoids are derived from membrane lipids, particularly arachidonic acid. Arachidonic acid can be enzymatically converted to the various eicosanoids, including the prostaglandins, leukotrienes, and thromboxanes. Synthesis of prostaglandin and leukotrienes involves the enzyme cyclo-oxygenase, which can be blocked by anti-inflammatory agents such as aspirin. Synthesis of leukotrienes requires the enzyme 5-lipoxygenase. As with steroid hormones, the relative distribution and abundance of the synthetic enzymes determines which eicosanoids will be produced by a given cell.

Secretion

The production and secretion of chemical signals usually are coordinated, although some signals can be stored. Peptide hormones often are stored temporarily in secretory vesicles and may be released by neural or hormonal stimuli impacting the secretory cell. For example, action potentials from the hypothalamus stimulate release of neuropeptides from the posterior pituitary while hormonal signals from the hypothalamus stimulate release of peptide hormones from the anterior pituitary. Similarly, catecholamines are stored in the nerve cells and their secretion is usually controlled by neural input, with action potentials serving as the proximal stimulus for release (Hadley 1996; Norris 1997).

Steroid-producing cells appear to secrete their hormones by diffusion as they are produced. There is no convincing evidence for any storage or specialized secretory mechanisms for steroid hormones. Eicosanoids also are released as they are produced, as is nitric oxide. Thyroid hormones are unusual in being the only hormones for which there is extracellular storage within an endocrine gland of a significant hormone supply. The hormones are contained in the structure of the precursor protein thyroglobulin, which is stored in colloid in the center of thyroid follicles, for later mobilization and enzymatic release by the follicular cells. It is speculated that the evolutionary development of this hormone storage relates to the trace element iodine being a critical component of these hormones (McNabb 1992; Hadley 1996; Norris 1997).

Patterns of secretion

Many hormones are released continuously, but some are secreted in discrete bursts at either regular (pulsatile) or irregular (episodic) intervals. Both modes of secretion can result in regulated concentrations of circulating hormones. Rates of synthesis and release may differ for different hormones. In some cases, for example when hormones are synthesized and secreted rapidly, it may be very difficult to determine "basal" hormone concentrations. This is a common concern in the measurement of stress-related corticosteroids, especially when animals are caught and sampled in the field, and it puts constraints on the protocols that are considered to yield reliable measurements. For example, to be reflective of the general level of corticosteroids in a wild bird (rather than the immediate stress of handling), blood samples for corticosterone analysis must be drawn within 5 minutes of catching the bird. In addition, it is now recognized that dynamic measurements of the stress response give the most reliable information about the capability of the bird to respond to stress. Such dynamic measurements require not only initial sampling immediately after catch, but also a follow-up sampling typically at 30 and 60 minutes later to assess the magnitude of the stress response (Wingfield 1994; Wingfield et al. 1995).

For hormones that are synthesized and released in pulses, the pulses themselves may be important in the signaling process. For example, GnRH is normally released in a pulsatile fashion from the hypothalamus. It has been shown that the pulses are a critical element in the stimulation of FSH and LH release from the anterior pituitary (Catt and Dufau 1991).

Hormones may show regular daily (diurnal or circadian) patterns of secretion or may show regular annual (circannual) patterns of secretion. These patterns are reflective of hormonal control of functions such as an animal's daily metabolic patterns by thyroid hormones or the coordination of the reproductive cycle (with the time of the year most appropriate for reproduction) by the hypothalamic–pituitary–gonadal axis (Hadley 1996; Norris 1997).

Transport

Signals may travel in the blood or interstitial fluid in their free form or they may be bound to carrier- binding proteins. These proteins are critical to the transport of the lipid or lipophilic hormones by the aqueous medium, blood. The carrier proteins may be specific for that signal, or they may be more general and carry a variety of molecules. Carrier molecules are thought be important in several ways in regulating hormone action:

- by protecting the hormone from degradation, especially by the liver;
- by determining the amount of "free" signal available to interact with receptors (since only unbound signal can interact with the receptor and cause biological action); and
- by acting as a "storage pool" of available hormone. There is evidence that some binding proteins may help deliver the signal to the target cell (Siiteri et al. 1982).

Peptide signals generally travel in their free form in the circulation because they contain charged amino acids, are usually quite polar, and, thus, are water soluble. Some peptide hormones (e.g., IGFs) are largely bound to carrier proteins, and it appears that this association functions in facilitating hormone supply to cell receptors (Hadley 1996).

Lipophilic hormones, such as thyroid and steroid hormones, are transported in the blood by binding proteins, and phylogenetic differences in these binding proteins have important implications for physiology and for the responses to pollutant chemicals. Thyroid hormones travel bound to thyroxine-binding globulin, transthyretin (thyroxine-binding prealbumin), and albumin. These binding proteins are thought to be instrumental in regulating the availability of "free" or unbound thyroid hormones to enter tissues (Robbins and Bartelena 1986). Thyroid hormone binding proteins of vertebrates differ between species and classes (Larsson et al. 1985), so knowledge of the binding protein types and concentrations in the species being investigated can be important in examining the disruption of hormonal control systems by chemical pollutants. In addition to transport of thyroid hormone in the circulation, transthyretin is also important in transporting T_4 across some membranes, e.g., possibly in T_4 entry across the blood-brain barrier into the central nervous system (DeGroot et al. 1996; Robbins 1996). Binding proteins can modulate the free thyroid hormone concentrations during embryonic avian development (McNabb and Hughes 1983), and some specific forms of binding proteins are important in environmental effects on hormonal control systems in reptiles (Licht et al. 1990). Some chemical pollutants are known to displace thyroid hormones from their binding proteins and alter the circulating free hormone concentrations in mammals (Darnerud et al. 1996; Morse et al. 1996). Pollutant binding also can interfere with the transport of retinoids (Brouwer and Van den Berg 1986).

Steroid hormones also travel bound to one of several proteins (Westphal 1986). Albumin serves as a general carrier for steroid hormones. Although albumin has a relatively low affinity for these hormones, it is so abundant that it can transport large amounts of hormone. Sex steroids (testosterone, estrogen and progestins) often are bound to a sex-hormone-binding globulin (SHBG), which also is called testosterone-estrogen-binding globulin (TEBG). Some form of sex-hormone-binding globulin is found in most oviparous vertebrates, although the affinity and specificity varies (Callard and Callard 1987). Interestingly, birds appear to lack a sex-hormone-binding globulin (Wingfield et al. 1984). Corticosteroids often are carried by a corticosteroid-binding globulin (CBG), also called transcortin.

Metabolism

Many signals are secreted in an inactive or less active form and are activated in peripheral tissues before they interact with receptors. Peptide signals may be secreted as larger prohormones that are activated by selective cleavage. For the thyroid hormones, thyroxine can be enzymatically converted to the more potent triiodothyronine in peripheral tissues, thus the presence of the appropriate enzymes is important in the quantitative supply of the most active hormone to different target tissues (Leonard and Visser 1986).

Steroid hormones undergo a wide variety of metabolic transformations. Many active steroid hormones are transformed from circulating precursors by the target tissues themselves. For example, the circulating androgens (testosterone and androstenediol) are converted to the estrogens (estradiol 17β and estrone) in target tissues by the action of the enzyme cytochrome P450 aromatase. Testosterone also can be converted to the more potent androgen dihydrotestosterone (DHT) in tissues possessing the enzyme 5α-reductase. In teleostean fishes, testosterone can be converted to 11-keto testosterone, which appears to be the major androgen of males (Kime 1987).

Degradation

The concentration of a signal at any given moment is dependent on the balance between its synthesis and degradation. Thus, mechanisms to remove signals from the circulation are an important aspect of signaling control. Degradation pathways exist for each of the various types of signaling molecules. The rate of degradation varies tremendously between signal types, between organisms, and between different physiological states. However, one generalization that can be made is that steroid hormones tend to be degraded more slowly than peptide hormones and therefore remain in circulation for a longer period of time. A typical steroid hormone half-life in circulation may be hours to days, compared with minutes to hours for a peptide hormone.

The means of degradation varies for each signal type. Proteins are degraded by an extensive complement of specific and nospecific proteases, catecholamines are degraded by specific deamination, and thyroid hormones are degraded enzymatically by deiodination or are conjugated to facilitate excretion (McNabb 1992; Hadley 1996; Norris 1997).

Steroids are catabolized by a series of reactions, including hydroxylations at various carbons and reductions of double bonds and ketones (Kime 1987). These reactions usually remove active sites important for receptor binding. In mammals, steroids are generally catabolized by the liver, but in oviparous vertebrates, they also may be catabolized by peripheral tissues including steroidogenic tissues. A key type of steroid conversion is conjugation to sulfate and glucuronide moieties. These groups make the steroids more water soluble and thus more likely to be removed from circulation by the kidneys. Conjugation of thyroid hormones and their release in bile is often considered a disposal mechanism. However, conjugation is reversible and this may lead to recycling from the gut. In mammals, conjugation may facilitate deiodination (see review in McNabb 1992). In teleost fishes, glucuronides (e.g., progesterone glucuronide) may serve as pheromones. They are water soluble and therefore can diffuse through the water to act on another individual (Kime 1987). However, glucuronides are de-conjugated very rapidly in water (Panter, personal communication); therefore the possibility is raised that it is the free steroid, rather than conjugates, that is acting as a pheromone.

Hormone concentrations in blood and tissues

Physiological chemical control (by hormones, growth factors, and their target organ receptors) is critical to the progression of normal events in the development and reproduction of vertebrates. There is a considerable body of knowledge in the endocrine literature on hormone profiles in different species, in different life stages, and under different environmental conditions that can be used to illustrate the changing physiological background against which pollutant effects will occur. This section illustrates some patterns of hormone changes during development and seasonal reproductive cycles. Although some basic patterns of hormones in the circulation or in tissues are similar among the vertebrates, there are phylogenetic differences between major taxa. In addition, there are also differences within and between individuals, depending on such factors as season, developmental stage, reproductive state, and life history. In addition, although circulating hormone concentrations are most investigated, they are only one of the types of background information needed to understand the control of development and reproduction. In an endocrine context, other types of necessary information are tissue hormone concentrations, hormone conversions, metabolic turnover of hormones, binding of hormones to their receptors, critical responsive developmental periods, and triggering of physiological actions.

Hormone concentration changes during development

Changes in circulating hormones or tissue availability of hormones during development are critical triggers of differentiation and maturation effects in tissues. Disruption of hormone signals that control developmental processes are the key to much of the vulnerability during transitional stages. Figures 3-5 through 3-7 present some examples of patterns of circulating hormones during key developmental and reproductive events in fishes and birds. However, the tissue content of hormones and tissue-specific metabolism of hormones also can be important.

The dramatic developmental changes that occur during amphibian metamorphosis have long been known to be triggered by thyroid hormones (Figure 3-5), and the maturation of deiodinase systems that convert T_4 to the more active T_3 are important in tissue-specific development (Galton 1988a, 1988b). Fish that undergo metamorphic changes (e.g., flounder) show circulating thyroid hormone patterns similar to those of amphibian metamorphosis, and other hormones such as corticosteroids are also important in these events (de Jesus et al. 1990). Likewise, the developmental process of smoltification in salmonid fishes is orchestrated by a suite of hormonal changes (Figure 3-6). In birds, which are endothermic as adults, the development of endothermy in precocial species appears suddenly during the perinatal period, and there are bursts of thyroid hormone activity that are associ-

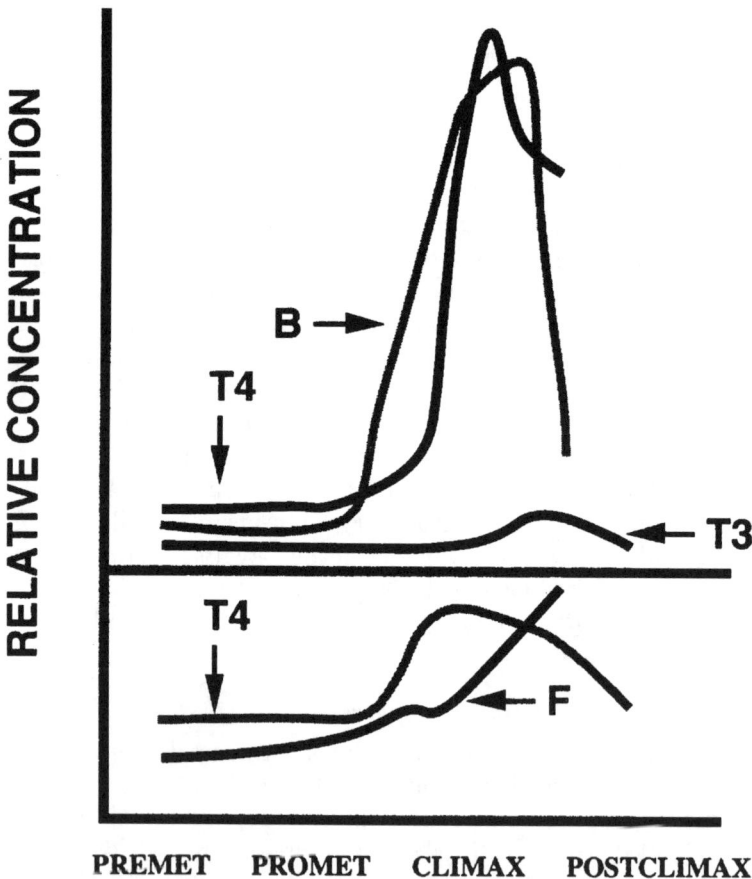

Figure 3-5 Circulating thyroid and corticosteroid hormone concentrations during metamorphosis in anuran amphibians. T_4, thyroxine; T_3, triiodothyronine; B, corticosterone; F, cortisol. PREMET, premetamorphosis; POSTMET, postmetamorphosis; CLIMAX, metamorphic climax. Redrawn from Dickhoff (1993) with permission from Academic Press, Inc.

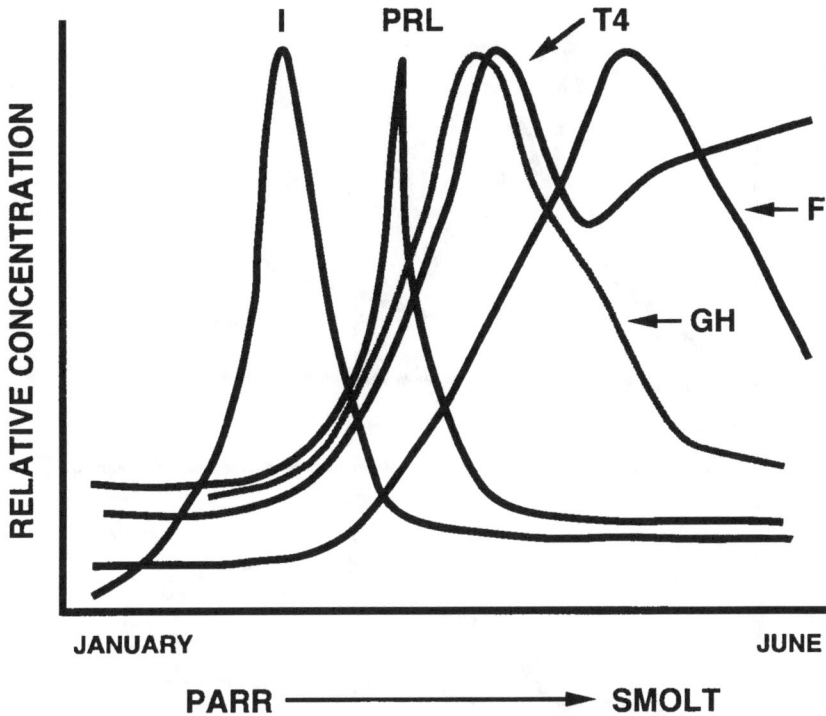

Figure 3-6 Circulating hormone concentrations during smoltification in salmonid fish. I, insulin; PRL, prolactin; GH, growth hormone; T_4, thyroxine; F, cortisol. Redrawn from Dickhoff (1993) with permission from Academic Press, Inc.

ated with this increase in metabolic capacity (Figure 3-7). In contrast, altricial birds are hatched in a less developed state, have little development of thyroid function at hatch, and develop most thyroid function and endothermy some time after hatching. Another striking example of hormone effects during development is the picture of gonadal steroid effects on brain development during critical periods of brain responsiveness.

Hormone concentration changes throughout the year

Reproductive cycles in vertebrates are under the control of the hypothalamic–pituitary–gonadal axis, which responds to environmental cues about season or maturity. Figure 3-8 illustrates a generalized pattern of pituitary gonadotropic hormones and gonadal steroid hormones during the period of vitellogenesis, oocyte maturation, and ovulation in teleosts. It needs to be recognized that there can be considerable variation in this pattern among the fishes and potentially even among different strains of the same species. Basically, gonadotropin I (GTH-I) (i.e., fish FSH) stimulates hepatic vitellogenin production via estrogen (estradiol) and

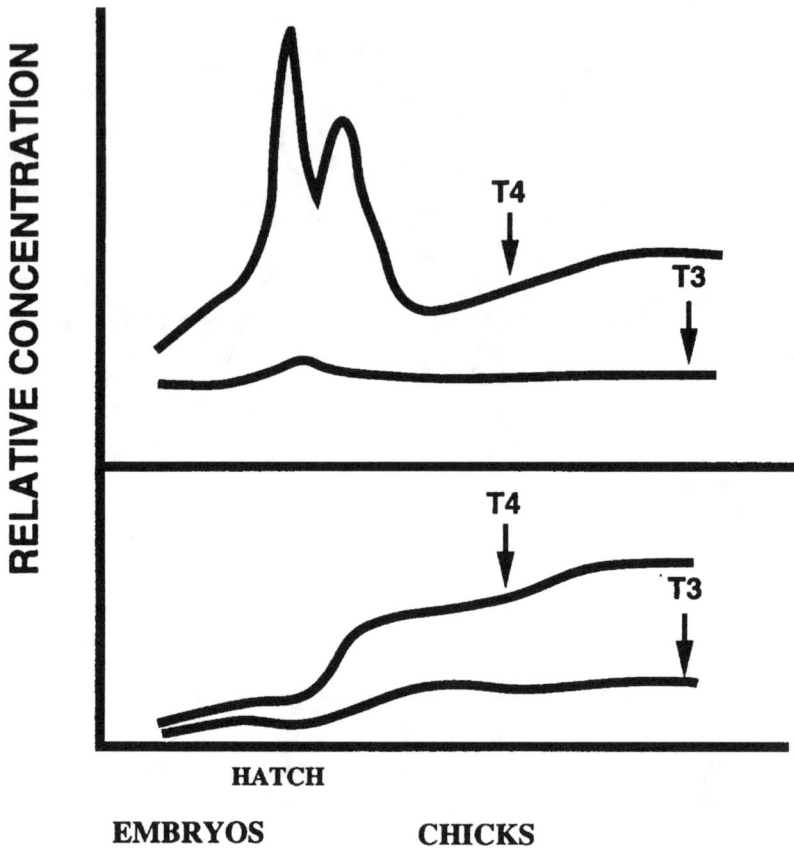

Figure 3-7 Circulating thyroid hormone concentrations in precocial and altricial birds from mid-incubation to 2 weeks posthatch. Data are representative of precocial Japanese quail or chickens (upper panel) and altricial ring doves or European starlings (lower panel). Redrawn from McNabb and Olson (1996) with permission from Science and Technology Letters.

consequent ovarian growth. Another gonadotropin, GTH-II, is more involved subsequently with production of maturation steroids and ovulation. The gonadally synthesized maturation steroids are usually progestins. Production of androgens such as testosterone also increase significantly in female fish near ovulation. For examples of temporal patterns of various reproductive hormones see Dodd and Sumpter (1984), Fitzpatrick et al. (1986), Prat et al. (1996), Planas et al. (1996) and Breton et al. (1998). Cycles such as these may differ with a number of factors related to reproductive life history as well as genetics.

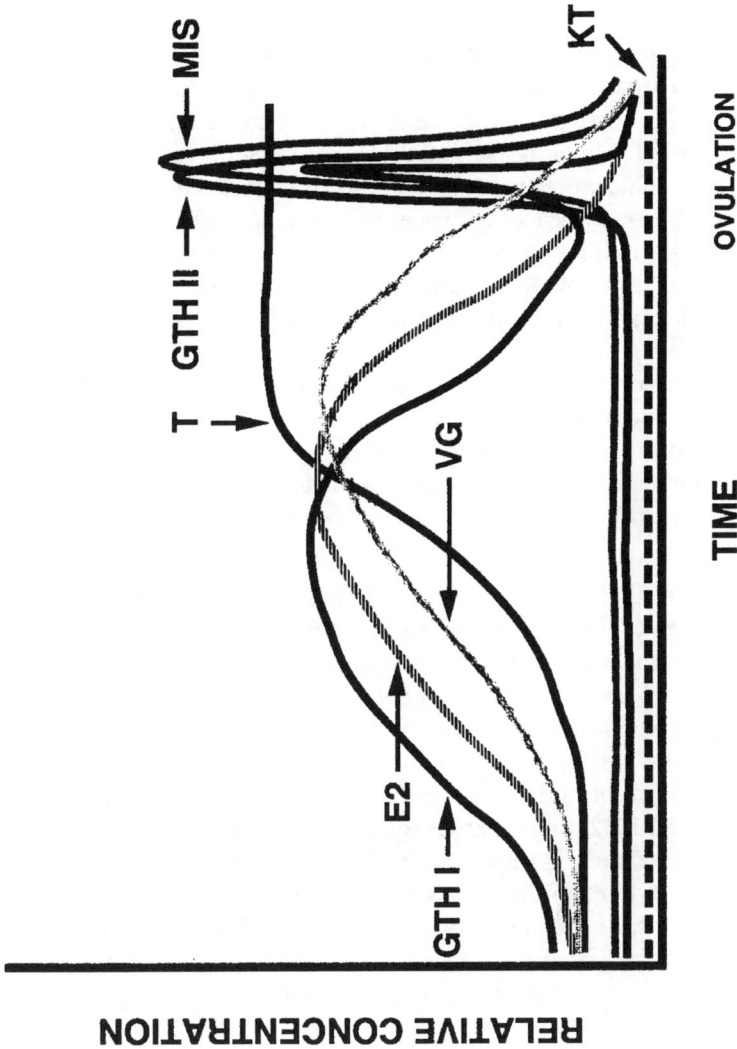

Figure 3-8 Generalized schematic of circulating hormone and vitellogenin profiles in a female teleost fish during vitellogenesis, final oocyte maturation, and ovulation. MIS, maturation inducing steroid (these include 17α,20β-dihydroxy- and 17α,20β, 21 trihydroxy forms of progesterone and are species-specific); E2, 17β-estradiol; GTH I (i.e.,fish FSH) and GTH II, gonadotropin I and II, repectively; KT, 11-ketotestosterone; T, testosterone; VG, vitellogenin. Drawn from data in Fitzpatrick et al. (1986), Prat et al. (1986), Planas et al. (1996), and Breton et al. (1998).

Development

Maternal materials in eggs

The females of oviparous vertebrates deposit a variety of materials in their eggs; pollutant chemicals from the mother's body may be among these materials. Most important nutritionally are yolk lipoproteins, which serve as the main source of nutrition for the developing embryo. Yolk quantity varies widely in different species within each of the oviparous vertebrate classes, and the availability of materials required by the embryo, as well as the exposure to egg contaminants, needs to be known to assess chemical pollutant effects on development. Some egg components are important in embryonic protection (e.g., immune components in albumin) and in stimulating developmental events (e.g., maternal hormones and growth factors). Input into the egg by the mother is complete at the time of egg laying, or oviposition. This fact seems to argue that the situation in oviparous vertebrates is dramatically different from that in mammalian embryonic and fetal development where there is continual exchange in utero between the mother and fetus. However, the situation is much more complex than this suggests, and the possibilities of additional maternal to embryo transfer and/or environment-to-embryo transfer have important implications for development. Some fish and reptile species, retain eggs in the body of the mother for varying periods of time before oviposition, and the eggs of some oviparous species are permeable to environmental materials after laying.

In fish and reptiles, the range from oviposition prior to embryonic development through varying degrees of egg retention, forms a continuum between oviparity and viviparity. At the viviparous end of this continuum, placenta-like associations are formed. In fish, large yolk size is associated with lecithotrophy (i.e., dependence on the yolk for nutrition), whether or not eggs are retained in the female during part of embryonic development. Small yolk size is associated with egg retention and varying degrees of matrotrophy (i.e., dependence on the mother for nutrition) (Wourms and Lombardi 1992). In viviparous squamate reptiles, which typically have eggs with large yolks, the yolk appears to be the primary source of organic nutrition; however, some inorganic materials may be supplied by the mother (Stewart 1992). Possible further complexity is added by some lizards with facultative viviparity, in which the eggs are retained for varying amounts of time. However, in one investigation of this phenomenon, there were no differences in embryonic size or development in embryos from eggs retained in the female for different lengths of time. These data suggest that the length of retention time is related to the egg-laying conditions rather than to nutritional benefits or detriments to variable times of maternal egg retention (Mathies and Andrews 1996).

In lecithotrophy, in addition to lipoprotein nutrients, lipophilic materials (such as steroid and thyroid hormones and many lipophilic pollutants known to act as hormone analogs or antagonists) enter eggs during yolk formation. In those species that retain eggs and form structures that permit the transfer of nutrients to embryos, transfer of lipophilic materials may continue during embryonic development. Whether hydrophilic materials can enter eggs depends on the permeability of the shell and the nature of its surroundings (whether in the mother's body or outside).

An array of materials, including vitamins, metals, and hormones, have to be sequestered in an oocyte to produce a viable egg and, in turn, a viable offspring. Some of these substances may enter adventitiously in the fluid phase during receptor-mediated uptake of vitellogenin (VTG). Others may be bound to VTG and other lipoproteins, but most probably enter quite specifically, bound to specific receptors (Perry et al. 1984; Shen et al. 1993). In chickens, both Vitamin A and thyroid hormones appear to be transported into oocytes, attached to a transport protein, transthyretin, which has its own specific receptor (Vieira et al. 1995). In oviparous animals, we know relatively little about the transport systems and the regulatory systems for the uptake of vitamins and metals into the growing oocyte.

After oviposition, the potential for exchange of materials between eggs and their environment varies with vertebrate class and egg types. In avian eggs, which are calcified, aside from the metabolic activities of the embryo, only gas exchange with, and evaporative water loss to, the environment can alter the in ovo environment after laying. For birds, the calcified shell is an important source of mineral elements (primarily calcium) for development of the embryonic skeleton (see reviews of eggshell and egg composition in birds by Noble 1991 and Carey 1996). Because of the shell's importance, any interference with normal shell formation or any pollutant materials that might appear in the shell (and be mobilized during embryonic development) could affect embryos. Reptiles lay eggs of a variety of shell types. Most lizard and all snake eggs have relatively thin, extensible eggs that may take up water from moist environments; turtles have flexible eggshells with a calcified layer; and crocodiles, some turtles, and a few lizards have rigid calcified shells that do not permit water uptake (Pough et al. 1998).

Materials such as hormones, growth factors and RNAs that encode for the production of hormones, growth factors, and receptors are deposited in the yolk of eggs and may have important influences on embryogenesis (Gilbert 1997). Fish larvae are generally physiologically immature with little or no capacity to produce certain enzymes, vitamins, and hormones, and they are dependent on exogenous sources (mother and/or live food) for the supply of these regulatory factors (Ayson and Lam 1993; Dabrowski and Blom 1994; Lam 1994). Thus, egg stores of maternal hormones may fill some of the regulatory needs of the developing embryo for growth, development, osmoregulation, and other physiological functions prior to the functional development of their own endocrine glands. In fish, thyroid hormones of maternal origin are deposited in egg yolk at high levels, and they can have significant effects on embryo development and survival (Brown et al. 1988, 1989; Tagawa and Hirano 1987, 1991). Some experiments that have manipulated maternal thyroid and corticosteroid hormone concentrations in fish eggs have demonstrated developmental effects on larvae (Kim and Brown 1997; Poholek et al. 1997), whereas other studies have shown no effect of thyroid hormone deficiency on early development (Tagawa and Hirano 1991). Some work has shown that hormones and growth factors deposited in avian eggs may play a role in early differentiation and growth of chicken (*Gallus domesticus*) embryos (insulin and IGFs; dePablo 1989), in skeletal tissue differentiation of Japanese quail (*Coturnix japonica*) embryos (thyroid hormones; Wilson and McNabb 1997), and in aggressive behavior of canary (*Serinus canaria*) nestlings (testosterone; Schwabl 1993). A number of growth factors are present and are important in early organization prior to gastrulation in avian development (Petitte and Karagenc 1996), but it is not known whether these factors or mRNAs for their production are present in yolk.

Chemical pollutants that are deposited in eggs and that act as hormone analogs or antagonists are likely to exert effects on differentiation of developing organ systems during embryonic life. Toxicological studies of embryos from bird eggs containing high concentrations of contaminants (e.g., Herring gulls, *Larus argentatus* in the Great Lakes; review, Fox 1993), correlative studies of the levels of pollutants in avian eggs with toxic effects (Tillitt et al. 1991), and experimental studies that have injected PCBs or TCDDs have documented teratogenic effects on avian embryonic development (Powell et al. 1996), as have in ovo studies using fish (Nacci et al. 1998). Although these studies have not provided mechanistic answers about how such disruption occurred, the types of defects observed are consistent with disruption of the hormonal control of embryogenesis and organogenesis by hormones that regulate differentiation and growth.

Life cycles

Many physiological processes and their control systems change ontogenically, and the animal's prior history and its current environment influence its progression through developmental stages (Schreck 1981, 1990). Likewise, underlying ontogenic "clocks" may play important roles in development, and the possibility that these clocks may be reset must be considered in any examination of transitions during development. The developmental stages at which chemical pollutant exposure occurs obviously is crucial to the type of effects that results. Assessments of developmental progression are complicated by differences in developmental rates and a variety of "life-history strategies" results. Thus, not only the stage at which exposure to a chemical pollutant occurs but also the way in which the animal achieved that stage may affect the physiological outcome of the exposure.

Developmental rates or patterns may be more important indicators of developmental stage than are the absolute numbers measured for physiological variables. For instance, activity levels of the enzyme Na/K-ATPase are used as indicators of the transition of anadromous salmonids from parr to smolt, i.e., the transition from fresh water to sea water during development (Zaugg and McLain 1972). However, there can be considerable variation in this indicator between individuals, generations, or populations, so the numeric value is not a powerful indicator. For example, the highest enzyme activity levels observed in one generation may be equal to the lowest values observed in another generation of that population or in some other population. Thus, what appears to be indicative of developmental progression is an increase in ATPase activity, not some absolute level of activity (Ewing and Birks 1982). Likewise biomarkers of smoltification such as hormones (T_4, T_3, growth hormone, and cortisol), morphological characters (body shape), skin pigment deposition (guanine), and behavior (migratory disposition) are affected by the nature of the growth pattern of the salmon. In the case of chinook salmon, *Oncorhynchus tshawytscha*, an individual that has limited growth in winter and grows rapidly in the spring will smolt, while one that grows fast in the winter and slow in the spring will not (Dickhoff et al. 1995).

Oviparous animals have a variety of life history "strategies." The life cycle of individuals within a population may differ, or in cases when individuals in a population are developing in synchrony, even geographically close populations may not be synchronous. Thus, there are many different developmental trajectories through which adulthood can be achieved (Thorpe 1989, 1994a, 1994b). Unfortunately, our understanding of how physiological mechanisms induce or regulate the developmental patterns is very limited. We speculate that the "strategy" for development into the next life-cycle stage is dependent on prior developmental trajectory coupled with bioenergetics (i.e., energy acquisition and utilization rates as suggested by Thorpe 1994a) and the physiological nature of growth. The key stages of development and the vulnerable control factors within them are discussed in detail in subsequent parts of this section.

The general paths of animal development are termed indirect and direct. Indirect development includes the broad stages of fertilized egg, embryo, larva, juvenile, and adult. Direct development differs in that there is no larval stage and the embryo develops into the juvenile, whose body form is essentially the same as that of the adult (Hanken et al. 1997). Among the classes of oviparous vertebrates, only fishes and amphibians include species that undergo indirect development, whereas no reptiles or birds have a larval stage. In this section, the emphasis is on the processes of transition between successive developmental stages. These transitional stages and the processes within them are highly vulnerable to chemical perturbation because the transitions are controlled by hormones and other chemical signals, nervous signals, and the ensuing chains of events that guide development. It has been suggested that the most sensitive of these transitions, with respect to chemical stressors, appears to be gametes, the highly plastic, early embryo prior to formation of eye pigmentation, hatching, and the transition to self-feeding (Rosenthal and Alderdice 1976). However, disruption of any normal developmental events can have permanent effects on the anatomy and physiology of the organism.

The following processes and transitions are described in this section:

- cleavage: the transition from the fertilized egg to the embryo;
- embryogenesis and organogenesis: the differentiation and growth of tissues and organs that occur primarily during the embryonic stage and also in the larval stage if there is one;
- hatching: the passage from the embryo in a protective egg casing to a free-living larval or juvenile form;
- initiation of feeding;
- fledging: a special case of a vulnerable transition when altricial birds leave the nest;
- metamorphosis: the transformation from the larval to the juvenile body form;
- migration and habitat shifts: movements that can occur at various life stages, often in association with moving to the site for reproduction;
- sexual maturity: the transition into reproductive potential of the adult stage; and
- senescence: the passage from reproductive potential to the end of life.

Adulthood and the production of gametes are covered in the next section.

Cleavage

Following fertilization, there are a number of mitotic divisions that divide the cytoplasm of the egg into smaller, nucleated cells (blastomeres). In general, in vertebrates, these divisions and the arrangement of the blastomeres that result are controlled by the maternal mRNAs and proteins stored in the egg. Cleavage divisions result in a marked increase in cell numbers without an increase in volume, and the resultant change in cytoplasmic-to-nuclear ratio is important to the activation of some embryonic genes (Gilbert 1997). Thus, maternal influences during cleavage, and potentially any chemical pollutant effects on egg formation, could play a major role in the initial stages of embryo development. However, if this type of effect reduces viability at very early stages it would be apparent only in a laboratory setting; if the embryo survived, effects of this sort might appear as teratogenicity.

Embryogenesis and organogenesis

Formation of the vertebrate body plan requires the orderly expression of genes that organize differentiation and growth (see Blaxter 1988; Raff 1996). Pattern formation—the positional identity of limbs and the antero-posterior axis—is controlled, in part, by products of a family of genes acting as nuclear transcription factors called the HOM-C/Hox homeotic genes (Krumlauf 1993). The homeotic gene complex is evolutionarily very old, highly conserved, and is found in plants as well as in animals. Consequently, disruption of the expression of this complex, or disruption of other transcription factors involved in regulating developmental differentiation, can result in teratogenic effects.

Retinoids are chemical signals, all derived from Vitamin A, that are natural morphogens in animals ranging from tunicates to mammals. They play roles in the development of the antero-posterior axis of the body, the central nervous system, and the heart (Means and Gudas 1995). An active, new area of investigation suggests that retinoids (which have receptors that are part of the nuclear steroid receptor superfamily) exert their actions, at least in part, through their actions on homeotic gene expression (Marshall et al. 1996). In the last decade, 2 families of retinoic acid receptors, the RAR and RXR isotypes, have been discovered. Additional levels of combinatorial complexity in this signaling system are generated by both RAR and RXR having 3 isotypes that in turn can homo- or heterodimerize and be ligand-dependent or independent (Chambon 1996). The complexity of this signaling system is enormous, with "cross-talk" among cell-surface signaling pathways, and multiple coactivators and corepressors playing a role in the effects of these receptors and their ligands in regulating gene expression. Of special interest is the formation of heterodimers of these retinoid receptors with thyroid hormone receptors and the implications for regulation and vulnerability to perturbation of this complex system. In addition, the metabolism of retinoic acid is complex (Napoli 1996); this means there are many potentially vulnerable sites in this signaling system. Various derivatives of Vitamin A act as signaling molecules that trigger developmental events in a variety of tissues. Disruption of enzymes involved in metabolizing Vitamin A may account for some of the deleterious effects of some polyhalogenated aromatic hydrocarbons (Zile 1992); other effects may result from retinoid displacement from binding proteins (Brouwer et al. 1986).

The teratogenic effects of exogenous retinoid exposure are well characterized and depend on the stage at which the exposure occurs (Means and Gudas 1995). Abnormalities in the neural tube occur with early exposure, as evidenced by malformations of the head, heart, thymus, and central nervous system. In special cases, limb formation can be induced by exogenous retinoic acid. Deficiency of Vitamin A also causes teratogenicity, characterized in mammals by absence of eyes and by problems in the urinogenital tract, the heart, and the lung.

Important work on 2 model systems in fish and amphibians, the zebrafish (*Brachydanio rerio*) and the African clawed frog (*Xenopus laevis*), indicates that different forms of retinoic acid predominate in embryos rather than in adults (Kraft et al. 1994; Costaridis et al. 1996). The retinoid-receptor type, RAR versus RXR, that enables the action of the ligand differs in these 2 species (Minucci and Ozato1996; Minucci et al. 1996), indicating differences between species or classes in the exact identity of the molecules involved. In neurulating embryos of zebrafish, inhibition of retinoic acid synthase activity results in larvae with wavy notochords, shortened spinal cords, and deformed pectoral fins (Marsh-Armstrong et al. 1995); all end-points with potential for quantifying disruption of this signaling system. An important finding is the requirement for both thyroid hormone and retinoic acid receptors in mediating the effects of thyroid hormones on embryonic differentiation in *Xenopus* (Puzianowska-Kuznicka et al. 1997).

Organogenesis occurs during the embryonic stage and, in the case of indirect developers, continues in the larva. Hormones and growth factors play important roles in the control of differentiation and growth of tissues and organs in embryonic development of oviparous vertebrates. In this regard the key hormones that are of importance are thyroid hormones, adrenal steroids, retinoids, growth hormone (through its stimulation of the production of insulin-like growth factors), and a variety of other growth factors. Typically there are "windows" during which critical events involved in differentiation and maturation of tissues occurs. Thus, factors that disrupt the control of these events can have permanent effects on the later structure and/or function of those tissues and organs. For example, thyroid hormones are necessary for the differentiation of brain, muscle, and skeletal tissues in vertebrates (McNabb and King 1993). In the case of brain differentiation, the formation of cell layers and the establishment of synaptic connections in appropriate patterns are critical to the attainment of brain function (Bernal and Nunez 1995). These events typically occur within a narrow "window" in the development of an animal; the timing of that window depends on the tissue or organ involved and the species under consideration. Molecular studies that identify the specific genes activated by hormones are providing information about whether oviparous and non-oviparous vertebrates have common mechanisms of hormone effects on tissue development. See, for example, the work of Denver (1997) comparing thyroid hormone effects on gene expression in the brain of rats and *Xenopus*.

There is considerable evidence that several types of polyhalogenated aromatic pollutants alter a number of aspects of thyroid function in oviparous vertebrates and in mammals. Several different types of studies conducted on different species suggest the following theory, although it should be noted that much more research is needed to confirm the accuracy of this assumption for any individual species. Studies in mammals (primarily rats) have shown that exposure to TCDD and some PCBs is associated with decreases in plasma T_4 concentrations, decreased binding of T_4 to the transport protein transthyretin caused by competition for binding with the pollutant (Darnerud et al. 1996, Morse et al. 1996), and increased hepatic uridine 5'-diphophate-glucuronosyltransferase (UDP-GT), a microsomal enzyme involved in T_4 glucuronidation, a process which results in enhanced T_4 excretion (Barter and Klassen 1994). The inference presented in these studies suggests that alterations in T_4 transport in the circulation plus enhanced glucuronidation increase T_4 excretion from the body, thereby disturbing thyroid function. Increases in the UDP-GT microsomal enzymes are consistent with these pollutant effects being mediated through Ah receptors. The decreases in circulating T_4 have negative feedback effects on the hypothalamic–pituitary–thyroid axis resulting in compensatory increases in pituitary release of thyrotropin (Darnerud et al. 1996). An explanation of how some polyhalogenated aromatic chemicals may displace thyroid hormones from their binding proteins is indicated by the structural similarities of PCBs and TCDD to T_4 (McKinney et al. 1985). If this structural similarity also leads to binding of these pollutants to thyroid hormone receptors, this binding could have either agonistic or antagonistic effects on events mediated by thyroid receptors.

Studies in adult and developing rats indicate some compensatory or adaptive responses to the effects of these pollutants on thyroid function. Specifically, in rats exposed to PCBs, although circulating T_4 is decreased, the activity of 5'deiodinase that converts T_4 to T_3 in the brain is increased (Morse et al. 1993, 1996). This increase in enzyme activity is likely a response to the decrease in circulating hormones (Leonard and Visser 1986) rather than a direct response to the pollutants, and it may or may not return brain T_3 concentrations to control levels (Morse et al. 1993, 1996).

Studies of fish and birds from environments contaminated by these pollutants provide evidence of disruption of thyroid function and of developmental abnormalities that could be caused by thyroid disruption (reviews, Leatherland 1992, 1993; Fox 1993). Wild birds (herons, cormorants, and eagles) exposed environmentally to chemical contaminants and laboratory studies of chicken embryos exposed to TCDD show increases in brain asymmetry with increasing exposure (Henshel 1998). Recent studies of alligators (*Alligator mississippiensis*) in Florida suggest alterations in thyroid hormone concentrations and thyroid hormone to growth relationships in severely polluted lakes compared to a reference lake (Crain et al. 1998). Overall, if the above disruptions of thyroid function result in either transient or long-range disturbances in hormone supply, such changes can affect a variety of hormone-mediated developmental events.

In fishes, the skeletal, cardiovascular, and digestive systems all exhibit developmental errors known to result from disruption of the endocrine control of their formation during embryonic life. The section above on embryogenesis described how disruption of retinoic acid control of some aspects of zebrafish skeletal development is being used as a model system to study developmental errors. The formation of the cardiovascular system also is sensitive to chemical perturbation and to chemicals acting through the AhR, which cause a number of characteristic lesions. The differentiation of the circulatory system in the estuarine mummichog, *Fundulus heteroclitus*, as well as in other species, is known to be vulnerable to chemical disruption, as evidenced by pericardial edema and tail hemorrhages (Prince and Cooper 1995a, 1995b). A non-destructive bioassay of cytochrome P450 induction, mediated by the AhR, has been developed and is shown to reflect an increased incidence of lesions in the cardiovascular system (Nacci et al. 1998).

Errors in embryogenesis or organogenesis in fish-eating birds from pollutant laden regions have been observed in studies that have monitored herring gull populations in the Great Lakes region for many years. Congenital malformations also have been seen in a number of other fish-eating birds from the Great Lakes during the same period. Because such malformations are considered to be uncommon among wild birds, these observations have been considered to be sensitive biomarkers of pollutant exposure (review, Fox 1993). Studies of birds in their native environments, where they are exposed to a complex mixture of pollutants and resultant breakdown products, do not reveal which compounds are responsible for developmental defects. However, laboratory studies of specific compounds and their effects on development are revealing some of the picture. For example, when PCB-126 or TCDD (both known to be common contaminants in the Great Lakes) were injected into the yolk of chicken eggs, both these compounds caused embryonic mortality but only PCB exposure resulted in significant increases in developmental abnormalities in hatchlings (Powell et al. 1996). However, other studies have demonstrated tetratogenic effects of TCDD (on the cardiovascular system) in chicken embryos (Cheung et al. 1981).

Hatching

The environment of the egg and the nature of the barrier between the embryo and the environment differs among and within the classes of oviparous vertebrates. During the perihatch period, the organism is leaving a relatively closed and protected system for life in the larger environment. This transition requires the functional maturation of a number of organs and multiple homeostatic mechanisms and often is associated with increased mortality.

In fish, hatching is critical because larvae are very sensitive to environmental factors, including contaminants, during this process (Rosenthal and Alderdice 1976; VonWesternhagen 1988). Hatching is accomplished by the softening of the chorionic membrane by enzymes located on the surface of various parts of the body, particularly the head, yolk sac, or pharynx. The nature of these enzymes may differ between species, and there may be several different enzymes or allozymes involved within individuals of some species. They range in molecular weight from 8 to 25 kilodaltons for fishes and appear to be several times larger in amphibians. Initiation of hatching may be under neural and hormonal control, and there have been suggestions that epinephrine, corticosteroids, and prolactin are involved (see review by Yamagami 1988). Hatching is facilitated by the thrashing movements of the embryo. The enzymes involved are pH- and temperature-sensitive; hatching under unnatural acidities or temperatures can be problematic, resulting in disruption of the normal process and exposure to solubilized contaminants.

In birds, significant steps in the hatching process are internal pipping of the beak into the air cell, pipping through the shell, pulmonary inflation/respiration, and emergence from the shell. Physiological processes that must be functional or attain function at this time are the use of lungs for respiration, the capability for high metabolic activity and muscle coordination associated with external pipping and exiting of the egg, gastrointestinal function for posthatch food processing, and control system maturation for regulation of these processes (Gill 1994). Prior to hatching, the lung undergoes significant anatomical and physiological changes that permit effective pulmonary gas-exchange to replace the role played by the chorioallantoic circulation during embryonic life (Wittmann et al. 1983).

There are 2 avian developmental modes—precocial and altricial—and the timing and patterns of differentiation/maturation of a number of tissues and systems differ markedly at hatch between these modes, which are the ends of a continuum (for a classification of intermediate patterns see Starck and Ricklefs 1998). In precocial species, chicks are hatched in a relatively advanced state and most systems (sensory/nervous, skeletal, locomotor, gastrointestinal) are functionally mature. Thermoregulatory responses that require endothermic heat production and thermoregulatory control functions also appear during the perihatch period. In contrast, altricial species are hatched in a much less advanced state, and some organs and organ systems remain relatively immature for some period after hatch. In general altricial nestlings are incapable of coordinated locomotion, their eyes and ears are closed, and they are completely dependent on parental care for feeding, brooding, and protection. The gastrointestinal system is well developed and functional at hatch, but many sensory/nervous, skeletal, and muscle tissues undergo considerable differentiation/maturation during nestling life (Gill 1994). Much is known about the hormonal control of development of these systems in precocial birds, but knowledge of such control in altricial species is limited (review, McNabb et al. 1998).

The attainment of endothermy (i.e., metabolic heat production for temperature regulation) and homeothermy (i.e., maintenance of constant body temperature) occur at different times in the 2 modes of avian development. In precocial birds, endothermic responses to cooling begin at hatch, and chicks begin to thermoregulate and maintain homeothermy from hatching onward. In contrast, altricial nestlings are ectothermic (i.e., dependent on external heat) and poikilothermic (i.e., having body temperatures that vary with ambient temperature) for some period after hatching. The peak in thyroid hormones during the perihatch period appears to be associated with the initiation of thermoregulatory responses and the capability for metabolic heat production (endothermy) that is needed for thermoregulation (McNabb and Olson 1996; Visser 1998). In general, altricial nestlings with many immature tissues grow more rapidly than precocial chicks with relatively mature function in most tissues and high energy expenditures for thermoregulation (Konarzewski et al. 1998).

Many of the developmental events during the perihatch period are under endocrine control. For example, in precocial birds there are increases in circulating glucocorticoid hormones prior to hatching; circulating thyroid hormones increase during the latter third of embryonic life and peak dramatically during the perihatch period (Figure 3-6). Control axes (hypothalamic–pituitary–other-endocrine-gland axes) play important roles in many events during this time. Examples of hormones involved in critical events during the perihatch period are glucocorticoid and thyroid hormones, which are known to be necessary for lung maturation and the initiation of the surfactant production required for lung inflation. Effective pulmonary respiration during the period of internal pipping depends on this maturation of lung function. Thyroid hormones are also necessary for the maturation of intestinal function (review, McNabb et al. 1998). Growth hormone, IGFs, and a number of other growth factors are critical to general growth, differentiation, and maturation of many tissues (Scanes et al. 1996). General knowledge of thyroid hormone actions in tissue differentiation and maturation suggests that the increases in thyroid hormones during the latter part of incubation are critical to the maturation of other functional systems (sensory/nervous, skeletal, muscle/locomotory; McNabb and King 1993) in these relatively independent hatchlings.

Events during the perihatch period are used as measurement endpoints in avian reproductive toxicology primarily because of the vulnerability to disruption and the consequent increases in mortality at this time. Some commonly used observations include embryo viability, pipping, successful hatching, and morphological development. Disruption of any of these attributes is the primary observation considered to indicate pollutant effects in many toxicological studies (see for example, the review by Fox [1993], on Great Lakes birds). The mechanisms of developmental disruption by pollutants generally are not determined, and much research is needed to confirm the causality of these effects.

Initiation of feeding

This transition, during which oviparous young shift from lecithotrophy or matrotrophy to exotrophy (i.e., obtaining nutrition from the external environment), requires the functional maturation of a number of organ systems. In nature and in human rearing of oviparous organisms, the failure of many individuals to succeed in feeding themselves and in obtaining adequate nutrition from food reveals the dangers of this transition.

In precocial birds and in some fishes, there is sufficient yolk present in the yolk sac to meet nutritional needs for several days. In altricial birds, relatively little yolk remains at hatch (e.g., 8% of hatchling mass for 31 altricial species compared to 15% in 18 precocial species; review, Carey 1996). In some fishes no yolk remains at hatch (Wourms and Lombardi 1992), so the young are largely or entirely dependent on their own capability of obtaining food.

In most teleost fishes at the time of hatching, the gastrointestinal tract is a simple undifferentiated tube. At this time, digestion is intracellular (Watanabe 1982; Govoni et al. 1986; Kishida et al. 1998). During larval development, in most cases, the stomach differentiates and the intestine also undergoes further differentiation. Differentiation of the intestine is regulated by hormones; the role of thyroid hormones has received the most attention, along with some work on cortisol (Miwa et al. 1992; Brown and Kim 1995; Huang et al. 1998). The differentiation of the stomach is a vulnerable process with great importance to the survival and growth of the larval fish (Tanaka et al.1995; Tanaka, Kawai et al. 1996; Tanaka, Kaji et al. 1996). The rapidity of epithelial tissue remodeling, the direct contact of the intestinal epithelium with environmental medium, and the regulation of epithelial tissue differentiation by hormones and growth factors make this a potentially vulnerable transition and a possible marker of perturbation.

Some fish brood their young in their mouth. Both maternal and paternal mouthbrooding tilapia have been shown to exude VTG into their surface mucus, and the young are known to contain VTG in their digestive tracts (Kishida and Specker 1993, 1994; Kishida et al. 1998), suggesting a route for transgenerational transfer of chemicals in fishes even after hatching. In tilapia, the transfer of material from parent to hatched offspring includes immunoglobulins (Takemura and Takano 1997).

Among birds, precocial hatchlings seeking their own food are subject to potential predation and are vulnerable to cold during the early posthatch period when they are imperfect thermoregulaters. Altricial avian young remain in the nest for some period of time and are completely dependent on parental care. If there is disruption of parental behavior (e.g., inadequate incubation attentiveness and brooding), as has been observed in association with high pollutant loads in fish eating birds from the Great Lakes, and demonstrated to occur in doves fed organochlorine mixtures, altricial nestlings are extremely vulnerable during this time (McArthur et al. 1983; Fox 1993).

Fledging

Altricial avian nestlings, after some period of remaining in a relatively immature state, go through a rapid period of maturation of many organs and systems, develop plumage, and gain the capability for flight from the nest. This transition involves leaving the relatively sheltered and protected nest and the care of the parents for the much harsher, larger environment. Young birds practice flight movements before leaving the nest. There may be some parental care for a brief period after fledging. However, this is a period of high mortality in many cases because developmental, physiological, and behavioral inadequacies are likely to make the fledglings especially vulnerable to environmental conditions or predation (review, Gill 1994).

Metamorphosis

Metamorphosis is an irreversible set of processes that results in changes in body form and physiology accompanied by a habitat shift (see Youson 1988). Among vertebrates, true metamorphosis occurs only in amphibians (most frogs) and fishes. The process of metamorphosis has rich potential for detecting chemical perturbation of development for the following reasons:
- the process is regulated by hormones,
- there is a robust ecological literature on the topic,

- the changes in body form are easy to quantify,
- the larval forms are aquatic and readily exposed to chemical agents, and
- processes at this stage are relatively independent of maternally derived chemical influences. In addition, the effects of chemical perturbation on morphology can be studied in culture.

Metamorphosis in amphibians (Kaltenbach 1996) and flounder (Miwa and Inui 1987) is triggered by increased thyroid hormone concentrations (Figure 3-5) and studies of thyroid endocrinology on this developmental transition date back to early this century. Steroid hormones including gonadal steroids (Hayes 1997) play important roles, with corticosteroids having received the most attention recently (de Jesus et al. 1990; Denver 1996; Rollins-Smith and Cohen 1996; Hayes 1997). The rate of metamorphosis can be chemically regulated; addition of exogenous thyroid hormone accelerates the rate, whereas the addition of goitrogenic chemical agents that interfere with thyroid hormone synthesis or metabolism can delay or prevent metamorphosis (the patterns of circulating thyroid hormones are shown in Figure 3-5). In an upcoming section, the physiology of metamorphosis of frogs and fishes will be reviewed with focus on model experimental species. The regulation of metamorphosis in eels and lampreys is less well understood (see Youson 1997).

During amphibian metamorphosis almost every tissue is remodeled. The most obvious changes are the emergence of limbs and the disappearance of the tail. Because tadpoles are herbivorous and frogs are carnivorous, the structure and function of the gastrointestinal system changes radically. With movement onto land, respiration switches from the gills and skin to the lungs. The molecular endocrinology of thyroid hormone function in amphibians is known in some detail, especially for *Xenopus*. The spatial and temporal expression of thyroid hormone receptors and specific responses to thyroid hormones have been described (Tata 1996; Shi 1996), lending this process to study of chemical perturbances on development at the molecular, biochemical, and cellular levels. There are diverse responses to thyroid hormones during amphibian metamorphosis (Tata 1996). Included among these responses are quantifiable changes in gene expression, new synthesis of both structural proteins and enzymes, programmed cell death, and appearance of new components of the immune system.

External and internal factors regulate metamorphosis and thyroid status during metamorphosis. Temperature and photoperiod influence the rate of metamorphosis (Kaltenbach 1996), as does food availability (Leips and Travis 1994). The timing and nature of neuroendocrine control over thyroid hormone production is not well understood, and what is known is perhaps unexpected. The hypothalamic, neuro-peptide-exerting influence over the pituitary thyrotropes is corticotropin-releasing factor (CRF) (Denver 1993). This may suggest that the effect of density on rate of metamorphosis is mediated through the neuroendocrine system, implying that disruption of chemical signaling could affect rate of metamorphosis through several endocrine pathways.

Flatfish metamorphosis has many parallels with amphibian metamorphosis. A key distinction is the prevalence of amphibians in freshwater systems and of flatfishes in marine and estuarine systems, although there are exceptions to both. Endocrine correlates and dependence on thyroid hormones of flatfish metamorphosis were determined in the last decade (de Jesus et al. 1993; Tanaka et al. 1995; Schreiber and Specker 1998). Water-soluble hydrocarbons from an oil spill have been shown to affect hormones known to play a role in flatfish metamorphosis (Stephens et al. 1997); however, impacts of environmental chemicals on rate of metamorphosis is an area which warrants consideration. Flatfish metamorphosis, like amphibian metamorphosis, lends itself to study because the change in body form is recognizable without technical skill. Thyroid hormone induction of specific proteins in muscles, of gastric glands and the stomach, and of immune system cells has been described (Inui et al. 1995; Huang et al. 1998).

Migration and habitat shifts
Some species of oviparous vertebrates migrate at particular stages during their life spans, while some migrate annually. Often the migration is to a new habitat where there is greater potential for energy acquisition for breeding activities and rearing of young. Preparation for migration involves physiological changes that are vulnerable to chemical perturbation and typically are very demanding energetically.

Among fishes, the parr-smolt transformation of anadromous salmonids has been well studied with respect to both physiological processes and their ecological significance (Hoar 1988). Regulation of the preparation for and migration to the sea includes photoperiodic cues (the principal environmental component) as well as many physiological signals from the endocrine system. Physiological processes that change with migration and are vulnerable to perturbation are osmoregulatory processes, metabolic processes including changes in body composition, lipid metabolism, gill osmoregulatory capabilities, and neurosensory functions critical for migration and homing. Migration that involves homing requires not only sensory (often olfactory) recognition of salient cues to use as guideposts for orientation, but also the ability to imprint these cues as juveniles for recall years later as adults. Hence pollutant disruption of sensory, memory, or innate homing mechanisms can reduce fitness (Schreck 1982b).

Migrant birds, which are homeothermic and have very high energy demands for flight, require very large amounts of energy for long distance migrations. Although some birds stop to feed during migration, other species fly nonstop for the entire distance of the migration. Energy storage is in the form of lipid and in nonstop migrants is in proportion to the demands of the migration distance (Hill and Wyse 1989). Birds that are long distance migrants may store as much as half their body weight as lipid in preparation for migration (Tucker 1971; review, Biebach 1996).

Any pollutant material that interferes with the hormonal control of migratory fat deposition, or with other aspects of the readiness for migration, can decrease the chances of migrant survival. Depending on the pollutant concentrations in the environment where migratory preparation occurs, these stored lipids may contain a large load of lipophilic pollutant chemicals, which will be mobilized within the body as lipid fuels are used during the migration.

Sexual maturity

The capability for sexual reproduction is known to be profoundly affected by many chemical pollutants that act as agonists or antagonists for reproductive steroids and alter embryogenesis or organogenesis in the reproductive organs of developing vertebrates. Such effects have been well documented in studies of both laboratory animals and wildlife (McLachlan 1985; Crain and Guillette 1997; Tyler, Jolding et al. 1998). Age at reproductive maturity has profound impacts on population structure and growth. Timing of this transition is influenced by rate of growth and accumulation of energy reserves in the juvenile stage; many species must reach a certain body weight and proportion of body fat before females mature sexually and become capable of reproduction (for studies of domestic birds, see Zelenka et al. 1987). Nutrient availability in the environment and nutrient stores in females are also important in the initiation and patterns of breeding of wild birds (review, Carey 1996). When chemical contaminants perturb growth rate, this in turn will influence the onset of sexual maturity. Bioenergetics and metabolism come into play at this transition because of the switch from somatic to gonadal growth, especially in the female.

Senescence

Senescence appears to be tightly coupled with reproductive development. Aside from animals that reproduce only once, little is known about what determines when an individual goes into the process of recrudescence or senescence following reproduction. Knowledge of the physiology of senescence in vertebrates (see early review by Comfort 1979) could be of value for environmental assessment. It is important to establish, when monitoring animals in the wild, if the physiological phenotype observed is natural (i.e., the organism is dying due to "old age") or is due to disruption of homeostatic mechanisms by toxicants. For oviparous animals, extensive information on the physiology of senescence exists for Pacific salmon, *Oncorhynchus* spp. (Idler et al. 1959; Schmidt and Idler 1962; Idler and Truscott 1963; van Overbeeke and McBride 1967; Donaldson and Fagerlund 1972), and there is some information on cod, *Gadus morhua* (Idler and Freeman 1965), lamprey, *Petromyzon* (Larsen 1980), and other poikilotherms (Schreibman 1987).

Growth and energetics

Overall growth of the organism is an important aspect of development that has been assessed in many studies of chemical pollutants. Overall growth of the organism reflects both cell differentiation/maturation processes and cell proliferation and hypertrophy, so growth cannot simply be separated from embryogenesis and organogenesis. However, our emphasis in this section is on the continued somatic growth that increases body mass and on the energetic considerations for that growth. There is rich physiological literature on growth rates, growth processes, and their control in vertebrates that provides important frames of reference for the use of growth data in assessment practice (Weatherley 1972; Schreibman et al. 1993).

Bioenergetics

Somatic growth is dependent on energy availability and processing efficiency. The ability to transform exogenous energy sources into growth obviously depends on the nature of the food consumed. It is also dependent on the energetic costs associated with maintenance functions, i.e., the costs associated with carrying on life-maintaining processes in the resting animal (Warren 1971; Burggren and Roberts 1991; Hochachka 1991; Weathers 1996). This energetic cost is referred to as "basal metabolic rate" in humans and "standard metabolic rate" (SMR) in all other animals that cannot be monitored in a truly inactive state. "Specific dynamic energy" is an important component of energy costs and is the nonutilized energy released as heat resulting from amino acid deaminations and other metabolic processing following food intake (i.e., it is energy not available to do physiological work). In fish, key, high energy demanding physiological processes include respiration and osmoregulation; these can each account for 25 to 40% of the entire energy budget of the organism (Jobling 1993).

Energy available to the organism above SMR is divisible into activities needed to carry on all other life functions such as food acquisition and processing, general activity, and pathogen resistance, with the balance of energy being available for growth and development (Warren 1971). This "active metabolic rate" is critical to growth, development, and reproduction, all being high-energy-demanding processes (Schreck 1982a). Overall energy budgets of wild animals are difficult to construct, but such budgets should be very useful in assessing the energetic restrictions on animals and ways in which such restrictions may interact with other stressors, such as chemical pollutants. For examples of energy budgets for birds, see Dawson and O'Connor 1996; Wiens and Farmer 1996.

Ectothermy/endothermy

An important difference among the oviparous vertebrate classes that can affect overall growth rates is that at some time during development, birds become endothermic (homeothermic) and expend large amounts of energy on temperature regulation, while the other 3 vertebrate classes are ectothermic (poikilothermic) throughout their life spans. The energy demands of producing endothermic heat to maintain homeothermy are very high; the advantage is that the animal gains independence of environmental temperature for the "price" of energy expenditure on heat production. Thus, endothermic vertebrates can remain active at low environmental temperatures but ectotherms cannot. It should be noted that under some conditions, some ectothermic vertebrates may maintain relatively high and constant body temperatures (e.g., "warm" fish like tuna and large lizards that behaviorally manipulate ectothermic heat) (Withers 1992; Schmidt-Nielsen 1997). For purposes of this discussion, these cases can be ignored because they typically involve manipulation and conservation of environmental (i.e., ectothermic) heat rather than metabolic energy use for endothermic heat production.

In birds, precocial species are capable of endothermic responses to cold from hatching onward. However, their ability to maintain a constant body temperature, and to do so at body temperatures characteristic of adults, is perfected during the early posthatching period. In contrast, altricial birds are ectothermic (i.e., do not respond metabolically to cold) at hatch and during early posthatching life, then develop homeothermy rapidly, often just prior to fledging from the nest (McNabb and Olson 1996). In many species, posthatching growth rates appear to be strongly influenced by the timing of development of homeothermy. Thus, precocial young, which are partitioning their energy between growth and endothermy, typically have lower growth rates than do altricial young that are ectothermic (McNabb et al. 1998).

A number of the effects of chemical pollutants that have been mentioned in the sections above could disrupt the development of endothermy and thermoregulatory control, although this subject has received little attention. For example, disruption of endothermic development in birds could result from alterations in thyroid hormones, which are involved in the development of endothermy and the differentiation and maturation of a number of tissue and organ types (McNabb and King 1993; McNabb and Olson 1996). Alterations in thyroid function of birds exposed to a number of chemical pollutions are known and disruption of heat production has been shown to result from thyroid effects of TCDD in adult rats (Potter et al. 1983) and of PCBs in mourning doves (Tori and Meyer 1981). Likewise, any alteration in growth-controlling hormones which are needed for the attainment of a body size adequate for a favorable surface-to-volume ratio could affect endothermic development. Effects on parental behavior, such as the decrease in parental attentiveness that has been observed in pollutant exposed birds from the Great Lakes, (review, Fox 1993) also could be important because it may lead to increased mortality in nestlings not capable of thermoregulation.

Lipid stores and utilization
Oviparous vertebrates have 2 phases of lipid metabolism during their development. Sheridan and Kao (1998) have characterized these as an initial lipogenic and lipid storage phase, followed by a lipolytic or mobilization phase. They have reviewed the literature on lamprey and anuran amphibian metamorphosis and salmonid smoltification to describe the general picture about lipid storage and subsequent lipid mobilization and utilization in the context of development and growth. This

sequence of events is under endocrine control by hormones that regulate the supply of energy for metabolism, development, and growth (Sheridan and Kao 1998). Chemical pollutants that disrupt the function of any of these endocrine components could interfere with growth by altering energy supply. Secondly, many chemical pollutants previously shown to interfere with development are lipid-soluble and stored in body fat. Thus, the pattern of lipid storage and mobilization will play an important role in the timing of exposure of the organism to such pollutants.

Homeostatic systems

Adaptational physiology

Physiological systems in animals have evolved in a background of environmental change (Calow and Sibly 1990). These changes have led to evolutionary adaptations in anatomy and physiology that are fixed in the genetics of the species (Barber et al. 1990; Diamond et al. 1991; Kramer et al. 1992), as well as acclimatory adaptations that are dynamic adjustments to change, the limitations of which also are fixed by genetics. Acclimatory adaptations occur in a time scale of weeks and provide considerable ability to adjust to environmental changes that impinge on the function of the animal (Precht 1958; Prosser 1986). Most physiological systems also possess redundancy, i.e., in many cases there are multiple ways in which an animal can adjust to both internal and external change. For example, if blood glucose decreases, there are 4 different types of hormones (glucagon, adrenaline, glucocorti- coids, growth hormone) that can increase blood glucose (Hadley 1996).

Vertebrate animals maintain chemical homeostasis in their blood and body fluids. This is a dynamic regulation by many physiological processes that maintain the constancy of internal chemical composition, thereby providing stable conditions for enzymatic, electrochemical, and other chemical processes in the body (Withers 1992; Schmidt-Nielsen 1997). With respect to temperature, only birds maintain temperature homeostasis or constancy; the other oviparous vertebrate classes typically have body temperatures that vary with the environment, i.e., they are poikilothermic and ectothermic.

Physiological adaptation to environmental stressors is mediated by neural and endocrine responses. Whether and how the animal is able to adapt to the stressor depends on a number of factors (e.g., exposure time, nature of the stressor, concen- tration of the stressor if it is chemical in nature, etc.). Assuming the stressor is not lethal, if the exposure is short-range, the organism is likely to make transient physiological adjustments that may or may not have lasting effects. If there is sustained exposure, the animal may be able to regain homeostatic control, i.e., it may achieve an adapted state in which there is some physiological compensation for the effects of the stressor. This new, adapted state may or may not be similar to that prior to exposure to the stressor (Precht 1958) and it may be a different energetic state because of the metabolic cost of dealing with the stressor. Some toxicological tests capitalize on adaptive responses (e.g., increases in liver enzymes involved in detoxification) as biomarkers for assessing pollutant exposure. It should be noted that continued exposure to a stressor or additional stressors added to the original situation may exceed the capacity of the organism to make adaptive responses, leading to irreversible impairment (Fox 1993; Heath 1995).

A key issue in considering adaptive physiological changes and chemical pollutants is at what point (or at what pollutant concentration) there is a shift from an adaptive response to a condition where the animal cannot cope physiologically with the pollutant. Thus, with some pollutants, at some concentrations, the animal's response may represent an adaptation that is not stressful but may be energetically expensive. However, as exposure continues in time there is a shift to a stressful condition that is damaging to the animal. Physiologists and endocrinologists often assess the stress of animals by measurements of circulating adrenal corticosteroid hormones that are released in response to stress.

Neurosensory function and behavior
In some cases the effects of chemical pollutants on animals are assessed by behavioral endpoints. For example, breeding behavior and parental care of nestlings have been shown to be altered in herring gulls exposed to high levels of pesticide mixtures in the Great Lakes (Peakell and Fox 1987; Fox 1993). Confirmation that the types of chemical pollutant mixtures found in the Great Lakes can cause such effects was obtained by experimental feeding of pollutant mixtures to doves (McArthur et al. 1983). What is unknown is the mechanism whereby these behavioral problems arise. However, it seems reasonable to assume that they are mediated through endocrine/central nervous interactions and could result from disruption at a number of possible points. For example, both thyroid and reproductive steroid hormone abnormalities are known to cause developmental and behavioral problems. There is a need for research that reveals more about the mechanisms of pollutant action that have behavioral consequences. In addition, the possibility that some systems may adapt to pollutant exposure needs to be considered. For example, avoidance of elevated metal concentrations by fathead minnows is reduced by previous exposure to those metals. This suggests that sensory system adaptation is responsible for this change (Hartwell et al. 1987).

Health
The general health of animals is important in determining their physiological capabilities and their vulnerability to environmental insults. Health is affected by both infectious agents and non-infectious agents.

Non-infectious determinants of health include physical habitat quality variables as well as nutrition (Adams 1990). Organisms living under environmental conditions that place them near their physiological tolerance limits are very different physiologically than those living near their preferred, more optimal conditions. Both the availability and quality of food affect general health. While important in limiting physiological performance, it is beyond the scope of this chapter to review how general nutrition affects vulnerability to contaminants; however, lipids and yolk are discussed in relation to their importance in the processes of maternal transfer and oogenesis.

Pathogenic organisms obviously affect physiological function directly as well as by inducing a general physiological stress response. Underlying infection by micro- and macroparasites can affect the capabilities of an animal to adapt to environmental change such as introduction to chemical pollutants. The degree of infection and the presence or absence of disease may differ between populations of the same species in nature, even within close geographic proximity or between generations at the same developmental stage (Currens 1997; Currens et al. 1997).

Stress

When vertebrates experience stress, neuroendocrine mechanisms are activated that help resist or adapt to the insult and then facilitate, if possible, recovery (Selye 1936, 1976; Moberg 1985; for reviews on fishes see Iwama et al. 1997). In fishes, the general stress response is elicited by many contaminants (Thomas 1990a). Briefly, the stress response entails the production and release of catecholamines and corticosteroids which result in osmoregulatory, energetic, immunological, and behavioral changes. Ways in which stress hormones are involved in physiological responses and affect population level responses has been outlined by Fausch et al. (1990), Shuter (1990), and Bartell (1990). These physiological responses cause shifts in energy compartmentalization (Schreck 1982a; Schreck and Li 1991) and retard or inhibit growth, development, and reproduction (Schreck 1993). While general aspects of this stress response are common throughout the vertebrates, there are some aspects that are unique to oviparous vertebrates. Specifically, egg quality or numbers may be affected by stress in the female. This is best understood with regard to alterations in yolk quality, yolk energy, or egg numbers in fishes (see reviews in Bromage and Roberts 1995). Circulating concentrations of free fatty acids can be decreased and the concentration and nature of blood proteins can be affected by stress. The concentration of sex hormones, particularly the androgens (Safford and Thomas 1987; Pickering et al. 1987; Carragher et al. 1989) in adult fishes of both sexes can be altered by stressors. There certainly is the possibility that substances such as stress-induced steroid hormones or neuropeptides are transferred to the oocyte, thereby affecting subsequent development of progeny (Schreck et al. 1991).

Stress can retard or inhibit growth and sometimes development in all vertebrate classes (Schreck 1993). However, the opposite effect, namely enhanced growth rates and enhanced indications of health, also can result from stress due to pollutant chemicals. Such overcompensation following the stress of pollutants is referred to as hormesis (Heath 1995). Stress can either accelerate or slow the timing of metamorphosis. Amphibians may undergo premature metamorphosis under conditions of crowding, reduced food, and decreased growth rate (Leips and Travis 1994). This acceleration of metamorphosis is consistent with the demonstration that cortisol interacts synergistically with thyroid hormones in larval flatfish and amphibians in triggering metamorphic changes (deJesus et al. 1993; Hayes 1995, 1997). In contrast, it is known that stress can retard or prevent the transition from parr to smolt in anadromous salmonids (Schreck 1982b; Schreck et al. 1985; Patiño et al. 1986). Therefore, environmental toxicants that either elicit a physiological stress response, or are agonists or antagonists of the stress response factors, could affect development and major organizational shifts in the body. In addition, multiple exposures to stressors have been shown to have cumulative effects in juvenile salmon, and diseased fish were more severely affected than healthy ones (Barton et al. 1986). It also should be noted that stress in fish can, and does, affect gamete quality— defined as the ability to produce viable offspring—in rainbow trout (Campbell et al. 1994).

Immune function

Development of the immune system of oviparous vertebrates has received increased attention in recent years. Oocytes have antibodies provided by maternal sources, thereby supplying the embryo with some degree of immune competence. The time at which specific components of the immune system form during early development varies widely between species and between higher taxa (Tatner 1996). There are differences in the ontogeny of different parts of the immune system, such as the major histocompatibility complex (Slater-Cid and Flajnik 1995) and T-cells (Rollins-Smith et al. 1992), during metamorphosis in anuran amphibians and development in birds (LeDouarin 1991). Suppression of T-cell-mediated immune function (by 30 to 45%) has been observed in prefledglings of 2 species of birds that eat fish contaminated with organochlorine pollutants in the Great Lakes (Grasman et al. 1996). The responsiveness of lymphocytes to suppression by corticosteroids is depressed during the period of parr-smolt transformation in salmonid fishes (Maule et al. 1993). In addition, sex steroids such as the androgens are extremely potent suppressers of antibody production (Slater and Schreck 1993). In fishes, both corticosteroids (Maule and Schreck 1991) and androgens (Slater et al. 1995a, 1995b) operate through receptor-mediated effects on lymphocytes. Together, both these classes of steroid hormones can have additive effects on immunosuppression, corticoids apparently operating by depression of a lymphokine such as an interleukin (Tripp et al. 1987) and androgens potentially operating by causing cell death (Slater and Schreck 1997). Thus, the immune system is vulnerable to perturbation during metamorphic, transitional, and reproductive stages (Schreck 1996).

Regulation of cellular and humoral immunity is a function of the interactions of the immune and endocrine systems as well as underlying effects from prior exposure to pathogens or other antigens. Studies in mammals illustrate the broad and interactive effects of chemical pollutants. For example, 3 agricultural chemicals (the herbicide metribuzin and 2 insecticides, aldicarb and methomyl) in combination have been shown to change immune function, impair certain neurological functions, and alter plasma concentrations of metabolism and growth-controlling hormones (Porter et al. 1993). While less is known about oviparous vertebrates than about mammals, our review of the literature suggests that virtually all hormones that have been examined in mammalian systems affect the immune system in some way. It is clear that numerous substances, including metals, aromatic hydrocarbons (including polychlorinated biphenyls), and pesticides, can affect the immune system of these vertebrates (Anderson 1996).

Reproduction

Control of reproduction

The complex events that occur during the reproductive cycles of oviparous vertebrates are under precise physiological regulation. Hormones and other chemical messengers secreted by the hypothalamic–pituitary–gonad axis control the timing and sequence of reproductive processes. Basic features of the reproductive endocrine system are highly conserved among the major classes of oviparous vertebrates and are shown in Figure 3-9. However, there can be differences in the timing of hormonal events between and within species because of the extreme variation in reproductive life histories.

Seasonal changes in photoperiod, temperature and other environmental factors and physiological changes such as nutritional state and social stimuli are detected by peripheral and internal sensory systems; the information is relayed by the nervous system to the endocrine system. Neurons of the brain regions where information is integrated release specific neurotransmitters and neuropeptides that influence the activities of hypothalamic neurons in the medial basal hypothalamus and preoptic region, which synthesizes gonadotropin-releasing hormone (GnRH), the primary neurohormone that controls reproduction. The neuronal system controlling the activity of GnRH neurons is highly complex, involving multiple pathways and neurotransmitters, and is not well understood for most oviparous vertebrates species. Several teleost species have 3 forms of GnRH (Powell et al. 1994), while most vertebrates have at least 2 forms of GnRH; a great deal of information is emerging about the significance of these forms and their receptor binding in the evolution of different vertebrate classes (Sherwood et al. 1993; King and Millar 1995; Millar et al. 1997).

In tetrapods, the axons of GnRH neurons and other hypothalamic neurons containing other releasing hormones terminate in the median eminence, a specialized neurohemal organ at the base of the hypothalamus. GnRH is released from axon terminals in the median eminence into the blood and is transported the short distance to the anterior pituitary by the hypothalamo-hypophyseal portal system.

SENSE ORGANS

BRAIN

LIVER

FEEDBACK

Neurotransmitters

Steroids

GnRH

Vitellogenin

GTH I & II
or
FSH & LH

PITUITARY

OVARY

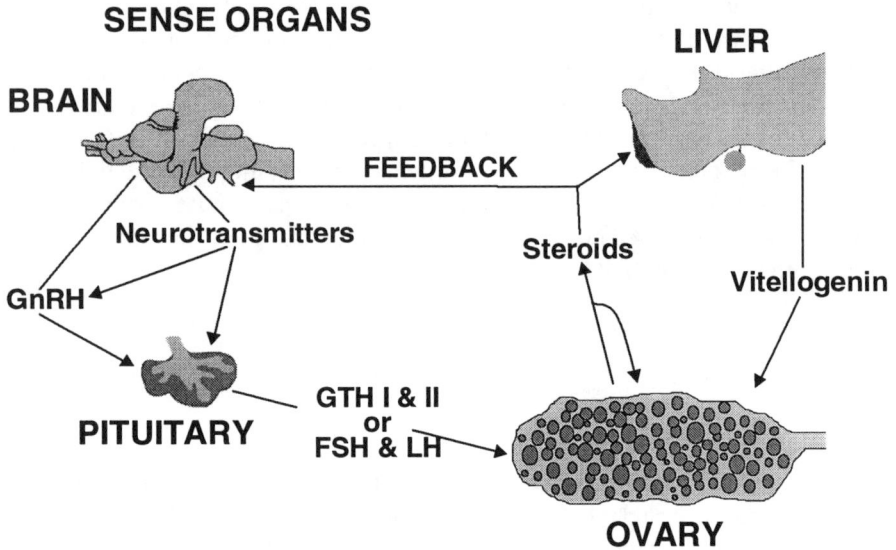

Figure 3-9 Potential sites of chemical interference with reproductive endocrine function in female oviparous vertebrates. Redrawn from "Teleost model for studying the effects of chemicals on female reproductive endocrine function," PT Thomas, *J Experimental Zool Suppl* 4:126–128, ©1990 Wiley-Liss, Inc. Adapted by permission of Wiley-Liss, Inc., a subsidary of John Wiley & Sons, Inc.

An extensive capillary system in the anterior pituitary gland delivers the GnRH to the gonadotropin-producing cells, the gonadotropes (Gorbman et al. 1983; Norris 1997). Teleosts lack a hypothalamo-hypophyseal portal system. Instead, the neurons containing the releasing hormones have long axons that extend down the pituitary stalk and release their contents in the vicinity of the pituitary hormone producing cells. GnRH binds to a specific receptor on the plasma membrane of gonadotropes and activates second messenger systems that regulate the synthesis and release of gonadotropins (Chang and Jobin 1994). Other neuropeptides and neurotransmitters also have been shown to influence gonadotropin secretion in a variety of oviparous species by exerting direct effects on the gonadotropes. For example, dopamine exerts a direct inhibitory influence on gonadotropin secretion in cyprinid fishes. Other pituitary hormones such as prolactin, growth hormone, thyroid stimulating hormone, and adrenocorticotropic hormone have been shown to have important reproductive functions at particular stages of the reproductive cycles in some oviparous vertebrates (reviewed in Norris and Jones 1987; Harvey et al. 1995).

It is widely accepted that the reproductive cycles of all major classes of vertebrates are under dual gonadotropin (GTH) control by FSH and LH in tetrapods and their equivalents, GTH-I and GTH-II, in teleosts. Recent data suggest that in salmonid fish GTH-I is FSH, based on both chemical and functional evidence (Prat et al. 1996; Tyler et al. 1997). The patterns of gonadotropin secretion during the reproductive cycle show marked differences among oviparous vertebrates (see review in Lamming

1984); therefore, few generalizations on the functions of the 2 gonadotropins can be made across all the vertebrate classes. Although the early stages of the reproductive cycle, which are characterized by growth and development of the gonads and of developing follicles in ovaries, are predominantly controlled by FSH or GTH-I in many vertebrate species, GTH I has not been found in all teleost species (Schultz et al. 1995), and GTH-II may have important functions during this stage of the reproductive cycle in some teleost species (Kobayashi et al. 1997). Final stages of the reproductive cycle, including final gamete maturation, ovulation, and spermiation are controlled by LH or the teleostean equivalent, GTH-II (Scott and Sumpter 1983; Prat et al. 1996).

Gonadotropins act in the gonads primarily to influence steroid hormone production and gametogenesis. Gonadotropins bind to specific receptors on the plasma membranes of somatic cells in the gonad, resulting in increases in cyclic-AMP and other second messenger systems, which regulate the production and secretion of steroid hormones, growth factors, inhibin, activin and other regulatory peptides (Van Der Kraak and Wade 1994). In females, the theca and granulosa cells of ovarian follicles are the sites of steroidogenesis, whereas in males steroid hormone production in the testis is confined to the Leydig cells (Hadley 1996) or interstitial cells in most species. During early stages of the gonadal cycle, gonadotropins stimulate the production of estrogens in females and androgens in males. In males FSH acts on Sertoli cells (which support the functions of spermatogenic cells), to initiate spermatogenesis and enhance the synthesis of androgen binding protein; LH regulates steroid production by Leydig cells. In females, both the granulosa and thecal cells of the ovarian follicle are involved in steroid production in some fish such as salmonids (2 cell model, Young et al. 1986). The sex steroids exert paracrine actions within the gonads to influence gametogenesis and also act outside the gonads to stimulate the development of accessory sexual structures (e.g., the oviduct) and secondary sexual structures (e.g., male plumage in birds). Regulation of the hypothalamic–pituitary–gonadal axis is provided by negative and positive feedback effects of the sex steroids and other gonadal hormones such as inhibin and activin. Feedback effects on gonadotropin secretion result from hormone binding to specific receptors in the pituitary and/or hypothalamus and possibly extrahypothalamic areas of the brain. Aromatization of androgens to estrogens appears to be necessary for androgens to exert their effects on gonadotropin secretion in many oviparous species, and high levels of aromatase enzyme have been found in the hypothalamus in several species (Callard et al. 1990). However, in several oviparous species, the mechanism by which estrogens affect hypothalamic neurons is indirect. The receptors for estrogens are not located in the hypothalamic GnRH neurons, but are present in adjacent neurons that release neurotransmitters such as catecholamines and gamma-amino-butyric acid, which in turn trigger GnRH secretion. Sex steroid hormone receptors also are present in other parts of the brain and are intermediaries in androgen and estrogen effects on reproductive and other behaviors (Norris 1997; Ottinger and Abdelnabi 1997).

In oviparous vertebrates the liver is a major site of estrogen action for stimulating the production of egg components (Tata 1986). At the beginning of the ovarian cycle, an increase in circulating estrogen up-regulates the hepatic estrogen receptor and a variety of estrogen responsive genes, including those that produce important egg proteins such as vitellogenin (the yolk percursor lipoprotein; Specker and Sullivan 1994), egg envelope proteins, and albumin. Metabolic hormones (such as growth hormone), are also involved in the dramatic changes in the partitioning of energy reserves and lipid utilization which are required for vitellogenin production, but their precise roles are poorly understood. Vitellogenin is transported in the circulation to the ovary, it binds to specific receptors on the oocyte plasma membrane, and it is sequestered by the growing oocytes (Perry et al. 1984; Barber et al. 1991; Tyler and Lancaster 1993; Elkin et al. 1995; Prat et al. 1998).

Toward the end of the reproductive cycle, when the gametes are fully grown, a surge in LH or GTH-II secretion stimulates the final maturation of the gametes, ovulation, and spermiation. In female birds and reptiles, LH acts directly to trigger final oocyte maturation, whereas in amphibians and fish, gonadotropins act by stimulating the synthesis of maturation-inducing steroids (MIS) (Nagahama et al. 1994; Thomas 1994). Increased production of MISs also occurs in male teleosts and is associated with maturation of sperm and the production of milt in which the sperm is contained. In oocytes, the actions listed above are triggered by progestin receptors that are on the plasma membrane (Nagahama et al. 1994). In teleost sperm, a receptor for the MIS also has been identified on the plasma membrane. This receptor is thought to be involved in sperm activation, as has been demonstrated for mammalian sperm (Thomas et al. 1997). The MISs and their conjugated metabolites are excreted by female fish during the periovulatory period and act as priming pheromones in males of several species to stimulate milt production (Stacey and Caldwell 1995). In addition to plasma membrane receptors, nuclear progestin receptors also are present in the ovaries of oviparous vertebrates and are believed to be involved in the control of prostaglandin synthesis. Prostaglandins are important mediators of ovulation and also may act as pheromones in stimulating courting behavior in some fish species (Stacey and Caldwell 1995).

The complexity of reproductive function in oviparous vertebrates, involving the precise physiological coordination of a wide variety of reproductive and endocrine processes by diverse tissues, makes it particularly vulnerable to interference by pollutant chemicals and other environmental factors. The integrative nature of the reproductive system complicates investigations of the primary sites and mechanisms of chemical interference on the hypothalamic–pituitary–gonad axis because changes in the activity of one level of the axis will ultimately affect the activities of other levels (Thomas 1990b). Thus, the consequences of endocrine disruption may be extensive and may involve many tissues and functions. Hormones act at more sites than just their primary target tissues, further complicating the interpretation of results. For example, estrogens and estrogen receptors are present in the testis and androgens, and androgen receptors are present in ovaries and presumably have important physiological functions in those locations, although such functions remain unclear (Sharpe 1997; Hess et al. 1997).

Pollutant chemicals potentially can disrupt endocrine function by altering the secretion, action, or metabolism of hormones. There is evidence for all 3 of these types of endocrine disruption. In birds and fish, exposure to polycyclic aromatic hydrocarbons, organochlorines, and heavy metals can lead to alterations in plasma gonadotropin levels in vivo or alterations in gonadotrope secretion in vitro (Cavanaugh and Holmes 1987; Thomas 1989; Van Der Kraak et al. 1992; Richie and Peterle 1979). These results suggest that the neuroendocrine axis is an important site of chemical pollutant interference with reproductive function. Gonadotropin secretion is triggered by GnRH binding to plasma membrane receptors and a calcium-dependent signal transduction system, which is disrupted in gonadotropes in vitro by heavy metals such as cadmium (Thomas 1993; Thomas and Khan 1997). Other heavy metals, which alter calcium homeostasis (see review by Viarengo and Nicotera 1991), may have a similar mechanism of action. Chemicals that impair hypothalamic neurotransmitter function can interfere with the neuroendocrine control of gonadotropin secretion. Long term exposure of Atlantic croaker to the PCB mixture Aroclor 1254, or to lead, causes decreases in hypothalamic concentrations of serotonin (a neurotransmitter which stimulates gonadotropin secretion), and a decline in gonadotropin secretion and reproductive impairment (Khan and Thomas 1996; Thomas and Khan 1997). Finally, estrogenic xenobiotics such as *o,p'*-DDT also can influence gonadotropin levels by interfering with the steroid negative feedback control at the hypothalamus and pituitary (Khan and Thomas 1998).

Chemical pollutant effects on gonadal steroid secretion have been demonstrated using in vitro steroidogenesis bioassays. Direct stimulatory effects of cadmium on steroidogenesis have been shown in experiments in which the metal was incubated with ovarian fragments in vitro (Thomas and Khan 1997). Interestingly, the increase in steroidogenesis was accompanied by elevations in ovarian cyclic-AMP levels, suggesting an effect on a mechanism mediated by plasma membrane receptors. In the majority of studies, however, steroidogenesis or the activity of steroidogenic enzymes in vitro have been assessed after in vivo exposure to chemicals (Crain et al. 1997; Sangalang and O'Halloran 1973; Thomas 1989; Van Der Kraak et al. 1992). With in vivo exposure, alterations in steroidogenesis could be mediated through disruption of hormones higher in the hypothalamic–pituitary–gonadal axis, i.e., they could reflect secondary effects. Thus, it is frequently difficult to differentiate between direct chemical effects on the gonads and secondary effects mediated by changes in gonadotropin secretion.

A wide variety of compounds have been shown to bind to hepatic estrogen receptors and mimic the stimulatory actions of estrogens on vitellogenin production by the liver (Sumpter and Jobling 1995; Jobling, Reynolds et al. 1995; Thomas and Smith 1993; Thomas and Khan 1997; MacLatchy and Van Der Kraak 1995). Several of these compounds also bind to nuclear progestin receptors in fish, birds, and reptiles (Pinter and Thomas 1997; Lundholm 1988; Vonier et al. 1997) and therefore have the potential to interfere with functions regulated by progestins. In addition, certain estrogenic chemicals are capable of displacing the MIS from its receptor on the plasma membrane on fish oocytes and sperm, of inhibiting germinal vesicle (oocyte nucleus) breakdown in oocytes and of decreasing sperm motility (Ghosh and Thomas 1995; Thomas et al. 1997; Thomas 1999).

Finally, there is evidence that exposure to organic compounds such as PCBs, which induce hepatic cytochrome P-450 enzymes, can result in increased hepatic clearance and metabolic inactivation of steroids (Yano and Matsuyama 1986). Moreover, the activity of a major steroid conjugating enzyme, glucuronyltransferase, also may increase after chemical exposure (Sivarajah et al. 1978). However, because physiological feedback systems may be able to compensate for increased steroid metabolism and excretion by stimulating steroid synthesis, the reproductive consequences of increased conjugation are unclear.

Sex determination and sex differentiation

Sex determination and sex differentiation are particularly sensitive to perturbation and thus are critical transitions vulnerable to environmental contaminants (Colborn and Clements 1992; Colborn et al. 1993). Most alterations in the processes of sex determination and/or sexual differentiation are likely to interfere with reproductive function, although detrimental effects should not categorically be assumed to occur. Both genotypic and phenotypic sex must be considered in sex determination; sex differentiation includes not only the gonads but also accessory sex structures and secondary sex characteristics.

Oviparous vertebrates have a variety of patterns of gonad differentiation during sex determination. These include gonochorism (in which the gonad develops into either a testis or an ovary), the most common sex determining mechanism in vertebrates, hermaphroditism (in which the gonad is comprised of both male and female tissue simultaneously), protogyny (where the gonad is ovarian early in the animal's life, but later becomes a testis), and protandry (in which the gonad is a testis early in the animal's life, but later becomes an ovary). Birds are gonochoristic, as are most reptiles and amphibians. Protogyny and protandry are most evident in fish, particularly marine fish species (Atz 1964; Breeder and Rosen 1966) Relatively little is known about the mechanisms, either environmental and/or physiological, that affect gonadal development in non-gonochoristic oviparous vertebrates, so it is premature at this time to try to generalize about the effects contaminants might have on the processes of sex determination and differentiation in these animals. Basic research is needed to unravel the sex determining processes in these species.

The genotypic sex of vertebrates, i.e., the sex chromosome complement of the individual, plays different roles in the determination of phenotypic sex and sexual differentiation among the oviparous vertebrate classes. Thus gonadal differentiation can be influenced by genotypic sex, hormones, or environmental factors, depending on the vertebrate class and species. In birds, the process of sex determination is genetic, but it differs from that of mammals (which have XX females and XY males) in that the male bird is the homogametic sex (ZZ) while the female is heterogametic (ZW). Genotypic sex normally determines early gonad differentiation, followed by gonadal steroid production, which in turn leads to phenotypic sex determination and differentiation of accessory sex structures and secondary sex characteristics. As in mammals, birds have a SRY-type gene in the heterogametic sex (on the W chromosome of genetically female birds, ZW), that seems to be important for the initial step of heterogametic sex differentiation (i.e., ovarian differentiation in birds, testes differentiation in mammals). The differentiated avian ovary then produces estrogens, of which estradiol is the primary one, that influence the development of accessory reproductive organs and secondary sex characteristics of females. The homogametic sex (ZZ males in birds) typically does not require gonadal steroid hormones for gonadal differentiation and is often referred to as the default sex (Hadley 1996; Norris 1997). Reptiles have sex chromosomes, but sex determination is primarily under the control of environmental factors, particularly temperature (Crews 1993; Crews et al. 1996; Lance 1997). In these animals, the development of ovaries or testes during about the middle third of embryonic life depends on the temperature at which the egg is incubated (Wibbels et al. 1993, 1994). However, gonadal steroids also are involved in reproductive development, and Crews (1993) has presented a hypothesis for the nature of the interactions of temperature and steroid effects (also see Pough et al. 1998 for a discussion of temperature-dependent sex determination in the context of reptilian life cycles). The sex determining mechanisms of amphibians have been reviewed recently by Hayes (1998). In fish, it is generally accepted that sex is determined by a pair of sex chromosomes, although few teleosts have sex chromosomes that are sufficiently distinctive to be easily identified. The heterogametic sex can be the male or the female, depending on the fish species. Minor sex determining genes also may be important in sex determination in some fish species. For example, in carp (*Cyprinus carpio*) it is known that genetic sex is controlled by more than one pair of alleles. Female (XX) carp, which also carry the recessive allele "mas–" develop either as males ($XX^{mas-\,mas-}$) or as hermaphrodites ($Xx^{mas+\,mas-}$) (Komen et al. 1992). In amphibians and fish, environmental factors also may play a role in sex determination; there are known examples of both male and female heterogamety and temperature-dependent sex determination in these species (Baroiller et al. 1996; Craig et al. 1996). Acidity also has been shown to affect sex determination in cichlid fish (Rubin 1985).

Following their initial differentiation, the gonads produce steroids which in turn control further reproductive development, including the development of accessory sex structures and secondary sex characteristics. In general, there is a "critical window" of time during which steroid hormones can alter fundamental aspects of gonadal differentiation. Oviparous animals, including fish, amphibians, and birds, differ from mammals in that the female phenotype develops under estrogen control, whereas the male phenotype develops in the absence of estrogen or in the presence of elevated androgens (Norris 1997). This generalization reflects the predominant hormone picture but it should be recognized that both estrogenic and androgenic hormones are present in both males and females, and there are many subtleties to the hormone balance that this generality does not take into account (Ottinger and Abdelnabi 1997). In addition, the timing of the critical window may vary with different species within a vertebrate class (see review of sexual differentiation in birds by Feyk and Giesy 1998).

In salmonid fish the critical window of sexual differentiation is for a few days around hatching. During this time brief exposure to steroidogenic compounds can result in phenotypic sex reversal of the subsequent offspring (Piferrer and Donaldson 1989). In common carp, however, the gonad remains undifferentiated for up to 120 days post-hatch (Gimeno et al. 1996, 1997) suggesting these fish are likely to have a broader critical window of vulnerability to perturbations in sexual development. In reptiles, although temperature is the key influence in sex determination, estrogens play a role in the differentiation of gonads, and additional exposure to exogenous estrogens or estrogenic chemicals can affect a variety of processes. Thus, it has been possible to assess the effects of putative estrogen mimics by spotting or painting them on turtle or alligator eggs and noting which compounds result in feminization or an intersex condition of the hatchlings from eggs reared at male-producing temperatures (Bergeron et al. 1994; Wibbels et al. 1994).

Because the differentiation of tissues associated with reproduction is plastic during a critical period, this differentiation can be influenced directly by alterations in steroid exposure (reviewed in Hunter and Donaldson 1983; Chang et al. 1996; Donaldson 1996) and indirectly by changes in the enzymes that control steroid synthesis and metabolism (Piferrer et al. 1994). Likewise, chemical pollutants that act as hormone mimics or antagonists and are present during the critical period can cause permanent, potentially deleterious alterations in the reproductive system. In birds, the critical period for much of reproductive development is during the latter part of embryonic life, so contaminants deposited in the egg are of particular concern (Ottinger and Abdelnabi 1997).

It is difficult to generalize about the effects of steroids and their analogs and antagonists on sexual differentiation because different species may have different responses to the same hormone, and the same hormone administered at different times can produce the opposite effects on sex phenotype in males versus females of the same species. For example, DDT and DDE, which are estrogenic and anti-androgenic (respectively) in mammals, have opposite effects on gonadal duct growth in salamanders; DDT, which was expected to be estrogenic instead antagonized the effects of estrogens. DDE, which was expected to be anti-androgenic, was estrogenic in some salamander tissues (Clark et al. 1998). The effects of hormones or pollutants also may be delayed; exposure at early life stages (e.g., egg or early larval stage of fish) may not be manifest until sexual maturity is reached many years later (Donaldson 1996). Studies of gull and tern populations have shown that exposure to PCBs in ovo can affect the sexual behavior of those individuals when they become adults (Fry et al. 1987). Such delays in effects make it difficult to establish exposure/ effect relationships for some environmental contaminants in wild populations.

Exposure to steroids at stages outside the critical windows of sex determination and differentiation also may affect sexual development, but greater exposure times and/ or higher concentrations of the compounds are generally required to cause an effect (Feist et al. 1996). For example, stimulation of reproductive tract growth and inhibition of testes development appear to be characteristics of exposure of adult fish to exogenous estrogens and their mimics. Induction of VTG production in males, reduction in testicular growth, liver hypertrophy, gonadotrope hyperplasia and even the occurrence of intersex phenotypes all have been shown to occur as a result of exogenous exposure of male fish to estrogens, estrogen mimics, and other chemical contaminants (Billard et al. 1981; Bortone and Davis 1994; Jobling, Sheahan et al. 1995; Tyler, Jolding et al. 1998). Exposure to exogenous androgens can cause sterility in fish of either genetic sex. In fish, transient sex reversal also can occur, for example, when phenotypic changes have been induced by contaminant chemicals, the fish may spontaneously revert back to its original phenotypic sex (Donaldson 1996). This phenomenon indicates that some effects of contaminants on sexual development may be transient and, therefore, may not necessarily impact the long term reproductive performance.

In some fish species, precocial sexual maturation occurs as a "natural" phenomenon. For example, male Atlantic and Pacific salmon, in the freshwater river where they are hatched, sometimes mature sexually during an early life stage weighing about 100 g without going to sea. Sexual maturation in salmon typically occurs during the period of 3 years or more that they spend at sea before returning to freshwater to spawn. The precocious males that have developed sexually at a small size in freshwater are essentially the same endocrinologically as adult male fish that may have spent 3 years or more in the ocean, weigh many kilograms, and are returning to freshwater to spawn. The cause of precocious maturation is uncertain, but it may involve a decreased growth rate during early life stages, suggesting that energy supply or compounds that interfere with growth may affect the timing of sexual development (Thorpe 1994a, 1994b; Dickhoff et al. 1995).

Secondary sex characteristics

In adults of some oviparous vertebrates, there are striking differences in secondary sexual characteristics between males and females. Some secondary sex characteristics develop very early; others, however, do not develop until puberty or even the final stages of sexual maturation, just prior to gamete production. Differences in secondary sex characteristics of males and females include differences in body size, body coloration, and behavior. Male oviparous vertebrates commonly have certain physical features that are sex-specific, e.g., the enlarged comb in chickens, the throat patch and thumb pads of frogs, the dewlap in some male lizards, the elongated claws of slider turtles, the development of a hooked jaw, or kipper, in some salmonid fish, and the appearance of tubercles and/or fat pads in cyprinid fish before spermiation (Hadley 1996; Norris 1997). Males of many species also show seasonal cycles of coloration and behavior that correspond with the breeding season. The observed differences between males and females are largely determined by differences in steroid hormone secretion from the gonads in each sex. Any contaminant that interferes with or mimics these steroid hormones could affect the development of these secondary sexual characteristics and may affect the reproductive performance of the animal (Bortone et al. 1989; Bortone and Davis 1994). Although secondary sex characteristics could be used to assess the affects of contaminants such assessments could be difficult and somewhat subjective.

Gametogenesis in females

Gamete growth and development, and subsequent embryonic development of the fertilized egg, are very complex processes. For this reason, success in producing viable offspring may be affected at many developmental stages and by many environmental and genetic factors. Interference with the formation of gametes at any stage during their synthesis is likely to affect the quality of the gamete and thus reproductive performance. A knowledge of the mechanisms underlying the processes of gamete development is essential for fully understanding how contaminants may affect gamete formation.

Oogenesis

Ovarian function differs among oviparous vertebrates according to the growth pattern of the oocytes. In some animals a single cohort of oocytes develops concurrently to form the season's batch of eggs. In others, eggs are recruited from a heterogeneous population of developing oocytes and are subsequently ovulated in several batches during each reproductive season. Whatever the pattern of ovarian function, reproduction imposes large metabolic demands, and, therefore, any additional stressors placed on the maturing female may affect egg production. In fish, within the limits set by genetic constraints, egg size varies between individuals within a species and between populations of the same species (Beacham and Murray

1985, 1987, 1993). In fish, age at maturity also may affect egg size (Sargent et al. 1987), with larger females producing larger eggs (Bagenal 1969; DeMartini 1991). Egg size also can be affected by the nutritional status of the female during ovarian recrudescence (Wootton 1979; Bromage and Jones 1991).

Despite differences in the dynamics of oocyte growth between and among the different classes of oviparous vertebrates, there appears to be the same general pattern of oocyte development (Tokarz 1978; Tata 1986) . Several staging systems for oocyte development that have been proposed incorporate physiological, bio-chemical, morphological and histological criteria, including follicle size, somatic tissues types and the presence of yolk. In fish, the major developmental events that occur can be classified broadly into the following phases or periods according to the state of oocyte growth: oogenesis, primary oocyte growth, secondary oocyte growth, maturation, and ovulation (reviewed in Wallace 1985; Tyler and Sumpter 1996). The genetic changes and ultrastructural events accompanying oocyte development have been described in detail for some oviparous animals (salmon, Nakamura and Nagahama 1993; zebrafish, Selman et al. 1993). Briefly, during the early stages of oocyte development, the following events occur during meiotic division: DNA is replicated (leptotene), homologous chromosomes pair (zygotene), and the chromo-some pairs shorten and thicken (pachytene). Chromosomes then separate and form lampbrush configurations (diplotene) just prior to the long period of cytoplasmic growth which is characterized by an enormous accumulation of yolk. Finally, a hormonal signal initiates the resumption of meiosis, whereby the nucleus, which has been arrested in meoitic prophase, breaks down and the chromosomes enter the first meoitic metaphase and continue with meiosis into the second meoitic metaphase. This process, termed oocyte maturation, converts the oocyte into an egg ready for fertilization (Nakamura and Nagahama 1993; Selman et al. 1993).

Vitellogenesis

Vitellogenesis, which encompasses both vitellogenin synthesis and its sequestration into the oocyte, is responsible for the enormous growth of oocytes during the secondary growth phase in most oviparous vertebrates and may account for as much as 90% of the final egg size (Tyler 1991). During this period extraovarian proteins are sequestered, processed, and packaged into oocytes; therefore, this period is of particular importance when considering maternal to oocyte transport of contaminants. In some fish, e.g., the stickleback and the Atlantic cod (*Gadus morhua*) , oocytes can pass through the vitellogenic phase of development in as little as 5 days (stickleback, Wallace and Selman 1979; cod, Kjeski et al. 1989); in others (e.g., salmonids) vitellogenesis may take 9 months or more to complete (Tyler and Sumpter 1996). Similarly, in birds, reptiles, and amphibians the period of time occupied by vitellogenic growth also may show considerable variation between species with different reproductive strategies. The duration of the vitellogenic growth process may be an important consideration with respect to the magnitude of accumulation of contaminants in the developing oocyte after contaminant exposure by adult females.

In many oviparous vertebrates, a hepatically derived plasma glycolipoprotein, VTG, is the principle precursor of yolk proteins. In birds, although very low density lipoprotein (VLDL) is the primary source of yolk protein, VTG is a secondary but major precursor of yolk protein. In some fish species (e.g., winter flounder, *Pseudopleuronectes americanus*) yolk protein precursors that differ from both VTG and VLDL are sequestered in oocytes during vitellogenesis (Nagler and Idler 1989). VTG and VLDL are selectively sequestered by receptor-mediated endocytosis (fish, Tyler and Lancaster 1993; birds, Shen et al. 1993; Bujo et al. 1994). Any factor that interferes with the function of this receptor system is likely to markedly affect the deposition of yolk reserves and, therefore, oocyte growth. During vitellogenesis concentrations of these lipoproteins in the plasma may be considerable. In salmonid fish, for example, blood VTG increases from concentrations of a few hundred mg/ml to levels that may exceed 50 mg/ml during vitellogenesis. In asynchronous ovulating fish, maximal levels of VTG tend to be much lower (generally a few hundred mg per ml to about 1 mg per ml). The levels of plasma VTG and the rates of VTG uptake in teleosts tend to reflect the pattern and mode of oocyte development. Studies have shown that circulating VLDL and VTG act as important transport proteins, binding metals and divalent cations as well as lipophilic hormones. The high rates of sequestration of VTG and VLDL into the developing oocyte and their ability to bind physiologically important materials and contaminants means they may act as major carriers of contaminants from the maternal circulation into the developing oocyte (reviewed in Specker and Sullivan 1994; McNabb and Wilson 1997). Some pollutants, such as organochlorines, have been shown to enter the developing oocyte associated with these lipoproteins (Ungerer and Thomas 1996). In some oviparous vertebrates, lipids from sources other than VTG and VLDL enter the developing oocyte during the secondary growth phase. In fish, there appears to be considerable variation in the amount of lipid in oocytes of different teleosts; some appear to be totally devoid of lipid droplets (Wallace and Selman 1981), whereas in others, such as the gourami (*Trichogaster cosby*, Belontiidae) more than one-third of the weight of an oocyte may be wax ester (Kaitaranta and Ackman 1981).The source and nature of these lipids, for the most part, have to be determined, but they may be of relevance when considering the passage of lipophilic contaminants into the developing oocyte.

Yolk proteins derived from VTG and VLDL include a variety of lipoproteins and phosphoproteins that differ between species (reviewed in Specker and Sullivan 1994). The yolk proteins derived from VTG, VLDL and other plasma precursors, and the lipid droplets serve as energy sources and protein building blocks for mobilization and use during embryonic development. Contaminants that enter the oocyte with VLDL or VTG may be stored and mobilized as the embryo starts to grow, a critical and vulnerable stage of development. However, we do not yet know how contaminants are partitioned within the egg, and this will clearly have an impact on how they may affect subsequent embryonic development.

The dependence of the embryo on the deposition, processing and storage of the correct complement of protein building blocks, energy supplies, minerals, vitamins, and hormones from the maternal circulation highlights the vulnerability of the developing oocyte to exposure to environmental contaminants.

Vitellogenin as a biomarker for estrogens

The production of VTG is estrogen-dependent and, therefore, generally restricted to females; little, if any, VTG can be detected in plasma of "normal" male oviparous animals (Copeland et al. 1986). However, males do carry the VTG gene, and exposure to natural, synthetic, or pollutant estrogens can trigger its expression in males (Chen 1983). Exposure of male rainbow trout to estrogenic chemicals can induce concentrations of plasma VTG that are far higher than those normally found in fully mature females (Purdom et al. 1994). These features of the vitellogenic process (estrogen dependency, the normal absence of VTG in male fish plasma, and the possible magnitude of change in the plasma VTG concentration) have provided the basis for a very sensitive bioassay of exposure to estrogenic chemical pollutants in fish. Radioimmunoassays and enzyme-linked immunoassays are available for measuring VTG in a wide variety of oviparous animals, (e.g., Benfey et al. 1989; Tyler and Sumpter 1990; Palmer and Palmer 1995; Palmer et al. 1998; Tyler, van Aerle et al. 1999) and there has been considerable progress toward the development of universal antibodies for VTG (Denslow et al. 1997). However, a limitation of VTG synthesis as a "tool" for monitoring exposure to environmental estrogens is the very steep dose-response curve; in fish only a 10-fold increase in the concentration of estrogen in the water can take the response curve from a non-estrogenic to a maximal response, i.e., a million-fold increase in vitellogenin concentration (Desbrow et al. 1996). Furthermore, the presence of VTG in the plasma of male fish is an indicator of estrogen exposure alone; it is not known if high VTG concentrations in male fish affect their reproductive performance. However, high rates of VTG production require a great deal of energy, so this may result in depletion of energy resources. Very high concentrations of VTG in the plasma have been shown to cause renal failure (Herman and Kincaid 1988). Other markers for exposure to estrogenic compounds are the zona radiata proteins (also called vitelline envelope proteins) that are synthesized in the liver of both sexes in response to estradiol (Oppen-Bersten et al. 1992; Hyllner et al. 1991).

Oocyte maturation

At the end of the period of oocyte growth, the post-vitellogenic oocytes of most oviparous vertebrates cannot be fertilized because the first meiotic division is incomplete and is arrested at prophase-I. (Nagahama 1987). A surge in gonadotropin (LH or GTH-II) secretion induces the resumption of meiosis and structural changes in the ooplasm that enable the oocytes to be fertilized after ovulation. This process, termed final oocyte maturation (FOM), oocyte maturation, or meiotic maturation, involves breakdown of the germinal vesicle (the oocyte nucleus), chromosome condensation, formation of the first meiotic spindle and polar body, completion of the first meiotic division, and continuation to metaphase II. The appearance of the ooplasm usually changes dramatically during FOM because of proteolytic cleavage and mobilization of yolk reserves (Nagahama 1987).

The entire process of FOM is completed within 24 hours, is highly synchronized in the ovaries of many teleost species (Brown-Peterson et al. 1988), and can be readily observed under a microscope. The details of FOM have been studied in spotted sea trout (*Cynoscion nebulosus*) oocytes and will be described here as an illustration. The first visible sign of FOM in sea trout oocytes is coalescence of the yolk granules around the nucleus (lipid coalescence), which begins around dawn on the day of spawning. Lipid coalescence is complete, the ooplasm has become clear by mid-morning, and the nucleus has begun to migrate to the animal pole (germinal vesicle migration). Germinal vesicle breakdown has occurred by mid-afternoon, and the oocyte has began to swell due to uptake of water by a Na^+,K^+ ATPase-dependent process. At dusk the oocyte has swollen to 8 times its original volume, and an oil droplet has formed. Shortly afterwards the follicular layer surrounding the oocyte ruptures and ovulation occurs (the oocyte is now referred to as an egg). Large numbers of oocytes (about 300,000) are released at once into the ovarian cavity. A minute opening (micropyle) forms at the animal pole through which the sperm will later enter and fertilize the egg. Spawning occurs within a 2 to 3 hour period and is completed by midnight (Brown-Peterson et al. 1988).

There is general agreement that gonadotropin induces FOM in teleosts and amphibians by stimulating the production of 21-carbon steroids, called MIS, by the follicular cells (Nagahama 1987; Jalabert et al. 1991). Current evidence indicates that progesterone is the sole MIS in amphibians, whereas a variety of progestins and deoxycorticosteroids have been identified as MISs in teleosts. The progestin 17,20α-dihydroxy-4-pregnen-3-one (17,20α-P) has been positively identified as the MIS in rainbow trout (Nagahama 1987) and is the likely MIS in other salmonids, some carp species, catfish, and several other teleost orders (Jalabert et al. 1991). Another 21-carbon steroid, 17,20α, 21-trihydroxy-4-pregnen-3-one (20α-S), has been positively identified as the MIS in Atlantic croaker (*Micropogonias undulatus*) and spotted sea trout (*Cynoscion nebulosus*) (Thomas 1994) and appears to be the major MIS in perciform fishes. Other 21-carbon steroids have been proposed as MISs in several species of flatfish (Scott and Canario 1987). The MISs bind to specific steroid receptors located on the oocyte plasma membrane to induce FOM (Sadler and Maller 1982; Patiño and Thomas 1990). In some species, the post-vitellogenic oocytes have to be "primed" with gonadotropin to become responsive to the MIS and undergo FOM. Up-regulation of the MIS receptor on oocyte membranes appears to be an essential component of this priming process by gonadotropin (Thomas and Patiño 1991). In contrast, convincing evidence that steroid hormones mediate the induction of meiotic maturation by gonadotropin is lacking for birds and reptiles (Jalabert et al. 1991). Increased plasma levels of progesterone have been observed in several avian and reptilian species accompanying the periovulatory surge of LH. However, detailed investigations on the actions of steroids on oocyte maturation in vitro are required to demonstrate that they function as MISs in these vertebrate groups.

The signal transduction pathways and intracellular mechanisms of MIS action have been investigated in fish and amphibians (Nagahama et al. 1994). A decrease in cAMP is required for MIS induction of FOM which appears to involve a guanine nucleotide-binding G-protein in the signal transduction pathway across the oocyte plasma membrane to the cytoplasm (Yoshikuni et al. 1995). A cytoplasmic factor, named maturation-promoting factor, is formed and is the intracellular mediator of FOM. Recent studies indicate that the structure of maturation-promoting factor is similar among vertebrates and composed of cdc 2 kinase and cyclin B (Nagahama et al. 1994). Interestingly, maturation-promoting factor appears to be a universal regulator of mitosis in addition to its function in initiating meiosis.

Ovulation

Once oocyte maturation is complete, the oocyte is released from the ovarian follicle by a process termed ovulation and is transported to an extraovarian site, where fertilization occurs (Jones and Schreibman 1987). The ovulatory process includes several preparatory changes in the oocyte and ovarian follicle prior to expulsion of the oocyte. Microvillar connections between the oocyte and surrounding follicle cells are broken (follicular separation), parts of the follicular wall decrease in thickness and an opening forms in the follicle wall (follicular rupture) due to digestion of tissue by proteolytic enzymes (Goetz et al. 1991). The oocyte is released by contraction of smooth muscles in the follicular wall. This process is regulated in part by prostaglandins in mammals and fish. The action of prostaglandins in the brook trout appears to be mediated by an increase in ovarian cAMP levels (Goetz et al. 1991).

The structural and biochemical changes in the ovary during ovulation are controlled by gonadotropin (LH or GTH-II) which is elevated in the blood prior to ovulation (Jones and Schreibman 1987; Norris 1997). Gonadotropin induces the synthesis of progestins and 11-deoxycorticosteroids by the ovarian follicle and these steroids are implicated in the regulation of ovulation in a variety of vertebrate species. Progester-one is the principal progestin synthesized by the ovaries of tetrapod vertebrates, whereas the MISs 17,20α-P and 20α-S are the major products of teleost ovaries and induce ovulation in vitro (Goetz 1997; Pinter and Thomas 1995a). One likely action of the MISs in teleost fish is to regulate ovarian prostaglandin synthesis (Goetz 1997). The induction of ovulation in teleost ovaries is mediated by a nuclear receptor presumed to be in follicular cells (Pinter and Thomas 1995a), but the induction of FOM by MISs is mediated by plasma membrane receptors on the oocyte (Pinter and Thomas 1995b). Nuclear progestin receptors also have been identified in the ovaries of avian species (Peddie et al. 1993).

Fecundity and atresia

The number of eggs produced (fecundity) is species-specific. Some oviparous vertebrates produce thousands of eggs (e.g., many fish species), whereas others may produce only one in any reproductive season (e.g., some bird species). Fecundity has been widely used in oviparous vertebrates as a measure of reproductive performance. Many factors affect fecundity, and high variance in fecundity may result from differences in intraspecific genetic variation, age, body size, environmental conditions, and nutrition (for fish, see Thorpe et al. 1984; Bromage et al. 1990; Flemming and Gross 1990; reviewed in Tyler and Sumpter 1996). Pollutant chemicals that impact egg production can potentially have significant effects at the population level, but they are working against a multiplicity of background influences. Thus, the use of fecundity for assessing effects of contaminants in wild populations requires a considerable amount of knowledge about the animal population of concern.

In mammals, the fecundity gene, (FecB gene) has been identified as playing a role in regulating fecundity (Braw-tal et al. 1993; McNatty et al. 1994), and this regulation is the subject of intensive study. Very little research on the genetics of fecundity has been conducted in birds, fish, reptiles, or amphibians, but it is likely that, as in mammals, there is a strong genetic basis. Any chemical that impacts the transcription and expression of the genes that control fecundity will in turn affect reproductive output.

The number of eggs produced is a result of the balance between recruitment of oocytes into the maturing pool and oocyte degeneration (referred to as atresia) (Ryan 1981). In mammals, atresia may occur at any stage of oocyte development and may play a major role in determining the numbers of oocytes that are recruited into the successive stages of development, thereby affecting the number of developing oocytes that eventually form mature eggs (Ryan 1981). Atresia appears to play an important role in oocyte recruitment in some fish species (particularly, batch-spawning fish that spawn a number of times each season), but appears to be an uncommon event in "synchronous" ovulators in which a single batch of eggs is produced and ovulated together in a season. The level of atresia in some fish has been shown to correlate with environmental stress (Johnson et al. 1997). In birds, the regulation of recruitment and atresia in the avian ovarian hierarchy of oocyte maturation has been studied in detail only in domestic chickens (Johnson 1996).

Gametogenesis in males

Spermatogenesis

Successful sperm production is an essential aspect of reproductive success for all but a few parthenogenic species. Interference with sperm formation by chemical pollutants can seriously affect reproductive performance (Garside and Harvey 1992; Working and Chellman 1993; Sundaram and Witorsch 1995). A chemical need not completely block spermatogenesis in order to have a detrimental effect; even a slight alteration in the quality of the sperm (e.g., a decrease in motility) may be sufficient to decrease or prevent fertilization of the egg.

The anatomy of the testes differs among the vertebrates (Roosen-Runge 1977), with the most obvious difference occurring between the anamniotes (fishes and amphibians) and the amniotes (mammals, birds, and reptiles). In anamniotes sperm development usually occurs in discrete cysts, sometimes called a spermatocyst, a follicle, or an ampulla. Typically, there is a clone of developing cells derived from a single stem cell within a cyst and the stages of spermatogenesis are synchronized among the clonal cells. The spermatocysts of anamniotes are usually compartmentalized into lobules or tubules within the testes. The exact arrangement differs among classes and among species within a class. Elasmobranchs have spheroidal cysts (ampulla) that are arranged in concentric zones representing successive stages of spermatogenesis (Dodd 1983). In contrast, most teleosts exhibit a lobular arrangement of the spermatocysts, although some show a tubular arrangement (Nagahama 1983). Anuran amphibians (frogs and toads) have spermatocysts arranged in tubules. Interestingly, some urodele amphibians (salamanders) have a longitudinal arrangement of cysts that form distinct lobes appearing almost like multiple testes (Roosen-Runge 1977). The lobes form a spermatogenic wave along the length of the testis, with the more cranial lobes in the earlier stages of spermatogenesis and the more caudal lobes in the later stages of spermatogenesis.

Amniote testes lack the cyst structure common to fishes and amphibians. Instead, they have a network of seminiferous tubules that contain the germinal tissues along their walls. The developing spermatogenic cells are arranged along the tubules in several layers that represent successive generations. Thus, the spermatogonia are located peripherally, the spermatocytes are located in the middle layers, and the spermatids located closest to the lumen of the tubule. The seminiferous tubules are long, usually open at both ends, and have varying degrees of anastomoses, depending on the species. The number of tubules also varies between species (Roosen-Runge 1977).

In both amniotes and anamniotes, there are nongerminal tissues associated with the developing germ tissues. In the amniotes, Sertoli cells (also called sustentacular cells or nurse cells) are in intimate association with multiple spermatogenic cells in all stages of development. The Sertoli cells appear to perform a variety of functions associated with spermatogenesis, including structural support for the developing gametes, sustenance for spermatogenic cells, steroidogenesis, phagocytosis of materials from the lumen, involvement (along with the basal lamina) in the blood/ testis barrier, and secretion of fluids and proteins into the lumen (Fawcett 1975; Setchell 1978). Similar cells, probably homologous, are present in anamniotes and form the walls of the spermatocysts or the boundaries of the lobules. Testes of both amniotes and anamniotes also have steroidogenic, nongerminal cells (Leydig cells) that are typically located in the interstitial spaces. The steroids (e.g., testosterone) produced by these cells are important in regulation of spermatogenesis.

The process of spermatogenesis is similar in most vertebrates despite wide variation in the size and anatomical arrangement of the testes. The timing of the spermatogenic stages and the details of the events of each stage differ by species, but the following cell types can be recognized in most species: primordial germ cells, spermatogonia, spermatocytes, spermatids, and spermatozoa. Primordial germ cells, which originate outside the testes, are the precursor cells to the gametes. They migrate to the site of the developing gonad by amoeboid movement. Primordial germ cells can divide mitotically and are not sexually differentiated, except for possessing specific sex chromosomes. Spermatogonia are diploid, mitotically active cells located within the gonad. The number of mitoses the spermatogonia undergo is fixed by species. The range of mitotic cycles varies between 4 and 12 among the vertebrates, with 4 to 5 cycles most common (Roosen-Runge 1977). Daughter cells from the successive mitoses are increasingly differentiated, especially with regard to the arrangement of the chromatin, which increasingly condenses during successive mitotic cycles. It is thought that the spermatogonial divisions serve to prepare the cells for meiosis. Spermatocytes are diploid, meiotically active cells. They are classified as primary or secondary, depending on the stage of meiosis in which they are engaged (meiosis I and meiosis II, respectively). Once meiosis is completed, the resulting haploid cells are called spermatids. These undergo extensive differentiation into spermatozoa through a process referred to as spermiogenesis. Spermiogenesis involves condensation of the nucleus, formation of the acrosome, elimination of the cytoplasm, development of a tail, and rearrangement of the mitochondria (Setchell 1978). Even after a spermatid has undergone the complex rearrangement into a spermatozoan, the cell may not be a competent gamete. The spermatozoan may need further maturation and activation prior to achieving the ability to fertilize an egg. Sperm activation (capacitation) may occur in the tubules of the testis, in the sperm ducts, or even in the female reproductive tract.

Spermiation

At the end of spermatogenesis, mature spermatozoa or late spermatids are released from their attachment to supporting cells (Sertoli cells) and transported from the testis by a process called spermiation (Morisawa and Morisawa 1990). In teleosts spermiation involves considerable thinning or hydration of the milt. The sperm are immotile when they leave the testis but mature and acquire the potential for motility as they pass along the male reproductive tract. The acquisition of the potential for sperm motility and thinning of the milt are controlled by gonadotropins in teleosts and in other oviparous vertebrates. In teleosts the action of gonadotropin is mediated by increased production of the MIS. Elevated circulating levels of 17,20α-P have been reported in male fish at spermiation (Scott and Baynes 1982). Treatment of rainbow trout with 17,20α-P increases the potassium content and pH of the seminal fluid (Scott and Baynes 1982; Yoshikuni and Nagahama 1991) and the sperm cAMP content. It has been proposed that the increase in cAMP, as the sperm are exposed to higher pHs along their passage down the sperm duct, confers the potential for motility (Morisawa and Morisawa 1990). In the spotted sea trout, the MIS probably acts on the sperm duct via the recently characterized nuclear progestin receptor (Pinter and Thomas 1995b). Sperm are stored in the vas deferens, seminal sacs (passerine birds), or sperm ducts (teleosts) and are immotile until they are ejaculated or spawned. In fish and amphibians with external fertilization, sperm motility is initiated when the sperm are released into the environment. A decrease in external osmolality or potassium ions initiates motility in fresh water species, whereas exposure to a hyperosmotic environment triggers motility in marine teleosts. In some species with internal fertilization, the sperm become motile in the female reproductive tract (Morisawa and Morisawa 1990).

At the time of insemination, the sperm of many vertebrate species are incapable of penetrating and fertilizing the oocyte. Capacitation, which involves structural modification of the plasma membrane of the sperm, includes loss of coating factors and occurs during residency of sperm in the oviduct. Hydrolytic enzymes, which aid in the penetration of the oocyte by sperm by weakening the zona pellucida, are released by exocytosis from the acrosome, an organelle underlying the plasma membrane. Progesterone exerts rapid nongenomic actions on mammalian sperm, most likely via a membrane receptor, to cause rapid calcium influx and the acrosome reactivation. The influx of calcium also results in hyperactivation, characterized by rapid circular movement in a proportion of the sperm. A plasma membrane receptor for the MIS on the sperm of spotted seatrout has been characterized (Thomas et al. 1997). Teleosts lack an acrosome. However, treatment of sea trout sperm with the MIS increases their velocity and turning rate, a response similar to hyperactivation in mammals. Thus, it has been concluded that gamete maturation in both male and female teleosts is mediated by plasma membrane receptors as well nuclear steroid receptors (Thomas et al. 1997).

There are several general ways in which chemical pollutants may interrupt or block spermatogenesis. These include interfering with cell division, the hypothalamic/ pituitary axis, Sertoli cell function, Leydig (interstitial) cell function, or sperm activation.

Many pharmaceutical and industrial chemicals inhibit cell division by interfering with protein or DNA synthesis, i.e., they act as anti-neoplastic agents. These agents target actively dividing tissues such as bone marrow, epithelium, and spermatogenic tissues (Garside and Harvey 1992). Other agents are known to be specifically toxic to spermatogenic tissues, e.g., busulfan, which is toxic to spermatogonia; 2-methoxyethanol, which is toxic to spermatocytes; and methyl chloride, which has toxic effects on spermatids (Working and Chellman 1993; Sundaram and Witorsch 1995).

There also are compounds that are directly toxic to the Sertoli cells. These chemicals have a direct effect on spermatogenesis by interfering with the important functions Sertoli cells perform in support of spermatogenic cells. Sertoli cell toxicity is usually irreversible because Sertoli cells typically do not divide in the adult. Examples of chemicals that are toxic to Sertoli cells are 1,3-dinitrobenzene, phthalate esters, and 2,5,-hexanedione (Working and Chellman 1993; Sundaram and Witorsch 1995).

Disruption of the hypothalamic–pituitary–gonadal axis also can disrupt gonadal function. Compounds that affect LH production will have an effect on spermatogenesis because LH is the primary stimulus for androgen (e.g., testosterone) production by the Leydig cell. The androgens, along with FSH, act on Sertoli cells and spermatogenic cells to stimulate sperm production. LH production can be altered at the level of the hypothalamus, where the stimulus for LH release GnRH is produced and secreted, or at the level of the anterior pituitary, where LH is produced and secreted. Similar disruptions of spermatogenesis result from agents that are directly toxic to Leydig cells, such as ethane-1,2 dimethanesulfonate and TCDD (Working and Chellman 1993; Sundaram and Witorsch 1995).

Disruption of spermatogenesis also can occur through agents acting at sites outside of the testes proper and the testicular endocrine axis. Depending on the species, one or more accessory structures may be involved in sperm storage and maturation and in secretion of seminal fluids. These accessory structures include the epididymis, the seminal vesicles, and the prostate gland. Methyl chloride is an example of a chemical that is toxic to the epididymis (Working and Chellman 1993; Sundaram and Witorsch 1995).

Stress is another important factor in inhibition of spermatogenesis. Stress is increased by exposure to environmental agents. The effects of stress on reproduction are mediated through multiple pathways. The adrenal axis is involved, as corticosteroids have been shown to inhibit LH production (Kamel and Kubajak 1987), but higher brain involvement also seems likely.

Control of gametogenesis

Any compound that interferes with hypothalamic–pituitary–gonadal axis control of gonad function (including gonad recrudescence in seasonally reproducing adults) or release of mature gametes may have an impact on reproduction. These effects may occur at the level of hormone synthesis, hormone secretion, or at the sites of action of these "reproductive" hormones with the target site. Several hormones, in addition to those of the hypothalamic–pituitary–gonad axis, have been implicated in regulating oocyte and sperm production in oviparous vertebrates; they include thyroid hormones, growth hormone, and insulin and insulin-like growth factors (IGFs). However, with the exception of the sex steroids, there has been little progress in understanding the precise roles of non-steroidal hormones in gonadal function. Some of these hormones act indirectly; for example, growth hormone, which influences the rate of body growth, can modulate female fecundity and egg size. Growth hormone also may act synergistically with gonadotropin in regulating steroidogenesis, possibly through IGFs (Hadley 1996), and there is some evidence that other growth factors may influence steroidogenesis (Saito and Shimada 1997). Although perturbations of the hormonal systems that effect gonadal development may potentially have a major impact on reproductive performance, these interactions are complex and generally have not been investigated.

Gamete quality

Successful reproduction in oviparous animals is dependent on the production of eggs and sperm of good quality. Good quality gametes may be defined as those that produce a viable embryo, although this is only a crude measure of gamete quality. In fish and amphibians, morphological characteristics of the larvae have been used as indicators of gamete quality (fish, Kjorsvik 1994). In eggs, the appearance of the chorion, the shape of the egg, egg transparency, and the distribution of oil globules can be related to quality of the egg, i.e., the chance that the egg will produce a viable embryo (Kjorsvik et al. 1990 and Bromage et al. 1994). For mammalian sperm, motility, direction of swimming, and duration of swimming activity have all been used to assess sperm quality, and these criteria also have been applied to sperm quality in some oviparous vertebrates (Kime et al. 1996). To date, however, there is little agreement regarding reliable methods for "quality" assessment of sperm or eggs in oviparous animals, and the methods of assessment need to be developed further .

The quality of gametes is determined by the intrinsic properties of the gametes themselves that have been provided by the parent (i.e., by genes, the maternal/ paternal mRNA transcripts stored in the gamete and, in the egg, nutrients contained within the yolk). Studies on mammals have shown that reproductive age of females can affect egg quality (Archibong et al. 1992; Koenig and Stormshak 1993; Navot et al. 1994), but studies of this nature in oviparous vertebrates have not been forthcoming. Only very recently have some attempts been made to address the genetic factors which underlie egg and sperm quality in oviparous vertebrates (Lam 1994; Nagahama 1994). However, in addition to inherent gamete quality, the environment in which the fertilized eggs are incubated can, and does, affect success in producing viable offspring. The conditions to which fertilized eggs are exposed, therefore, often become encompassed in the term "egg quality." Our knowledge of the processes affecting egg and sperm quality in oviparous vertebrates is very limited, but a few studies in fish show that egg quality may vary considerably from season to season in the wild (Kjorsvik et al. 1990). In multiple-spawning fish, there also are considerable variations in the quality of eggs produced in different batches over a single spawning season, even when the batches of eggs are maintained under apparently equivalent culture conditions (Kjesbu et al. 1996; Manning and Crim 1995). Environmental factors that are known to affect egg and sperm quality include the physio-chemical conditions in which the eggs and sperm are released and in which the embryo is subsequently incubated (temperature, pH, exposure to pollutant contaminants, etc.). Physical features of the spawning/nesting environment can also affect oviposition and sensitivity of the adults to hormone signals (Soyano et al. 1993).

There is a great deal of general information about the major determinants of egg and sperm quality. This has led to the widespread belief that the nutrient materials sequestered by the oocyte, and their processing during growth and maturation, are key factors affecting egg quality (Craig and Harvey 1984; Kjorsvik et al. 1990; Bromage et al. 1992). Diet has received the greatest attention with respect to its affect on egg quality, and the major influences on quality are exerted by just a few dietary constituents (Watanabe et al. 1991; Harel et al. 1994; Blom and Dabrowski 1995). Despite the considerable efforts that have been directed towards unraveling the importance of dietary components in determining egg quality, however, the evidence that diet can directly affect egg quality is very limited. The physiology of the adults and their hormonal status, which affects the incorporation of compounds, including hormones, into eggs and sperm, have some effects on gamete quality because stressing fish or exposing them to an environmental contaminant is deleterious to reproduction (Campbell et al. 1994; Brooks et al. 1995; Contreras-Sanchez et al. 1998).

Reproductive behavior

Alterations in neuroendocrine and sensory processes that influence behavior can affect reproduction and development. Studies of birds from Great Lakes colonies with high pollutant exposure exhibit disruption of mating, incubation behavior, and parental care in comparison to colonies from more pristine areas (see reviews by Peakall and Fox 1987; Fox 1993; Feyk and Giesy 1998). Such disruption could result from chemical pollutants that disrupt the gonadal steroid-directed development of sex-specific brain function and behavior. Estrogen (primarily estradiol) is the steroid hormone responsible for the differentiation of reproductive behaviors in birds, in which the "default" sex is male (Ottinger and Abdelnabi 1997). For example, exposure of the embryonic brain to chemicals that create an inappropriate estrogenic signal at a critical time for organization and activation of sex-specific behaviors may cause permanent sex reversal of behavior in male birds. Likewise treatment of female embryos with antiestrogens can result in behavioral masculinization (reviews, Schlinger and Arnold 1995; Ottinger and Abdelnabi 1997). Another example involves the skew in sex ratios with an increased proportion of females in populations of several bird species from contaminated areas in the Great Lakes. In this case, female-female pairs or associations of 2 females with one male appear to be a response of the birds to attempt to rear young despite inadequate numbers of males compared to the numbers of nests. Thus, in this case, the behavioral abnormality may be an indirect result of the chemical pollutant effects on the sex ratios in the population (Feyk and Giesy 1998). It should be noted that affects on brain development and its later behavioral consequences also involve a variety of other hormonal influences and are intimately involved with monoamine neurotransmitters. Studies of avian brain function and development in both precocial and altricial birds have recently added a great deal of basic information about these functions (reviews, Schlinger and Arnold 1995; Ottinger and Abdelnabi 1997) and illustrate the potential for vulnerability to disruption by chemical influences.

Alterations in behavior of fish can be affected by external stressors. Stress induced behavioral alterations in fish and some reptiles may lead to disruption in orientation that will impair reproductive migration leading to reduced reproductive success or "genetic contamination." Other facets affected include food acquisition and predator evasion (Olla et al. 1996, Birge et al. 1993; Brown et al. 1985) and territoriality, which may result in reduced reproductive success (Rand 1985; Beitinger 1990). For an overview of the responses of fish to contaminants, see Schreck et al. (1997).

Conclusions

There are some fundamental aspects of the physiology of development and reproduction of oviparous animals that are well understood and that are important to the assessment and management of pollutant chemicals. Thus, much is known about the control of development, and this knowledge has allowed the identification of the critical ontogenic stages. For a limited number of oviparous species, the basics of the reproductive process have been fairly well described. We also now have fundamental understanding of the receptor and second messenger systems of hormones that control development and reproduction. There also is considerable work in the area of how vertebrates adapt to environmental changes and respond physiologically to those changes (including many chemical pollutants) that elicit stress responses.

We require better understanding of how normal physiology differs from abnormal physiology. We also do not understand how different regulatory systems relate to each other, e.g., how the stress response affects reproduction. Unfortunately, we do not have much information in the area of regulation of some processes involved in development. For example, we do not fully understand sex determining mechanisms in some oviparous vertebrates. Concerning environmental consequences of toxicants, we do not know how changes in gamete production (i.e., number and quality of gametes) affect production (quality and fitness) of offspring, nor do we have much information on how stress in general affects survivorship.

To fill these critical information gaps we need more research in the area of basic physiology. Natural ranges in variability need to be described. More model systems and keystone species need to be made available that provide access for assessment over ranges of environments.

References

Adams SM. 1990. Status and use of biological indicators for evaluating the effects of stress on fish. *Am Fish Soc Symp* 8:1–8.

Adkins-Regan E. 1987. Hormones and sexual differentiation. In: Norris DO, Jones RE, editors. Hormones and reproduction in fishes, amphibians and reptiles. New York: Plenum. p 1–29.

Anderson DP. 1996. Environmental factors in fish health: immunological aspects. In: Iwama G, Nakanishi T, editors. The fish immune system organism, pathogen, and environment. New York: Academic Press. p 289–310.

Ankley G, Bradbury S, Hermens J, Mekenyan O, Tollefsen K-E. 1997. Current approaches to the use of structure-activity relationships (SARs) in identifying the hazards of endocrine modulating chemicals to wildlife. In: Tattersfield L, Matthiessen P, Campbell P, Grandy N, Lage R, editors. SETAC-Europe/OECD/ED expert workshop in endocrine modulators and wildlife: assessment and testing. Brussels, Belgium: SETAC. p 19–40.

Anstead GM, Carlson KE, Katzenellenbogen JA. 1997. The estradiol pharmacophore: ligand structure-estrogen receptor binding affinity relationships and a model for the receptor binding site. *Steroids* 62:268–303.

Anzick SL, Kononen J, Walker RL, Azorsa DO, Tanner MM, Guan XY, Sauter G, Kallioniemi OP, Trent JM, Meltzer PS. 1997. AIB1, a steroid receptor coactivator amplified in breast cancer and ovarian cancer. *Science* 277:965–968.

Archibong AE, Maurer RR, England DC, Stormshak F. 1992. Influence of sexual maturity of donor on in vivo survival of transferred porcine embryos. *Biol Reprod* 47:1026–1030.

Arnold AP, Gorski, RA. 1984. Gonadal steroid induction of sexual differences in the CNS. *Ann Rev Neurosci* 7:413–442.

Atz JW. 1964. Intersexuality in fishes. In: Amstrong CN, Marshall AJ, editors. Intersexuality in vertebrates including man. New York: Academic Press. p 145–232.

Ayson FG, Lam TJ. 1993. Thyroxine injection of female rabbitfish *Siganus guttatus* broodstock—changes in thyroid-hormone levels in plasma, eggs, and yolk-sac larvae, and its effect on larval growth and survival. *Aquaculture* 109:83–93.

Bagenal T. 1969. The relationship between food supply and fecundity in brown trout *Salmo trutta. J Fish Biol* 1:167–182.

Bandyopahyay A, Bandyopadhyay J, Choi HH, Choi HS, Kwon HB. 1998. Plasma membrane mediated action of progesterone in amphibian (*Rana dybowskii*) oocyte maturation. *Gen Comp Endocrinol* 109:293–301.

Bannister R, Safe S. 1987. Synergistic interactions of 2,3,7,8-TCDD and 2,2',4,4',5,5'-hexachlorobiphenyl in C57BL/6J and DBA/2J mice: role of the Ah receptor. *Toxicology* 44:159–169.

Barber DL, Sanders EJ, Aebersold R, Schneider WJ. 1991. The receptor for yolk lipoprotein deposition in the chicken oocyte. *J Biol Chem* 266:18761–18770.

Barber I, Baird DJ, Calow P. 1990. Clonal variation in general responses of *Daphnia magna* Straus to toxic stress. II. Physiological effects. *Functional Ecol* 4:409–414.

Barnard JA, Jyons RM, Moses HL. 1990. The cell biology of transforming growth factor β. *Biochim Biophys Acta* 1032:79–87.

Baroiller JF, Nakayama I, Foresti F, Chourrout D. 1996. Sex determination studies in 2 species of teleost fish *Oreochromis-Niloticus* and *Leporinus-Elongatus*. *Zool Stud* 35(4):279–285.

Bartell SM. 1990. Ecosystem context for estimating stress-induced reductions in fish populations. *Am Fish Soc Symp* 8:167–182.

Barter RA, Klassen CD. 1994. Reduction of thyroid hormone levels and alteration of thyroid function by four representative UDP-glucuronosyltransferase inducers in rats. *Toxicol Appl Pharmacol* 128:9–17.

Barton BA, Schreck CB, Sigismondi LA. 1986. Multiple acute disturbances evoke cumulative physiological stress responses in juvenile chinook salmon. *Trans Am Fish Soc* 115:245–251.

Beacham TD, Murray CB. 1985. Effect of female size, egg size and water temperature on chum salmon *Oncorhynchus keta* from the Nitinat River, British Columbia. *Can J Fish Aquat Sci* 42:1755–1765.

Beacham T, Murray C. 1987. Adaptive variation in body size, age, morphology, egg size, and developmental biology of chum salmon *Oncorhynchus keta* in British Columbia. *Can J Fish Aquat Sci* 44:244–261.

Beacham TD, Murray CB. 1993. Fecundity and egg size variation in North American Pacific salmon *Oncorhynchus*. *J Fish Biol* 42:485–508.

Beitinger TI. 1990. Behavioral reactions for the assessment of stress in fishes. *J Great Lakes Res* 16:495–528.

Benfey TJ, Donaldson EM, Owen TG. 1989. An homologous radioimmunoassay for coho salmon *Oncorhynchus kisutch* vitellogenin, with general applicability to other Pacific salmonids. *Gen Comp Endocrinol* 75:78–82.f

Bergeron JM, Crews D, McLachlan JA. 1994. PCBs as environmental estrogens: turtle sex determination as a biomarker of environmental contamination. *Environ Health Perspect* 102:780–781.

Bernal J, Nunez J. 1995. Thyroid hormones and brain development. *Eur J Endocrinol* 133:390–398.

Berridge MJ. 1993. Inositol trisphosphate and calcium signaling. *Nature* 361:315–325.

Biebach H. 1996. Energetics of winter and migratory fattening. In: Carey C, editor. Avian energetics and nutritional ecology. New York: Chapman and Hall. p 280–323.

Billard R, Breton B, Richard M. 1981. On the inhibitory effect of some steroids on spermatogenesis in adult rainbow trout (*Salmo gairdneri*). *Can J Zool* 59:1479–1487.

Birge WJ, Hoyt RD, Black JA, Kercher MD, Robinson WA. 1993. Effects of chemical stresses on behavior of larval and juvenile fishes and amphibians. In: Fuiman LA, editor. Water quality and the early life stages of fishes: selected papers from the 16th Annual Larval Fish Conference; 16–20 June 1992; Kingston RI. Bethesda MD: American Fisheries Society.

Blackshear PJ, Nairn AC, Kuo JF. 1988. Protein kinases 1988: a current perspective. *FASEB J* 2:2957–2969.

Blaxter JHS. 1988. Pattern and variety in development. In: Hoar WS, Randall DJ, editors. Fish physiology. Volume 11A. San Diego CA: Academic Press. p 1–58.

Blom JH, Dabrowski K. 1995. Reproductive success of female rainbow trout *Oncorhynchus mykiss* in response to graded dietary ascorbyl monophosphate levels. *Biol Reprod* 52:1073–1080.

Bolander FF. 1994. Molecular endocrinology, 2nd ed. San Diego CA: Academic Press. 601 p.

Bortone SA, Davis WP. 1994. Fish intersexuality as an indicator of environmental stress. *Bioscience* 44:165–172.

Bortone SA, Davis WP, Bundrick CM. 1989. Morphological and behavioral characters in mosquitofish as potential bioindication of exposure to kraft mill effluent. *Bull Environ Contam Toxicol* 43:370–377.

Bouck GR, Johnson DA. 1979. Medication inhibits tolerance to seawater in coho salmon smolts. *Trans Am Fish Soc* 108:63–66.

Bradford CS, Fitzpatrick MS, Schreck CB. 1992. Evidence for ultra-short-loop feedback in ACTH-induced interrenal steroidogenesis in coho salmon: acute self-suppression of cortisol secretion in vitro. *Gen Comp Endocrinol* 87: 292–299.

Braw-tal R, McNatty KP, Smith P, Heath DA, Hudson NL, Philips DJ, McCleod BJ, Davis GH. 1993. Ovaries of ewes homozygous for the X-linked inverdale gene (Fec XI) are devoid of secondary and tertiary follicles but contain many abnormal structures. *Biol Reprod* 49:895–907.

Breeder CM, Rosen DE. 1966. Modes of reproduction in fishes. Garden City NY: Natural History Press. 941 p.

Breton B, Govoroun M, Mikolajczyk T. 1998. GTH I and GTH II secretion profiles during the reproductive cycle in female rainbow trout: relationship with pituitary responsiveness to GnRH-A stimulation. *Gen Comp Endocrinol* 111:38–50.

Bromage N, Jones J. 1991. The effects of seasonal alterations in ration on fecundity and maturation in rainbow trout *Oncorhynchus mykiss* . In: Scott AP, Sumpter JP, Kime DE, Rolfe MS, editors. Proceedings of the Fourth International Symposium on the Reproductive Physiology of Fish. Sheffield, UK: Fish Symp 91.

Bromage NR, Roberts RJ, editors. 1995. Broodstock management and egg and larval quality. Oxford, UK: Blackwell Science Ltd. 424 p.

Bromage NR, Hardiman P, Jones J, Springate J, Bye V. 1990. Fecundity, egg size and total egg volume differences in 12 stocks of rainbow trout *Oncorhynchus mykiss*. *Aquaculture Fish Manage* 21:269–284.

Bromage N, Jones J, Randall C, Thrush M, Davies B, Springate J, Duston J, Barker G. 1992. Broodstock management, fecundity, egg quality and the timing of egg-production in the rainbow-trout *Oncorhynchus mykiss. Aquaculture* 100:141–166.

Bromage N, Bruce M, Basavaraja N, Rana K. 1994. Egg quality determinants in finfish: the role of overripening with special reference to the timing of stripping in the Atlantic halibut *Hippoglossus hippoglossus. J World Aquaculture Soc* 25:13–21.

Brooks S, Pottinger TG, Tyler CR, Sumpter J. 1995. Does cortisol influence egg quality in the rainbow trout *Oncorhynchus keta*. In: Goetz F, Thomas P, editors. Proceedings of the Fifth International Symposium on the Reproductive Physiology of Fish. Austin TX: Fish Symposium '95. p 180.

Brouwer A, van den Berg KJ. 1986. Binding of a metabolite of 3,4,3',4'-tetraochlorobiphenyl to transthyretin reduces serum vitamin A transport by inhibiting the formation of protein complex carrying both retinol and thyroxin. *Toxicol Appl Pharmacol* 85:301–312.

Brown CL, Doroshov SI, Nunez JM, Hadley C, Vaneenennaam J, Nishioka RS, Bern HA. 1988. Maternal triiodothyronine injections cause increases in swimbladder inflation and survival rates in larval striped bass *Morone saxatilis. J Exp Zool* 248:168–176.

Brown CL, Doroshov SI, Cochran MD, Bern HA. 1989. Enhanced survival in striped bass fingerlings after maternal triiodothyronine treatment. *Fish Physiol Biochem* 7:295–299.

Brown CL, Kim BG. 1995. Combined application of cortisol and triiodothyronine in the culture of larval marine finfish. *Aquaculture* 135:79–89.

Brown JA, Johansen PH, Colgan PW, Mathers, RA 1985. Changes in the predator-avoidance behavior of juvenile guppies (*Poecilia reticulata*) exposed to pentachlorophenol. *Can J Zool* 63:2001–2005.

Brown-Peterson N, Thomas P, Arnold CR. 1988. Reproductive biology of the spotted seatrout *Cynoscion nebulosus* in south Texas. *Fish Bull* 86(2):373–388.

Brzozowski AM, Pike ACW, Dauter Z, Hubbard RE, Bonn T, Engstrom O, Ohman L, Greene GL, Gustafsson J-A, Carlquist M. 1997. Molecular basis of agonism and antagonism in the oestrogen receptor. *Nature* 389:753–758.

Bujo H, Hermann M, Kaderli MO, Jaconsen L, Sugawara S, Nympf J, Yamamotot T, Schneider WJ. 1994. Chicken oocyte growth is mediated by an eight ligand binding repeat member of the LDL receptor family. *EMBO* 13:5165–5175.

Burbach KM, Poland A, Bradfield CA. 1992. Cloning of the Ah-receptor cDNA reveals a distinctive lingan-activated transcription factor. *Proc Natl Acad Sci USA* 89:8185–8189.

Burggren W, Roberts JL. 1991. Respiration and metabolism. In: Ladd Prosser C, editor. Environmental and metabolic animal physiology. NY: Wiley-Liss. p. 353–436.

Byskov AG, Andersen CY, Nordholm L, Thøgersen H, Guoliang X, Wassmann O, Andersen JV, Guddal E, Roed T. 1995. Chemical structure of sterols that activate oocyte meiosis. *Nature* 374:559–562.

Callard IP, Callard GV. 1987. Sex steroid receptors and non-receptor binding proteins. In: Norris DO, Jones RE, editors. Hormones and reproduction in fishes, amphibians, and reptiles. New York: Plenum Press. p 355–384.

Callard GV, Schlinger B, Pasmanik M. 1990. Nonmammalian vertebrate models in studies of brain-steroid interactions. *J Exp Zool Suppl* 4:6–16.

Calow P, Sibly RM. 1990. A physiological basis of population processes: ecotoxicological implications. *Functional Ecol* 4:283–288.

Campbell PM, Pottinger TG, Sumpter JP. 1994. Preliminary evidence that chronic confinement stress reduces the quality of gametes produced by brown and rainbow trout. *Aquaculture* 120:151–169.

Carey C. 1996. Female reproductive energetics. In: Carey C, editor. Avian energetics and nutritional ecology. New York: Chapman and Hall. p 324–374.

Carragher JF, Sumpter JP, Pottinger TG, Pickering AD. 1989. Deleterious effects of cortisol implantation on reproductive function in two species of trout *Salmo trutta L.* and *Salmo gairdneri Richardson. Gen Comp Endocrinol* 16:310–321.

Catt KJ, Dufau ML. 1991. Gonadotropic hormones: biosynthesis, secretion, receptors, and actions. In: Yen SSC, Jaffe RB, editors. Reproductive endocrinology. Philadelphia PA: W.B. Saunders Co. p 105–155.

Cavanaugh KP, Holmes WN. 1987. Effects of ingested petroleum on the development of ovarian endocrine function in photostimulated mallard ducks *Anas platyrhynchos. Arch Environ Contam Toxicol* 16:247–253.

Chambon P. 1996. A decade of molecular biology of retinoic acid receptors. *FASEB J* 10:940–954.

Chang JP, Jobin RM. 1994. Regulation of gonadotropin release in vertebrates: a comparison of GnRH mechanisms of action. In: Davey KG, Peter RE, Tobe SS, editors. Perspectives in comparative endocrinology. Ottawa, ON, Canada: National Research Council of Canada. p 41–51.

Chang LT, Yu NW, Hsu CY, Liu HW. 1996. Gonadal transformation in male *Rana-catesbeiana* tadpoles intraperitoneally implanted with estradiol capsules. *Gen Comp Endocrinol* 102:299–306.

Chen TT. 1983. Identification and characterization of oestrogen-responsive gene products in the liver of the rainbow trout. *Can J Biochem Cell Biol* 61:802–810.

Cheung MO, Gilbert EG, Peterson RE. 1981. Cardiovascular teratogenicity of 2,3,7,8-tetracholorodibenzo-*p*-dioxin in the chick embryo. *Toxicol Appl Pharmacol* 61:197–204.

Clark EJ, Norris DO, Jones DE. 1998. Interactions of gonadal steroids and pesticides (DDT, DDE) on gonaduct growth in larval tiger salamanders, *Ambystoma tigrinum. Gen Comp Endocrinol* 109:94–105.

Cohen P. 1988. Protein phosphorylation and hormone action. *Proc Royal Soc London B* 234:115–144.

Colborn T, Clement C, editors. 1992. Chemically induced alterations in sexual and functional development: the wildlife/human connection. Volume 21, Advances in modern environmental toxicology. Princeton NJ: Princeton Scientific. 403.

Colborn T, Vom Saal FS, Soto AM. 1993. Developmental effects of endocrine-disrupting chemicals in wildlife and humans. *Environ Health Perspect* 101:378–383.

Comfort A. 1979. The biology of senescence. New York: Elsevier. 414 p.

Contreras-Sanchez W, Schreck CB, Fitzpatrick MS, Pereira CB. 1998. Effects of stress on the reproductive performance of rainbow trout *Onchorhynchus mykiss. Biol Reprod* 58:439–447.

Cooper RL, Barrett MA, Goldman JM, Rehnberg GL, McElroy WK, Stoker TE. 1994. Pregnancy alterations following xenobiotic-induced delays in ovulation in the female rat. *Fund Appl Toxicol* 22:474–480.

Copeland P, Sumpter JP, Walker TK, Croft M. 1986. Vitellogenin levels in male and female rainbow trout *Salmo gairdneri Richardson* at various stages of the reproductive cycle. *Comp Biochem Physiol* 83B:487–493.

Costa LG, Murphy SD. 1987. Interaction of the pesticide chlordimeform with adrenergic receptors in mouse brain: an in vitro study. *Arch Toxicol* 59:323–327.

Costa LG, Olibet G, Murphy SD. 1988. Alph$_2$-adrenoreceptors as a target for formamidine pesticides: in vitro and in vivo studies in mice. *Toxicol Appl Pharmacol* 93:319–328.

Costaridis P, Horton C, Zeitlinger J, Holder N, Maden M. 1996. Endogenous retinoids in the zebrafish embryo and adult. *Dev Dyn* 205:41–51

Craig JCA, Harvey SM. 1984. Egg quality in rainbow trout. The relation between egg viability, selected aspects of egg composition, and time of stripping. *Aquaculture* 40:115–134.

Craig JK, Foote CJ, Wood CC. 1996. Evidence for temperature-dependent sex determination in sockeye-salmon *Oncorhynchus-nerka*. *Can J Fish Aquat Sci* 53:141–147.

Crain DA, Guillette Jr LJ. 1997. Endocrine-disrupting contaminants and reproduction in vertebrate wildlife. *Rev in Toxicol* 1:207–231.

Crain DA, Guillette Jr LJ, Rooney AA, Pickford DB. 1997. Alterations in steroidogenesis in alligators *Alligator mississippiensis* exposed naturally and experimentally to environmental contaminants. *Environ Health Perspect* 105:528–533.

Crain DA, Guillette Jr LJ, Pickford DB, Percival HF, Woodward AR. 1998. Sex-steroid and thyroid hormone concentrations in juvenile alligators *Alligator mississippiensis* from contaminated and reference lakes in Florida. *Environ Toxicol Chem* 17:446–452.

Crews D. 1993. The organizational concept and vertebrates without sex chromosomes. *Brain Behav Evol* 42:202–214.

Crews D, Cantu AR, Rhen T, Vohra R. 1996. The relative effectiveness of estrone, estradiol-17-beta, and estriol in sex reversal in the red-eared slider, *Trachemys scripta*, a turtle with temperature-dependent sex determination. *Gen Comp Endocrinol* 102:317–326.

Currens KP. 1997. Evolution and risk in conservation of Pacific salmon [Ph.D. dissertation]. Corvallis OR: Oregon State University.

Currens KP, Hemmingsen AR, French RA, Buchanan DV, Schreck CB, Li HW. 1997. Susceptibility to disease and introgression in a wild population of rainbow trout. *N Amer J Fish Manage* 17:1065–1078.

Dabrowski K, Blom JH. 1994. Ascorbic acid deposition in rainbow trout *Oncorhynchus mykiss* eggs and survival of embryos. *Comp Biochem Physiol* A108:129–135.

Darnerud PO, Morse D, Klasson-Wehler E, Brouwer A. 1996. Binding of a 3,3',4,4'-tetracholorobiphenyl (CB-77) metabolite to fetal transthyretin and effects on fetal thyroid hormone levels in mice. *Toxicology* 106:105–114.

Dawson WR, O'Connor TP. 1996. Energetic features of avian thermoregualtory responses. In: Carey C, editor. Avian energetics and nutritional ecology. New York: Chapman and Hall. p 85–124.

DeGroot LJ, Larsen PR, Hennemann G. 1996. Thyroid hormone transport, cellular uptake, metabolism, and molecular action. In: The thyoid and its diseases. Sixth ed. New York: Churchill Livingstone. p 61–111.

de Jesus EG, Inui Y, Hirano T. 1990. Cortisol enhances the stimulating action of thyroid hormones on dorsal fin ray resorption of flounder larvae in vitro. *Gen Comp Endocrinol* 79:167–173.

de Jesus EG, , Hirano T., Inui Y. 1993. Flounder metamorphosis: its regulation by various hormones. *Fish Physiol Biochem* 11:323–328.

DeMartini EE. 1991. Annual variations in fecundity, egg size, and the gonadal and somatic conditions of queenfish *Seriphus politus* (Sciaenidae). *Fish Bull* 89:9–18.

Denslow ND, Chow M, Chow MM, Bonomelli S, Folmar LC, Heppell SA, Sullivan CV. 1997. Development of biomarkers for environmental contaminants affecting fish. In: Rolland RM, Gilbertson M, Peterson RE, editors. Chemically induced alterations in functional development and reproduction of fishes. Pensacola FL: SETAC. p 73–86.

Denver RJ. 1993. Acceleration of anuran metamorphosis by corticotropin-releasing hormone-like peptides. *Gen Comp Endocrinol* 91:38–51.

Denver RJ. 1996. Neuroendocrine control of amphibian metamorphosis. In: Gilbert LI, Tata JR, Atkinson BG, editors. Metamorphosis: postembryonic reprogramming of gene expression in amphibian and insect cells. San Diego CA: Academic Press. p 434–464.

Denver RJ. 1997. Thyroid hormone action in brain development: molecular biological aspects. In: Kawashima S and Kikuyama S, editors. Advances in comparative endocrinology. Proceedings of the XIIIth International Congress of Comparative Endocrinology. p 1071–1078.

Desbrow C, Routledge E, Sheehan D, Waldock M, Sumpter JP. 1996. The identification and assessment of oestrogenic substances in sewage treatment works effluents. Bristol, UK: Environment Agency.

DeVito MJ, Thomas T, Martin E, Umbreit TH, Gallo MA. 1992. Antiestrogenic action of 2,3,7,8-tetrachlorodibenzo-*p*-dioxin: tissue specific regulation of estrogen receptor in CD1 mice. *Toxicol Appl Pharamcol* 113:284–292.

Diamond SA, Newman MC, Mulvey M, Guttman SI. 1991. Allozyme genotype and time-to-death of mosquitofish *Gambusia holbrooki*, during acute inorganic mercury exposure: a comparison of populations. *Aquat Toxicol* 21:119–134.

Dickhoff WD. 1993. Hormones, metamorphosis and smolting. In: Schreibman MP, Scanes CG, Pang PKT, editors. The endocrinology of growth, development, and metabolism in vertebrates. San Diego CA: Academic Press. p 519–540.

Dickhoff WW, Beckman BR, Larsen DA, Mahnken CVW, Schreck CB, Sharpe C, Zaugg WS. 1995. Quality assessment of hatchery-reared spring chinook salmon smolts in the Columbia River Basin. In: Schramm HL, Piper RG, editors. Uses and effects of cultured fishes in aquatic ecosystems. Bethesda MD: American Fisheries Society. p 292–302.

Dodd JM. 1983. Reproduction in cartilaginous fishes *Chondrichthyes*. In: Hoar WS, Randall DJ, Donaldson EM, editors. Fish physiology. Volume IX, Reproduction. Part A. Endocrine tissues and hormones. New York NY: Academic Press. p 31–96.

Dodd JM, Sumpter JP. 1984. Fishes. In: Lamming GE, editor. Marshall's physiology of reproduction. Fourth ed. Volume 1, Reproductive cycles of vertebrates. Edinburgh, Scotland: Churchill-Livingstone. p 1–126.

Donaldson EM. 1996. Manipulation of reproduction in farmed fish, *Animal Reproduction Science* 42:381–392.

Donaldson EM, Fagerland UHM. 1972. Corticosteroid dynamics in Pacific salmon. *Gen Comp Endocrinol* 3:254–265.

Elkin RG, MacLachlan HM, Schneider WJ. 1995. Characterization of the Japanese quail oocyte receptor for very low density lipoprotein and vitellogenin. *J Nutr* 125:1258–1266.

Ellis L, Travere JM, Levine BA. 1991. Insulin receptor tyrosine kinase structure and function. *Biochem Soc Trans* 19:426–432.

Endler JA. 1986. Natural selection in the wild. Princeton NJ: Princeton University Press.

Ewing RD, Birks EK. 1982. Criteria for parr-smolt transformation in juvenile chinook salmon *Oncorhynchus tschawytscha*. *Aquaculture* 28:185–194.

Fantl WJ, Escobedo JA, Martin GA, Turck CW, del Rosario M, McCormick F, Williams LT. 1992. Distinct phosphotyrosines on a growth factor receptor bind to specific molecules that mediate different signaling pathways. *Cell* 267:413–423.

Fausch KD, Lyons J, Karr JR, Angermeier PL. 1990. Fish communities as indicators of environmental degradation. *Am Fish Soc Symp* 8:123–144.

Fawcett DW. 1975. Ultrastructure and function of the Sertoli cell. In: Hamilton DW, Greep RO, editors. Volume 5, Handbook of physiology. Male reproductive system. Washington DC: American Physiological Society. p 21–55.

Fawell SE, Lees JA, White R, Parker MG. 1990. Characterization and colocalization of steroid binding and dimerization activities in the mouse estrogen receptor. *Cell* 60:953–962.

Feist G, Schreck CB, Gharrett AJ. 1996. Controlling the sex of salmonids. Corvallis OR: Oregon Sea Grant. ORESH-H-96-001.

Feyk LA, Giesy JP. 1998. Xenobiotic modulation of endocrine function in birds. In: Kendall RJ, Dickerson RL, Giesy JP, Suk WP, editors. Principles and processes for evaluating endocrine disruption in wildlife. Pensacola FL: SETAC. p 121–140.

Fitzpatrick MS, Van der Kraak G, Schreck CB.1986. Profiles of plasma sex steroids and gonadotropin in coho salmon, *Oncorhynchus kisutch*, during final maturation. *Gen Comp Endocrinol* 62:437–451.

Flemming IA, Gross MR. 1990. Latitudinal clines: a trade off between egg number and size in Pacific salmon. *Ecology* 71:1–11.

Fox GA. 1993. What have biomarkers told us about the effects of contaminants on the health of fish-eating birds in the Great Lakes? The theory and a literature review. *J Great Lakes Res* 19:722–736.

Fry FEJ. 1947. Effects of the environment on animal activity. Toronto, Ontario, Canada: Fish. Res. Lab. No. 68. Toronto Univ. Press. Biol. Ser. No. 55.

Fry DM, Toone CK, Speich SM, Peard RJ. 1987. Sex ratio skew and breeding patterns of gulls: demographic and toxicological considerations. *Stud Av Biol* 10:26–43.

Galzi JL, Revah F, Bessis A, Changeux JP. 1991. Functional architecture of the nicotinic acetylcholine receptor: from electric organ to brain. *Annu Rev Pharmacol* 31:37–72.

Galton VA 1988a. The ontogeny of iodothyronine 5'-mondeiodinase activity in *Rana cateshiana* tadpoles. *Endocrinology* 122:640–645.

Galton VA. 1988b. Iodothyronine 5'-deiodinase activity in the amphibian *Rana catesbeiana* at different stages of the life cycle. *Endocrinology* 122:1746–1750.

Gandini O, Kohno H, Curtis S, Korach KS. 1997. Two transcription activation functions in the amino terminus of the mouse estrogen receptor that are affected by the carboxy terminus. *Steroids* 62:508–515.

Garside DA, Harvey PW. 1992. Endocrine toxicology of the male reproductive system. In: Atterwill CK, Flack JD, editors. Endocrine toxicology. Cambridge, UK: Cambridge University Press. p 285–312.

Ghosh S, Thomas P. 1995. Antagonistic effects of xenobiotics on steroid-induced final maturation of Atlantic croaker oocytes in vitro. *Mar Environ Res* 39:159–163.

Gilbert SF. 1997. Developmental biology. Fifth ed. Sunderland MA: Sinauer Associates. 918 p.

Gill FB. 1994. Ornithology. Second ed. New York: W. H. Freeman and Co. 720 p.

Gimeno S, Gerritsen A, Bowmer T, Komen H. 1996. Feminization of male carp. *Nature* 384:221–222.

Gimeno S, Komen H, Venderbosch PM, Bowmer T. 1997. Disruption of the sexual differentiation in genetic male common carp *Cyprinus carpio* exposed to an alkyphenol during different life stages. *Environ Sci Technol* 31:2884–2890.

Glass CK. 1994. Differential recognition of target genes by nuclear receptor monomers, dimers and heterodimers. *Endocrine Rev* 15:391–407.

Goetz FW. 1997. Follicle and extrafollicular tissue interaction in $17\alpha,10\beta$ Dehydroxy-4-pregnen-3-one-stimulated ovulation and prostaglandin synthesis in the yellow perch *Perca flavescens* ovary. *Gen Comp Endocrinol* 105:121–126.

Goetz FW, Berndtson AK, Ranjan M. 1991. Ovulation: mediators at the ovarian level. In: Pang PKT, Schreibman M, editors. Vertebrate endocrinology: fundamental and biomedical implications. San Diego CA: Academic Press. p 127–203.

Goldman JM, Cooper RL. 1993. Assessment of toxicant-induced alterations in the luteinizing hormone control of ovulation in the rat. In: Heindel JJ, Chapin RE, editors. Methods in toxicology, Volume 3B. Female reproductive toxicology. San Diego CA: Academic Press. p 79–91.

Goldman JM, Cooper RL, Edwards TL, Rehnberg GL, McElroy WK, Hein J. 1991. Suppression of the luteinizing hormone surge by chlordimeform in ovariectomized, steroid-primed female rats. *Pharmacol Toxicol* 68:131–136.

Goodman HM. 1994. Basic medical endocrinology. 2nd ed. New York: Raven Press. 332 p.

Gorbman A, Dickhoff WW, Vigna SR, Clark NB, Ralph CL. 1983. Comparative endocrinology. New York: Wiley. 592 p.

Govoni JJ, Boehlert GW, Watanabe Y. 1986. The physiology of digestion in fish larvae. *Env Biol Fishes* 16:59–77.

Grasman KA, Fox GA, Scanlon PF, Ludwig JP. 1996. Organochlorine-associated immunosuppression in prefledgling Caspian terns and herring gulls from the Great Lakes: an ecoepidemiological study. *Environ Health Perspect* 104 (suppl 4):829–842.

Hadley ME. 1996. Endocrinology. 4th ed. Englewood Cliffs NJ: Prentice Hall. 502 p.

Halachmi S, Marden E, Martin G, MacKay H, Abbondanza C, Brown M. 1994. Estrogen receptor-associated proteins: possible mediators of hormone-induced transcription. *Science* 264:1455–1458.

Hanken J, Jennings DH, Olsson L. 1997. Mechanistic basis of life-history evolution in anuran amphibians: direct development. *Amer Zool* 37:160–171.

Harel M, Tandler A, Kissil GW, Applebaum SW. 1994. The kinetics of nutrient incorporation into body-tissues of gilthead seabream *Sparus aurata* females and the subsequent effects on egg composition and egg quality. *British J Nutr* 72:45–58.

Harrison SC. 1991. A structural taxonomy of DNA-binding domains. *Nature* 353:715–719.

Harrison S. 1993. Metapopulations and conservation. In: Edwards PJ, May RM, Webb NR, editors. Large-scale ecology and conservation biology. The 35th Symposium of the British Ecological Society with the Society for Conservation Biology University of Southampton. p 111–128.

Hartwell SI, Cherry DS, Cairns Jr J. 1987. Field validation of avoidance of elevated metals by fathead minnow (*Pimephales promelas*) following in situ acclimation. *Environ Toxicol Chem* 6:189–200.

Harvey S, Scanes CG, Daughaday WH. 1995. Growth hormone. Basic, clinical and applied aspects. Boca Raton FL: CRC Press. 523 p.

Hayes TB. 1995. Interdependence of corticosterone and thyroid hormones in larval toads (*Bufo boreas*). I. Thyroid hormone-dependent and independent effects of corticosterone on growth and development. *J Exp Zool* 271:95–102.

Hayes TB. 1997. Steroids as potential modulators of thyroid hormone activity in anuran amphibians. *Amer Zool* 37:185–194.

Hayes TB. 1998. Sex determination and primary sex differentiation in amphibians: Genetic and developmental mechanisms. *J Exp Zool* 281:373–399.

Heath AG. 1995. Physiological energetics, Chapter 8, Section IX, Effects of pollutants on larval and juvenile growth. In: Water pollution and fish physiology. 2nd ed. Boca Raton FL: Lewis. p 191–195.

Henshel DS. 1998. Developmental neurotoxic effects of dioxin and dioxin-like compounds on domestic and wild avian species. *Environ Toxicol Chem* 17:88–98.

Herman RL, Kincaid HL 1988. Pathological effects of orally administered estradiol to rainbow trout. *Aquaculture*. 72:165–172.

Hess RA, Bunick D, Lee K-H, Bahr J, Taylor, JA., Korach KS, Lubahn DB. 1997. A role for oestrogens in the male reproductive system. *Nature* 390:509–512.

Hill RW, Wyse GA. 1989. Animal physiology. 2nd ed. New York: Harper and Row. 656 p.

Hoar WS. 1988. The physiology of smolting salmonids. In: Hoar WS, Randall DJ, editors. Fish physiology. New York NY: Academic Press. p 275–343.

Hochachka PA. 1991. Design of energy metabolism. In: Ladd Prosser C, editor. Environmental and metabolic animal physiology. New York: Wiley-Liss. p 325–352.

Huang L, Schreiber AM, Soffientino B, Bengtson DA, Specker JL. 1998. Metamorphosis of summer flounder (*Paralichthys dentatus*): thyroid status and the timing of gastric gland formation. *J Exp Zool* 280:413–420.

Hunter GA, Donaldson EM. 1983. Hormonal sex control and its application to fish culture. In: Hoar WS, Randall DJ, Donaldson EM, editors. Fish physiology. Volume 9B. New York: Academic Press. p. 223–303.

Hunter T. 1987. A thousand and one protein kinases. *Cell* 50:822–829.

Hunter T. 1989. Protein-tyrosine phosphatases: the other side of the story. *Cell* 58:1013–1016.

Hyllner SJ, Oppen-Berntsen DO, Helvik JV, Walther BT, Haux C. 1991. Oestradiol-17β induces the major vitelline envelope proteins in both sexes in teleosts. *J Endocrinol* 131:229–236.

Idler DR, Freeman HC. 1965. A demonstration of an impaired hormone metabolism in moribund Atlantic cod *Gadus morhua L. Can J Biochem Physiol* 43:620–623.

Idler DR, and Truscott B. 1963. In vivo metabolism of steroid hormones by sockeye salmon. A. Impaired hormone clearance in mature and spawned Pacific salmon *Oncorhynchus nerka*. B. Precursors of 11-ketotestosterone. *Can J Biochem Physiol* 41:875–887.

Idler DR, Ronald AP, Schmidt PJ. 1959. Biochemical studies on sockeye salmon during spawning migration. VII. Steroid hormones in plasma. *Can J Biochem Physiol* 37:1228–1238.

Ignar-Trowbridge DM, Pimentel M, Parker MG, McLachlan JA, Korach KS. 1996. Peptide growth factor cross-talk with the estrogen receptor requires the A/B domain and occurs independently of protein kinase C or estradiol. *Endocrinol* 137:1735–1744.

Inglese J, Freedman NJ, Koch WJ, Lefkowitz RJ. 1993. Structure and mechanism of the G protein-coupled receptor kinases. *J Biol Chem* 268:23735–23738.

Inui Y, Yamano KK, Miwa S. 1995. The role of thyroid hormone in tissue development in metamorphosing flounder. *Aquaculture* 135:87–98.

Iwama GK, Pickering AD, Sumpter JP, Schreck CB. 1997. Fish stress and health in aquaculture. Society for Experimental Biology Seminar Series 62.

Jalabert B, Fostier A, Breton B, Weil C. 1991. Oocyte maturation in vertebrates. In: Pang PKT, Schredibuan MP, editors. Vertebrate endocrinology. Volume 4A, Fundamentals and biomedical implications. San Diego CA: Academic Press. p 23–90.

Janowski BA, Willy PJ, Devi TR, Falck JR, Mangelsdorf DJ. 1996. An oxysterol signaling pathway mediated by the nuclear receptor LXRa. *Nature* 383:728–731.

Jobling M. 1993. Bioenergetics: feed intake and energy partitioning. In: Rankin JC, Jensen FB, editors. Fish ecophysiology. London: Chapman and Hall. p 1–44.

Jobling S, Reynolds T, White R, Porter MG, Sumpter JP. 1995. A variety of environmentally persistent chemicals, including some phthalate plasticizers are weakly estrogenic. *Environ Health Perspect* 103:582–588.

Jobling S, Sheahan DA, Osborne JA, Matthiessen P, Sumpter JP. 1995. Inhibition of testicular growth in rainbow trout *Oncorhynchus mykiss* exposed to environmental estrogens. *Environ Toxicol Chem* 15:194–202.

Johnson AL. 1996. The avian ovarian hierarchy: a balance between follicle differentiation and atresia. *Poult Avian Biol Rev* 7:99–110.

Johnson LL, Sol SY, Lomax DP, Nelson GM, Sloan CA, Casillas E. 1997. Fecundity and egg weight in English sole *Pleuronectes vetulus* from Puget Sound, Washington: influence of nuritional status and chemical contaminants. *Fish Bull* 95:231–249.

Jones RE, Schreibman M. 1987. Ovulation: insights about the mechanisms based on a comparative approach. In: Norris DO, Jones RE, editors. Hormones and reproduction in fishes, amphibians and reptiles. New York: Plenum Press.

Joyeux, Balaguer AP, Gagne D, Nicolas JC. 1996. In vitro and in vivo interactions between nuclear receptors and estrogen response elements. *J Steroid Biochem* 58:507–515.

Kahn CR, Smith RJ, WW Chin. 1992. Mechanism of action of hormones that act at the cell surface. In: Wilson JD, Foster DW, editors. Textbook of endocrinology. 8th ed. Philadelphia PA: Saunders. p 91–134.

Kaitaranta JK, Ackman RG. 1981. Total lipids and lipid classes of fish roe. *Comp Biochem Physiol* B69:725–729.

Kaltenbach JC. 1996. Endocrinology of amphibian metamorphosis. In: Gilbert LI, Tata JR, Atkinson BG, editors. Metamorphosis: postembronic reprogramming of gene expression in amphibian and insect cells. San Diego: Academic Press. p 403–431.

Kamel F, Kubajak CL. 1987. Modulation of gonadotropin secretion by corticosterone: interaction with gonadal steroids and mechanism of action. *Endocrinology* 121:561–568.

Kato S, Endoh H, Masuhiro Y, Kitamoto T, Uchiyama S, Sasaki H, Masushige S, Gotoh Y, Nishida E, Kawashima H, Metzger D, Chambon P. 1995. Activation of the estrogen receptor through phosphorylation by mitogen-activated protein kinase. *Science* 270:1491–1494.

Kelce WR, Stone CR, Laws SC, Gray LE, Kemppainen JA, Wilson EM. 1995. Persistent DDT metabolite p,p'-DDE is a potent androgen receptor antagonist. *Nature* 375:581–585.

Khan I, Thomas P. 1996. Disruption of neuroendocrine function in Atlantic croaker exposed to Aroclor 1254. *Mar Environ Res* 42:145–149.

Khan IA, Thomas P. 1997. Aroclor 1254-induced alterations in hypothalamic monoamine metabolism in the Atlantic croaker *Micropogonias undulatus*: correlation with pituitary gonadotropin release. *Neurotoxicology* 18:553–560.

Khan IA, Thomas P. 1998. Estradiol-17 and o,p'-DDT stimulate gonadotropin release in Atlantic croaker. *Mar Environ Res* 46:149–152.

Khara I, Saatcioglu F. 1996. Antiestrogenic effects of 2,3,7,8-tetrachlorodibenzo-p-dioxin are mediated by direct transcriptional interference with the liganded estrogen receptor. *J Biol Chem* 271:10533–10537.

Kim BG, Brown CL. 1997. Interaction of cortisol and thyroid hormone in the larval development of pacific threadfin. *Am Zool* 37:470–481.

Kime DE. 1987. The steroids. In: Chester-Jones I, Ingleton PM, Phillips JG, editors. Fundamentals of comparative vertebrate endocrinology. NY: Plenum Press. p. 3–56.

Kime DE, Ebrahimi M, Nysten N, Roelants I, Rurangwa E, Morre HDM, Ollevier F. 1996. Use of computer assisted sperm analysis (CASA) for monitoring the effects of pollution on sperm quality of fish; application to the effects of heavy metals. *Aquatic Toxicol* 36:223–237.

King JA, Millar RP. 1995. Evolutionary aspects of gonadotropin-releasing hormone and its receptor. *Cell Molec Neurobiol* 15:5–23.

Kishida M, Specker JL. 1993. Vitellogenin in tilapia (*Orecochromis mossambicus*): Induction of two forms of estradiol, quantification in the plasma and characterization in oocyte extract. *Fish Physiol Biochem* 12:171–182.

Kishida M, Specker J L. 1994. Vitellogenin in the surface mucus of tilapia (*Oreochromis mossambicus*): Possibility for uptake by the free-swimming embryos. *J Exp Zool* 268:259–268.

Kishida M, Johanning KM, Bengtson DA, Specker JL. 1998. Intestinal uptake of lipovitellin from brine shrimp (*Artemia franciscana*) by larval inland silversides (*Menidia beryllina*) and striped bass (*Morone saxatilis*). *Comp Biochem Physiol* 119A:415–421.

Kjesbu OS, Solemdal P, Bratland P, Fonn M. 1996. Variation in annual egg-production in individual captive Atlantic cod *Gadus morhua*. *Can J Fish Aquat Sci* 53:610–620.

Kjeski OS, Koyvi H, Solemdal PL, Walker MG. 1989. Oogenesis in cod *Gadus morhua L.* studied by light and electron microscopy. *J Fish Biol* 34:735–746.

Kjorsvik E. 1994. Egg quality in wild and broodstock cod *Gadus morhua L. Journal-World Aquaculture Society* 25:22–31.

Kjorsvik E, Mangorjensen A, Holmefjord I. 1990. Egg quality in fishes. *Advances In Marine Biology* 26:71–113.

Klinge CM, Bambra RA, Hilf R. 1992. What differentiates anti-estrogen-liganded vs. Estradiol-liganded estrogen receptor action? *Oncol Res* 4:137–144.

Kobayashi M, Amano M, Kim MH, Yoshiura Y, John YC, Suetake H, Aida K. 1997. Gonadotropin-releasing hormone and gonadotropin in goldfish and masu salmon. *Fish Physiol Biochem* 17:1–8.

Koenig JLF, Stormshak F. 1993. Cytogenetic evaluation of ova from pubertal and 3rd-estrous gilts. *Biol Reprod* 49:1158–1162.

Komen J, Yamashita M, Nagahama Y.1992. Testicular development induced by a recessive mutation during gonadal differentiation of female common carp *Cyprinus carpio L. Develop Growth Diff* 34:535–544.

Konarzewski M, Kooijman SALM, Ricklefs RE. 1998. Models for avian growth and development. In: Starck JM, Ricklefs RE, editors. Avian growth and development. Evolution within the altricial-precocial spectrum. New York: Oxford University Press. p. 340–365.

Kraft JC, Schuh T, Juchau M, Kimelman D. 1994. The retinoid X receptor ligand, 9-cis-retinoic acid, is a potential regulator of early *Xenopus* development. *Proc Natl Acad Sci USA* 91:3067–3071.

Kramer VJ, Newman MC, Mulvey M, Ultsch GR. 1992. Glycolysis and Krebs cycle metabolites in mosquitofish, *Gambusia holbrooki*, Girard 1859, exposed to mercuric chloride: Allozyme genotype effects. *Environ Toxicol Chem* 11:357–364.

Kramer VJ, Helferich WG, Bergman A, Klasson-Wehler E, Giesy JP. 1997. Hydroxylated polychlorinated biphenyl metabolites are anti-estrogenic in a stably transfected human breast adenocarcinoma (MCF7) cell line. *Toxicol Appl Pharmacol* 144:363–376.

Krumlauf R. 1993. Hox genes and pattern formation in the branchial region of the vertebrate head. *Trends Genet* 9:106–12.

Kuiper GGJM, Enmark E, Pelto-Huikko M, Nilsson S, Gustaffson J-A. 1997. Cloning of a novel estrogen receptor expressed in rat prostate and ovary. *Proc Natl Acad Sci USA* 93:5925–5930.

Kumar V, Chambon P. 1988. The estrogen receptor binds tightly to its responsive element as a ligand-induced homodimer. *Cell* 55:145–156.

Kurebayashi S, Miyashita Y, Hirose T, Kasayama S, Akira S, Kishimoto T. 1997. Characterization of mechanisms of interleukin-6 gene repression by estrogen receptor. *J Steroid Biochem Molec Biol* 60:11–17.

Lam TJ. 1994. Hormones and egg/larval quality in fish. *Journal-World Aquaculture Society* 25:2–12.

Lamming GE. 1984. Marshall's physiology of reproduction. Volume 1: Reproductive cycles of vertebrates. Edinburgh, Scotland: Churchill Livingston. 842 p.

Lance VA. 1997. Sex determination in reptiles: an update. *Am Zool* 37:504–513.

Landers JP, Bunce NJ. 1991. The Ah receptor and the mechanism of dioxin toxicity. *Biochem J* 276:273–287.

Larsen LO. 1980. Physiology of adult lampreys, with special regard to natural starvation, reproduction, and death after spawning. *Can J Fish Aqat Sci* 37:1762–1779.

Larsson M, Pettersson T, Carlstrom A. 1985. Thyroid hormone binding serum of 15 vertebrates species: isolation of thyroxine-binding globulin and prealbumin analogs. *Gen Comp Endocrinol* 58:360–375.

Latchman DS 1990. Eukaryotic transcription factors. *Biochem J* 270:281–289.

Leatherland JF. 1992. Endocrine and reproductive function in Great Lakes Salmon. In: Colborn T, Clement C, editors. Chemically induced alterations in sexual and functional development: the wildlife/human connection. Princeton NJ: Princeton Scientific. p 129–145.

Leatherland JF. 1993. Field observations on reproductive and developmental dysfunction in introduced and native salmonids from the Great Lakes. *J Great Lakes Res* 19:737–751.

LeDouarin N. 1991. Studies on the ontogeny of the immune function in birds. *Adv Exp Med Biol* 292:19–30.

Lee AV, Weng C-N, Jackson JG, Yee D. 1997. Activation of estrogen receptor-mediated gene transcription by IGF-1 in human breast cancer cells. *J Endocrinol* 152:39–47.

Leips J, Travis J. 1994. Metamorphic responses to changing food levels in two species of hylid frogs. *Ecology* 75:1345–1356.

Leonard JL, Visser TJ. 1986. Biochemistry of deiodination. In: G. Hennemann, editor. Thyroid hormone metabolism. New York: Marcel Dekker. p 189–229.

Licht P, Pavgi S, Denver RJ. 1990. The role of hormone binding in the cold suppression of hormone stimulation of the pituitary, thyroid and testis of the turtle. *Gen Comp Endocrinol* 80:381–392.

Lin HY, Wang XF, Ng-Eaton E, Weinberg RA, Lodish HF. 1992. Expression cloning of the TGF-β type II receptor, a functional transmembrane serine/threonine kinase. *Cell* 68:775–785.

Lundholm CE. 1988. The effects of DDE, PCB and chlordane on the binding of progesterone to its cytoplasmic receptor in the eggshell gland mucosa of birds and the endometrium of mammalian uterus. *Comp Biochem Physiol* 89C:361–368.

MacLatchy DL, Van Der Kraak GJ. 1995. The phytoestrogen-sistosterol alters reproductive endocrine status of goldfish *Carassius auratus*. *Toxicol Appl Pharmacol* 134:305–312.

Maniatis T, Goodbourn S, Fischer JA. 1987. Regulation of inducible and tissue-specific gene expression. *Science* 236:1237–1244.

Manning AJ, Crim LW. 1995. Variability in egg quality and production in a batch spawning flounder *Pleuronectes ferrugineus*. In: Goetz FW, Thomas P, editors. Proceedings of the Fifth International Symposium on the Reproductive Physiology of Fish. Austin TX: University of Texas. 238 p.

Margolis B, Li N, Koch A, Mohammadi M, Hurwitz DR, Zilberstein A, Ullrich A, Pawson T, Schlessinger J. 1990. The tyrosine phosphorylated carboxyterminus of the EGF receptor is a binding site for GAP and PLC-g. *EMBO J* 9:4375–4380.

Margolis B. 1992. Proteins with SH2 domains: transducers in the tyrosine kinase signaling pathway. *Cell Growth Diff* 3:73–80.

Marletta MA. 1989. Nitric oxide: biosynthesis and biological significance. *Trends Biochem Sci* 14:488–492.

Marshall H, Morrison A, Studer M, Popperl H, Krumlauf R. 1996. Retinoids and hox genes. *FASEB J* 10:969–978

Marsh-Armstrong N, McCaffery P, Hyatt G, Alonso L, Vowling JE, Gilbert W, Drager WC. 1995. Retinoic acid in the anteroposterior patterning of the zebrafish trunk. *Arch Dev Biol* 205:103–113.

Mathies T, Andrews RA. 1996. Extended egg retention and its influence on embryonic development and egg water balance: Implications for the evolution of viviparity. *Physiol Zool* 69:1021–1035.

Mattick S, Glenn K, de Haan G, Shapiro DJ. 1997. Analysis of ligand dependence and hormone response element synergy in transcription by estrogen receptor. *J Steroid Biochem Molec Biol* 60:285–294.

Maule AG, Schreck CB. 1991. Stress and cortisol treatment changed affinity and number of glucocorticoid receptors in leukocytes and gill of coho salmon. *Gen Comp Endocrinol* 84:83–93.

Maule AG, Schreck CB, Sharpe C. 1993. Seasonal changes in cortisol sensitivity and glucocorticoid receptor affinity and number in leukocytes of coho salmon. *Fish Physiol Biochem* 10:497–506.

McArthur MLB, Fox GA, Peakall DB, Philogene BJR. 1983. Ecological significance of behavioral and hormonal abnormalities in breeding ring doves fed an organochlorine mixture. *Arch Environ Contam Toxicol* 12:343–353.

McDonnell DP, Clemm DL, Hermann T, Goldman ME, Pike JW. 1995. Analysis of estrogen receptor function in vitro reveals three distinct classes of antiestrogens. *Mol Endocrinol* 9:659–669.

McKinney JD, Fawkes J, Jordan S, Chae K, Oatley S, Coleman RE, Briner W. 1985. 2,3,7,8-tetrachlorodibenzo-*p*-dioxin (TCDD) as a potent and persistent thyroxine agonist: a mechanistic model for toxicity based on molecular reactivity. *Environ Health Perspectives* 61:41–53.

McLachlan JA, editor. 1985. Estrogens in the environment II. Influences on development. New York: Elsevier Science. 435 p.

McLachlan JA. 1993. Functional toxicology: a new approach to detect biologically active xenobiotics. *Environ Health Perspect* 101:386–387.

McNabb FMA. 1992. Thyroid hormones. Englewood Cliffs NJ: Prentice Hall. 283 p.

McNabb FMA, Hughes TE. 1983. The role of serum binding proteins in determining free thyroid hormone concentrations during development in quail. *Endocrinology* 113:957–963.

McNabb FMA, King DB. 1993. Thyroid hormone effects on growth, development and metabolism. In: Schreibman MP, Scanes CG, Pang PKT, editors. The endocrinology of growth, development, and metabolism of vertebrates. New York: Academic Press. p 393–417.

McNabb FMA, Olson JM. 1996. Development of thermoregulation and its hormonal control in precocial and altricial birds. *Poult Avian Biol Reviews* 7:111–125.

McNabb FMA, Wilson CM. 1997. Thyroid hormone deposition in avian eggs and effects on embryonic development. *Am Zool* 37:553–560.

McNabb FMA, Scanes CG, Zeeman M. 1998. Endocrine control of development. In: Starck JM, Ricklefs RE, editors. Avian growth and development. Evolution within the altricial-precocial spectrum. New York: Oxford University Press. p. 174–202.

McNatty KP, Hudson NL, Shaw L, Moore L. 1994. Plasma concentrations of FSH, LH, thyroid stimulating hormone and growth hormone after exogenous stimulation with GnRH, TRH and GHRH in Booroola ewes that are homozygous carriers or non-carriers of the Fec B gene. *J Reprod Fertility* 102:177–183.

Means AL, Gudas LJ. 1995.The roles of retinoids in vertebrate development. *Annu Rev Biochem* 64:201–33.

Meffe G, Carroll CR. 1994. Principles of conservation biology. Sunderland MA: Sinauer Associates. 600 p.

Migliaccio A, Rotondi A, Aurichio F. 1986. Estradiol receptor: phosphorylation on tyrosine in uterus and interaction with anti-phosphotyrosine antibody. *EMBO J* 5:2867–2872.

Migliaccio A, Di Domenico M, Green S, de Falco A, Kajtaniak EL, Blasi F, Chambon P, Aurichio F. 1989. Phosphorylation on tyrosine of in vitro synthesized human estrogen receptor activates its hormone binding. *Mol Endocrinol* 3:1061–1069.

Migliaccio A, Pagano M, DeGoeij CCJ, DiDomenico M, Castoria G, Sluyser M, Auricchio F. 1992. Phosphorylation and estradiol binding of estrogen receptor in hormone-dependent and hormone-independent GR mouse mammary tumors. *Int J Cancer* 51:733–739.

Millar RP, Troskie B, Sun Y-M, Ott T, Wakefield I, Myburgh D, Pawson A, Davidson JS, Flanagan C, Katz A, Hapgood J, Illing N, Weinstein H, Sealfon SC, Peter RE, Terasawa E, King J. 1997. Plasticity in the structural and functional evolution on GnRH: a peptide for all seasons. In: Kawashima S and Kikuyama S, editors. Advances in comparative endocrinology. Proceedings of the XIIIth International Congress of Comparative Endocrinology. p 15–27.

Minucci S, Ozato K. 1996. Retinoid receptors in transcriptional regulation. *Curr Opin Genet Dev* 6:567–574.

Minucci S, Saint-Jeannet JP, Toyama R, Scita G, DeLuca LM, Tiara M, Levin AA, Ozato K, Dawid IB.1996. Retinoid X receptor-selective ligands produce malformations in *Xenopus* embryos. *Proc Natl Acad Sci USA* 93:1803–1807.

Mitchell PJ, Tijan R. 1989. Transciptional regulation in mammalian cells by sequence-specific DNA binding proteins. *Science* 245:371–378.

Miwa S, Inui Y. 1987. Effects of various doses of thyroxine and triiodothyronine on the metamorphosis of the flounder *Paralichthys olivaceus. Gen Comp Endocrinol* 67:356–63.

Miwa S, Yamano K, Inui Y. 1992. Thyroid hormone stimulates gastric development in flounder larvae during metamorphosis. *J Exp Zool* 261:424–430.

Moberg GP. 1985. Biological response to stress: key to assessment of animal well-being. In Moberg GP, editor. Animal stress. Bethesda MD: American Physiological Society. p 27–49.

Morisawa M, Morisawa S. 1990. Acquisition and initiation of sperm motility. In: Gagnon C, editor. Controls of sperm motility: biological and clinical aspects. Boca Raton FL: CRC Press. p 137–151.

Morse DC, Groen D, Veerman M, Van Amerongen CJ, Koëter DBWM, Smits Van Prooije AE, Visser TJ, Koeman JH, Brouwer A. 1993. Interference of polychlorinated biphenyls in hepatic and brain thyroid hormone metabolism in fetal and neonatal rats. *Toxicol Appl Pharmacol* 122:27–33.

Morse DC, Wehler EK, Wesseling W, Koeman JH, Brouwer A. 1996. Alterations in rat brain thyroid hormone status following pre- and postnatal exposure to polychlorinated biphenyls (Aroclor 1254). *Toxicol Appl Pharmacol* 136:269–279.

Murdoch FE, Meier DA, Furlow JD, Grunwold KAA, Gorski J. 1990. Estrogen receptor binding to a DNA response element in vitro is not dependent upon estradiol. *Biochemistry* 29:8377–8385.

Nacci D, Coiro L, Kuhn A, Champlin D, Munns Jr W, Specker J, Cooper K. 1998. Nondestructive indicator of ethoxyresorufin-o-deethylase activity in embryonic fish. *Environ Toxicol Chem* 17:2481–2486.

Nagahama N. 1994. Molecular biology of oocyte maturation in fish. In: Davey KG, Peter RE, Tobe SS, editors. Perspectives in comparative endocrinology. Ottawa, Ontario, Canada: National Research Council of Canada. p 193–198.

Nagahama Y. 1983. The functional morphology of teleost gonads. In: Hoar WS, Randall DJ, Donaldson EM, editors. Fish physiology. Volume IX , Reproduction. Part A, Endocrine tissues and hormones. New York: Academic Press. p 223–275.

Nagahama Y. 1987. Endocrine control of oocyte maturation. In: Norris DO, Jones RE, editors. Hormones and reproduction in fishes, amphibians and reptiles. New York: Plenum Press. p 171–202.

Nagahama Y, Yoshikuni M, Yamashita M, Tanaka M. 1994. Regulation of oocyte maturation in fish. In: Sherwood NM, Hew CL, editors. Fish physiology, Volume XIII. New York: Academic Press. p 393–440.

Nagler JJ, Idler DR. 1989. Ovarian uptake of vitellogenin and another high density lipoprotein in winter flounder *Psuedopleuronectes americanus* and their relationship to yolk proteins. *Biochem Cell Biol* 68:330–335.

Nakamura M, Nagahama Y. 1993. Ultrastructural-study on the differentiation and development of steroid-producing cells during ovarian-differentiation in the amago salmon *Oncorhynchus rhodurus*. *Aquaculture* 112:237–251.

Napoli JL. 1996. Biochemical pathways of retinoid transport, metabolism, and signal transduction. *Clinical Immunology and Immunopathology* 80:S52–S62.

Nardulli AM, Greene GL, O'Malley BW, Katzenellenbogen BS. 1988. Regulation of progesterone receptor messenger ribonucleic acid and protein levels in MCF-7 cells by estradiol: Analysis of estrogen's effect on progesterone receptor synthesis and degradation. *Endocrinology* 122:935–944.

Navot D, Drews MR, Bergh PA, Guzman I, Karstaedt A, Scott RT, Garrisi GJ, Hofmann GE. 1994. Age-related decline in female fertility is not due to diminished capacity of the uterus to sustain embryo implantation. *Fertility and Sterility* 61:97–101.

Nishizuka Y. 1992. Intracellular signaling by hydrolysis of phospholipids and activation of protein kinase C. *Science* 258:607–614.

Noble RC. 1991. Comparative composition and utilization of yolk lipid by embryonic birds and reptiles. In: Deeming DC, Ferguson MWJ, editors. Egg incubation: its effects on embryonic development in birds and reptiles. Cambridge, UK: Cambridge University Press. p. 17–28.

Norris DO. 1997. Vertebrate endocrinology. 3rd ed. San Diego CA: Academic Press. 634 p.

Norris DO, Jones RE, editors. 1987. Hormones and reproduction in fishes, amphibians, and reptiles. New York: Plenum. 613 p.

Nuñez SB, Medin JA, Braissant O, Kemp L, Wahli W, Ozato K, Segars JH. 1997. Retinoid X receptor and peroxisome proliferator-activated receptor activate an estrogen responsive gene independent of the estrogen receptor. *Molec Cell Endocrinol* 127:27–40.

Olla BL, Davis MW, Ryer CH, Sogard SM. 1996. Behavioral determinants of distribution and survival in early stages of walleye pollock, *Theragra chaleogramma*: a synthesis of experimental studies. *Fishery Oceanography* Supplement 1:167–178.

Oppen-Berntsen DO, Gram-Jensen E, Walther BT. 1992. Zona radiata proteins are synthesized by rainbow trout (*Oncorhynchus mykiss*) hepatocytes in response to oestradiol-17β. *J Endocrinol* 135:293–302.

Ottinger MA, Abdelnabi MA. 1997. Neuroendocrine systems and avian sexual differentiation. *Am Zool* 37:514–523.

Pabo CO, Sauer RT. 1992. Transcription factors: Structural families and principles of DNA recognition. *Annu Rev Biochem* 61:1053–1095.

Paech K, Webb P, Kuiper GGJM, Nilsson S, Gustafsson J-A, Kushner PJ, Scanlan TS. 1997. Differential ligand activation of estrogen receptors ERα and ERβ at AP1 sites. *Science* 277:1508–1510.

Palmer BD, Palmer SK. 1995. Vitellogenin induction by xenobiotic estrogens in the red-eared turtle and African clawed frog. *Environ Health Perspect* 103(Suppl 4):19–25.

Palmer BD, Huth LK, Pieto DL, Selcer KW. 1998. Vitellogenin as a biomarker for xenobiotic estrogens in an amphibian model system. *Environ Toxicol Chem* 17:30–36.

Patiño R, Schreck CB, Banks JL, Zaugg WS. 1986. Effects of rearing conditions on the developmental physiology of smolting of coho salmon. *Trans Am Fish Soc* 115:828–837.

Patiño R, Thomas P. 1990. Characterization of membrane receptor activity for 17, 20, 21-trihydroxy-4-pregnen-3-one in ovaries of spotted seatrout *Cynoscion nebulosus*. *Gen Comp Endocrinol* 78:204–217.

Pawson T, Schlessinger J. 1993. SH2 and SH3 domains. *Curr Biol* 3:434–442.

Peakall DB, Fox GA. 1987.Toxicological investigations of pollutant-related effects in Great Lakes gulls. *Environ Health Perspect* 71:187–193.

Peddie M, Onagbesan M, Woolveridge I. 1993. The role of epidermal growth factor and other factors in the paracrine and autocrine control of ovarian follicular development in the domestic hen. In: Sharp PJ, editor. Avian endocrinology. Bristol, UK: Journal of Endocrinology, Burgess Science Press. p 321–330.

Perry MM, Griffin HD, Gilbert AB. 1984. The binding of very low density lipoproteins to the plasma membrane of the hen's oocyte. *Exp Cell Res* 151:433–446.

Petitte JN, Karagenc L. 1996. Growth factors during early events in avian embryo. *Poult Avian Biol Rev* 7:75–87.

Pickering AD, Pottinger TG, Carragher J, Sumpter JP. 1987. The effects of acute and chronic stress on the levels of reproductive hormones in the plasma of mature male brown trout, *Salmo trutta* L. *Gen Comp Endocrinol* 68:249–259.

Piferrer F, Donaldson EM. 1989. Gonadal ontogenesis in coho salmon *Oncorhynchus kisutch* after a single treatment with androgen or estrogen during ontogenesis. *Aquaculture* 77:251–262.

Piferrer F, Zanuy S, Carrilo M, Solar II, Devlin RH, Donaldson EM. 1994. Brief treatment with an aromatase inhibitor during sex differentiation causes chromosomally female salmon to develop as normal, functional males. *J Exp Zool* 270:255–262.

Pinter J, Thomas P. 1995a. Characterization of a progestogen receptor in the ovary of the spotted seatrout *Cynoscion nebulosus*. *Biology of Reproduction* 52:667–675.

Pinter J, Thomas P. 1995b. Studies of the nuclear progestogen receptor in the ovary of the spotted seatrout: regulation of the final stages of ovarian development. In: Goetz FW, Thomas P, editors. Proceedings of the fifth international symposium on the reproductive physiology of fish. Austin TX: University of Texas. p. 302–304.

Pinter J, Thomas P. 1997. The ovarian progestogen receptor in the spotted seatrout, *Cynoscion nebulosus*, demonstrates steroid specificity different from progesterone receptors in other vertebrates. *J Steroid Biochem Molec Biol* 60:113–119.

Planas JV, Athos J, Swanson P. 1996. Regulation of ovarian steroidogenesis in vitro by gonadotropins during sexual maturation in coho salmon (*Oncorhynchus kisutch*). In: Goetz FW, Thomas P, editors. Proceedings of the fifth international symposium on the reproductive physiology of fish. Austin TX: University of Texas. p. 296–298.

Poholek AJ, Schreiber AM, Specker JL. 1997. Effects of thyroid hormones and cortisol on the early development of summer flounder. *Am Zool.* 37:157A.

Porter WP, Green SM, Debbink NL, Carlson I. 1993. Groundwater pesticides: Interactive effects of low concentrations of carbamates aldicarb and methomyl and the triazine metribuzin on thyroxine and somatotropin levels in white rats. *J Toxicol Environ Health* 40:15–34.

Potter CL, Sipes IG, Russell DH. 1983. Hypothyroxinemia and hypothermia in rats in response to 2,3,7,8-tetracholordibenzo-*p*-dioxin administration. *Toxicol Appl Pharamacol* 69:89–95.

Pottinger TG, Pickering AD, Hurley MA. 1992. Consistency in the stress response of individuals of two strains of rainbow trout *Oncorhynchus mykiss. Aquaculture* 103:275–298.

Pough FH, Andrews RG, Cadle JE, Crump ML, Savitzky AH, Wells KD. 1998. Herpetology. Englewood Cliffs NJ: Prentice Hall. 577 p.

Powell DC, Aulerich RJ, Meadows JC, Tillitt DE, Giesy JP, Stromborg KL, Bursian SJ. 1996. Effects of 3,3',4,4',5-pentachlorobiphenyl (PCB 126) and 2,3,7,8-tetrachlorodibenzo-*p*-dioxin (TCDD) injected into the yolks of chicken *Gallus domesticus* eggs prior to incubation. *Arch Environ Contam Toxicol* 31:404–409.

Powell JFF, Zohar Y, Elizur A, Park C, Fischer WH, Craig AG, Rivier JE, Lovejoy PA, Sherwood NM. 1994. Three forms of gonadotropin-releasing hormone characterized from brain of one species. *Proc Natl Acad Sci USA* 91:12081–12085.

Prat F, Sumpter JP, Tyler CR. 1996. Validation of radioimmunoassays for two salmon gonadotropins (GTH I and GTH II) and their plasma concentrations throughout the reproductive cycle in male and female rainbow trout (*Oncorhynchus mykiss*). *Biol Reprod* 54:1375–1382.

Prat F, Coward K, Sumpter JP, Tyler CR. 1998. Molecular characterization and expression of two lipoprotein receptors in the rainbow trout, *Onchrhynchus mykiss. Biol Reprod* 58:1146–1153.

Precht I. 1958. Concepts of the temperature adaptation of unchanging reaction systems of cold-blooded animals. In: Prosser CL, editor. Physiological adaptations. Washington DC: American Physiology Society. p 50–78.

Prince R, Cooper KR. 1995a. Comparisons of the effects of 2,3,7,8-tetrachlorodibenzo-*p*-dioxin on chemically impacted and nonimpacted subpopulations of *Fundulus heteroclitus*: I. TCDD toxicity. *Environ Toxicol Chem* 14:579–587.

Prince R, Cooper KR. 1995b. Comparisons of the effects of 2,3,7,8-tetrachlorodibenzo-*p*-dioxin on chemically impacted and nonimpacted subpopulations of *Fundulus heteroclitus*: II. Metabolic considerations. *Environ Toxicol Chem* 14:589–595.

Prosser CL. 1986. Adaptational biology. Molecules to organisms. New York: Wiley. 784 p.

Ptashne M. 1988. How eukaryotic transcriptional activators work. *Nature* 335:683–689.

Purdom CE, Hardiman PA, Bye VJ, Eno NC, Tyler CR, Sumpter JP. 1994. Estrogenic effects of effluents from sewage treatment works. *Chem Ecol* 8:275–285.

Putney JW, Bird G St J. 1993. The inositol phosphate-calcium signaling system in nonexcitable cells. *Endocr Rev* 14:610–631.

Puzianowska-Kuznicka M, Damjanovski S, Shi YB. 1997. Both thyroid hormone and 9-cis retinoic acid receptors are required to efficiently mediate the effects of thyroid hormone on embryonic development and specific gene regulation in *Xenopus laevis*. *Mol Cell Biol* 17:4738–4749

Raff RA. 1996. The shape of life: genes, development, and the evolution of animal form. Chicago IL: University of Chicago Press. 520 p.

Rand GM. 1985. Behavior. In: Rand GM and Petrocelli SR, editors. Fundamentals of aquatic toxicology methods and applications. New York: Hemisphere. p 221–263.

Revelli A, Massobrio M, Tesarik J. 1998. Nongenomic actions of steroid hormones in reproductive tissues. *Endocrine Rev* 19:3–17.

Richie PJ, Peterle TJ. 1979. Effects of DDE on circulating luteinizing hormone levels in ring doves during courtship and nesting. *Bull Environ Contam Toxicol* 23:220–226.

Robbins J. 1996. Thyroid hormone transport proteins and the physiology of hormone binding. In: Braverman LE, Utiger RE, editors. Werner and Ingbar's the thyroid. A fundamental and clinical text. 7th ed. Philadelphia PA: Lippincott-Raven. p 96–110.

Robbins J, Bartelena L. 1986. Plasma transport of thyroid hormones. In: Hennemann G, editor. Thyroid hormone metabolism. New York: Marcel Dekker. p 3–38.

Rollins-Smith LA, Cohen N. 1996. Metamorphosis: an immunologically unique period in the life of the frog. In: Gilbert LI, Tata JR, Atkinson BG, editors. Metamorphosis: postembryonic reprogramming of gene expression in amphibian and insect cells. San Diego: Academic Press. p 626–646.

Rollins-Smith LA, Blair PJ, Davis AT. 1992. Thymus ontogeny in frogs: T-cell renewal at metamorphosis. *Dev Immunol* 2:207–213.

Roosen-Runge EC. 1977. The process of spermatogenesis in animals. Cambridge, UK: University of Cambridge Press. 214 p.

Rosenthal H, Alderdice DF. 1976. Sublethal effects of environmental stressors, natural and pollutional, on marine fish eggs and larvae. *J Fish Res Board Can* 33:2047–2065.

Rubin DA. 1985. Effects of pH on sex ratio in chichlids and a poecilliid *Teleostei*. *Copeia* 1985:233–235.

Rudas P, Bartha T. 1993. Thyroxine and triiodothyronine uptake by the brain of chickens. *Acta Vet Hung* 41:395–408.

Rudas P, Bartha T, Frenyo VL. 1993. Thyroid hormone deiodination in the brain of young chickens acutely adapts to changes in thyroid status. *Acta Vet Hung* 41:381–393.

Rudas P, Bartha T, Frenyo VL. 1994. Elimination and metabolism of triiodothyronine depend on the thyroid status in the brain of young chickens. *Acta Vet Hung* 42:218–230.

Ryan RJ. 1981. Follicular atresia. Some speculations of biochemical marks and mechanisms. In: Schwartz NB, Hunziker-Dunn M, editors. Dynamics of Ovarian Function. New York: Raven Press. p 1–12.

Sadler SE, Maller JL. 1982. Identification of a steroid receptor on the surface of Xenopus oocytes by photoaffinity labeling. *J Biol Chem* 257:355–361.

Safford SE, Thomas P. 1987. Effects of capture and handling on circulating levels of gonadal steroids and cortisol in the spotted seatrout *Cynoscion nebulosus*. In: Reproductive physiology of fish. St. John's, Newfoundland, Canada: Memorial University of Newfoundland. p 312–314.

Saito N, Shimada K. 1997. Control of ovarian steroidogenesis: gene expression of steroidogenic enzymes. In: Harvey S, Etches RJ, editors. Perspectives in avian endocrinology. Bristol, UK: Journal of Endocrinology Ltd. p 193–200.

Sanderson MJ, Charles AC, Boitano S, Dirksen ER. 1994. Mechanisms and function of intercellular calcium signaling. *Mol Cell Endocrinol* 98:173–187.

Sangalang GB, O'Halloran MJ. 1973. Adverse effects of cadmium on brook trout testis and on in vitro testicular androgen synthesis. *Biol Reprod* 9:394–403.

Sargent RC, Taylor PD, Gross MR. 1987. Parental care and the evolution of egg size in fishes. *Am Natur* 129:32–46.

Scanes CG, Radecki SV, Aramburo C. 1996. Growth hormone and growth factors in avian development. *Poult Avian Biol Rev* 7:89–98.

Schlessinger J, Ullrich A. 1992. Growth factor signaling by receptor tyrosine kinases. *Neuron* 9:383–391.

Schlinger BA, Arnold AP. 1995. Estrogen synthesis and secretion by the songbird brain. In: Micevych PE, Hammer Jr RP, editors. Neurobiological effects of sex steroid hormones. New York: Cambridge University Press. p 297–323.

Schmidt PJ, Idler DR. 1962. Steroid hormones in the plasma of salmon at various states of maturation. *Gen Comp Endocrinol* 2:204–214.

Schmidt-Nielsen K. 1997. Animal physiology. Adaptation and environment. Fifth edition. Cambridge, UK: Cambridge University Press. 607 p.

Schreck CB. 1981. Stress and compensation in teleostean fishes: response to social and physical factors. In: Pickering AD, editor. Stress and fish. London: Academic Press. p 295–321.

Schreck CB. 1982a. Stress and rearing of salmonids. *Aquaculture* 28:241–249.

Schreck CB. 1982b. Parr-smolt transformation and behavior. In: Brannan EL and Salo EO, editors. Proceedings salmon and trout migratory behavior symposium. Seattle WA: Univ Washington Press. p 164–172.

Schreck CB. 1990. Physiological, behavioral, and performance indicators of stress. *Am Fish Soc Symp* 8:29–37.

Schreck CB. 1993. Glucocorticoids: Metabolism, growth, and development. In: Schreibman MP, Scanes CG, Pang PKT, editors. The endocrinology of growth, development and metabolism in vertebrates. NY: Academic Press. p 367–392.

Schreck CB. 1996. Immunomodulation: endogenous factors. In: Iwama G, Nakanishi T, editors. The fish immune system. Fish physiology. Volume 15. New York: Academic Press. p 311–337.

Schreck CB, Fitzpatrick MS, Feist GW, Yeoh C-G. 1991. Steroids: developmental continuum between mother and offspring. In: Scott AP, Sumpter JP, Kime DE, Rolfe MS, editors. Proc. 4th Inter. Symp. Reprod. Physiol Fish. Sheffield, UK: Fish Symposium '91. p 256–258.

Schreck CB, Li HW. 1991. Performance capacity of fish: Stress and water quality. In: Brune DE, Tomasso JR, editors. Aquaculture and water quality. Baton Rouge LA: World Aquaculture Society. *Advances in World Aquaculture* 3:21–29.

Schreck CB, Patino R, Pring CK, Winton JR, Holway JE. 1985. Effects of rearing density on indices of smoltification and performance of coho salmon *Oncorhynchus kisutch*. *Aquaculture* 45:345–358.

Schreck CB, Olla BL, Davis MW. 1997. Behavioral responses to stress. In: Iwama GW, Sumpter J, Pickering AD, Schreck CB, editors. Fish stress and health in aquaculture. Cambridge UK: Cambridge University Press. p 145–170.

Schreiber AM, Specker JL. 1998. Metamorphosis in the summer flounder (*Paralichthys dentatus*): stage-specific developmental response to altered thyroid status. *Gen Comp Endocrinol* 111:156–166.

Schreibman MP. 1987. Aging of the neuroendocrine system. In: Norris DO, Jones RE, editors. Hormones and reproduction in fishes, amphibians, and reptiles. New York: Plenum. p 563.

Schreibman MP, Scanes CG, Pang PKT, editors. 1993. The endocrinology of growth, development, and metabolism of vertebrates. New York: Academic Press. 607 p.

Schulz RW, Bogerd J, Bosma PT, Peute J, Rebers FEM, Zanbergen MA, Goos HJT. 1995. Physiological, morphological, and molecular aspects of gonadotropins in fish with special reference to the African catfish, *Clarias gariepinus*. In: Goetz FW, Thomas P, editors. Proceedings of the fifth international symposium on reproductive physiology of fish. Austin TX: University of Texas. p 2–6.

Schwabl H. 1993. Yolk is a source of maternal testosterone for developing birds. *Neurobiology* 90:11446–11450.

Scott AP, Baynes SM. 1982. Plasma levels of sex steroids in relation to ovulation and spermiation in rainbow trout *Salmo gairdneri*. In: Richter CJJ, Goos HJT editors. Proceedings of the international symposium on reproductive physiology of fish. Pudoc Wageningen. p 103–106.

Scott AP, Canario AVM. 1987. Status of oocyte maturation inducing steroids in teleosts. In: Idler DR, Crim LW, Walsh JM, editors. Proceedings of the 3rd international symposium on the reproductive physiology of fish. St. John's, Newfoundland, Canada: Memorial University of Newfoundland. p 224–234.

Scott AP, Sumpter JP. 1983. A comparison of the female reproductive cycles of autumn-spawning and winter-spawning strains of rainbow trout (*Salmo gairdneri*). *Gen Comp Endocrinol* 52:79–85.

Selcer KW, Leavitt WW. 1988. Progesterone down-regulation of nuclear estrogen receptor: A fundamental mechanism in birds and mammals. *Gen Comp Endocrinol* 72:443–452.

Selcer KW, Leavitt WW. 1991a. Progesterone downregulates progesterone receptor, but not estrogen receptor, in the estrogen-primed oviduct of a turtle *Trachemys scripta*. *Gen Comp Endocrinol* 83:316–323.

Selcer KW, Leavitt WW. 1991b. Estrogens and progestins. In: Schriebman MP, Jones R, editors. Volume 4, Vertebrate Endocrinology: Fundamentals and Biomedical Implications. Reproduction. New York: Academic Press. p 67–114.

Selman K, Wallace RA, Sarka A, Xiaoping Q. 1993. Stages of oocyte development in the zebrafish *Brachydanio rerio. J Morphol* 218 :203–224.

Selye H. 1936. A syndrome produced by diverse nocuous agents. *Nature Lond* 138:32.

Selye H. 1976. Stress in health and disease. Boston MA: Butterworths. 1256 p.

Setchell BP. 1978. The mammalian testis. Ithaca NY: Cornell University Press. 450 p.

Sharpe RM. 1997. Do males rely on female hormones? *Nature* 390:447–448.

Shen X, Steyer E, Retzek H, Sanders EJ, Schneider WJ. 1993. Chicken oocyte growth: receptor-mediated yolk deposition. *Cell Tissue Res* 272:459–471.

Sheridan MA, Kao Y-H. Regulation of metamorphosis-associated changes in the lipid metabolism of selected vertebrates. *Am Zool* 38:350–368.

Sherwood NM, Lovejoy DA, Coe IR. 1993. Origin of mammalian gonadotropin-releasing hormones. *Endocr Rev* 14:241–254.

Shi YB, Ishizuya-Oka A. 1996. Biphasic intestinal development in amphibians: embryogenesis and remodeling during metamorphosis. In: Current topics in developmental biology. Volume 32. San Diego CA: Academic Press. p 205–235.

Shuter BJ. 1990. Population-level indicators of stress. *Am Fish Soc Symp* 8:145–166.

Sibley DR, Benovic JL, Caron MG, Lefkowitz RJ. 1987. Regulation of transmembrane signaling by phosphorylation. *Cell* 48:913–917.

Siiteri PK, Murai JT, Hammond GL, Nisker FA, Raymoure WJ, Kuhn RW. 1982. The serum transport of steroid hormones. *Recent Prog Horm Res* 38:457–510.

Sivarajah K, Franklin CS, Williams WP. 1978. The effects of polychlorinated biphenyls on plasma steroid levels and hepatic microsomal enzymes in fish. *J Fish Biol* 13:401– 409.

Slater-Cid L, Flajnik MF. 1995. Evolution and developmental regulation of the major histocompatibility complex. *Crit Rev Immunol* 15:31–75.

Slater CH, Schreck CB. 1993. Testosterone alters the immune response of chinook salmon *Oncorhynches tshawytschu. Gen Comp Endocr* 89:291–298.

Slater CH, Schreck CB. 1997. Physiological levels of testosterone kill salmonid leukocytes, in vitro. *Gen Comp Endocrinol* 106:113–119.

Slater CH, Fitzpatrick MS, Schreck CB. 1995a. Characterization of an androgen receptor in salmonid lymphocytes: possible link to androgen induced immunosuppression. *Gen Comp Endocrinol* 100:218–225.

Slater CH, Fitzpatrick MS, Schreck CB. 1995b. Androgens and immunocompetence in salmonids: specific binding in and reduced immunocompetence of salmonid lymphocytes exposed to natural and synthetic androgens. *Aquaculture* 136:363–370.

Specker JL, Sullivan CV. 1994. Vitellogenesis in fishes: status and perspectives. In: Davey KG, Peter RE, Tobe SS, editors. Perspectives in comparative endocrinology. Ottawa, Ontario, Canada: National Research Council of Canada. p 304–315.

Soyano K, Saito T, Nagae M, Yamauchi K. 1993. Effects of thyroid hormone on gonadotropin-induced steroid-production in medaka *Oryzias latipes* ovarian follicles. *Fish Physiol Biochem* 11:265–272.

Stacey NE, Caldwell JR. 1995. Hormones as sex pheromones in fish: widespread distribution among freshwater species. In: Goetz FW, Thomas P, editors. Proceedings of the fifth international symposium on the reproductive physiology of fish. Austin TX: University of Texas Press. p 244–248.

Stabel S, Parker PJ. 1991. Protein kinase C. *Pharmacol Ther* 51:71–95.

Starck JM, Ricklefs RE. 1998. Patterns of development: the altricial-precocial spectrum.In: Starck JM, Ricklefs RE, editors. Avian growth and development. Evolution within the altricial-precocial spectrum. New York: Oxford University Press. p. 3–30.

Stephens SM, Brown JA, Frankling SC. 1997. Stress responses of larval turbot, *Scophthalmus maximus L.*, exposed to sub-lethal concentrations of petroleum hydrocarbons. *Fish Physiol Biochem* 17:433–439.

Stewart JR. 1992. Placental structure and nutritional provision to embryos in predominantly lecithotrophic viviparous reptiles. *Am Zool* 32:303–312.

Sumpter JP, Jobling S. 1995. Vitellogenesis as a biomarker for estrogenic contamination of the aquatic environment. *Environ Health Perspectives* 103:173–178.

Sundaram K, Witorsch RJ. 1995. Toxic effects on the testis. In: Witorsch J, editor. Reproductive toxicology. 2nd ed. New York: Raven Press. p 99–121.

Suzuki M. 1993. Common features in DNA recognition helices of eukaryotic transcription factors. *EMBO J* 12:3221–3226.

Suzumoto BK, Schreck CB, McIntyre JD. 1977. Relative resistances of three transferring genotypes of coho salmon *Oncorhynchus kisutch* and their hematological responses to bacterial kidney disease. *J Fish Res Board Can* 34:1–8.

Tagawa M, Hirano T. 1987. Presence of thyroxine in eggs and changes in its contents during early development of chum salmon *Oncorhynchus keta*. *Gen Comp Endocrinol* 68:129–135.

Tagawa M, Hirano T. 1991. Effects of thyroid-hormone deficiency in eggs on early development of the medaka *Oryzias latipes*. *J Exp Zool* 257:360–366.

Takemura A, Takano K. 1997. Transfer of maternally derived immunoglobulin (IgM) to larvae in tilapia, *Oreochromis mossambicus*. *Fish Shellfish Immunol* 7:417–427.

Tanaka M, Tanangonan JB, Tagawa M, de Jesus EG, Nishida H, Isaka M, Kimura R, S., Hirano T. 1995. Development of the pituitary, thyroid and interrenal glands and applications of endocrinology to the improved rearing of marine fish larvae. *Aquaculture* 135:111–126.

Tanaka M, Kawai S, Seikai T, Burke JS. 1996. Development of the digestive organ system in Japanese flounder in relation to metamorphosis and settlement. *Mar Fresh Behav Physiol* 28:19–31.

Tanaka M, Kaji T, Nakamura Y, Takahashi Y. 1996. Developmental strategy of scombrid larvae: high growth potential related to food habits and precocious digestive system development. In: Survival strategies in early life stages of marine resources. Rotterdam, The Netherlands: A.A. Balkema. p 125–139.

Tata JR. 1986. Coordinated assembly of the developing egg. *Bio Essays* 4:197–200.

Tata JR. 1996. Metamorphosis: an exquisite model for hormonal regulation of post-embryonic development. *Biochem Soc Symp* 62:123–136.

Tatner MF. 1996. Natural changes in the immune system of fish. In: Iwama G, Nakanishi T, editors. The fish immune system organism, pathogen, and environment. New York: Academic Press. p 255–269.

Taylor CW. 1990. The role of G proteins in transmembrane signaling. *Biochem J* 272:1–13.

Thomas P. 1989. Effects of Aroclor 1254 and cadmium on reproductive endocrine function and ovarian growth in Atlantic croaker. *Mar Environ Res* 28:499–503.

Thomas P. 1990a. Molecular and biochemical responses of fish to stressors and their potential use in environmental monitoring. *Am Fish Soc Symp* 8:9–28.

Thomas P. 1990b. Teleost model for studying the effects of chemicals on female reproductive endocrine function. *J Exp Zool* (Suppl.) 4:126–128.

Thomas P. 1993. Effects of cadmium on gonadotropin secretion from Atlantic croaker pituitaries incubated in vitro. *Mar Environ Res* 35:141–145.

Thomas P. 1994. Hormonal control of final oocyte maturation in sciaenid fishes. In: Davey KG, Peter RE, Tobe SS, editors. Perspectives in comparative endocrinology. Ottawa, Ontario, Canada: National Research Council of Canada. p 619–625.

Thomas P. 1999. Nontraditional sites of endocrine disruption by chemicals on the hypothalamus-pituitary-gonadal axis: interactions with steroid hormone membrane receptors, monoaminergic pathways and signal transduction systems. In: Naz RK, editor. Endocrine distruptors: effects on male and female reproductive systems. Boca Raton FL: CRC Press. p 3–38.

Thomas P, Khan IA. 1997. Mechanisms of chemical interference with reproductive endocrine function in sciaenid fishes. In: Rolland RM, Gilbertson M, and Peterson RE, editors. Chemically induced alterations in functional development and reproduction of fishes. Pensacola FL: SETAC. p 29–52.

Thomas P, Patiño R. 1991. Changes in 17,20β,21-trihydroxy-4-pregnen-3-one membrane receptor concentrations in ovaries of spotted seatrout during final oocyte maturation. In: Scott AP, Sumpter J, Kime D, Rolge MS, editors. Proceedings 4th international symposium on reproductive physiology of fish. East Anglia, UK: University of East Anglia. p 122–124.

Thomas P, Smith JS. 1993. Binding of xenobiotics to the estrogen receptor of spotted seatrout: A screening assay for potential estrogenic effects. *Mar Environ Res* 35:147–151.

Thomas P, Breckenridge-Miller D, Detweiler C. 1997. Binding characteristics and regulation of the 17,20β,21-trihydroxy-4-pregnen-3-one (20β-S) receptor on testicular and sperm plasma membranes of spotted seatrout *Cynoscion nebulosus*. *Fish Physiol Biochem* 17:109–116.

Thomas P, Breckenridge-Miller D, Detweiler C. 1998. The teleost sperm membrane progestogen receptor: interactions with xenoestrogens. *Mar Environ Res* 46:163–167.

Thorpe JE. 1989. Developmental variation in salmonid populations. *J Fish Biol* 35:295–303.

Thorpe JE. 1994a. An alternative view of smolting in salmonids. *Aquaculture* 121:105–113.

Thorpe JE. 1994b. Salmonid fishes and the estuarine environment. *Estuaries* 17:76–93.

Thorpe JE, Miles MS, Keay DS. 1984. Development rate, fecundity and egg size in Atlantic salmon *Salmo salar* L. *Aquaculture* 43:289–306.

Timbrell JA. 1991. Principles of biochemical toxicology. 2nd ed. London, UK: Taylor and Francis. 415 p.

Tillitt DE, Ankley GT, Verbrugge DA, Giesy JP, Ludwig JP, Kubiak TJ. 1991. H4IIE rat hepatoma cell bioassay-derived 2,3,7,8-tetrachlorodibenzo-*p*-dioxin equivalents in colonial fish-eating waterbird eggs from the Great Lakes. *Arch Environ Contam Toxicol* 21:91–101.

Tokarz RR. 1978. Oogonial proliferation, oogenesis, and folliculogenesis in non-mammalian vertebrates. In: Jones RE, editor. The vertebrate ovary. New York: Plenum Press. p 145–179.

Torchia J, Rose DW, Inostroza J, Kamei Y, Westin S, Glass CK, Rosenfeld MG. 1997. The transcriptional co-activator p/CIP bind CBP and mediates nuclear-receptor function. *Nature* 387:677–684.

Tori GM, Mayer LP. 1981. Effects of polychlorinated biphenyls on the metabolic rates of mourning doves exposed to low ambient temperatures. *Bull Environ Contam Toxicol* 27:678–682.

Tripp RA, Maule AG, Schreck CB, Kaattari SL. 1987. Cortisol-mediated suppression of salmonid lymphocyte responses in vitro. *Develop Comp Immunol* 11:565–576.

Tsai MJ, O'Malley BW. 1994. Molecular mechanisms of action of steroid/thyroid receptor superfamily members. *Annual Rev Biochem* 63:451–486.

Tucker VA. 1971. Flight energetics in birds. *Am Zool* 11:115–124.

Tyler CR. 1991. Vitellogenesis in salmonids. In: Scott AP, Sumpter JP, Kime DE, Rolfe MS, editors. Proceedings of the 4th international symposium on the reproductive physiology of fish. Sheffield, UK. p 295–299.

Tyler CR, Lancaster PM. 1993. Isolation and characterization of the receptor for vitellogenin from follicles of the rainbow trout *Oncorhynchus mykiss. J Comp Physiol* 163:219–224.

Tyler CR, Sumpter JP. 1990. The development of a radioimmunoassay for carp *Cyprinus carpio* vitellogenin. *Fish Physiol Biochem* 8:129–140.

Tyler CR, Sumpter JP. 1996. Oocyte growth and development in teleosts. *Rev Fish Biol Fish* 6:287–318.

Tyler CR, Pottinger TG, Coward K, Prat F, Beresford N, Maddix S. 1997. Salmonid follicle-stimulating hormone (GtH I) mediates vitellogenic development of oocytes in the rainbow trout, *Oncorhynchus mykiss. Biol Reprod* 57:1238–1244.

Tyler CR, Jolding S, Sumpter JP. 1998. Endocrine disruption in wildlife: a critical review of the evidence. *Crit Rev Toxicol* 28:319–361.

Tyler CR, van Aerle R, Hutchinson T, Maddix S, Trip H. 1999. An in vitro testing system for endocrine disruptors in fish early life stages using induction of vitellogenin. *Environ Toxicol Chem* 18:337–347.

Unwin N. 1993. Neurotransmitter action: opening of ligand-gated ion channels. *Cell* 72:31–41.

Ungerer J, Thomas P. 1996. Transport and accumulation of organochlorines in the ovaries of Atlantic croaker *Mar Environ Res* 42:167–171.

Van Der Kraak G, Wade MG. 1994. A comparison of signal transduction pathways mediating gonadotropin actions in vertebrates. In: Davey KG, Peter RE, Tobe SS, editors. Perspectives in comparative endocrinology. Ottawa, Ontario, Canada: National Research Council of Canada. p 59–63.

Van Der Kraak GJ, Munkittrick KR, McMaster ME, Portt CP, Chang JP. 1992. Exposure to bleached pulp mill effluent disrupts the pituitary-gonadal axis of white sucker at multiple sites. *Toxicol Appl Pharmacol* 115:224–233.

van Overbeeke AP, McBride JR. 1967. The pituitary gland of the sockeye *Oncorhynchus nerka* during sexual maturation and spawning. *J Fish Res Board Can* 24:1791–1810.

Vega QC, Cochet C, Filhol O, Chang CP, Rhee SG, and Gill GN. 1992. A site of tyrosine phosphorylation in the C-terminus of the epidermal growth factor receptor is required to activate phospholipase C. *Mol Cell Biol* 12:128–135.

Viarengo A, Nicotera R. 1991. Possible role of Ca^{++} in heavy metal cytotoxicity. *Comp Biochem Physiol* 100E:81–84.

Vieira AV, Sanders EJ, Schneider WJ. 1995. Transport of transthyretin into chicken oocytes. *J Biol Chem* 7:2952–2956.

Visser GH. 1998. Development of temperature regulation. In: Starck JM, Ricklefs RE, editors. Avian growth and development. Evolution within the altricial-precocial spectrum. New York: Oxford University Press. p 117–156.

Vonier PM, Crain DA, McLachlan JA, Guillette Jr. LJ, Arnold SF. 1997. Interaction of environmental chemicals with the estrogen and progesterone receptors from the oviduct of the American alligator. *Environ Health Perspect* 104:1318–1322.

VonWesternhagen H. 1988. Sublethal effects of pollutants on fish eggs and larvae. In Hoar WS, Randall DJ, editors. Fish physiology. Volume 11A. San Diego CA: Academic Press. p 253–346.

Wallace RA. 1985. Vitellogenesis and oocyte growth in non-mammalian vertebrates. In: Browder LW, editor. Developmental biology. Volume 1, Oogenesis. New York: Plenum Press. p 127–177.

Wallace RA, Selman K. 1979. Physiological aspects of oogenesis in two species of sticklebacks, *Gasterosteus aculeatus* (L.) and *Apeltes quadracus* (Mitchill). *J Fish Biol* 14:551–564.

Wallace RA, Selman K. 1981. Cellular and dynamic aspects of oocyte growth in teleosts. *Am Zool* 21:325–345.

Waller CL, Oprea TI, Chae K, Park H-K, Korach KS, Laws SC, Wiese TE, Kelce WR, Gray Jr. LE. 1996. Ligand-based identification of environmental estrogens. *Chem Res Toxicol* 9:1240–1248.

Walton KM, Dixon JE. 1993. Protein tyrosine phosphatases. *Annu Rev Biochem* 62:101–120.

Waples RS. 1991. Definition of "species" under the Endangered Species Act: application to Pacific salmon. National Oceanographic and Atmospheric Administration Technical Memo. Washington DC: National Marine Fisheries Service. F/NWC-194. 29 p.

Warren CE. 1971. Biology and water pollution control. London UK: W.B. Saunders Company.

Watanabe Y. 1982. Intracellular digestion of horseradish peroxidase by the intestinal cells of teleost larvae and juveniles. *Bull Jap Soc Sci Fish* 48:37–42.

Watanabe T, Fujimura T, Lee MJ, Fukusho K, Satoh S, Takeuchi T. 1991. Nutritional studies in the sees production of fish. Effect of polar and nonpolar lipids from krill on quality of eggs of red sea bream, *Pagrus major*. *Nippon Suisan Gakkaishi-Bulletin of the Japanese Society of Scientific Fisheries* 57:695–698.

Weatherley AH. 1972. Growth and ecology of fish populations. London, UK: Academic Press. 293 p.

Weathers WW. 1996. Energetics of postnatal growth. In: Carey C, editor. Avian energetics and nutritional ecology. New York: Chapman & Hall. p 461–496.

Westphal U. 1986. Steroid-protein interactions II. Volume 27, Monographs on endocrinology. New York: Springer-Verlag. 603 p.

Whitlock Jr JP, Okino ST, Dong L, Ho HP, Clarke-Katzenburg R, Ma O, Li H. 1996. Induction of cytochrome P4501A1: a model for analyzing mammalian gene transcription. *FASEB J* 10:809–818.

Wibbels T, Gideon P, Bull JJ, Crews D. 1993. Estrogen-induced and temperature-induced medullary cord regression. *Differentiation* 53:149–154.

Wibbels T, Bull JJ, Crews D. 1994. Temperature-dependent sex determination—a mechanistic approach. *J Exp Zool* 270:71–78.

Wiens JA, Farmer AH. 1996. Population and community energetics. In: Carey C, editor. Avian energetics and nutritional ecology. New York: Chapman & Hall. p 497–526.

Willy PJ, Umesono K, Ong ES, Evans RM, Heyman RA, Mangelsdorf DJ. 1995. LXR, a nuclear receptor that defines a distinct retinoid response pathway. *Genes & Development* 9:1033–1045.

Wilson JD, Foster DW, editors. 1992. Williams' textbook of endocrinology. 8th ed. Philadelphia PA: Saunders. 1712 p.

Wilson CM, McNabb FMA. 1997. Maternal thyroid hormones in Japanese quail eggs and their influence on embryonic development. *Gen Comp Endocrinol* 107:153–165.

Wingfield JC. 1994. Modulation of the adrenocortical response to stress in birds. In: Davey KG, Peter RE, Tobe SS, editors. Perspectives in comparative endocrinology. Ottawa, Ontario, Canada: National Research Council of Canada. p 520–528.

Wingfield J, Matt KS, Farner DS. 1984. Physiologic properties of steroid hormone-binding proteins in avian blood. *Gen Comp Endocrinol* 53:281–292.

Wingfield JC, O'Reilly KM, Astheimer LB. 1995. Modulation of the adrenocortical responses to acute stress in Arctic birds: a possible ecological basis. *Am Zool* 35:285–294.

Winter GW, Schreck CB, McIntyre JD. 1979. Resistance of different stocks and transferring genotypes of coho salmon and steelhead trout to bacterial kidney disease and vibriosis. *Fish Bull* 77:795–802.

Withers PC. 1992. Comparative animal physiology. Fort Worth TX: Saunders College Publishing. 949 p.

Witschi E. 1971. Mechanisms of sexual differentiation. In: Hamburgh M, Barrington EJW, editors. Hormones in development. New York: Appleton-Century-Crofts. p 601–618.

Wittmann J, Kluger W, Petry H. 1983. Motility pattern and lung respiration of embryonic chicks under the influence of L-thyroxine and thiourea. *Comp Biochem Physiol* 75A:379–384.

Wootton RJ. 1979. Energy costs of egg production and environmental determinants of fecundity in teleost fishes. *Symp Zool Soc Lond* 44:133–159.

Working PK, Chellman GJ. 1993. The testis, spermatogenesis, and the excurrent duct system. In: Scialli AR, Zinaman MJ, editors. Reproductive toxicology and infertility. New York: McGraw-Hill. p 55–76.

Wourms JP, Lombardi J. 1992. Reflections on the evolution of piscine viviparity. *Am Zool* 32:276–293.

Yamagami K. 1988. Mechanisms of hatching in fish. In Hoar WS, Randall DJ, editors. Fish physiology. Volume 11A. San Diego CA: Academic Press. p 447–499.

Yano T, Matsuyama H. 1986. Stimulatory effect of PCB on the metabolism of sex hormones in carp hepatopancreas. *Bull Jap Soc Scient Fish* 52:1837–1852.

Yoshikuni M, Nagahama Y. 1991. Endocrine regulation of gametogenesis in fish. *Bull Inst Zool Acad Sin Monogr* 16:139–162

Yoshikuni M, Oba Y, Nagahama Y. 1995. A pertussis toxin sensitive GTP-binding protein is involved in the signal transduction pathway of the maturation-inducing hormone (17α, 20β-dihydroxy-4-pregnen-3-one) of rainbow trout *Oncorhynchus mykiss* oocytes. In: Goetz FW, Thomas P, editors. Proceedings 5th international symposium of the reproductive physiology of fish. Austin TX: University of Texas Press. p 248–350.

Young G, Adachi S, Nagahama Y. 1986. Role of ovarian thecal and granulosa layers in gonadotropin-induced synthesis of salmonid maturation-inducing substance (17α,20β-dihydroxy-4-pregnen-3-one). *Dev Biol* 118:1–8.

Youson JH. 1988. First metamorphosis. In: Hoar WS, Randall DJ, editors. Fish Physiology, Vol. XI. San Diego CA: Academic Press. p 135–196.

Youson JH. 1997. Is lamprey metamorphosis regulated by thyroid hormones? *Am Zool* 37:441–460.

Zacharewski T, Harris M, Safe S. 1991. Evidence for the mechanism of action of the 2,3,7,8-tetrachlorodibenzo-*p*-dioxin mediated decrease of nuclear estrogen receptor levels in wild type and mutant mouse Hepa 1c1c7 cells. *Biochem Pharmacol* 41:1931–1939.

Zaugg WS, McLain LR. 1972. Changes in gill adenosine-triphosphatase activity associated with parr-smolt transformation in steelhead, coho and spring chinook salmon. *J Fish Res Board Can* 29:161–171.

Zelenka DJ, Jones DE, Dunnington EA, Siegel PB. 1987. Selection for body weight at eight weeks of age. 18. Comparisons between mature and immature pullets at the same live weight and age. *Poult Sci* 66:41–46.

Zile MH. 1992. Vitamin A homeostasis endangered by environmental pollutants. *Proc Soc Exp Biol Med* 201:141–53.

CHAPTER 4

Ecological Responses of Oviparous Vertebrates to Contaminant Effects on Reproduction and Development

Kenneth A. Rose, Larry W. Brewer, Lawrence W. Barnthouse, Glen A. Fox, Nicholas W. Gard, Mary Mendonca, Kelly R. Munkittrick, Laurie J. Vitt

Recent evidence that very low-level exposures to endocrine disrupting contaminants can result in significant toxic effects to organisms has renewed attention to the more general issue of the ecological consequences of contaminants. Oviparous vertebrates, such as fish, reptiles, and birds may be especially susceptible to many contaminants. Eggs of oviparous vertebrates undergo external development and are therefore vulnerable to contaminant exposure via direct contact with the environment. Contaminants can affect the viability of early life stages or exert effects on adults that ultimately can lead to reduced survival or reproductive success. Contaminants that affect reproduction and early life stages are a major concern because the dynamics of many populations are very sensitive to changes in early-life-stage survival.

Contaminants are one of a multitude of anthropogenic-related stressors that affect oviparous vertebrate populations. Marine fisheries are declining on a worldwide basis (Botsford et al. 1997) and, although overfishing is a major cause (Garcia and Newton 1997), habitat alteration and contaminants have also been implicated as contributing factors. According to Temple (1986), habitat loss threatens 82% of the world's endangered avian species, overharvest threatens 44%, invasive exotic species threaten 35%, and less pervasive variables such as toxic chemicals and natural events threaten 12%. The primary causes of habitat loss include agriculture conversion, urban and industrial development, and deforestation. Pesticides, particularly herbicides, also can reduce habitat and alter habitat quality (Gard and Hooper 1995). The worldwide declines in amphibian populations (Vitt et al. 1990; Wake 1991) and increased frequency of malformed frogs (Ouellet et al. 1997) have brought the debate over the role of anthropogenic stressors in population declines of

CHAPTER PREVIEW

Reproductive and Developmental Effects of Contaminants in Oviparous Vertebrates. Richard T. Di Giulio and Donald E. Tillitt, editors.
©1999 Society of Environmental Toxicology and Chemistry (SETAC). ISBN 1-880611-37-6

oviparous vertebrates into the general public arena. Although some cases of population decline and species endangerment are caused by natural factors, the vast majority in modern times ultimately can be traced directly or indirectly to human activities.

Contaminants can affect individuals, populations, communities, ecosystems, and derived properties of these biological units (e.g., diversity, rates of energy cycling). Contaminant effects on levels of biological organization beyond the individual have ecologically relevant consequences. We focus on population-level responses in this chapter because of pragmatic considerations. Although we recognize that all levels of biological organization can be important in different situations, our focus is on scaling individual-level effects to the population level. It is often difficult to establish the ecological relevance of effects at lower levels than the population, and data are typically inadequate to project individual-level effects to organization levels beyond the population level. Furthermore, any community-level change necessarily employs population level changes, which for vertebrates are significant.

A major theme throughout this chapter is the use of life-history theory to attempt to generalize population responses to contaminants across taxa. The consequences of contaminant effects can vary greatly depending on the life-history strategy of the species of interest. Oviparous vertebrates exhibit a vast array of complex life cycles and reproductive strategies. Although these taxa obviously differ in the specifics of their life styles, they can be placed into a common life history strategy framework. Life-history theory permits generalizations across taxa and helps avoid the trap that every species and every contaminant must be studied in detail before population-level consequences can be inferred.

In this chapter, we present issues related to extrapolating individual-level effects of contaminants to the population level. In the "Ecological Effects" section, we present our rationale for focusing on the population level and a general overview of the empirical and modeling approaches for population analyses. In the "Life-History Framework" section, we present a life-history theory framework that uses both life cycles and life-history strategies. The "Population Monitoring and Modeling" section goes into more detail on empirical monitoring and modeling approaches appropriate for assessing population-level effects of contaminants. We advocate a life history-based monitoring approach and structured population models. In the "Contaminant Effects" section, we present examples of empirical and modeling studies of population level effects of contaminants on oviparous vertebrates. Examples are grouped into 5 categories: modeling examples that emphasize life-history strategies, persistent contaminants, nonpersistent contaminants, essential compounds, and large-scale episodic events. The section entitled "Issues Confounding Population Analyses" illustrates, with examples, 5 issues that can confound detecting and quantifying the effects of contaminants on populations. These issues are: density-dependence, spatial scales and dispersal, community interactions, cumulative effects, and stochasticity. In the final section, we synthesize our results under the themes of reptiles and amphibians, populations, monitoring, toxicity endpoints, taxa versus contaminants, and confounding factors. We conclude with a statement concerning the current state of affairs in assessing population-level effects of contaminants and a plea for multidisciplinary studies.

Ecological Effects

Individual and population-level responses

Population-level responses bridge the gap between exposure and toxicological response on one hand and risk assessment on the other hand. We view ecological effects as all effects beyond the level of the individual. Toxicology is the study of contaminant effects at the individual level or lower. We realize that our focus on the population level may be controversial, especially in light of recent emphasis on higher-order levels such as species diversity, environmental quality, and ecosystem integrity. However, current methods for extrapolating individual-level effects on oviparous vertebrates to higher levels of biological organization generally are not developed sufficiently to permit realistic assessments at levels higher than the population. For many species, even population-level assessments are tenuous because of high uncertainty in basic demographic rates (reproduction, growth, mortality) and in contaminant chemistry and toxicological information from laboratory studies or field measurements. Population responses of oviparous vertebrates are often a basic indicator of ecological effects, and some taxa provide important economic and social values. We anticipate that the feasibility of population-level assessments is rapidly increasing due to advances in modeling and monitoring methods.

For the purposes of this chapter, we define a population as a distinct group of interbreeding, or potentially interbreeding, individuals occupying a specific space and time. Important population characteristics include: genetic composition, demographic attributes (e.g., age structure, abundance), and occupation of an ecological niche. Individual-level effects can vary from biochemical to whole-organism measurements. Examples commonly measured in vertebrate studies include impaired maturation, behavioral anomalies, reduced fertility, and reduced egg viability (see Chapter 5).

Two types of population-level responses can be used to infer ecological effects. Some analyses may stop at the individual level and use the distribution of individual-level responses as the ecological endpoint. For example, estimation of the percentage of offspring that exhibit some morphological deformity may be a sufficient endpoint in some situations. Extrapolation of individual-level effects to the dynamics of the population of organisms is more ecologically relevant, but it is also more difficult. Examples of population responses to contaminants include reductions in total abundance or recruitment (survival to adult), alterations in age or size structure, distortion of sex ratio, changes in interannual variability in abundance (i.e., changes in stability), and reductions in population resilience.

Methods for scaling individual-level effects to the population

Two general approaches for scaling individual-level effects to the population level are empirical data collection and mathematical or computer modeling. Empirical data collection tends to be more credible and easier to explain because the variables of interest are directly measured using methods that are generally agreed upon by the scientific community. Advantages of modeling approaches are that they tend to be cheaper to perform, generate results more quickly, and can predict for unobserved conditions and cumulative effects. Modeling and empirical approaches are not mutually exclusive; modeling helps understand cause and effect in highly variable natural systems, thus permitting extrapolation of empirical data, and empirical data provide the basis for constructing and validating models. The most powerful and convincing population-level analyses utilize both approaches. In the "Population Monitoring and Modeling" section, we argue for the use of life-history-based monitoring and structured (matrix projection and individual-based) population modeling for assessing contaminant effects on oviparous vertebrate populations.

Life-History Framework

Life cycles

Oviparous vertebrates generally exhibit complex life cycles that vary greatly among species. The life cycle can be viewed as progression of basic life stages (egg, embryo/larva, juvenile, subadult or immature adult, sexually mature adult, senescent adult) linked by major developmental events. Many species exhibit dramatic developmental changes at various stages in their life cycle (e.g., aquatic to terrestrial; tadpoles metamorphosing to frogs; increasing in weight by several orders of magnitude; freshwater to saltwater). These developmental changes often are accompanied by major changes in habitat or ecological niche. An example of a life cycle using striped bass (*Morone saxatillis*) is shown in Figure 4-1.

Life-history strategies of fish vary greatly among species (Winemiller and Rose 1992). The striped bass example typifies the many long-lived species that utilize rivers and inshore waters as nursery grounds and reside offshore as adults. Striped bass migrate each spring into the freshwater portions of rivers where they release millions of tiny eggs to the water, with each egg having less than a 1 in a million chance to survive to spawn. Other species release large numbers of eggs, but spread their release over protracted spawning seasons. Many other species (e.g., smallmouth bass) release a few large eggs into nests, which are guarded by the males to ensure high survival of their progeny. Longevity and age of maturation can vary

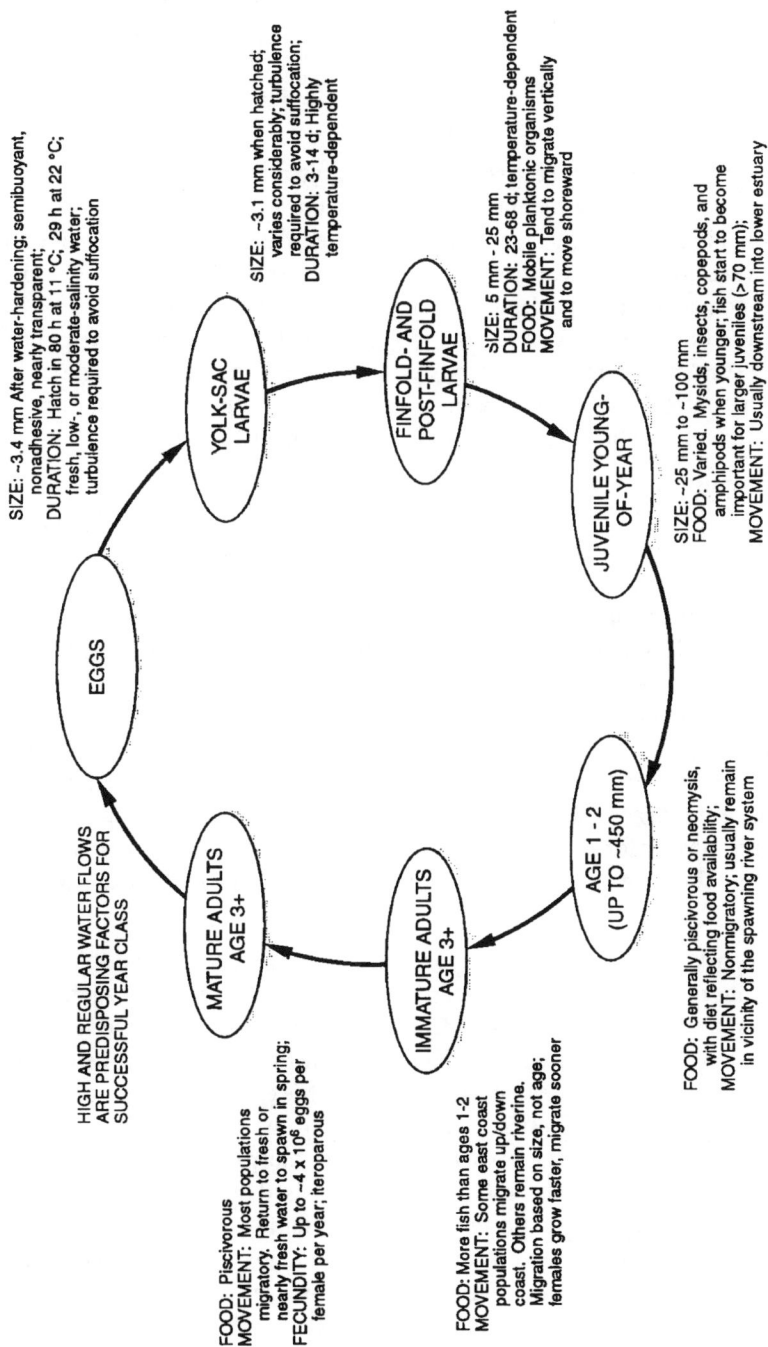

Figure 4-1 Striped bass as an illustration of a complex life cycle

EGGS

SIZE: ~3.4 mm After water-hardening; semibuoyant, nonadhesive, nearly transparent; DURATION: Hatch in 80 h at 11 °C; 29 h at 22 °C; fresh, low-, or moderate-salinity water; turbulence required to avoid suffocation

YOLK-SAC LARVAE

SIZE: ~3.1 mm when hatched; varies considerably; turbulence required to avoid suffocation; DURATION: 3-14 d; Highly temperature-dependent

FINFOLD- AND POST-FINFOLD LARVAE

SIZE: 5 mm - 25 mm DURATION: 23-68 d; temperature-dependent FOOD: Mobile planktonic organisms MOVEMENT: Tend to migrate vertically and to move shoreward

JUVENILE YOUNG-OF-YEAR

SIZE: ~25 mm to ~100 mm FOOD: Varied. Mysids, insects, copepods, and amphipods when younger; fish start to become important for larger juveniles (>70 mm); MOVEMENT: Usually downstream into lower estuary

AGE 1 - 2 (UP TO ~450 mm)

FOOD: Generally piscivorous or neomysis, with diet reflecting food availability; MOVEMENT: Nonmigratory; usually remain in vicinity of the spawning river system

IMMATURE ADULTS AGE 3+

FOOD: More fish than ages 1-2 MOVEMENT: Some east coast populations migrate up/down coast. Others remain riverine. Migration based on size, not age; females grow faster, migrate sooner

MATURE ADULTS AGE 3+

FOOD: Piscivorous MOVEMENT: Most populations migratory. Return to fresh or nearly fresh water to spawn in spring; FECUNDITY: Up to ~4 x 10⁶ eggs per female per year; iteroparous

HIGH AND REGULAR WATER FLOWS ARE PREDISPOSING FACTORS FOR SUCCESSFUL YEAR CLASS from months to decades among species. Even species in lakes and reservoirs use a variety of habitats (e.g., littoral zone versus open water) that change with ontogeny. Diets can range from tiny zooplankton to other fish, often showing this progression from larvae to adults within a species. The complex life-history strategies of fish result in great variation in the life stages exposed to contamination among different species.

A wide diversity of life-history strategies is also exhibited by amphibians (e.g., Duellman and Trueb 1986) and reptiles (e.g., Dunham and Miles 1985). Reproduction is seasonal in most species of amphibians and reptiles, and social behavior and its correlates are hormonally controlled. Social systems of adult amphibians and reptiles vary considerably depending on species, habitat characteristics, and population attributes such as density. Many species are territorial and have polygynous mating systems in which reproductive success of individual males varies greatly. Many species rely on chemical cues for social communication or for prey identification, and the chemosensory system is hormonally mediated.

Oviparous amphibians deposit eggs in water and less commonly on land or on surfaces (e.g., leaves overhanging the water, tree trunks) (Crump 1995). The eggs of most species hatch into tadpoles (frogs) or larvae (salamanders) that remain in water until metamorphosis; metamorphosis usually involves an abrupt ontogenetic change in morphology, physiology, behavior, and habitat (Wassersug 1975). Because the entire larval period is spent in aquatic environments, the potential for exposure to contaminants in the larval period is primarily through the gills, skin, and diet. In most, but not all, species of amphibians tadpoles undergo drastic metamorphosis into a juvenile and enter the terrestrial or arboreal environment. This transition results in the activation of new physiological processes and the reabsorption and redistribution of large amounts of body mass. The major routes of contaminant exposure for juvenile and adult amphibians are contact with wetted surfaces during the terrestrial or arboreal portion of their lives and through food consumption.

For the most part, oviparous reptiles deposit their eggs in nests in moist soil, inside of decaying logs or tree trunks, or in piles of humic material (e.g., dense, decaying piles of leaves). Females of some species attend the nests, but in most species, females abandon the nests and eggs are exposed to the physical environment. For many species, particularly those that are short-lived, the egg stage is the most susceptible to introduction of contaminants directly from the environment. For the most part, eggs hatch into miniature juveniles similar morphologically to adults, except that some may retain substantial yolk reserves used during early growth. Juveniles and adults may be aquatic, terrestrial, or in some instances, marine. Long-lived reptiles, particularly turtles, are known to accumulate toxic substances (Bishop et al. 1991, 1994). The primary avenue of uptake of toxic substances for juveniles and adults is through food consumption.

Birds are the only homeothermic oviparous vertebrates, and only birds fly. Life-cycles of many bird species, particularly those in temperate climates, are characterized by annual migrations, often over great distances. However, unlike fishes, reptiles, and amphibians, where some species produce live young, all birds lay hard-shelled eggs that develop outside of the mother. Body size of adult birds varies little within populations and after onset of reproduction. Adult mortality and fecundity are approximately constant after the onset of reproduction. Reproduction is highly seasonal, and most species are highly territorial during the reproductive period. Pre-laying behaviors may be more complex in birds than in other oviparous wildlife, and behaviorally mediated courtship rituals are common in many species. Within species, clutch size is proximately adjusted to make efficient use of locally available food. Birds can be characterized broadly into those species that give birth to young that are blind, unfeathered, and incompletely developed (altricial species) and those that give birth to fully developed young and that possess locomotory and foraging skills shortly after birth (precocial species). Unlike many fishes, reptiles, and amphibians, birds are characterized by the production of a few eggs with large yolk sacs. Intensive parental care of eggs and young is common. Many larger species delay the onset of reproduction, thus allowing the acquisition of skills, behavior patterns, and social status necessary to successfully raise young and optimally use resources. Because avian eggs are hard-shelled and relatively well protected from the external environment, the primary exposure pathway to contaminants for developing embryos is through maternal transfer. The primary route of uptake for juvenile and adult birds is generally through food consumption.

Complex life cycles typical of oviparous vertebrates greatly complicate population-level analyses. Many oviparous vertebrates are long-lived and are affected by variation in natural factors and stresses other than contaminants at any number of life stages. Complete population-level manifestation of contaminant effects that operate on reproduction or early life stages requires multigenerational information. Population monitoring and modeling on the time scale of decades often are needed to fully realize population-level responses to contaminants and to partition population dynamics between contaminant effects and other sources of variation.

Population monitoring and modeling also require analyses over broad spatial regions and must deal with immigration and emigration among subpopulations (also called metapopulations). Large scale migration of birds and fishes greatly complicates sampling of some life stages. Many fish species that migrate spend some portion of their early life in quasi-separate nursery grounds only to intermix later in life with individuals from other nursery areas. Immigration and emigration of young among nursery areas and intermixing of subpopulations make determination of localized contaminant effects difficult. Many oviparous vertebrates, especially terrestrial species, operate on a landscape of habitat types whose spatial pattern (e.g., degree of fragmentation) can greatly affect their demographic rates.

The complex life cycles of oviparous vertebrates can make even simple statements about population-level losses to contaminants difficult. Monitoring data can generate estimates of the number of young killed by a contaminant spill. Recasting these numbers killed as a percentage of the population, however, requires agreement and estimation of what is the total number of individuals in the population. One could use total numbers ranging from the number of individuals in the subpopulation directly impacted by the contaminant spill to the total number of individuals in all subpopulations combined. At the extreme, total numbers could include all the individuals of the species on the east coast (e.g., striped bass) or in the western U.S. (e.g., California quail, *Callipepla californica*).

Life-history strategies

Life-history theory deals with the constraints among demographic variables and the manner in which these constraints shape strategies for dealing with different environments. Life-history strategies are generalizations of the different combinations of traits and vital rates that make up the diversity found in life cycles. Life-history characteristics are important determinants of population level changes in response to exogenous stressors, including contaminant exposure. A thorough knowledge of the life history of a species is essential for understanding and predicting population responses to chemical stressors. The priority considerations for choosing species for assessment should relate to exposure and life-history characteristics that determine their vulnerability. In general, populations with delayed maturity and low fecundity (few, large eggs) are especially susceptible to contaminant effects. Examples of other potentially important life-history characteristics include length of the endogenous feeding period, yolk volume, diet, feeding and growth rates, metamorphosis, and temperature-dependent sex determination.

Life-history patterns for oviparous vertebrates generally conform to linear or multidimensional continua. Theoretical and empirical analyses of life-history patterns typically follow the linear r-K continuum (Pianka 1970). The intrinsic rate of population increase (r) is a measure of the balance between recruitment and loss. The carrying capacity (K) of the environment represents the equilibrium population size that can be supported by available resources. K and r vary as the biotic forces impinging on the individuals of the population change, but their relative values reflect a species' reproductive potential, adult body size, longevity, the type and pattern of resource use, and other aspects of life-history characteristics. The r-strategy refers to short-lived populations characterized by individuals with short lives, early sexual maturity, large clutches/broods, and high annual reproductive effort. The K-strategy refers to long-lived populations characterized by individuals with long lives, delayed sexual maturity, small clutches/broods, and low annual reproductive effort. Some demographic and life-history attributes of r- and K-strategy species typical of oviparous vertebrates are listed in Table 4-1.

Table 4-1 Some demographic and life-history attributes associated with *r*- and *K*-strategy populations of oviparous vertebrates[1]

Attributes	r-strategy	K-strategy
Population size	Seasonally variable; highest after breeding season; lowest at beginning of breeding season; recolonization may be needed	High to low, but relatively stable from year to year; recolonization not necessary
Age structure	Seasonally and annually variable; most numerous in younger classes, least in adults; usually < 3 year classes	Adult age classes relatively stable; most numerous in adult classes; multiple year classes
Suvivorship pattern	Type III (high mortality in early life stages, then low for remainder of cohort's existence)	Type II (a constant death rate) or Type III
Mean generation time	Equivalent to age at sexual maturity	Often exceeds age at sexual maturity
Population turnover	Rapid	Variable, typically > 1.5 times age at sexual maturity to decades
Age at sexual maturity	Usually < 2 y	Usually > 4 y; can be much older
Longevity	Short, rarely > 1 y	Commonly from > 8 y to decades
Body size	Small, relative to taxonomic group	Small to large
Clutch size	Moderate to large	Small to large
Clutch frequency	Few breeding seasons; often multiple times within a season	Multiple breeding seasons; usually once per season
Annual reproductive effort	High	Low to moderate
Examples:		
Fish	Small minnows	Pike, salmon
Amphibians and reptiles	Anurans, small lizards	Turtles, alligators, snakes
Birds	Passerine birds	Large birds of prey, albatross

[1]Based on Pianka (1970) and Zug (1993)

Triangular continua have been developed for some groups of organisms, such as the one developed by Winemiller and Rose (1992) for North American fishes (Figure 4-2). Three extreme strategies (opportunistic, equilibrium, periodic) are noted, which differ in their values of juvenile survivorship, age of maturity, and fecundity. The equilibrium strategy is similar to the K-strategy of the linear r-K scheme. Spawners expend some effort in caring for their progeny, resulting in low fecundity and high juvenile survivorship. The opportunistic strategy involves early maturation, frequent reproduction over an extended spawning season, and rapid larval growth, all leading to a fast population growth rate. The opportunistic strategy differs from the r-strategy in having the smallest rather than the largest clutches. A periodic strategist exhibits delayed maturation associated with attaining sufficient size to produce many large clutches consisting of small eggs. Spawning tends to be synchronous, with optimal conditions every few years leading to high recruitments that sustain the population for the intervening years of low recruitment. Juvenile growth rates are high; small, unguarded eggs result in low juvenile survivorship.

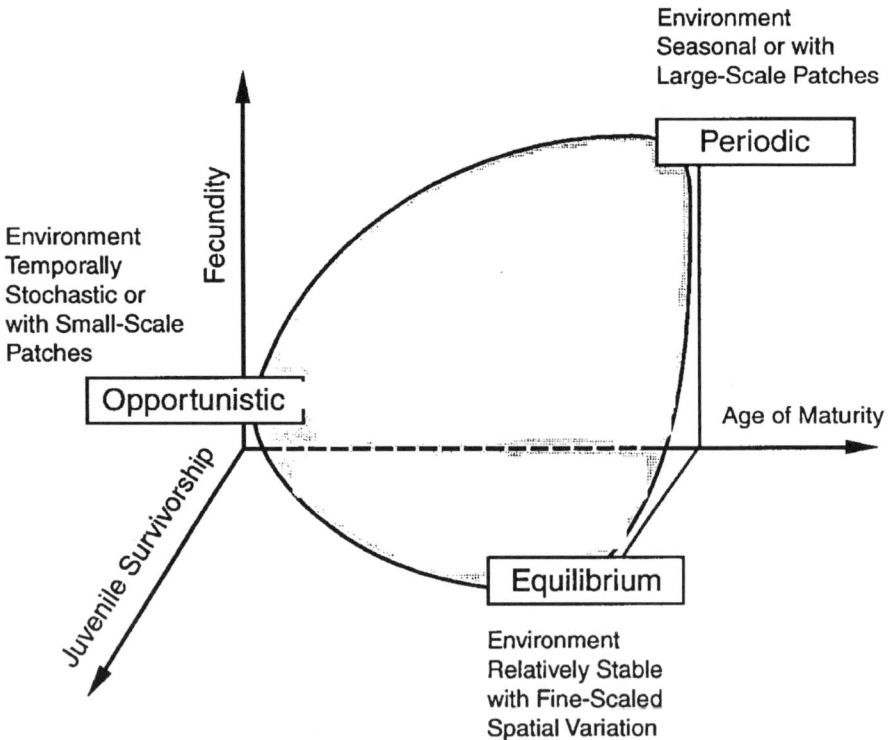

Figure 4-2 3-dimensional version of life-history strategies from Winemiller and Rose (1992). The scheme was determined by statistical analysis of the life-history traits of North American fish species. Individual species can fall anywhere on the shown surface.

While developed for fish, the triangular continuum developed by Winemiller and Rose (1992) also seems appropriate for many other taxa. Whether one uses a linear or 3-dimensional continua is not important; what is important is that some scheme is used so that the species of interest can be placed into a broader context. To illustrate, below we use the traditional r-K continua for birds, and the 3-dimensional scheme for fish, reptiles, and amphibians.

Small-bodied, short-lived bird species (r-strategists) often live at high population densities but because of greater vulnerability to various mortality factors, are prone to greater population fluctuations and hence to local variations in abundance. Many passerine bird species can be classified as r-strategists because they attempt to raise a large number of young in a year, breed when they reach 1 year old, and usually have short life-spans. Marked declines in reproductive success of r-strategy species are rapidly followed by declines in adult abundance, while increased adult mortality may have little negative effect on annual production and subsequent recruitment. The larger-bodied, long-lived species (K-strategists) generally have more stable populations and a larger non-breeding component and are more resilient to short-term reductions in reproductive rates. However, increased adult mortality may result in rapid population declines. The low population densities for many of these species make them especially vulnerable because of the large area required to maintain an adequate population. This is especially true for raptors, but also for the foraging areas of long-lived seabirds and the breeding areas of waterfowl and wading birds. The California condor (*Gymnogyps californianus*) has such a low reproductive rate that its population could not persist except for the extraordinary longevity of the adults (Mertz 1971). The species could not survive unless mortality of young adult condors is kept low and senescent death is deferred until a very late stage. Such demographic considerations point to the vulnerability of low reproducers in the face of predation and catastrophic environmental change. These larger-bodied species typically have delayed reproduction, and, therefore, populations contain a large proportion of "immatures," which may occupy different habitat from breeding individuals. Most importantly, the low reproductive rate and delayed sexual maturity of long-lived species means that they require a much longer time to recover from a population decline than do short-lived species. Regardless of life-history, however, adult mortality is usually lower than juvenile mortality in most bird species.

Amphibian life histories can be described by the triangular continuum proposed by Winemiller and Rose (1992). The life histories of many frog species roughly fit the opportunistic strategy (early maturity, low fecundity, low juvenile survival), although juvenile survival varies greatly among species. Many salamanders and caecilians, on the other hand, typically fit the equilibrium strategy (late maturity, low fecundity, high juvenile survival). Although they have very different life-history strategies, the majority of frogs and salamanders share life cycles that feature an aquatic larval stage adapted for rapid growth and a terrestrial phase adapted for dispersal (Wilbur 1980).

Reptile life histories vary both among and within major groups (Dunham and Miles 1985; Dunham et al. 1988). Crocodilians, turtles, snakes, large-bodied lizards, and the single living genus of Rhynchocephalians, *Sphenodon*, exhibit the periodic strategy of late maturity, high fecundity, and low juvenile survivorship. Adult crocodilians attend nests throughout egg development, offsetting some juvenile mortality. Most small-bodied lizards mature early, have high or low fecundity (depending on species), and have low juvenile survival.

Importance of size and environmental factors

Size is an important determinant of many of the reproductive, developmental, growth, and mortality rates that govern progression through a life cycle. Size has been the basis of scaling many biological rates (Peters 1983) and the focus of many population studies and models (Ebenman and Persson 1988; Tuljapurkar and Caswell 1997) and life-history analyses (Roff 1992). The representation of reproduction, growth, and survival in mathematical models almost always includes a component for the influence of size. Size has been shown to be important in fish for affecting their development rates, feeding efficiency, metabolic rates, fecundity, likelihood of survival, competitive interactions, and predator-prey interactions (e.g., Miller et al. 1988; Cowan et al. 1997; Sogard 1997). Sedinger et al. (1995) examined the relationship between size of female black brant (*Branta bernicla*) goslings late in their first summer (as a measure of early growth) and adult body size, breeding propensity, and fecundity. Larger goslings survived at a significantly higher rate, were significantly larger as adults, began breeding at a significantly earlier age, and laid significantly more and larger eggs than smaller birds. In snapping turtles (*Chelydra serpentina*), larger hatchlings are more successful in obtaining food, which may allow them to maximize their growth rate and, concurrently, their chances for survival (McKnight and Gutzke 1993). Size also determines time of maturation, egg size, and clutch size in snapping turtles (Gibbons et al. 1981; Congdon and Tinkle 1982), all of which affect generation time and fecundity. Size is invariably included, either explicitly or implicitly, on one or more of the axes used to distinguish among life-history strategies (e.g., Dunham and Miles 1985; Roff 1992; Winemiller and Rose 1992).

Environmental factors also can influence the life-history strategy of individual species. For example, Rowley and Russell (1991) indicate that for a given body size, tropical and subtropical bird species exhibit more K-strategy characteristics, with lower fecundity and higher survival, than temperate zone equivalents. This could be due to tropical ecosystems being more stable and less prone to environmental fluctuations than temperate ecosystems.

Even for species with closely related life histories, susceptibility to contaminant effects can vary because of unidentified physiological and environmental factors. For example, in a series of studies on pulp mill effluent impacts, lake whitefish (*Coregonus clupeaformis*) showed dramatic differences in size of mature fish, gonadal size, and circulating levels of steroid hormones, while white sucker (*Catostomus commersoni*) showed less dramatic, but still significant, differences (McMaster et al. 1991; Munkittrick, McMaster et al. 1992). In contrast, longnose sucker (*Catostamus catostomus*) and brown bullhead (*Ameiurus nebulosus*) also showed changes in levels of steroid hormones but no impacts on whole organism characteristics (Munkittrick, Van Der Kraak et al. 1992; McMaster et al. 1996) . The reasons for the differences among these species are not obvious, and both white sucker and longnose sucker show similar life-history characteristics. Differences in habitat preferences and mobility likely contributed to the difference in responses among these species.

Population Monitoring and Modeling

Empirical monitoring

We argue for a life-history-based monitoring approach, rather then the straightforward approach of directly measuring population abundances. Abundance maybe the ultimate metric of concern, but it is typically a poor early warning indicator of problems. Direct monitoring of population abundances is often not feasible and frequently is inconclusive. Ideally, one would measure population abundances over time under unimpacted and impacted conditions; differences could then be attributed to the impact. Summing over different species would yield community-level responses. In practice, however, population monitoring is often fraught with difficulties. The complex life cycles of oviparous vertebrates greatly complicates monitoring design. Many species are relatively long-lived and populations exhibit wide interannual fluctuations from natural causes, implying that years to decades of data would be needed to separate contaminant effects from other sources of variation. Many factors simultaneously affect reproduction, growth, and survival and therefore affect population abundances. These factors can include temperature variation, contaminant exposure, habitat change, species competition, and exploitation. Obtaining baseline data against which to assess the effects of any specific stressor is not easy. Monitoring does not typically begin until a population appears to have been already impacted. Substituting space for time by comparing abundances between different locations introduces the problem of determining whether sites differ due to the contaminant or due to other differences between the sites.

Life-history-based monitoring involves measurement of individual-level variables (e.g., fecundity, growth) that relate to population health. The assumptions underlying life-history-based monitoring are that individual organisms respond to stressors by altering their performance and that individuals from a variety of species with similar life histories respond similarly. Monitoring individual performance can show whether a population is stressed and, in some cases, indicate the life stage and type of stressor that is the likely cause. Individual performance measures include fecundity, age of maturity, nest success, growth rate, mean size at age, and condition. Most populations have a capacity for compensating for perturbations by density-dependent changes in their survival or reproductive rates, and changes in abundance can be buffered by the ability of individuals in the population to adjust their performance measures. Density-dependent responses can mask population-level changes, but they can be detected by changes in the life-history-related performance characteristics of individuals. Understanding the limits of how these life-history characteristics can change can have implications for determining cumulative effects of multiple stressors and the ability of the population to compensate for further perturbations. The pattern of changes in these performance characteristics sometimes also can provide clues as to the likely cause of the stress. For example, stress-induced increases in juvenile mortality can lead to different set of responses than stress-related reductions in adult growth.

In general, life-history-based monitoring is more powerful and sensitive than is abundance-based monitoring. Life-history-based monitoring can detect changes in the population, with relatively small sample sizes, that would be undetectable in noisy abundance estimates. Often, detectable changes in life-history variables are indicative of large changes in population dynamics. For most species, it is much easier and more accurate to measure individual fecundity or size than to estimate the total number of individuals in the population. This is especially true for many fish, amphibian, and reptile species that do not gather seasonally in highly restricted, predictable locations to breed and guard their young, thereby not providing a convenient method for censuring population numbers. When abundance monitoring is feasible, it is very useful. In many situations, when abundance monitoring is impractical, life-history-based monitoring offers a viable alternative.

Life-history-based monitoring has been used in many environmental studies, although it has not always been explicitly stated. Munkittrick and Dixon (1989a) developed a formal framework for life-history-based monitoring. We summarize their framework for fish monitoring here but readily acknowledge that others have used various aspects of it for other taxa. The original interpretive framework of this method was developed from concepts described by Colby (1984) to explain walleye (*Stizostedion vitreum*) population responses to stock rehabilitation efforts and was later adapted by Munkittrick, Van Der Kraak et al. (1992) to measure the potential effects of a bleached kraft mill effluent on fish. Munkittrick and colleagues generalized the framework to accommodate a wide variety of different stressors and to provide information on the likely type of stressor and life stage affected (Munkittrick and Dixon 1989a, 1989b; Gibbons and Munkittrick 1994).

The general approach uses 1 or 2 adult "sentinel" fish species as indicators of the health of fish populations (see Gibbons and Munkittrick 1994). In brief, fish characteristics were grouped according to age structure (mean age or age distribution), energy expenditure (growth rate, gonad weight, fecundity, age at maturity), and energy storage (condition, liver weight, tissue lipid levels). Based on comparisons with a reference area, a response characteristic for each grouping was assigned as either an increase (+), a decrease (–), or no change (0) for the sentinel species from the exposure area. The response pattern was then compared against generalized response patterns in order to provide direction and focus for research into the causal factors (Table 4-2). There is still some debate about the sample size of fish required to establish these measurements, but studies have shown that preliminary indications can be achieved with sample sizes of fewer than 20 adult individuals from each sex for the fish species commonly used (Munkittrick 1992). The success and accuracy of the sampling effort also depends on the reference sites selected.

The utility of the life-history-based monitoring framework is that it provides a theoretical framework to understand the stressors that determine how a population arrived at its present state. Life-history-based monitoring would be appropriate for many situations involving oviparous vertebrates. We would expect to be able to detect stress-related responses in individual performance before we could detect declines in many populations. Also, many of the variables measured as part of life-history-monitoring can be used as inputs to predictive population dynamics models. Life-history-based monitoring is especially useful when direct monitoring of abundances is clearly impractical because of long-lived species or high interannual variability in abundances.

Types of models

Population models have been widely used in the management of bird and fish populations. The motivation behind the development of many of these models was to evaluate the effects of harvest and hunting on population dynamics relative to other stresses. Recreational and commercial harvest of some of these taxa are important economically on a national scale. Scientists have been strongly motivated to develop methods for estimating the number of individuals that can be harvested without causing irreversible harm to populations and for understanding why some species can sustain more intensive harvesting than others. Both fundamental and applied questions have motivated the development of these population models. However, population models have not, as yet, been commonly used in evaluating the effects of contaminants on fish and avian populations. Because both harvesting and toxic chemicals can reduce the capacity of populations to sustain themselves, many of the methods used to model impacts of harvesting could be modified to evaluate the impacts of contaminants.

Population models can be crudely classified as aggregate, structured, or individual-based. Aggregate models treat population abundance or biomass as the state

Table 4-2 Generalized response patterns of fish populations to changes in populations[1,2]

Generalized pattern	Cause of changes	Follow-up study	Age distribution[3]	Energy utilization[4]	Energy storage[5]
Exploitation	Decreased competition between adults associated with mortality or eutrophication	Examine food resource availability and population density	Shift to younger	Increased	Increased
Recruitment failure	Shift to older age classes associated with decreased reproductive success	Detailed examination of spawning habitat, utilization, and reproductive development	Shift to older	No change	No change
Multiple stressors	Simultaneous impacts on food availability and reproductive success	Detailed studies of reproductive development and food resources	Shift to older	Decreased	Decreased
Food limitation	Increased competition associated with increased reproductive success or decreased food availability	Examine food resource availability and population density	No change	Decreased	Decreased
Niche shift	Modest increase in competition for forage base	Examine food base and competition aspects	No change	Decreased	No change
Metabolic redistribution	Inability to maximally utilize available food resources	Detailed physiological studies of energetics	Shift to younger	Mixed	Mixed
Chronic recruitment failure	Shift to population of older individuals	Detailed study of reproductive performance	Shift to older	Increased	Increased or decreased
Null response	No obvious changes	Check population size data to see if carrying capacity of the system has changed	No change	No change	No change

[1]Modified from Munkittrick and Dixon 1989a, 1989b; Gibbons and Munkittrick 1994
[2] The stressors could be chemical- or habitat-related, and the surviving populations integrate the conditions in the receiving environment (both chemical and habitat).
[3]Age distribution can be indicated by mean age or larger samples for ages of the population.
[4]Energy utilization can be reflected in growth rate, reproductive rates, or age at maturity.
[5]Energy storage can be reflected in condition factors or in lipid storage levels.

variable. The total number or biomass of individuals in the population is predicted. Structured models subdivide the total abundance into numbers in various classes. These classes are typically based on age, life stage, or size. Structured approaches can either be static or dynamic. Static models assume equilibrium conditions and generate population growth rates, or similar metrics, from class-specific survival and reproductive rates. An age-structured dynamic model would predict the number in each age class of the population through time. The ultimate structured approach is to represent each individual in the population. Individual-based modeling has been gaining popularity in ecology (DeAngelis and Gross 1992; DeAngelis et al. 1994). Below we briefly describe some commonly used static and dynamic structured modeling approaches.

A fundamental, static measure of populations is the intrinsic rate of population increase (r). The value of r under equilibrium can be determined from age-specific survival and fecundity:

$$\sum_{x=1}^{n} e^{-rx} l_x m_x = 1 \qquad \text{Equation 4-1,}$$

where l_x is age-specific probability of survival and m_x is the age-specific fecundity of an average organism of age x. Despite its simplicity, Equation 4-1 illustrates some very important aspects of population dynamics that are relevant to assessing ecological effects of contaminants that affect reproduction. First, note that the terms l_x and m_x are multiplicative, and the product of these terms determines the value of r, not the individual values. A given proportional decrease in survival at any given age is equivalent to the same proportional decrease in age-specific fecundity. In other words, from the population perspective, there is nothing qualitatively unique about fecundity effects as compared to effects on survival.

Cole (1954), followed by Lewontin (1965) and many others (reviewed by Stearns 1977), used this model to explore the consequences for population growth of different life-history patterns. One of the first results discovered was that changing the age at which a typical organism becomes sexually mature has a much greater influence on population growth than does changing its life span or its age-specific reproductive rate. Thus, for example, a contaminant that delayed sexual maturation could have a greater impact than one that simply reduced the number of eggs produced per individual. Most of these theoretical results are quite general and apply to any species; the life-history framework developed by Winemiller and Rose (1992) is a direct application of this concept to empirical data on the life histories of North American fish species.

A second static measure that is related to Equation 4-1 is the reproductive potential model. Reproductive potential P is defined as the expected contribution of a female recruit (a 1-year-old female individual) to future generations of recruits:

$$P = s_0 \sum_{i=1}^{n} s_i f_i m_i \qquad \text{Equation 4-2,}$$

where s_0 is the probability that an egg will hatch and survive to age 1, s_i is the annual probability of survival from age $i - 1$ to i, m_i is the probability of being sexually mature at age i, and f_i is the fecundity at age i. Equation 4-2 has been used by Goodyear (1977, 1988) and Barnthouse et al. (1987, 1989) to demonstrate impacts on fish populations from influences that affect early life stages. Goodyear used the reproductive potential model to explore the trade-off between power-plant-related mortality and fishing mortality. Given estimates of a minimum reproductive potential required to ensure long-term stability of the Hudson River striped bass population, Goodyear calculated the amount by which allowable harvest of the population would have to be reduced to balance a given rate of power plant mortality. Barnthouse et al. (1987, 1989) used the model to show how standard test endpoints used in fish toxicity testing could be linked to relevant life-history parameters of fish populations.

A third static measure related to reproductive potential that has been used with birds is lifetime reproductive success (LRS). Lifetime reproductive success is measured as the total number of young raised (fledgings or as recruits to the breeding population) by recognizable individuals through successive breeding attempts through most or all of their natural life span. Most species of birds are conspicuous, diurnal, and can be readily trapped and individually marked, so their survival can be monitored. For some species, eggs and young can be counted in successive breeding attempts. This type of monitoring has facilitated long-term demographic studies and investigations into the factors that influence LRS.

Clutton-Brock (1988) and Newton (1989) synthesized the major findings of longitudinal studies of LRS for 29 species of birds representing a wide array of genera, body sizes, life spans, life-styles, and mating strategies. They found that a large fraction of all fledgings produced die before they can breed, that successful individuals vary greatly in their productivity, and that the most successful individuals contribute disproportionately to the next generation. An estimated 62 to 87% of all fledgings do not contribute to the population; roughly 1 in 10 eggs is expected to produce a breeding adult. Analyses showed that breeding life span and survival of hatched young to fledging are important demographic determinants of LRS, while lifetime fecundity (total eggs laid) was relatively unimportant (Clutton-Brock 1988; Newton 1989). Individual variation in age at first breeding is an important determinant of breeding life span. Furthermore, environmental fluctuations, habitat quality, and phenotype of the reproductive adults (e.g., female body size) were particularly important in short-lived species.

Matrix projection models are the next step in complexity from the static models described above. Population growth rate (r), reproductive potential (P), and LRS can be used to explore general aspects of population dynamics but cannot provide detailed predictions concerning the dynamics of specific populations. Matrix projection models are much more flexible and suitable for population-specific applications. Matrix population models have been developed into a powerful framework for demographic analyses (Caswell 1989). Such models allow the classification of individuals by any variable of biological interest (e.g., age, size, stage); we use age for simplicity. In this approach, the number of individuals in each of an arbitrary number of age classes at time t is expressed in terms of the numbers present at time t − 1:

$$N_t = LN_{t-1}$$

Equation 4-3,

where N_t and n_{t-1} are vectors containing the numbers of organisms in each age class and L is the matrix defined by

$$L = \begin{bmatrix} s_0 f_1 & s_1 f_2 & s_2 f_3 & \cdots & s_{k-1} f_k & 0 \\ s_0 & 0 & 0 & \cdots & 0 & 0 \\ 0 & s_1 & 0 & \cdots & 0 & 0 \\ 0 & 0 & s_2 & \cdots & 0 & 0 \\ 0 & 0 & 0 & \cdots & s_k & 0 \end{bmatrix}$$

Equation 4-4,

where s_k is the age-specific probability of surviving from one time interval to the next, and f_k is the average fecundity of an organism of age k. Analogous versions of Equation 4-4 can be derived for populations divided into size classes and stage classes. Matrix models can be analyzed under equilibrium conditions (using eigenvector mathematics) to obtain metrics of population health such as growth rate and the stable age distribution. Eigenvalue analysis can also be used to compute the sensitivity, or elasticity, of the population growth rate to various elements of the projection matrix. In addition to equilibrium analyses, matrix projection models can be used to simulate time trajectories of population–age distributions. Variants of the basic matrix model can be derived by making the survival and reproduction parameters random variables, or functions of environmental parameters or population size. The matrix-projection approach has been widely used both in fisheries and avian studies (e.g., Barnthouse et al. 1990; Saila et al. 1991; McDonald and Caswell 1993). Provided realistic relationships among contaminant exposure and egg production, hatching success, survival, growth, and maturation can be derived, matrix-projection models are also applicable to the study of population-level effects of contaminants (Grant et al. 1983; Emlen 1989; Slade 1994; Landahl et al. 1997).

Individual-based models can be viewed as the ultimate structured approach. An individual-based model provides the opportunity to evaluate the influence of characteristics of individual organisms on the characteristics of populations. Two types of individual-based models are 1) distribution or p-state models that follow the probability distribution of individual characteristics (e.g., length) and 2) configuration or i-state models that involve the simulation of the activities of hundreds or thousands of individual organisms. The sum over all individuals yields population-level metrics. Distribution models are typically partial differential equations, while configuration models are often computer simulations that involve Monte Carlo methods. Distribution models allow for extensive mathematical analysis, whereas configuration models are much more flexible and allow practically any level of biological detail to be imposed on individuals (DeAngelis and Rose 1992; DeAngelis et al. 1993).

Both distribution and configuration individual-based models are being used to examine contaminant effects on animal populations. Kooijman and Metz (1984) did pioneering work on the influence of contaminants on metabolism and population growth using a distribution model of the copepod *Daphnia*. Hallam and Lassiter (Hallam et al. 1990; Lassiter and Hallam 1990) extended this approach to include a thermodynamically based model of the uptake of contaminants from aqueous media and a definition of death in terms of the internal dissolved contaminant concentration within an organism. Rose and Cowan (1993) developed an individual-based model of striped bass recruitment for the Potomac River that tracked the daily spawning, growth, and mortality of thousands of individual fish from birth to 1 year of age. Rose et al. (1993) used this model to examine both lethal and sublethal effects (e.g., reduced ability to capture prey) of a generic contaminant. Chronic and episodic exposures were simulated by modifying survival, growth, and feeding rates of individuals of various life stages. They concluded that toxic and chronic exposures generally had greater affect than temperature fluctuations and episodic exposures, juveniles were the least sensitive of the young-of-the-year (YOY) life stages, and priority research areas were realistic estimation of contaminant exposure and greater emphasis on sublethal effects.

Contaminant Effects

Below we present some examples of studies that examine the population-level consequences of contaminant exposure on oviparous vertebrates. We grouped the examples into 5 categories: life-history and general responses, persistent contaminants, nonpersistent contaminants, essential compounds, and large-scale episodic events. Studies under the life-history and general responses group mostly use models to examine how life-history strategies influence population responses to contaminants and other stressors. These studies all explicitly include life-history strategies in their analysis and do not discuss specific contaminants. The next 3 groups use studies that report the effects of specific contaminants, but these groups differ in the type of contaminant analyzed. Persistent contaminants are those that are highly resistant to environmental degradation (e.g., Hg, polychlorinated biphenyls [PCBs]). Many persistent contaminants of concern are lipophilic and tend to bioaccumulate in organisms and biomagnify in food chains. Nonpersistent compounds degrade or are metabolized quickly in the environment, are often not lipophilic, and tend to not bioaccumulate. Examples of nonpersistent compounds are some polycyclic aromatic hydrocarbons (PAHs) and organophosphorous pesticides. Essential compounds are those chemicals that are essential in small amounts for normal biological functions but that can have toxic effects at high exposures (e.g. Se, some retinoids). The final group of contaminant effects examines large-scale, episodic events. We use oil spills to illustrate the idea that large-scale, visually dramatic contaminant events do not necessarily translate into major population-level effects. Examples in each category are generally presented in order of empirical examples involving fish, birds, and reptiles and amphibians, followed by modeling examples.

We tried to maintain a balance in the cited studies between those that used modeling versus empirical approaches as well as among the different taxa. Our examples are not intended to be a comprehensive review of the literature but rather use selected examples to illustrate some general themes. We rely heavily on our own experiences. Because the life-history and generic contaminant group was dominated by studies that used modeling and other theoretical approaches, we emphasize empirical approaches in the other groups. Most of the empirical examples involve fish and birds because these species have been studied more than reptiles and amphibians. Field monitoring data on fish, and to some extent birds, tend to be from locations that have experienced a variety of anthropogenic stresses. Many fish and bird population studies are initiated because of declines in commercially or recreationally important species or species experiencing habitat declines from human activities (e.g., farming, logging). Data on reptiles and amphibians tend to come from pristine areas because investigators have tended to intentionally seek out undisturbed areas for study. Studies of amphibians and reptiles that include contaminants have focused on laboratory toxicity testing or measurement of levels of the contaminants in organism tissues. The predominance of fish and bird examples throughout this chapter reflects this difference in information availability and should not be interpreted as a statement on the relative importance of fish and birds versus reptiles and amphibians.

Life history and general responses

Barnthouse et al. (1990) developed density-dependent, age-structured models of 2 fish populations representing contrasting life histories: the Chesapeake Bay striped bass population and the Gulf of Mexico menhaden (*Brevoortia patronus*) population. This analysis used 2 well-studied species to clearly demonstrate how life-history strategy can influence population responses to contaminant stress. The striped bass is a long-lived species that fits the periodic life-history type as defined by Winemiller and Rose (1992) (see Figures 4-1 and 4-2). Individual adult striped bass can live for as long as 30 years. A typical adult female can produce more than 1 million eggs per year, but survival of those eggs is poor except for occasional years in which environmental conditions are favorable. The Gulf menhaden is a short-lived species typical of the opportunistic life-history strategy. Longevity of an individual menhaden is about 5 years; individuals reach maturity at age 2 years. The Gulf menhaden population has been relatively stable and supports one of the largest commercial fisheries in U.S. coastal waters.

To permit simulation of available data on fluctuations in abundance of both populations, the survival rate of YOY fish (s_0 in Equation 4-4) was made a function of both population density and environmental variability. The model was used to investigate the interaction between life-history and contaminant effects. Population abundances were simulated for 100 years, with and without contaminant exposure. Of interest was the interaction between contaminant toxicity to YOY fish, average population size, and probability of extinction. The menhaden was found to have a greater capacity to tolerate contaminant stress, measured in terms of both reduction in mean abundance and frequency of extinctions of model populations. Simultaneous exposure to both chemicals and harvesting was found to cause larger reductions in abundance and higher probabilities of extinction in striped bass than in menhaden.

Demographic models for turtles, which are late-maturing, long-lived, and relatively fecund species, have shown that maintenance of population numbers is primarily dependent on adult and juvenile survival, and less sensitive to changes to age at maturity, fecundity, or nest survival (Congdon et al. 1993, 1994). Model analyses indicated that the life-history traits of turtles result in severe constraints on the ability of populations to respond to chronic disturbance.

Young (1968) used a simple model to compare the population responses of a hypothetical robin with a hypothetical bald eagle. The robin is a short-lived, rapidly maturing, prolific species, while the bald eagle is a long-lived, late maturing, slow reproducing species. He performed 5-year simulations to examine the effects on population size of 10% reductions in survival from young to adult, fertility (10% less young per female), and sterility (no young from 10% of the matings). Under the assumption that 10% of the population is affected, sterility had a greater effect on population size for the robin, while reduced survival had the largest effect for the bald eagle.

Grier (1980) used a deterministic life-table model and a stochastic Monte Carlo model to simulate the effects of different rates of reproduction and survival on population outcomes of the bald eagle (*Haliaeetus leucocephalus*). Both models showed that survival was far more critical to the existence of the population than was reproduction and that short-term chances of extinction depended largely on survival rate, partially on the initial size of the population, and almost not at all on reproductive rate. Higher reproduction rates simply led to larger populations among those that survived. Contrary to intuition, rates of reproduction may be relatively inconsequential to this long-lived, slow breeding species. However, the focus of field studies has always been on reproduction, and we have almost no information on survival.

Life-history strategies have been used by Williams and Emlen (1994) in predictive models to determine which population parameter is most sensitive to alteration by chemicals or other human-induced stressors for different types of birds. Fecundity was considered to rise with age, reach a plateau, and then decline in older birds. Three life stages were considered: early juveniles that are not reproductively active, late juveniles and early adults with some reproduction, and older adults with higher reproductive rates. Species were grouped by size into small, medium, and large classes, with survival increasing and fecundity decreasing in progression from smaller to larger size classes. Life-stage-specific values for fecundity and survival were allowed to fluctuate randomly over time around mean values characteristic of each size class and life stage. The endpoints examined were population density over time and the number of times the population declined below 20%, which was considered a pseudoextinction level. Multiple simulations were performed to determine how these endpoints responded to change in each demographic parameter.

The analysis of Williams and Emlen (1994) showed that the relative importance of survival in different life stages depended on the life-history strategy. For small birds, early survival had the greatest influence on population density. A 10% decrease in survival resulted in a 30% decrease in population density. Adult survival was of second most importance, while changes in birth rates were of lesser importance. No single demographic variable was a critical determinant of population density for medium-sized birds, although changes in adult survival had the greatest effect on population size. Adult survival was the most critical parameter for large birds, with 10% increases or decreases resulting in 65% increases or 46% decreases in population size. Although this model uses a very simplistic categorization of life-history characteristics, it provides insight into which individual-level parameters are most likely to be important in influencing population-level responses to chemical stressors for different types of birds. In this way, the model can serve as a useful starting point in focusing studies on the most relevant demographic features when attempting to predict impacts on avian populations.

For many species whose life styles incorporate broad geographic areas, landscape-scale issues become important for assessing population effects of contaminants. Fahrig and Freemark (1995) regard a toxic event as just another type of ecological disturbance and use the existing ecological literature to argue by extension. They indicate that the landscape-scale effect of a toxic event on population survival depends on the landscape structure, the spatiotemporal distribution of the toxic event, and the dispersal characteristics and habitat-specific demographics of the organism. They infer that small, local events can have population-level consequences if the population is patchy and the scale of the disturbances affects an area that is a critical patch for regional survival. Frequent, local events can have significant effects if their frequency is high enough to inhibit recolonization. Large-scale events represented the most serious threat to species survival. The immediate effect of a large scale event is a reduction in numbers across all local populations. Some of these reductions, in combination with other stochastic factors, are likely to lead to local extinctions, which can translate into regional effects depending on the dispersal characteristics of the species. Fahrig and Freemark (1995) conclude that dispersal, which is greatly affected by habitat distribution, generally acts to increase population stability by decreasing the effects of disturbances such as toxic events.

Persistent contaminants

Significant exposure to persistent lipophilic compounds can occur for the early life stages of fish and other oviparous vertebrates. Differences in trophic position and life-history characteristics can interact to increase or decrease the vulnerability of a species to exposure. For example, lake trout (*Salvelinus namaycush*) are higher in the food chain, have a longer period to maturity, and transfer a smaller percentage of their contaminants to the egg than yellow perch *(Perca flavescens)* therefore showing greater bioaccumulation of these compounds.

Populations of lake trout began to decline in Lake Ontario prior to the 1940s, with virtual extinction occurring by the 1960s. This population decline has long been thought to be associated with overfishing, habitat loss, and predation by sea lamprey, *Petromyzon marinus* (Ankley and Giesy 1998). Throughout the 1970s and 1980s, attempts were made to maintain the lake trout populations by stocking, but success was limited and reproduction remained at very low levels into the early 1990s. The chronically poor reproduction suggested that environmental contaminants may have been playing a role. A series of studies examined the relationships between historical trends in contaminant levels and reproductive success in Lake Ontario lake trout (see Ankley and Giesy 1998)

These detailed studies clearly demonstrated that historical exposures of lake trout to compounds with dioxin (TCDD)-like activity were primary contributors to the adverse population level changes documented in Lake Ontario during the early 1950s. These studies were used to build evidence that allowed a logical progression through the following steps. First, it was established that lake trout early life stages were very sensitive to exposure to TCDD-like compounds that bound to the Ah-receptor and that the pathology observed in the 1980s and 1990s in lake trout larvae (i.e., blue sac disease) was consistent with laboratory exposure to Ah-receptor binding chemicals such dioxins, furans and PCBs. Then, toxicity equivalency values that were derived for converting chemical residue levels for a variety of compounds to TCDD equivalents showed that the effects of the compounds could reasonably be assumed to be additive. Next, dose-response relationships between the chemicals (as TCDD equivalents) and fry mortality were estimated. Then, experiments were performed that allowed determination of the maternal dose of TCDD toxic equivalents required to result in contaminant levels in eggs that would lead to lethal doses for fry. Next, back-calculated sediment concentrations required to elevate maternal body burdens to levels that would result in fry mortality after spawning were determined. Finally, data were used to establish retrospective levels of TCDD toxic equivalents in Lake Ontario sediment through dated sediment core samples and to predict retrospective survival levels of lake trout fry dating back to the 1940s. When taken together, this logical, stepwise progression led to a convincing argument that dioxins, furans, PCBs, and perhaps unidentified compounds with similar modes of toxicity played a major role in the decline in lake trout. The lake trout experience suggests that contaminants may have played a greater role in many of the fish population changes originally blamed on other factors over the past 50 years.

Persistent contaminants affecting bird populations have been documented for a variety of species. We use the decline and subsequent recovery of the sparrow hawk (*Accipiter nisus*) in response to organochlorine pesticides to illustrate the type of empirical evidence that links persistent contaminants to bird population dynamics. It was in the sparrowhawk that Ratcliffe (1967) first illustrated eggshell thinning and suggested a relationship between thinning and organochlorines; later empirical evidence suggested that adult mortality was also important in their decline (Newton and Wyllie 1992). The sparrowhawk is a year-round resident in Great Britain where it feeds in both woods and open country on small birds. Dichlorodiphenyltrichloro-ethane (DDT) was introduced into agricultural use in Britain during 1946 to 1947 and was used heavily throughout the country into the 1980s. Aldrin and dieldrin were introduced in 1956 and were widely used as seed treatments for cereals. Their use on spring-sown cereals was withdrawn in 1962 and on autumn-sown cereals in 1975.

Until the late 1950s, sparrow hawks were common and widespread throughout Britain, breeding wherever there was a suitable woodland. Then suddenly, over most of the country, the species showed a marked decline in numbers, disappearing almost completely from some districts. Newton et al. (1986) found a highly signifi-cant negative correlation in sparrow hawks between dichlorodiphenyldichloroethylene (DDE) content and both eggshell thickness and productivity. No significant relationship was detected between PCBs or HEOD (the stable metabolite of aldrin and dieldrin) and thinning or productivity. Production of young was reduced by an estimated 14 to 35% in different areas. Newton et al. (1992) analyzed the temporal trends in mortality based on 1029 sparrowhawk examined over the period 1963 to 1990 and concluded that HEOD probably accounted for 50% of all recorded deaths of this species.

Sparrow hawks recovered across Britain after these compounds were withdrawn from agricultural use. The proportion of sparrowhawk deaths attributable to HEOD declined from 1963 to 1975 and from 1976 to 1986 following a marked reduction in chemical use and fell to near zero during 1987 to 1990 (Newton et al. 1992). Reproductive success generally increased during the 1980s as eggshell thinning and DDE concentrations in eggs declined. The temporal correspondence between the decline and recovery of this sparrowhawk population and the introduction and withdrawal of dieldrin and aldrin as cereal seed dressings suggests that excessive adult mortality due to these pesticides was more important than DDE-related reproductive impacts (Newton and Wyllie 1992).

Other examples of population declines, and sometimes recoveries, attributable to persistent contaminants include affects of the insecticide dieldrin on Merlin (*Falco columbariu*) near Kindersley, Saskatchewan (Fox 1971; Houston and Hodson 1997) effects of hexachlorobenzene, PCBs, and 2,3,7,8-TCDD on herring gulls (*Larus argentatus*) in eastern Lake Ontario (Gilbertson 1974; Gilman et al. 1977; Fox 1993; Grasman et al. 1996), and effects of PCBs on Forster's terns (*Sterna forsteri*) in Green Bay in Lake Michigan (Kubiak et al. 1989; Harris et al. 1993). In contrast to the sparrowhawk example described above in which contaminant effects on adult survival were important, the dominant effect of contaminants on these species was related to reduced reproductive success.

The few available reptile and amphibian examples illustrate some interesting population-level effects of persistent contaminants. Elevated levels of DDE have been found in the juvenile body tissues and eggs of American alligators (*Alligator mississippiensis*) in some Florida lakes (Heinz et al. 1991). Gonadal abnormalities have been found in both males and females, as well as reduced penis size in males. Although viable eggs are still being produced in this population, they have reduced hatching success. Several studies of DDE effects on snakes have shown that higher residues of DDE tend to be found in snakes that feed aquatically rather than terrestrially, and residues levels increased with trophic level (Fleet and Plapp 1978; Ohlendorf et al. 1988; Burger 1992). The number of snake species at a given site increased 2 years after the ban of DDT, with apparent recolonization of oviparous species that had disappeared. Oviparous species have been found to be more affected by DDT, showing large decrease in abundance when DDT levels were high, than are viviparous species (Fleet and Plapp 1978).

DDT and its metabolites, other organochlorines, and pH have been found to have significant impacts on amphibian populations by repeatedly hindering hatching or early larval success. For example, a study in Point Pelee National Park in Canada found that, almost 30 years after the banning of DDT use, body-tissue levels of DDT and its metabolites in the short-lived spring peepers (*Pseudacris crucifer*) were higher than the allowable limit for fish (Russell et al. 1995). Since 1972, 3 species of frogs have disappeared at this site; a fourth disappeared while DDT was still in use. Russell et al. (1995) state that there has been no wetland loss, harvesting, or habitat acidification that can explain the loss of these species.

Jaworska, Rose, and Brenkert (1997) developed an individual-based model of largemouth bass (*Micropterus salmoides*) that incorporated effects of PCBs on the growth of individual fish larvae and juveniles. Individuals were followed daily through development, growth, feeding, and mortality processes from spawning to 1 year of age. Survival and growth decreased with increasing PCB tissue concentrations. Monte Carlo simulation was used to simulate the effects of inter-annual variations in environmental conditions. Predicted October largemouth bass abundances decreased with increasing PCB levels (Figure 4-3). Density-dependent processes largely, although not completely, offset the effects of PCBs at low exposure concentrations. PCB-related mortality imposed on eggs and larvae led to increased growth and improved survival of juveniles. The large overlap in the distributions of predicted abundances among baseline and 4 PCB concentrations implies that natural variability was large enough to effectively mask the adverse effects of the PCBs.

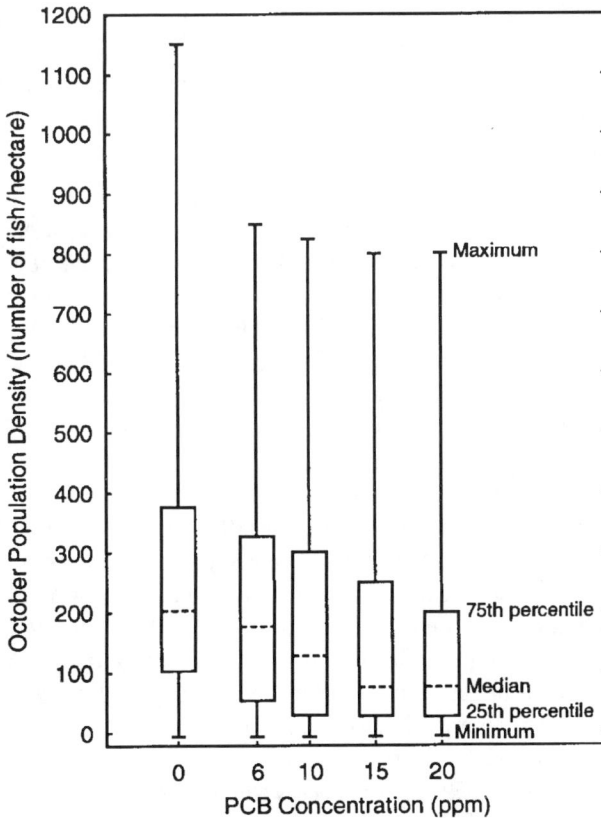

Figure 4-3 Predicted October abundances of largemouth bass under baseline (0 ppm), 6, 10, 15, and 20 ppm PCBs (Jaworska, Rose, and Brenkert 1997). Monte Carlo simulation was used to vary model inputs to mimic realistic interannual variation. The 500 Monte Carlo simulations performed for each PCB concentration are summarized by showing predicted percentiles, minimum, and maximum values of abundances.

Nonpersistent contaminants

Modern, nonpersistent pesticides vary in their specificity, toxicity, mode of action, and rate and frequency of use. Herbicides, for example, are relatively nontoxic to birds, but can have indirect effects related to habitat alteration. Insecticides, on the other hand, may have both direct, acutely toxic effects on birds and significant indirect effects related to alteration of their prey insect populations.

Effluents from some pulp mills contain chemicals that may act as endocrine disruptors. These chemicals are waterborne with short environmental half-lives and are not detected by traditional bioassays. The physiological changes associated with exposure to these compounds can be similar to those seen with persistent compounds; however, recovery from a single exposure can be quick, and these compounds are not expected to biomagnify. These chemicals are rapidly metabolized and maternal transfer may not be as pronounced. Exposures are associated with changes in fish due to the elimination and continuous uptake of new chemicals. Although binding to physiological receptors may be short-lived, the constant uptake of new chemicals from effluent discharges means that the receptors are constantly being bound with new chemicals.

During the past 10 years, there have been extensive studies of the impacts of effluents from pulp mills on fish populations (e.g., McMaster et al. 1991, 1996; Adams et al. 1992; Hodson et al. 1992; Munkittrick, Van Der Kraak et al. 1992; Munkittrick, McMaster et al. 1992; Gagnon et al. 1994a, 1994b; Sandström 1994). In these studies, fish exposed to effluents from some pulp mills have shown effects including delayed sexual maturity, lowered recruitment, reduced fecundity, and smaller egg size. These effects have been observed at some sites at very diluted (1 to 3%) effluent concentrations, when the effluent discharges themselves (not diluted) did not show acute toxicity. These effects were associated with changes in the levels of gonadal sex steroids. Life-cycle studies with fathead minnows (*Pimephales promelas*) have confirmed the capability of some effluents to delay maturity and reduce egg production (Kovacs et al. 1995). Other suggested effects of reduced gonadal steroids are alterations in secondary sex characteristics and in the many behaviors in fish that rely on chemical communication. The compounds associated with the physiological changes in the fish appear to be waterborne and are rapidly metabolized with a half-life for clearance of less than 7 days. Although the contaminants responsible for the changes have not been identified, some level of recovery has been seen observed in laboratory studies (Kovacs et al. 1996) and near many of the sites (Munkittrick et al. 1997) after process changes were implemented that affected the potency of the effluents. Detailed analysis of the composition of effluents has shown that none of the compounds tested so far can account for all of the observed effects (Van Der Kraak et al. 1998). The disappearance of TCDDs and polychlorinated dibenzofurans after process modernization has not resulted in improved reproduction at all sites studied.

There is a risk that other effluents will be found to contain nonpersistent, rapidly metabolized compounds that disrupt physiological systems and endocrine control of reproduction in fish. Physiological responses that may impact fish populations also recently have been seen with sewage effluents (Purdom et al. 1994), with compounds such as nonyl phenols (Jobling and Sumpter 1993; White et al. 1994), which are present in many effluents, and with previously unidentified contaminants in a pesticide formulation of trifluoro-nitromethyl-phenol (Munkittrick et al. 1994; Hewitt et al. 1996, 1998). Some of these compounds have been shown to be estrogenic and induce vitellogenin synthesis in male fish. Synthetic and natural estrogens in sewage effluent have also been implicated as affecting fish reproduction (Desbrow et al. 1998). Recent surveys have found hermaphrodism and other abnormalities in wild fish (Jobling et al. 1998). The effects on population dynamics of these physiological changes in male fish have not, as yet, been clearly established. Many of the ongoing programs mobilized to look at the "endocrine" issue are using biochemical markers as indicators of responses, which unfortunately will not provide the life-history information required to evaluate the significance of the changes. It is crucial that these surveys provide some information on growth, reproduction, and age distributions to evaluate the significance of the exposures.

Oviparous organisms such as amphibians and fish are especially at risk from these nonpersistent compounds because they are commonly restricted to living in discharge areas and, because they are ectothermic, their ability to metabolize compounds will be restricted by environmental temperatures. Rapid depuration will be reduced during winter when water flows and dilution are also reduced. The opposite effect of reduced toxicity can occur if the toxic agent is a metabolite. Furthermore, fish and amphibians have a less developed suite of detoxification enzymes than do birds and mammals, and they will have a reduced capability for dealing with some of these compounds.

Chemicals used in modern agriculture have had varying degrees of influence on bird populations. According to O'Connor and Shrubb (1986), pesticides can have 3 main ecological effects on bird populations:
- kill plants and animals directly influencing the agricultural landscape and its ecology,
- change food and habitat resources by removing plant species and altering invertebrate fauna, and
- allow the farmer much greater freedom to select his crops and rotations, thereby modifying farmland habitat.

Indirect effects of these chemicals may be more important than their direct effects. Direct effects of these nonpersistent, highly toxic chemicals on individual animals have been observed and measured during field and laboratory studies, but population-level responses have been difficult to quantify. Hoffman and Albers (1984) and Mineau et al. (1994) showed that a variety of commercial formulations of insecticides and herbicides cause embryotoxicity or eggshell effects in laboratory experiments. However, definitive links between direct effects in individuals (e.g., mortality, cholinesterase inhibition, reproductive effects) and population-level responses have been difficult to prove. Avian field studies of the effects of modern insecticides such as organophosphates and carbamates have had mixed results. In the last 40 years, large areas of forest have been repeatedly and intensively treated with a variety of insecticides to control defoliating insects. Field study results have been variable but, in general, massive mortalities have not been detected. Mortality may have been masked by immigration from large "floater" populations (DeWeese et al. 1979; Herman and Bulger 1979; Busby et al. 1990) or simply did not occur. Avian toxicity predicted by laboratory studies is not always detected in the field, and when detected, may or may not translate into observable population effects.

Indirect effects of agricultural chemicals on bird populations can be important. Agricultural chemicals kill both target and nontarget invertebrates and plants. Agricultural chemicals can reduce the abundances of insects or seeds eaten by birds and reduce the plant species composition, which acts as food for the insects and as nesting cover for the birds. In the United Kingdom, 24 of 40 bird species normally found in lowland farming districts are declining (RSPB 1997). Invertebrates are important in the diets of the declining species, and their population trends have temporally correlated with pesticide use. Most declines began between 1974 and 1985 when the area of crops sprayed with herbicides, insecticides, and fungicides increased greatly in this region. It is hypothesized that pesticide-induced reduction in invertebrates and alteration of habitats have played major roles in these population declines (RSPB 1997).

An excellent example of indirect effects of pesticides on avian reproductive performance involves the gray partridge (*Perdix perdix*). In this species, chick survival is a principal factor determining population changes (Southwood and Cross 1969), as is usually the case with short-lived species with high annual fecundity. Partridge chicks are insectivorous during the first 10 days after hatching, and their survival largely depends on the number of arthropods available in cereal grain crops in June (Potts 1980). The densities of 5 preferred insect taxa explained over 50% of the variation in chick survival (Green 1984). The preferred taxa represent about 20% of the local arthropod community. Routinely used herbicides were responsible for a reduction in the density of insect-rich plant species and a 50% reduction in the numbers of preferred insects. In a series of experiments in which the selectivity of the herbicide and insecticide regime applied to the preferred habitats (field margins and headlands) were manipulated, the densities of preferred insects and survival of partridge chicks were markedly increased (Rands 1985). Similar effects of changing agricultural landscape structure on the food resources and survival of gray partridge chicks in Poland were recently reported by Panek (1997).

Freemark and Boutin (1995) reviewed the existing literature to assess the extent to which wildlife living in terrestrial habitats has been affected by use of agricultural herbicides in temperate landscapes. They concluded that

- the weight of evidence indicates that the widespread and intensive use of agricultural herbicides have affected wildlife and that most effects have been interpreted as negative from ecological and socioeconomic points of view;
- habitat patterns in agricultural landscapes have changed dramatically in conjunction with chemically intensive farming in North America and western Europe; and
- strong evidence exists for adverse effects from changes in habitat pattern on beneficial insects and arthropods in the UK and on birds in North America and western Europe. Habitat alteration is a major factor in the loss of bird populations (Temple 1986; Freemark and Boutin 1995).

Because habitat alterations are often long-term, their potential for having long-lasting or permanent effects on bird populations is great. Quantifying the extent of habitat alteration resulting from contaminants is difficult because there are numerous non-chemical causes of habitat loss and alteration (e.g., urban and industrial expansion, road construction, deforestation by clear cutting, burning, and soil tillage). However, there is substantial evidence that pesticide induced habitat alteration is a significant contributor to population-level effects. Indirect, habitat-related effects of pesticides may be far more ecologically significant than acute or chronic direct effects.

Chemically induced production of intersexes was recently illustrated by laboratory experiments on the red-eared slider, *Trachemys scripta* (Bergeron et al. 1994). Sex determination in turtles is a temperature-dependent process. A change in threshold temperature results in an all-or-none response; intersexes are very rarely documented in nature. However, exposure to an environmental estrogen mimic (2,3,4,5-trichloro-biphenylol) resulted in 21% of the hatchlings being intersexes. These hatchlings would be perfectly viable in the environment, utilizing resources, but may have limited (if any) ability to reproduce. The production of intersexes can have major implications at the population level.

Few modeling studies have explicitly dealt with the effects on oviparous vertebrates of low-level exposure to nonpersistent contaminants. Modeling approaches tend to deal better with changes in mortality, growth, or reproductive success than with sublethal and behavioral effects. Changes in survival, growth, or reproduction can usually be simulated by altering one or more model parameters. Sublethal effects and behavioral changes first must be translated into changes in survival, growth, or reproduction that can then be imposed by changing model parameter values. Few population models simulate processes at a fine enough resolution to incorporate inputs that directly relate to sublethal and behavioral effects. We propose a modeling strategy that combines individual-based and matrix-projection approaches for simulating the effects of nonpersistent contaminants (Figure 4-4). The individual-based model serves a bridge between contaminant effects and inputs to the matrix-projection model. The individual-based approach is ideally suited to simulate

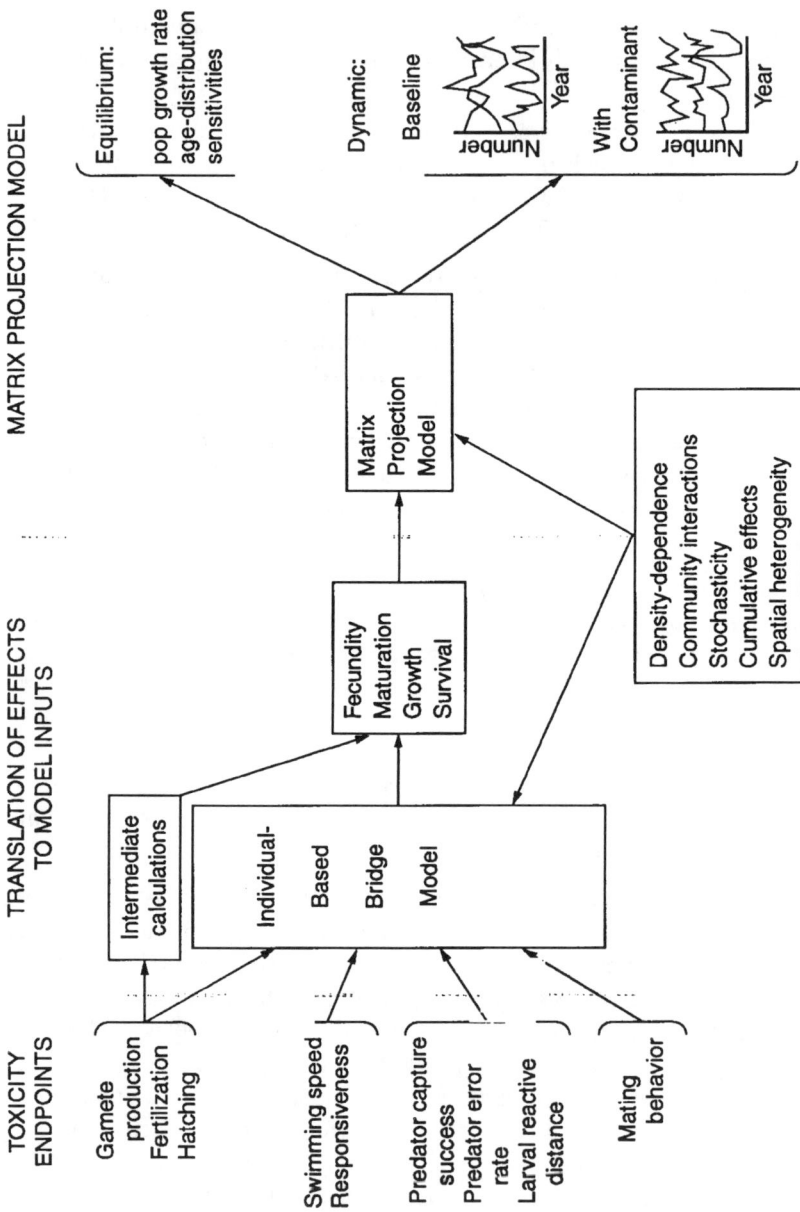

Figure 4-4 An example of using an individual-based model to bridge between sublethal toxicological effects and inputs to a matrix projection model

sublethal and behavioral effects and translate them to changes in reproductive, growth, and survival rates that can be used by the more aggregated matrix-projection approach.

Essential compounds

There are many chemicals that can be essential to performance and survival but that become toxic at high concentrations. Individual performance can be affected by changes in the circulating levels of these essential compounds in individuals or by changes in deposition rates of these compounds into eggs. For example, some essential compounds can cause decreases in the concentrations of important circulating hormones that affect homeostasis during embyrogenesis. Physiological responses have also been associated with substitution of exogenous metals for required nutrients (e.g., lead for calcium, Chang and Cockerham 1994). Calcium is a critical regulator of many physiological functions, and this substitution has been associated with impacts on key physiological processes controlling reproduction.

The essential metal selenium is a micronutrient that has been shown to be toxic at high concentrations and has been associated with dramatic changes in fish populations. Selenium exposure associated with fly ash from coal-fired power plants led to dramatic decreases in adult densities and nearly 99% declines in density of YOY bluegill (Cumbie and Van Horn 1979). The mechanism appeared to be larval toxicity associated with maternal transfer of selenium into eggs prior to spawning (Gillespie and Baumann 1986).

An elevated rate of embryotoxicity (dead and deformed embryos), reproductive failure, and adult mortality attributable to selenium toxicosis was observed in a number of waterfowl and shorebird species at the Kesterson National Wildlife Refuge in California during the 1980s (Ohlendorf et al. 1989; Williams et al. 1989). Agricultural drain water that had leached naturally occurring selenium from soils was used for marsh management at Kesterson Reservoir. Williams et al. (1989) estimated that prefledging brood mortality approached 100% for American avocets (*Recurvirostra americana*) and black-necked stilts (*Himantopus mexicanus*) at Kesterson in 1984 and 1985. They also concluded that other factors could not explain the complete recruitment failure during both years. Ohlendorf et al. (1989) studied the reproductive success of several species of waterfowl and shorebirds at Kesterson from 1983 to 1985. Nest success varied by species and between years. Predation, desertion, and water-level fluctuations were the most important causes of reproductive failure for most species, but embryotoxicosis (i.e., mortality, deformity, and lack of embryonic development) was the most important cause of nest failure in eared grebes (*Podiceps nigricollis*). Embryotoxicosis also reduced the hatchability of eggs of all other species at Kesterson in one or more years. These findings indicate that external factors such as predation and habitat quality were the most important factors affecting population dynamics but that selenium toxicosis also contributed to decreased reproductive success in these bird species.

Other metals have also been associated with problems in oviparous vertebrates. For example, acid deposition affecting the chemical speciation of aluminum was intensively studied as a cause for the loss of fish species in various regions of the U.S. (Baker et al. 1990). An estimated 1.5 to 3 million waterfowl die annually in the U.S. from lead gunshot poisoning (Sanderson and Bellrose 1986; USFWS 1986). However, it was difficult to directly equate this mortality to population declines in most species, as losses due to lead poisoning are only a small fraction of the total annual sport harvest. Other species, particularly raptors, died from secondary poisoning following consumption of prey tissues with embedded or ingested lead shot. Lead-poisoned birds lose mobility and become more vulnerable to predation and other sources of mortality, including hunting (Eisler 1988).

Stochastic contamination events

Oil spills illustrate that dramatic contaminant effects do not necessarily result in dramatic long-term population responses. Marine transport of petroleum along the Pacific and Atlantic coasts of North America, and elsewhere in the world, has long been recognized as an ecological hazard, and there have been several well-documented tanker spills. Because coastal North America areas are occupied by abundant forage fish, a large proportion of the world's seabird population seasonally resides there.

The ecological importance of the seabird populations at risk and the potential for a shipping disaster have lead to extensive studies of the toxicity of petroleum to birds. In fact, few contaminants have been characterized as well as oil. Oil is a variable and complex mixture that changes greatly when spilt on water. Different crudes and fractions vary in their toxicity. The water repellency and insulative properties of the plumage of birds are detrimentally affected by oil, and birds ingest oil both while trying to preen their plumage and through the water and food they ingest. Oil has been shown to have a variety of significant toxic effects (e.g., Leighton et al. 1983; Peakall et al. 1983; Butler et al. 1986; Holmes and Cavanaugh 1990; Culic et al. 1991). Several field studies of experimentally exposed individuals have been conducted which showed decreased clutch size, hatching success, chick survival, breeding success, and abnormal parental behavior (Butler et al. 1988; Fry et al. 1986).

Samuels and Lanfear (1982) created a demographic model to simulate the effects of an oil spill on populations of glaucous-winged gulls (*Larus glaurescens*) and common murres (*Uria aalge*) in the northern Gulf of Alaska. The model, which did not consider movement between colonies, predicted that a spill which reduced the population by 50% would require a recovery time of approximately 20 years for the gulls and 70 years for the murres. Ford et al. (1982) developed an extensive set of models to determine the sensitivity to oil spills of breeding guillemots and kittiwakes on the Pribilof Islands. They considered 2 important one-time mortality scenarios: increased adult mortality and increased chick starvation. Chicks are more likely to starve because of a reduction in the rate of food delivery that occurs when adults are forced to make longer foraging trips to avoid oil. Recovery time was longest under the adult mortality scenario; changes in chick survival rates had little effect. Kittiwakes seemed less sensitive than guillemots to perturbations. When a one-time mortality increase was superimposed on a chronic, low-level change in survival or fecundity, recovery times were markedly longer.

Given this knowledge, the supertanker *Exxon Valdez* grounding in Prince William Sound, Alaska in 1989 was expected to have a marked effect on the local bird populations. About 44 million liters of crude oil were released, contaminating roughly 2100 km of shoreline. An estimated 300,000 birds were killed. Some 35,000 carcasses were recovered representing 90 species; 74% of the carcasses were murres. However, despite the obvious acute impact on murres, investigators were unable to detect long-term changes in numbers, phenology, or egg or chick production. The effects of the oil spill were indistinguishable from the natural responses of the murres to long-term environmental changes. Any effects of the spill on murre colony attendance were relatively short-term and may have reflected recruitment of formerly non-breeding individuals at spill-affected colonies or immigration from unaffected colonies (Boersma et al. 1995; Erikson 1995; Piatt and Anderson 1996)

Oakley and Kuletz (1996) were conducting an ongoing study of the population dynamics, reproduction, and foraging of pigeon guillemots (*Cepphus columba*) on an island 30 km from the grounding site at the time of the *Exxon Valdez* spill. They detected a decline in numbers that lasted for 3 years after the spill, but no effects on reproduction or foraging were observed. Monitoring the size and productivity of 24 colonies (10 oiled by spill, 14 not) of black-legged kittiwakes in Prince William Sound between 1984 and 1994 revealed no changes in numbers in response to the spill, but productivity at oiled colonies was markedly reduced between 1989 and 1994 (Irons 1996). The black oystercatcher (*Haematopus bachmani*) feeds on intertidal invertebrates. The proportion of nonbreeding pairs increased, eggs were smaller, and chick mortality was higher on oiled sites than on the non-oiled site. Chick mortality was positively correlated with the amount of oil present in foraging territories. In 1990, oystercatcher chick production remained lower on oiled territories and failed completely on bioremediated shorelines (Sharp et al. 1996). Nesting surveys of bald eagles in 1989 and 1990 in coastal south-central Alaska revealed a large-scale reproductive failure, but that was limited to western Prince William Sound and only in 1990 (Bernatowicz et al. 1996).

Wiens et al. (1996) examined the effect of the oil spill on the marine bird community based on survey counts between June 1989 and 1991 from 10 study bays differing in the magnitude of initial oiling. Species richness, species diversity, and species occurrence were all decreased, especially in heavily oiled bays, but the differences had disappeared by 1991. Species richness of several guilds of birds feeding on, or close to, the shoreline was negatively related to initial oiling level until mid-1990. Of these guilds, the richness of a guild of winter visitant and resident species showed the greatest negative association with initial oiling. However, the richness of guilds of solitary or colonial species that dive or feed on fish showed no significant relationships with oiling at any time.

Despite the highly visible, dramatic impacts of the *Exxon Valdez* oil spill, few long-term population-level effects could be identified. Although not discussed here, similar results also were obtained with many of the fish populations. For example, despite the oil spill resulting in about a 30% loss in spawning and nursery habitat for pink salmon in the Prince William Sound, Geiger et al. (1996) estimated the percentages of wild salmon not returning at 28% in 1990, and dropping to 6% in 1991 and 1992. Population responses are not necessarily proportional to the impacts.

Issues Confounding Population Analyses

We presented above empirical and modeling examples of studies designed to detect and quantify contaminant effects on oviparous vertebrate populations. The relatively few definitive field studies that established convincing evidence of cause and effect generally resulted from long-term, extensive investigations. More common are the empirical studies leading to equivocal results that can neither confirm nor deny the importance of contaminant effects on population dynamics, and modeling results that fail to convince the many who distrust theoretical analyses. Below we describe 5 issues that confound population-level analyses and make the detection and quantification of contaminant effects so difficult and modeling results so controversial. These issues are density-dependence, spatial scales and dispersal, community interactions, cumulative effects, and stochasticity. These are general issues that apply to both modeling and empirical approaches and to a wide variety of taxa. The complex life cycles of many oviparous vertebrates play a contributing role in all of these issues.

Density dependence

Density dependence occurs when reproductive or survival rates vary with population abundances. Compensatory density dependence occurs when reproduction or survival increase with decreasing abundances. Compensatory density dependence acts to offset increases in mortality. Depensatory density dependence occurs when reproduction or survival decreases with increasing abundance, and, like most positive, regulatory feedback mechanisms, depensatory density dependence is destabilizing to the population. Population responses to contaminant-related stress can be masked by compensatory density dependence. Thus, no changes in population abundance under contaminant exposure can be misleading because the compensatory reserve of the population is lower under contaminant exposure and the population is less resilient to additional stresses.

Density dependence is fundamental to understanding population dynamics (Murray 1994) and, therefore, to predicting the effects of contaminants. The basis of life-history monitoring is that individuals adjust their reproductive or survival rates in response to stress-related changes in abundances. Yet, detecting and quantifying density dependence in field situations is difficult (Eberhardt 1970; Slade 1977; den Boer and Reddingius 1989), and its formulation in models is often at the root of controversy (Barnthouse et al. 1984). Krebs (1991), in the context of arguing that looking for density dependence does not justify long-term studies, even goes as far as to characterize the density-dependent paradigm as bankrupt. While this is an overstatement, the strong opinion expressed illustrates the importance and contro- versy of density dependence. Barnthouse et al. (1984) concluded that "After more than a decade of study and tens of millions of dollars, it was still not possible to draw definitive conclusions about the long-term effects of entrainment and im- pingement (mortality) on fish populations in the Hudson River. . . .We believe that lack of success occurred because of insufficient understanding of the underlying biological processes affecting population dynamics."

Density dependence is also important for understanding bird population dynamics. In a 15-year study of the population dynamics of the Swainson's thrush (*Catharus ustulatus*) in the scarce riparian woodlands of coastal California, numbers of new and total adults captured in a given year were significantly dependent on the number of hatching-year birds caught in the previous year (Johnson and Geupel 1996). In addition, per capita productivity was inversely density-dependent and may partially act to regulate adult abundance. The strong habitat specificity exhibited by this species on the breeding grounds may reinforce this density dependence. Dobson and Hudson (1994) provide several examples of density-dependence survival during the juvenile life stage. For example, they suggest that as much as 50% of the variation in the survival of YOY tawny owls (*Strix aluco*) can be explained by the number of nests from which they were produced. These authors also suggest that the assumption commonly made about birds that adult survival is density- independent while juvenile survival, and hence recruitment, are density-dependent is likely an oversimplification and more appropriate for short-lived than long-lived species.

Recruitment of non-breeding seabirds into a breeding colony is one of the most important factors in determining changes in breeding numbers from year to year. Studies of the kittiwake (Porter and Coulson 1987) and great skuas (Klomp and Furness 1992) have provided clear evidence for the existence of a pool of potential recruits that was temporarily restricted from breeding, suggesting a density- dependent aspect to recruitment to the colony. These authors concluded that the size of the non-breeding pool is the first indicator of a decline in the well-being of the population, declining several years before the breeding population declined. Monitoring of the number of non-breeders attending seabird colonies would provide a sensitive signal of adverse environmental stress affecting seabirds and would be especially valuable for the majority of populations, where adult survival rates are not assessed.

A series of experiments using 4 species of frog larvae (Wilbur 1987) demonstrated how population density can interact with natural disturbances to affect population dynamics. Wilbur (1987) simultaneously examined the effects of pond drying (as a measure of disturbance), predation, and population density on the structure of the tadpole community. In high density treatments, species interactions (competition) determined survival and differentially affected life-history characteristics. Predation by newts in high-density treatments affected relative abundance of tadpoles, but not biomass, because the competitively dominant species was most frequently eaten. At low densities, predation by newts drastically lowered both survival and biomass of tadpoles. Reduced growth rates resulting from competition in high-density treatments resulted in high mortality because the ponds dried before tadpoles could metamorphose, whereas, in the presence of predators, density-dependent growth resulted in more tadpoles metamorphosing before the ponds dried.

Spatial scales and dispersal

The complex life cycle of many oviparous vertebrate species includes dispersal of individuals over broad geographic areas. Such populations can be viewed in the context of a spatial mosaic of linked metapopulations. Within a relatively small geographic region, most mobile animal populations inhabit a variety of different habitats, and intraspecific variation in reproductive success and in survival occurs across habitat types. Succession and land use by humans result in temporal variation in habitat availability. In this case, population dynamics models that explicitly deal with spatial differences in habitat on landscape scales are required. Spatially explicit population models operating over broad regions have recently received attention (Dunning et al. 1995). These models consider both species-habitat relationships and the arrangement of the habitats in space and time, and can be used to predict the response of organisms to habitat change occurring at a variety of spatial and temporal scales. We present several analyses based on birds because the effects of the spatial arrangement of habitat on bird metapopulations has been the focus of much attention.

Immigration and emigration are probably crucial to the health and persistence of local populations of many bird species. Because populations often occupy a variety of local areas and may experience different local birth and death rates, in some areas production of young may offset mortality (source populations), while in other areas mortality may exceed production (sink populations) with population levels maintained by continual immigration. As Pulliam (1988) pointed out, 2 conditions follow from such a spatial population substructure: 1) a small source population may maintain a large sink population indefinitely, if the surplus production in the source populations is large and the per capita deficit in the sink populations is small, and 2) numbers in areas of sink populations may depend largely on conditions in the areas that contain source populations. A reproductively successful source population is essential if the metapopulation is to persist.

Howe et al. (1991) modeled temperate breeding passerines as source and sink metapopulations connected by density-dependent dispersal of adults and juveniles. In the absence of an available sink habitat, surplus individuals remain in the source population as "floaters." Sink populations augment metapopulation size, and may even prevent extinction when intrinsic dispersal occurs. A large fraction of a species' population may reside and reproduce in sink habitats. A reduction in the critical dispersal threshold through habitat deterioration, overharvest or other environmental perturbation may lead to the sudden collapse of the metapopulation. Howe et al. (1991) conclude that sink habitats appear to be most critical for metapopulations of r-strategy species (i.e., species with high rates of juvenile dispersal and low adult survivorship).

A landscape perspective is required to understand how species are distributed across complex mosaics of habitat patches that differ in their quality and use to the individual. Complex landscapes affect ecological processes which, in turn, affect populations. Landscapes are characterized by their physiognomy (i.e., features associated with the physical layout of habitat elements) and their composition (i.e., the relative amounts of each habitat that they contain), both of which can affect populations. Dunning et al. (1992) describe 4 types of ecological processes that operate at the landscape scale: complementation, supplementation, source/sink relationships, and neighborhood effects. Landscape complementation occurs when different patch types occur in close proximity within a landscape and thus can support a larger population than can landscapes in which these habitats are far apart. In the case of landscape supplementation, the population in a focal patch may be increased if that patch is located in a portion of the landscape that contains additional available resources. Source/sink relationships occur when relatively productive patches serve as sources of emigrants, which disperse to less productive patches called sinks. Neighborhood effects occur when a species' abundance in a particular focal patch may be more strongly affected by characteristics of contiguous patches than by those of more distant parts of the landscape. Any of the above processes also could indirectly affect an organism by directly impacting the organism's competitors or predators. If landscape-level processes affect the persistence of populations, then these processes must be considered when attempting to predict contaminant effects.

Lande (1988) employed a metapopulation model of the northern spotted owl (*Strix occidentalis caurina*) to predict the effect of future habitat alteration. The model represented local extinction and colonization, territoriality, and dispersal behavior in a patchy environment. Lande's analysis showed that populations at demographic equilibrium need not occupy all of the patches of suitable habitat available to it, and that a population may become extinct in the presence of suitable habitat because of the difficulties experienced by juveniles searching the landscape for available patches. Destruction of old-growth forest habitat by logging constituted a major threat to the continued existence of the northern spotted owl.

Pulliam et al. (1992) developed a model that incorporated dispersal, survivorship, reproductive success, and habitat information for the Bachman's sparrow (*Aimophila aestivalis*). Their simulations suggest that variations in demographic variables can affect population size more than variation in dispersal ability and that changes in adult and juvenile survivorship can have especially large impacts on the probability of population extinction. The presence of habitat types that served as permanent sources of dispersers increased the total population size in the landscape and lowered the probability of extinction.

Pulliam (1994) modified the Bachman's sparrow model to include agricultural habitats in the landscape. The model assumed that 10, 20, or 30% of the land was in agricultural use. Pesticide applications on agricultural land were assumed to reduce the fledging success of birds breeding in territories next to agricultural habitats by 50%. In the absence of pesticide-induced mortality, the breeding population decreased by 34%, although none of the simulated populations went extinct even at the highest rate of agricultural land use. However, when pesticide use was included, a 0 to 30% increase in agricultural land use reduced the population size by 54% and resulted in a 33% chance of extinction.

Akcakaya and Atwood (1997) presented a habitat-based metapopulation analysis of the dynamics of the federally threatened coastal California gnatcatcher (*Polioptila c. californica*) for an 850 km^2 region of Orange County, California. A habitat suitability model was developed and verified, and then was used to calculate the spatial structure of the metapopulation, including the size and spatial arrangement of habitat patches. Field collected data were used to estimate survival, fecundity, and dispersal rates and the impact of weather-related catastrophes. These results were combined to develop a stage-structured, stochastic, spatially explicit model. The model predicted a fast decline and high risk of population extinction with most combinations of parameters. Population parameters were most sensitive to density-dependent effects, catastrophes, and adult survival and fecundity.

Although many of the examples of spatially explicit models presented above did not consider the effects of contaminants, the basic structure exists. These models could be modified to incorporate contaminant effects. For example, if the dose-dependent effects of a contaminant on adult survival or fecundity were quantifiable, this information could be incorporated into the models to determine the sensitivity of population-level endpoints to fluctuations in these measures of individual reproductive performance. Although far less studied, the survival and reproductive success of individuals of many fish, reptiles, and amphibian species are also influenced by local habitat conditions, and their populations also can be viewed as metapopulations. For example, anadromous fish species such as striped bass migrate into rivers to spawn but then mix as adults in the coastal ocean. We see the coupling of contaminant effects to landscape-level population models, and the application of these models to fish, amphibians, and reptiles as ripe areas for further research and application.

Community effects

Most analyses are at the population level, even though populations rarely exist in isolation. Uncertainty in interspecific interactions usually deters performing analyses at the community level. There has also been the belief that we have enough difficulty modeling population dynamics, without attempting to also represent interspecific competition and predation. Yet, population responses can be greatly influenced by community-level interactions.

Rose (in press) presented an example of a modeling analysis that demonstrated how community-level interactions can distort population responses. In this analysis, the daily spawning, growth, and mortality of individuals of yellow perch and walleye were followed throughout their lifetimes (Rose et al., in press). Adult walleye predation is the major source of mortality on YOY juvenile and yearling yellow perch. The model was run in population and community modes (Jaworska, Rose, and Barnthouse 1997). The community mode corresponded to a 2-species model in which predation on yellow perch varied dependent on walleye dynamics. The population version removed walleye from the model; mortality rates of YOY juvenile and yearling life stages of yellow perch (usually eaten by walleye) were set to average values from the baseline simulation of the community version. Metabolic rate of juvenile yellow perch was increased by 20% to mimic sublethal effects of a contaminant. Inclusion of a dynamic predator (walleye) in simulations reversed the effects of increased metabolism on yellow perch population abundances. Increased metabolism caused a 10% decrease in long-term average adult yellow perch abundance under population conditions, and a 55% increase under community conditions. These results demonstrate that interspecific interactions can, in some situations, greatly affect population responses to stressors. Some caution is appropriate when interpreting predictions based on population-level analyses.

Cumulative effects

Populations are often faced with multiple stressors that act simultaneously to influence population dynamics. The cumulative effects of multiple stressors can be far greater than the sum of their independent effects.

Striped bass population dynamics in the Sacramento River–San Joaquin Delta were simulated using an individual-based, YOY model coupled to an age-structured adult model (Rose et al., in prep.; Rose, in press). The striped bass population in this system has been declining over the past several decades, and a variety of causes (including reduced food, increased diversion of water for agricultural use, and reduced adult survival rates) have been blamed for the decline (Stevens et al. 1985). Prey composition (zooplankton and *Neomysis*), YOY mortality due to diversions of water for agriculture, and adult survival rates were specified at values reflective of pre- and post-decline periods. Predicted population abundances remained stable when prey composition, diversion mortality, and adult survival were all at their pre-decline values, and exhibited a dramatic decline (similar to the observed decline) when all 3 were set to their post-decline values (Figure 4-5). However, whenever each was set to post-decline values singly, only small reductions in population abundances were predicted. Examination of each of these factors alone would not have indicated the dramatic population decline that would result when all 3 factors were changed. If these 3 factors were operating independently, one can take the product of their separate effects and predict that all 3 together would cause a 55% decline in average adult striped bass abundance. However, the model predicted a greater 85% decline when all 3 factors were set to their post-decline values. Thus, the cumulative effects of the 3 stressors were greater than would be expected from the examination of each of the stressors alone.

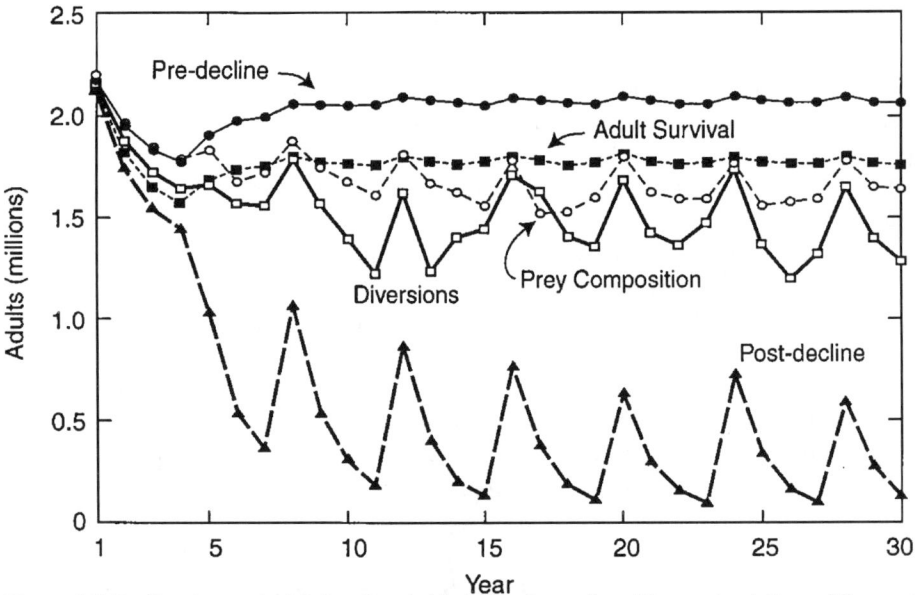

Figure 4-5 Predicted annual adult (age 3 and older) abundances from 30-year simulations of the Sacramento-San Joaquin River striped bass population model with prey composition, diversion mortality, and adult survival all at pre-decline values, each changed singly to its post-decline value, and all changed to their post-decline values (Rose et al., in prep.).

Stochasticity and power

High interannual variation in population abundances makes isolating contaminant effects difficult. Much of the variation appears as noise to investigators, resulting in comparisons between reference and impacted populations having low power of detection.

We use the previously cited modeling study by Jaworska, Rose, and Brenkert (1997) to illustrate the difficulties in separating the effects of a contaminant from other sources of variation. Although median predicted October abundances of large-mouth bass decreased with increasing concentrations of PCBs, predicted distributions overlapped substantially (Figure 4-3). Sample sizes needed to statistically detect significant differences in mean October abundances (t-test, α = 0.05, β = 0.2) between baseline (0 ppm PCB) and PCB simulations were 300 for 6 ppm, 92 for 10 ppm, 47 for 15 ppm, and 32 for 20 ppm. These sample sizes are the number of years that October abundances would have to measured, under both baseline and PCB conditions, to detect a statistically significant difference in mean values. The long data record required to detect differences is especially daunting when one considers that detection was attempted under ideal conditions. The only difference between baseline and PCB simulations was the effects induced by contaminant exposure; all other factors were identical between simulations.

Synthesis and Conclusions

We synthesize the major themes of this chapter below under the headings of reptiles and amphibians, populations, monitoring, toxicity endpoints, taxa versus contaminants, and confounding factors. The last section summarizes the current state of affairs in assessing population-level effects of contaminants on oviparous vertebrates and offers a plea for multidisciplinary studies.

Reptiles and amphibians

We would encourage more quantitative studies on the effects of contaminants on reptile and amphibian populations. Amphibians and reptiles are especially good predictors of the impact of environmental chemicals on populations because they
- exhibit great diversity in morphology, habitats utilized, and life histories;
- occur at high relative abundance and are important in many food webs;
- can be easily manipulated in experiments and monitored in the field;
- allow for contaminant effects to be measured under the extreme biological conditions of high competition and predation; and
- are highly vulnerable to contaminant exposure in both aquatic and terrestrial environments.

Exposure can be direct (e.g., through permeable eggs, skin, and gills in the case of salamanders) or indirect (e.g., through diet or maternal transfer of chemicals bioaccumulated in the yolk). Because of the intermediate phylogenetic position of amphibians and reptiles, they have the potential to shed insight into the link between the responses of ectothermic and endothermic oviparous vertebrates. Reptiles, especially those that are long-lived and late maturing (e.g., turtles and crocodilians) are also ideal models for the study of lipophilic, persistent compounds because they accumulate these contaminants over long periods of time.

Populations

A fundamental assumption of this chapter is that the population is the most relevant level of biological organization for assessing contaminant effects. We realize that this assumption is not necessarily in vogue, especially with the current emphasis on species diversity and ecosystem management. However, we are convinced that the population is the appropriate level for many analyses. Analyses performed at levels of organization below the population often face the difficult hurdle of establishing the ecological significance of lower-level effects. Performing analyses at higher levels of biological organization than the population are fraught with technical and measurement limitations. In some situations, community-level analyses are possible, but require extensive effort. Furthermore, community-level analyses may mask some important population responses. Sampling multiple populations often involves sacrificing some sampling effort on individual populations. Less intensive sampling of individual populations can miss some subtle, but important, population responses.

Monitoring

We advocate the use of life-history-based monitoring. Life-history-based monitoring involves measuring changes in a suite of individual-level characteristics, which can be used to indicate population status, to provide information on the type of stress involved, and to provide estimates of inputs to population dynamics models. These changes in individual characteristics can be detected more easily and with smaller sample sizes than can changes in abundances.

Two issues that repeatedly arise in population analyses are the selection of reference sites and the decision of which variables to measure. The complex life histories of many oviparous vertebrates make determination of the status of an unimpacted population practically impossible. Differences in population dynamics between a reference site and the impacted site cannot simply be attributed to the contaminant. Other differences between the sites can have potentially significant, but often unknown, effects on population dynamics. Measuring abundance is intuitively appealing but often has low power for detecting contaminant effects. Many taxa are long-lived and use a variety of habitats distributed in space. In addition, their population dynamics are affected by a multitude of abiotic and biotic factors. Measurements of abundance are simply too noisy to allow for detection of effects with a reasonable data record.

Toxicity endpoints

Data from laboratory assessments have to include some aspect of performance related to growth, reproduction, or survival, or should include measurements that can be translated to individual-level performance. There is presently a gap between commonly reported individual-level effects and those effects that can be used to infer population-level responses. If an empirical approach is used, individual-level effects need not be scalable if the individual-level effect is the endpoint. However, extrapolation to the population level requires toxicological endpoints be related to whole organism performance. This is not to say that a laboratory toxicology study needs to measure survival or lifetime egg production. Rather, the endpoint reported must be able to be ultimately related to growth, survival, or reproduction. An intermediate model or set of calculations may be needed to link the reported toxicological endpoint to a change in the inputs of a population dynamics model (see Figure 4-4).

More attention should be paid to measuring and reporting sublethal effects of contaminants that can be scaled to the population level. Most predictions of population-level consequences of contaminant effects have focused on lethality; increased mortality is easily understood as having population-level implications. Reproduction, growth, and mortality of oviparous vertebrates also include complex behavioral components that can be very sensitive to contaminant effects. This sensitivity is exemplified by the recent attention paid to endocrine disrupting chemicals, which can cause sublethal effects at low exposure levels. Laboratory assessments of sublethal effects must be designed in close collaboration with population ecologists in order to ensure that reported data includes adequate information for use in population-level assessments, either directly or via bridge models.

Taxa versus contaminants

We suggest that greater differences in population responses arise from exposures to different contaminants than from different taxa exposed to the same contaminant. The number of commonalities among the taxa in terms of their population dynamics was surprising. The similarities became especially clear when species were viewed in a general life-history framework of life cycles and life-history strategies. However, in our review of contaminant effects, we had difficulty in making generalizations across the different types of contaminants, even for studies performed on the same taxonomic group. This lack of obvious relationships between population responses and contaminant type could be because the nature of the available information. Many of the analyses of specific contaminant effects are site-specific, making comparison to other studies difficult. Also, there is a great deal of variation among individuals and among species in toxicant metabolism and sensitivity to contaminant effects, and the timing and duration of exposure can vary considerably even among closely related taxa.

We advocate the use of individual-based and matrix projection models imbedded in a life-history framework to analyze population effects of contaminants. These modeling approaches are flexible enough to handle many different taxa and contaminants, are readily accepted in the scientific community, and can be adapted to the quantity and quality of the available data.

Confounding factors

Quantifying contaminant effects at the population level is confounded by dispersal, density dependence, stochasticity, community-level interactions, and cumulative effects of multiple stressors. We illustrated each of these with empirical or modeling examples. Greater awareness of these issues in population-level studies should improve our ability to detect and separate contaminant effects from other sources of population variation. We believe that consideration of these issues can help explain some of the ecological responses in nature that surprise us from time to time.

Prognosis

Surprisingly few definitive studies have unequivocally documented the role of contaminants in causing population declines in oviparous vertebrates. While there are some clear-cut examples, especially those involving bird populations and organochlorine pesticides, there are also many examples that were suggestive but inconclusive. The cases where contaminants were clearly linked to population declines required an extensive series of studies that tediously provided the pieces to a complex puzzle. A major question facing us today is whether the lack of wide-spread examples of contaminants causing declines is because we are not looking in enough detail at specific sites or because such situations are the exception rather than the rule. As the pool of potential contaminants continues to increase, and populations are subjected to increasing pressures from man's activities, it is impera-tive that we have confidence in the tools we use for detecting population-level effects of contaminants or other anthropogenic stressors. We must be certain that when we do not detect a contaminant effect on the population, it is truly because there is no effect and not because our methods are inadequate. The situation is only going to get more complicated. The technology is available for performing analyses of the population-level effects of contaminants: chemical fate and exposure methods, toxicity databases, life-history-based monitoring, individual-based and matrix-projection life-cycle modeling, and life-history theory. The limiting factor is our ability as scientists to effectively and efficiently work together. Toxicologists, physiologists, chemists, population ecologists, and risk assessors must collaborate in true multidisciplinary efforts if we hope to effectively manage and maintain our natural resources.

References

Adams SM, Crumby WD, Greeley MS, Shugart LR, Saylor CF. 1992. Responses of fish populations and communities to pulp mill effluents: a holistic approach. *Ecotoxicol Environ Safety* 24:347–360.

Akcakaya HR, Atwood JL. 1997. A habitat-based metapopulation model of the California gnatcatcher. *Conserv Biol* 11:422–434.

Ankley GT, Giesy JP. 1998. Endocrine disruptors in wildlife: a weight-of-evidence perspective. In: Kendall R, Dickerson R, Suk W, Giesy J, editors. Principles and processes for assessing endocrine disruption in wildlife. Pensacola FL: SETAC. p 349–367.

Baker JP, Bernard DP, Christensen SW, Sale MJ, Freda J, Heltcher K, Marmorek D, Rowe L, Scanlon P, Suter G, Warren-Hicks W, Welbourn P. 1990. Biological effects of changes in surface water acid-base chemistry. Washington, DC: National Acid Precipitation Assessment Program. NAPAP State of Science and Technology Report 13.

Barnthouse LW, Boreman J, Christensen SW, Goodyear CP, Van Winkle W, Vaughan DS. 1984. Population biology in the courtroom: the Hudson River controversy. *BioScience* 34:14–19.

Barnthouse LW, Suter II GW, Rosen AE, Beauchamp JJ. 1987. Estimating responses of fish populations to toxic contaminants. *Environ Toxic Chem* 6:811-824.

Barnthouse LW, Suter II GW, Rosen AE. 1989. Inferring population-level significance from individual-level effects: an extrapolation from fisheries science to ecotoxicology. In: Suter GW, Lewis MA, editors. Aquatic toxicology and environmental fate: 11th volume. Philadelphia: American Society for Testing and Materials. ASTM STP 1007. p 289-300.

Barnthouse LW, Suter GW II, Rosen AE. 1990. Risks of toxic contaminants to exploited fish populations: influence of life history, data uncertainty and exploitation intensity. *Environ Toxicol Chem* 9:297–311.

Bergeron JM, Crews D, McLachlan JA. 1994. PCBs as environmental estrogens: turtle sex determination as a biomarker of environmental contamination. *Environ Health Perspec* 102:780–781.

Bernatowicz JA, Schemp PF, Bowman TD. 1996. Bald eagle productivity in south-central Alaska in 1989 and 1990 after the *Exxon Valdez* oil spill. *Am Fish Soc Symp* 18:785–797.

Bishop CA., Brooks RJ, Carey JH, Ng P, Norstrom RJ, Lean DRS. 1991. The case for a cause-effect linkage between environmental contamination and development in eggs of the common snapping turtle (*Chelydra s. serpentina*) from Ontario, Canada. *J Toxicol Environ Health* 33:521–547.

Bishop CA, Brown GP, Brooks RJ, Lean DRS, Carey JH. 1994. Organochlorine contaminant concentrations in eggs and their relationship to body size, clutch characteristics of the female snapping turtle (*Chelydra serpentina serpentina*) in Lake Ontario, Canada. *Arch Environ Contam Toxicol* 27:82–87.

Boersma PD, Parrish JK, Kettle AB. 1995. Common murre abundance, phenology, and productivity on the Barren Islands, Alaska: the *Exxon Valdez* oil spill and long-term environmental change. In: Wells PG, Butlerand JN, Hughes JS, editors. *Exxon Valdez* oil spill: fate and effects in Alaskan waters. Philadelphia: American Society for Testing and Materials. ASTM STP 1219. p 820–853.

Botsford LW, Castilla JC, Peterson CH. 1997. The management of fisheries and marine ecosystems. *Science* 277:509–515.

Burger J. 1992. Trace element levels in pine snake hatchlings: tissues and temporal differences. *Arch Environ Contam Toxicol* 22:209–213.

Busby DG, White LM, Pearce PA. 1990. Effects of aerial spraying of fenitrothion on breeding white-crowned sparrows. *J Appl Ecol* 27:743–755.

Butler RG, Harfenist A, Leighton FA, Peakall DB. 1988. Impact of sublethal oil and emulsion exposure on the reproductive success of Leach's storm-petrels: short and long-term effects. *J Appl Ecol* 25:125–143.

Butler RG, Peakall DB, Leighton FA, Borthwick J, Harmon RS. 1986. Effects of crude oil exposure on standard metabolic rate of Leach's storm-petrel. *Condor* 88:248–249.

Caswell H. 1989. Matrix population models: construction, analysis, and interpretation. Sunderland MA: Sinauer Associates. 328 p.

Chang LW, Cockerham L. 1994. Toxic metals in the environment. In: Cockerham LG, Shane BS, editors. Basic environmental toxicology. Boca Raton FL: CRC Press. p 109–132.

Clutton-Brock TH. 1988. Reproductive success. Chicago IL: University of Chicago Press. 538 p.

Colby PJ. 1984. Appraising the status of fisheries: rehabilitation techniques. In: Cairns VW, Hodson PV, Nriagu JO, editors. Contaminant effects on fisheries. *Adv Environ Sci Tech* 16:233–257.

Cole LC. 1954. The population consequences of life history phenomena. *Quarterly Review of Biology* 29:103–137.

Congdon JD, Tinkle DW. 1982. Reproductive energetics of the painted turtle (*Chrysemys picta*). *Herpetologica* 38:228–237.

Congdon JD, Dunham AE, Van Loben Sels RC. 1993. Demographics of Blanding's turtles (*Emydoidea blandingii*): implications for conservation and management of long-lived organisms. *Conserv Biol* 7:826–833.

Congdon JD, Dunham AE, Van Loben Sels RC. 1994. Demographics of common snapping turtles (*Chelydra serpentina*): implications for conservation and management of long-lived organisms. *Am Zool* 34:397–408.

Cowan JH, Rose KA, Houde ED. 1997. Size-based foraging success and vulnerability to predation: selection of survivors in individual-based models of larval fish populations. In: Chambers R, Trippel E, editors. Early life history and recruitment in fish populations. New York: Chapman and Hall. p 357–386.

Crump ML. 1995. Parental care. In: Heatwole H, Sullivan BK, editors. Amphibian biology, Volume 2. NSW Australia: Surrey Beatty and Sons.

Culik BM, Wilson RP, Woakes AT, Sanudo FW. 1991. Oil pollution of Antarctic penguins: effects on energy metabolism and physiology. *Marine Pollut Bull* 22:388–391.

Cumbie PM, Van Horn SL. 1979. Selenium accumulation associated with fish mortality and reproductive failure. *Proc Annual Conf Southeastern Assoc Fish Wildlife Agencies* 32:612–624.

DeAngelis DL, Gross LJ. 1992. Individual-based models and approaches in ecology: populations, communities, and ecosystems. Boca Raton FL: CRC Press. 544 p.

DeAngelis DL, Rose KA. 1992. Which individual-based approach is most appropriate for a given problem? In: DeAngelis DL, Gross LJ, editors. Individual-based models and approaches in ecology: populations, communities, and ecosystems. Boca Raton FL: CRC Press. p 67–87.

DeAngelis DL, Rose KA, Crowder L, Marschall E, Lika D. 1993. Fish cohort dynamics: application of complementary modeling approaches. *Am Natur* 142:604–622.

DeAngelis DL, Rose KA, Huston MA. 1994. Individual-oriented approaches to modeling populations and communities. In: Levin SA, editor. Frontiers in mathematical biology, Lecture Notes in Biomathematics, Vol. 100. New York: Springer-Verlag. p 390–410.

den Boer PJ, Reddingius J. 1989. On the stabilization of animal numbers. Problems of testing 2. Confrontation with data from the field. *Oecologia* 79:143–149.

DeWeese LR, Henny CJ, Floyd RJ, Bobal KA, Schulz AW. 1979. Responses of breeding birds to aerial sprays of trichlorfon (dylox) and carbaryl (sevin-4-oil) in Montana Forests. Washington DC: U.S. Department of Interior. Special Scientific Report Wildlife 224.

Desbrow C, Routledge EJ, Brighty GC, Sumpter JP, Waldock M. 1998. Identification of estrogenic chemicals in STW effluent-I-chemical fractionation and in vitro biological screening. *Env Sci Tech* 32:1549–1558.

Dobson A, Hudson P. 1994. Assessing the impact of toxic chemicals: temporal and spatial variation in avian survival rates. In: Kendall RJ, Latcher TE, editors. Wildlife toxicology and population modeling: integrated studies of agriecosystems. Boca Raton FL: CRC Press. p 85–98.

Duellman WE, Trueb L. 1986. Biology of amphibians. New York: McGraw-Hill. 670 p.

Dunham AE, Miles DB. 1985. Patterns of covariation in life history traits of squamate reptiles: the effects of size and phylogeny reconsidered. *Amer Natur* 126:231–257.

Dunham AE, Miles DB, Resnick DN. 1988. Life history patterns in squamate reptiles. In: Gans C, editor. Biology of the reptilia, Vol. 16. New York: Alan R. Liss. p 441–522.

Dunning JB, Danielson BJ, Pulliam HR. 1992. Ecological processes that affect populations in complex landscapes. *Oikos* 65:169–175.

Dunning JB, Stewart DJ, Danielson BJ, Noon BR, Root TL, Lamberson RH, Stevens EE. 1995. Spatially explicit population models: current forms and future uses. *Ecol Applic* 5:3–11.

Ebenman B, Persson L. 1988. Size-structured populations. New York: Springer-Verlag. 284 p.

Eberhardt LL. 1970. Correlation, regression, and density dependence. *Ecology* 51:306–310.

Eisler R. 1988. Lead hazards to fish, wildlife, and invertebrates: a synoptic review. Washington DC: U.S. Fish and Wildlife Service. Biol. Rep. 85(1.14)

Emlen JM. 1989. Terrestrial population models for ecological risk assessment: a state-of-the-art review. *Environ Toxicol Chem* 8:831–842.

Erikson DE. 1995. Surveys of murre colony attendance in the northern Gulf of Alaska following the Exxon Valdez oil spill. In: Wells PG, Butler JN, Hughes JS, editors. *Exxon Valdez* oil spill: fate and effects in Alaskan waters. Philadelphia PA: American Society for Testing and Materials. ASTM STP 1219. p 780–819.

Fahrig L, Freemark K.1995. Landscape-scale effects of toxic events for ecological risk assessment. In: Cairns Jr J, Niederlehner BR, editors. Ecological toxicity testing. Boca Raton FL: Lewis. p 193–208.

Fleet RR, Plapp Jr FW. 1978. DDT residues in snakes decline since DDT ban. *Bull Environ Contam Toxicol* 19:383–388.

Ford RG, Wiens JA, Heinmann D, Hunt GL. 1982. Modelling the sensitivity of colonially breeding marine birds to oil spills: guillemot and kittiwake populations on the Pribilof islands, Bering Sea. *J Appl Ecol* 19:1–31.

Fox GA. 1971. Recent changes in the reproductive success of the pigeon hawk. *J Wildlife Manage* 35:122–128.

Fox GA. 1993. What have biomarkers told us about the effects of contaminants on the health of fish-eating birds in the Great Lakes? The theory and a literature view. *J Great Lakes Res* 19:722–736.

Freemark, K, Boutin C. 1995. Impacts of agricultural herbicide use on terrestrial wildlife in temperate landscapes: a review with special reference to North America. *Agric Ecosyst Environ* 52:67–91.

Fry DM, Swenson J, Addiego LA, Grau CR, Kang A. 1986. Reduced reproduction in wedge-tailed shearwaters exposed to weathered Santa Barbara crude oil. *Arch Environ Contam Toxicol* 15:453–463.

Gagnon MM, Dodson JJ, Hodson PV. 1994a. Ability of BKME (bleached kraft mill effluent) exposed white suckers (*Catostomus commersoni*) to synthesize steroid hormones. *Comp Biochem Physiol* 107C:265–273.

Gagnon MM, Dodson JJ, Hodson PV. 1994b. Seasonal effects of bleached kraft mill effluent on reproductive parameters of white sucker (*Catostomus commersoni*) populations of the St. Maurice River, Quebec, Canada. *Can J Fish Aquat Sci* 51:337–347.

Garcia S, Newton S. 1997. Current situation, trends, and prospects in world capture fisheries. In: Pikitch EK, Huppert DD, Sissenwine MP, editors. Global trends: fisheries management. Bethesda MD: American Fisheries Society. p 3–27.

Gard NW, Hooper MJ. 1995. An assessment of potential hazards of pesticides and environmental contaminants: a synthesis and review of critical issues. In: Martin TE, Finch DM, editors. Ecology and management of neotropical migratory birds. New York: Oxford University Press. p 294–310.

Geiger HJ, Bue BG, Sharr S, Wertheimer AC, Willeffe TM. 1996. A life history approach to estimating damage to Prince William Sound pink salmon caused by the *Exxon Valdez* oil spill. *Am Fish Soc Symp* 18:487–498.

Gibbons JW, Semlitsch RD, Greene JL, Schubauer JP. 1981. Variation in age and size at maturity of the slider turtle (*Pseudemys scripta*). *Am Natur* 117:841–845.

Gibbons WN, Munkittrick KR. 1994. A sentinel monitoring framework for identifying fish population responses to industrial discharges. *J Aquat Eco Health* 3:227–237.

Gilbertson M. 1974. Pollutants in breeding herring gulls in the lower Great Lakes. *Can Field Nat* 88:273–280.

Gillespie RC, Baumann PC. 1986. Effects of high tissue concentrations of selenium on reproduction by bluegills. *Trans Am Fish Soc* 115:208–213.

Gilman AP, Fox GA, Peakall DB, Teeple SM, Carrol TR, Haymes GT. 1977. Reproductive parameters and egg contaminant levels of Great Lakes herring gulls. *J Wildl Manage* 41:458–468.

Goodyear CP. 1977. Assessing the impact of power plant mortality on the compensatory reserve of fish populations. In: Van Winkle W, editor. Assessing the effects of power-plant-induced mortality on fish populations. New York: Pergamon. p 186–195.

Goodyear CP. 1988. Implications of power plant mortality for management of the Hudson River striped bass fishery. *Am Fish Soc Monogr* 4:245–254.

Grant WE, Fraser SO, Isakson KG. 1983. Effect of vertebrate pesticides on non-target wildlife populations: evaluation through modelling. *Ecol Model* 21:85–108.

Grasman KA., Fox GA, Scanlon PF, Ludwig JP. 1996. Organochlorine-associated immunosuppression in prefledgling Caspian terns and herring gulls from the Great Lakes: an ecoepidemiological study. *Environ Health Perspect* 104(Suppl 4):829–842.

Green RE. 1984. The feeding ecology and survival of partridge chicks (*Alectoris rufa* and *Perdix perdix*) on arable farmland in East Anglia. *J Appl Ecol* 21:817–830.

Grier JW. 1980. Modeling approaches to bald eagle population dynamics. *Widl Soc Bull* 8:316–322.

Hallam TG, Lassiter RR, Li J, McKinney W. 1990. Toxicant-induced mortality in models of Daphnia populations. *Environ Toxicol Chem* 9:597–621.

Harris HJ, Erdman TC, Ankley GT, Lodge KB. 1993. Measures of reproductive success and polychlorinated biphenyl residues in eggs and chicks of Forster's Terns on Green Bay, Lake Michigan, Wisconsin - 1988. *Arch Environ Contam Toxicol* 25:304–314.

Heinz GH, Glooschenko V, Haffner GD, Lazar R. 1991. Contaminants in American alligator eggs from Lake Apopka, Lake Griffin, and Lake Okeechobee, Florida. *Environ Monit Assess* 16:277–285.

Herman SG, Bulger JB. 1979. Effects of a forest application of DDT on nontarget organisms. Wildlife Monograph 69. Suppl. to *J Wildlife Manage* 43(4).

Hewitt LM, Carey JH, Munkittrick KR, Parrott JL, Solomon KR, Servos MR. 1998. Identification of chloro-nitro-trifluoromethyl-substituted dibenzo-*p*-dioxins in lampricide formulations of 3- trifluoromethyl-4-nitrophenol: assessment to induce mixed function oxidase activity. *Environ Toxicol Chem* 17:841–950.

Hewitt LM, Scott IM, Munkittrick KR, Van Der Kraak GJ, Solomon KR, Servos MR. 1996. Development of TIE procedures for complex mixtures using physiological responses in fish. In: Bengtson DA, Henshel DS, editors. Environmental toxicology and risk assessment: biomarkers and risk assessment, 5th Volume. Philadelphia PA: American Society for Testing and Materials. STP 1306. p 37–52.

Hodson PV, McWhirther M, Ralph K, Gray B, Thivierge D, Carey J, Van Der Kraak G, Whittle DM, Levesque MC. 1992. Effects of bleached kraft mill effluent on fish in the St. Maurice River, Quebec. *Environ Toxicol Chem* 11:1635–1651.

Hoffman DH, Albers PH. 1984. Evaluation of potential embryotoxicity and teratogenicity of 42 herbicides, insecticides, and petroleum contaminants to mallard eggs. *Arch Environ Contam Toxicol* 13:15–27.

Holmes WN, Cavanagh KP. 1990. Some evidence for an effect of ingested petroleum on the fertility of the mallard drake (*Anas platyrhynchos*). *Arch Environ Contam Toxicol* 19:898–901.

Houston CS, Hodson KA.1997. Resurgence of breeding merlins, *Falco columbarius richardsonii*, in Saskatchewan grasslands. *Can Field Nat* 111:243–248.

Howe RW, Davis GJ, Mosca V. 1991. The demographic significance of "sink" populations. *Biol Conserv* 57:239–255.

Irons DB. 1996. Size and productivity of black-legged kittiwake colonies in Prince William Sound before and after the *Exxon Valdez* oil spill. *Am Fish Soc Symp* 18:738–747.

Jaworska JS, Rose KA, Barnthouse LW. 1997. General response patterns of fish populations to stress: an evaluation using an individual-based simulation model. *J Aquat Ecosyst Stress Recovery* 6:15–31.

Jaworska JS, Rose KA, Brenkert AL. 1997. Individual-based modeling of PCB effects on young-of-the-year largemouth bass in southeastern U.S. reservoirs. *Ecol Model* 99:113–135.

Jobling S, Sumpter JP. 1993. Detergent components in sewage effluent are weakly estrogenic to fish: an in vitro study using rainbow trout (*Oncorhynchus mykiss*) hepatocytes. *Aquat Toxicol* 27:361–372.

Jobling S, Nolan M, Tyler CR, Brighty G, Sumpter JP. 1998. Widespread sexual disruption in wild fish. *Env Sci Tech* 32:2498–2506.

Johnson MD, Geupel GR. 1996. The importance of productivity to the dynamics of a Swainson's thrush population. *Condor* 98:133–141.

Klomp NI, Furness RW. 1992. Non-breeders as a buffer against environmental stress: declines in numbers of great skuas on Foula, Shetland, and prediction of future recruitment. *J Appl Ecol* 29:341–348.

Kooijman SALM, Metz JAJ. 1984. On the dynamics of chemically stressed populations: the deduction of population consequences from effects on individuals. *Ecotoxic Environ Safety* 8:254–274.

Kovacs TG, Gibbons JS, Martel PH, Voss RH. 1996. Improved effluent quality at a bleached kraft mill as determined by laboratory biotests. *J Toxicol Environ Health* 49:101– 129.

Kovacs TG, Gibbons JS, Tremblay LA, O'Connor BI, Martel PH,Voss RH. 1995. The effects of a secondary-treated bleached kraft pulp mill effluent on aquatic organisms as assessed by short-term and long-term laborator tests. *Ecotox Environ Safety* 31:7–22.

Krebs CJ. 1991. The experimental paradigm and long-term population studies. *IBIS* 133 (Supplement 1):3–8.

Kubiak TJ, Harris HJ, Smith LM, Schwartz TR, Stalling DL, Trick JA, Sileo L, Docherty DE, Erdman TC. 1989. Microcontaminants ansd reproductive impairment of the Forster's tern on Green Bay, Lake Michigan - 1983. *Arch Environ Contam Toxicol* 18:706–727.

Landahl, JT, Johnson LL, Collier TK, Stein JE, Varanasi U. 1997. Marine pollution and fish population parameters: English sole (*Pleuronectes vetulus*) in Puget Sound, WA. *Trans Am Fish Soc* 126:519–535.

Lande R. 1988. Demographic models of the northern spotted owl (*Strix occidentalis* caurina). *Oecologica* 75:601–607.

Lassiter RR, Hallam TG. 1990. Survival of the fattest: implications for acute effects of lipophilic chemicals on aquatic populations. *Environ Toxicol Chem* 9:585-596.

Leighton FA, Peakall DB, Butler RG. 1983. Heinz-body hemolytic anemia from ingestion of crude oil: a primary toxic effect in marine birds. *Science* 220:871–873.

Lewontin RC. 1965. Selection for colonizing ability. In: Baker RK, Stebbins GL, editors. The genetics of colonizing species. New York: Academic Press. p 77–91.

McDonald DP, Caswell H. 1993. Matrix methods for avian demography. In: Power DM, editor. Current ornithology, Vol. 10. New York: Plenum Press. p 139–185

McMaster ME, Van Der Kraak GJ, Munkittrick KR. 1996. An epidemiological evaluation of the biochemical basis for steroid hormonal depressions in fish exposed to industrial wastes. *J Great Lakes Res* 22:153–171.

McMaster ME, Van Der Kraak GJ, Portt CB, Munkittrick KR, Sibley PK, Smith IR, Dixon DG. 1991. Changes in hepatic mixed function oxygenase (MFO) activity, plasma steroid levels and age at maturity of a white sucker (*Catostomus commersoni*) population exposed to bleached kraft pulp mill effluent. *Aquat Toxicol* 21:199–218.

McKnight CM, Gutzke WHN. 1993. Effects of embryonic environment and of hatchling housing conditions on growth of young snapping turtles (*Chelydra serpentina*). *Copeia* (May 3):475–482.

Mertz DB. 1971. The mathematical demography of the California condor population. *Am Nat* 105:437–453.

Miller TJ, Crowder LB, Rice JA, Marschall EA. 1988. Larval size and recruitment mechanisms in fishes: toward a conceptual framework. *Can J Fish Aquatic Sci* 45:1657–1670.

Mineau P, Boersma DC, Collins B. 1994. An analysis of avian reproduction studies submitted for pesticide registration. *Ecotoxicol Environ Safety* 29:304–329.

Munkittrick KR. 1992. A review and evaluation of study design considerations for site-specifically assessing the health of fish populations. *J Aquat Eco Health* 1:283–293.

Munkittrick KR, Dixon DG. 1989a. An holistic approach to ecosystem health assessment using fish population characteristics. In: Munawar, Dixon DG, Mayfield CI, Reynoldson T, Sadar MH, editors. Environmental bioassay techniques and their application. Dordrecht The Netherlands: Kluwer Academic Publishers. p 122–135.

Munkittrick KR, Dixon DG. 1989b. Use of white sucker *(Catostomus commersonl)* populations to assess the health of aquatic ecosystems exposed to low-level contaminant stress. *Can J Fish Aquat Sci* 46:1455–1462.

Munkittrick KR, McMaster ME, Portt CB, Van Der Kraak GJ, Smith IR, Dixon DG. 1992. Changes in maturity, plasma sex steroid levels, hepatic mixed-function oxygenase activity, and the presence of external lesions in lake whitefish *(Coregonus clupeaformis)* exposed to bleached kraft mill effluent. *Can J Fish Aquat Sci* 49:1560–1569.

Munkittrick KR, Servos MR, Carey JH, Van Der Kraak GJ. 1997. Environmental pulp and paper wastewater: evidence for a reduction in environmental effects at North American pulp mills since 1992. *Water Sci Technol* 35:329–338.

Munkittrick KR, Servos MR, Parrott JL, Martin V, Carey JH, Flett PA, Van Der Kraak GJ. 1994. Identification of lampricide formulations as a potent inducer of MFO activity in fish. *J Great Lakes Res* 20:355–365.

Munkittrick KR, Van Der Kraak GJ, McMaster ME, Portt CB. 1992. Response of hepatic mixed function oxygenase (MFO) activity and plasma sex steroids to secondary treatment and mill shutdown. *Environ Toxic Chem* 11:1427–1439.

Murray BG. 1994. On density dependence. *Oikos* 69:520–523.

Newton I. 1989. Lifetime reproduction in birds. London, UK: Academic Press. 496 p.

Newton I, Bogan JA, Rothery P. 1986. Trends and effects of organochlorine compounds in sparrowhawk eggs. *J Appl Ecol* 23:461–478.

Newton I, Wyllie I. 1992. Recovery of a sparrowhawk population in relation to declining pesticide contamination. *J Appl Ecol* 29:476–484.

Newton I, Wyllie I, Asher A. 1992. Mortality from pesticides aldrin and dieldrin in British sparrowhawks and kestrels. *Ecotoxicology* 1:31–44.

O'Connor RJ, Shrubb M. 1986. Farming and birds. Cambridge, UK: Cambridge University Press. 290 p.

Oakley KL, Kuletz K. 1996. Population, reproduction, and foraging of pigeon guillemots at Naked Island, Alaska, before and after the *Exxon Valdez* oil spill. *Am Fish Soc Symp* 18:759–769.

Ohlendorf HM, Hothem RL, Aldrich TW. 1988. Bioaccumulation of selenium by snakes and frogs in the San Joaquin Valley, California. *Copeia* 3:704–710.

Ohlendorf HM, Hothem RL, Welsh D. 1989. Nest success, cause-specific nest failure, and hatchability of aquatic birds at selenium-contaminated Kesterson Reservoir and a reference site. *Condor* 91:787–796.

Ouellet M, Bonin J, Rodrigue J, DesGranes J-L, Lair S. 1997. Hindlimb deformities (ectromelia, ectrodactyly) in free-living anurans from agricultural habitats. *J Wildl Dis* 33:95–104.

Panek M. 1997. The effect of agricultural landscape structure on food resources and survival of grey partridge (*Perdix perdix*) chicks in Poland. *J Appl Ecol* 34:787–792.

Peakall DB, Miller DS, Kinter WB. 1983. Toxicity of crude oils and their fractions to nesting herring gulls - 1. Physiological and biochemical effects. *Marine Environ Res* 8:63–71.

Peters RH. 1983. The ecological implications of body size. Cambridge, UK: Cambridge University Press. 329 p.

Pianka ER. 1970. On *r*- and *K*-selection. *Am Nat* 104:592–597.

Piatt JF, Anderson P. 1996. Response of common murres to the *Exxon Valdez* oil spill and long-term changes in the Gulf of Alaska marine ecosystem. *Am Fish Soc Symp* 18:720–737.

Porter JM, Coulson JS. 1987. Long-term changes in recruitment to the breeding group, and the quality of recruits at a kittiwake (*Rissa tridactyla*) colony. *J Anim Ecol* 56:675–689.

Potts GR. 1980. The effects of modern agriculture, nest predation and game management on the population ecology of partridges (*Perdix perdix* and *Alectoris rufa*). *Adv Ecol Res* 11:1–82.

Pulliam HR. 1988. Sources, sinks and population regulation. *Am Nat* 132:652–661.

Pulliam HR. 1994. Incorporating concepts from population and behavioral ecology into models of exposure to toxins and risk assessment. In: Kendall RJ, Latcher TE, editors.

Wildlife toxicology and population modelling: integrated studies of agroecosystems. Boca Raton, FL: CRC Press. p 13–26.

Pulliam RH, Dunning JB, Liu J. 1992. Population dynamics in complex landscapes: a case study. *Ecol Applic* 2:165–177.

Purdom CE, Hardiman PA, Bye VJ, Eno NC, Tyler CR, Sumpter JP. 1994. Estrogenic effects of effluents from sewage treatment works. *Chem Ecol* 3:275–285.

Rands MRW. 1985. Pesticide use on cereals and the survival of grey partridge chicks: a field experiment. *J Appl Ecol* 22:49–54.

Ratcliffe DA. 1967. Decrease in eggshell weight in certain birds of prey. *Nature* 215:208–210.

Roff DA. 1992. The evolution of life histories. New York: Chapman, and Hall. 528 p.

Rose KA. Why are quantitative relationships between environmental quality and fish populations so elusive? *Ecolog Applic* (In press).

Rose KA, Cowan JH. 1993. Individual-based model of young-of-the-year striped bass population dynamics: I. Model description and baseline simulations. *Trans Am Fish Soc* 122:415–438.

Rose KA, Cowan JH, Houde ED, Coutant CC. 1993. Individual-based modeling of environmental quality effects on early life stages of fish: a case study using striped bass. *Am Fish Soc Sym* 14:125–145.

Rose KA, Cowan JH, Miller LW, Stevens DE, Kimmerer WJ, Brown R. A model analysis of the factors affecting the decline in the Sacramento-San Joaquin River striped bass population. (In prep).

Rose KA, Rutherford ES, McDermott D, Forney JL, Mills EL. 1999. Individual-based model of yellow perch and walleye populations in Oneida Lake. *Ecolog Monographs* 69:127–154.

Rowley I, Russell E. 1991. Demography of passerines in the temporate southern hemisphere. In: Perrins CM, Lebreton J-D, Hirons GJM, editors. Bird population studies: relevance to conservation and management. Oxford, UK: Oxford University Press. p 22–44.

[RSPB] Royal Society for the Protection of Birds. 1997. The indirect effects of pesticides on farmland birds. Bedfordshire, UK: RSPB.

Russell RW, Hecnar SJ, Haffner GD. 1995. Organochlorine pesticide resdiues in southern Ontario peepers. *Environ Toxicol Chem* 14:815–817.

Saila S, Martin B, Ferson S, Ginzburg L, Millstein J. 1991. Demographic modeling of selected fish species with RAMAS. Final Report, Research Project 2553. Palo Alto CA: Electric Power Research Institute. EPRI EN-7178.

Samuels WB, Lanfear KJ. 1982. Simulations of seabird damage and recovery from oilspills in the northern Gulf of Alaska. *J Environ Manage* 15:169–182.

Sanderson GC, Bellrose FC. 1986. A review of the problem of lead poisoning in waterfowl. Champaign IL: Illinois Natural History Survey. Spec. Publ. No. 4.

Sandström O. 1994. Incomplete recovery in a coastal fish community exposed to effluent from a modernized Swedish coastal bleached kraft mill. *Can J Fish Aquat Sci* 51:2195–2202.

Sedinger JS, Flint PL, Lindberg MS. 1995. Environmental influence on life-history traits: growth, survival, and fecundity in black brant (*Branta bernicla*). *Ecology* 76:2404–2414.

Sharp BE, Cody M, Turner R. 1996. Effects of the *Exxon Valdez* oil spill on the black oystercatcher. *Am Fish Soc Symp* 18:748–758.

Slade NA. 1977. Statistical detection of density dependence from a series of sequential censuses. *Ecology* 58:1094–1102.

Slade NA.1994. Models of structured populations: age and mass transition matrices. In: Kendall RJ, Lacher TE, editors. Wildlife toxicology and population modeling. Boca Raton FL: Lewis. p 189–199.

Sogard SM. 1997. Size-selective mortality in the juvenile stage of teleost fishes: a review. *Bull Marine Sci* 60:1129–1157.

Southwood TRE, Cross DJ. 1969. The ecology of the partridge. III. Breeding success and the abundance of insects in natural habitats. *J Anim Ecol* 38:497–509.

Stearns SC. 1977. The evolution of life history traits: a critique of the theory and a review of the data. *Ann Rev Ecol Syst* 8:145–171.

Stevens DE, Kohlhorst DW, Miller LW, Kelly DW. 1985. The decline of striped bass in the Sacramento-San Joaquin Estuary, California. *Tran Am Fish Soc* 114:12–30.

Temple SA 1986. The problem of avian extinctions. *Current Ornithol* 3:453–485.

Tuljapurkar S, Caswell H. 1997. Structured-population models in marine, terrestrial, and freshwater systems. New York: Chapman and Hall.

[USFWS] U.S. Fish and Wildlife Service. 1986. Use of lead shot for hunting migratory birds in the United States. Final Environmental Impact Statement. Washington DC: U.S. Government Printing Office.

Van Der Kraak GJ, Munkittrick KR, McMaster ME, MacLatchy DL. 1998. A comparison of bleached kraft pulp mill effluent, 17β-estradiol and β-sitosterol effects on reproductive function in fish. In: Kendall RJ, Dickerson R, Geisy JP, Suk W, editors. Principles and processes for evaluating endocrine disruption in wildlife. Pensacola FL: SETAC. p 249–265.

Vitt LJ, Caldwell JP, Wilbur HM, Smith DC. 1990. Amphibians as harbingers of decay. *Bioscience* 40:418.

Wake DB. 1991. Declining amphibian populations. *Science* 253:860.

Wassersug RJ. 1975. The adaptive significance of the tadpole stage with comments on the maintenance of complex life cycles in anurans. *Amer Zool* 15:405–417.

White R, Jobling S, Hoare SA, Sumpter JP, Parker JP. 1994. Environmentally persistent alkylphenolic compounds are estrogenic. *Endocrinology* 135:175–182.

Wiens JA, Crist TO, Day RH, Murphy SM, Hayward CD. 1996. Effects of the *Exxon Valdez* oil spill on marine bird communities in Prince William Sound, Alaska. *Ecol Applic* 6:829–841.

Wilbur HM. 1980. Complex life histories. *Ann Rev Ecol Syst* 11:67–93.

Wilbur HM. 1987. Regulation of structure in complex systems: experimental temporary pond communities. *Ecology* 68:1437–1452.

Williams B, Emlen J. 1994. Population models as research tools: an empirical perspective. In: Kendall RJ, Lacher TE, editors. Wildlife toxicology and population modeling: integrated studies of agroecosystems. Boca Ration, FL: CRC Press. p 501–508.

Williams ML, Hothem RL, Ohlendorf HM. 1989. Recruitment failure in American avocets and black-necked stilts nesting at Kesterson Reservoir, California, 1984–1985. *Condor* 91:797–802.

Winemiller KO, Rose KA. 1992. Patterns of life-history diversification in North American fishes: Implications for population regulation. *Can J Fish Aquat Sci* 49:2196–2218.

Young H. 1968. A consideration of insecticide effects on hypothetical avian populations. *Ecology* 49:992–994.

Zug GR. 1993. Herpetology: an introductory biology of amphibians and reptiles. San Diego CA: Academic Press. 527 p.

CHAPTER 5

Reproductive and Developmental Toxicology of Contaminants in Oviparous Animals

Anne Fairbrother, Gerald T. Ankley, Linda S. Birnbaum, Steven P. Bradbury,
Bettina Francis, L. Earl Gray, David Hinton, Lyndal L. Johnson, Richard E. Peterson,
Glen Van Der Kraak

Ecological risk assessments can utilize endpoints at any of a number of biological levels of organization (Figure 5-1). Except in the case of rare and endangered species, assessment endpoints of most concern in ecological risk assessment typically are at the population and community levels (Suter 1993). However, these endpoints generally are highly integrative and not reflective of any particular stressor. For example, fecundity can be affected adversely by a variety of non-chemical or chemical stressors, and even the latter could represent multiple toxic modes of action (MOA). From a risk assessment perspective, it is desirable that endpoints reflect the MOA of chemicals for a number of reasons. For example, in diagnostic or retrospective assessments, an endpoint reflective of a specific MOA can be critical in the identification of contaminants or classes of contaminants responsible for adverse effects (Ankley and Giesy 1998; Fairbrother et al. 1997). In prospective assessments, an understanding of MOA is critical as a basis for among-species and among-chemical extrapolations. Coupled with an understanding of reproductive and developmental physiology (see Chapter 3), predictions can be made regarding the potential for effects in species for which empirical toxicity data do not exist.

As one proceeds to lower levels of biological organization, it is increasingly possible to define chemical-specific MOAs as causes of observed effects. Unfortunately, however, it may not be possible to relate direct effects at lower levels of organization to impacts at the individual or population level. For example, induction of vitellogenin in male fish is a fairly specific indicator of estrogen receptor agonists, yet the consequences of this response at the population level are uncertain. The integration of individual-level responses into predictions of population-level effects is discussed more fully in Chapter 4.

CHAPTER PREVIEW

Reproductive and Developmental Effects of Contaminants in Oviparous Vertebrates. Richard T. Di Giulio and Donald E. Tillitt, editors.
©1999 Society of Environmental Toxicology and Chemistry (SETAC). ISBN 1-880611-37-6

Level of Organization	Measurement Endpoint
Ecosystems/Communities	Species Assemblages Species Diversity Keystone or Sentinel Species (Presence/Absence) Functional Indices (e.g., Productivity/Energy Flow)
Populations	Census Data Age Structure Size/Age Relationships Sex Ratios Recruitment
Individuals	Condition Factors Growth/Development Genetic vs. Phenotypic Sex Fecundity Functional Assays (e.g., Immune Challenge) Hormone/Receptor Concentrations Biomarkers (e.g., Vitellogenin, Enzyme Induction)
Suborganismal	Functional Responses Structural Responses Receptor Binding Assays

(Left axis, bottom to top arrow: Increasing specificity in terms of MOA)

(Right axis, bottom to top arrow: Increasing relevance in terms of ecological effects)

Figure 5-1 Schematic of biological levels of organization in the context of ecological risk assessment

In many of the case studies described in this chapter, initial demonstrations of impact were made at the individual level (e.g., observation of deformities), yet determination of causes of these impacts was achieved through collection of data at lower levels of organization (e.g., cellular or subcellular responses). And, in the most comprehensive case studies, confirmation of suspect causative agents was made by reconsideration of impacts at higher levels of organization after removal of the stressor (Schwabe et al. 1977). Hence, from a risk assessment perspective, information at all biological levels of organization (Figure 5-1) can be extremely useful and highly integrative.

In this chapter, we review the toxicological mechanisms by which chemicals can affect reproduction and development in oviparous species, paying particular attention to the sensitive life stages identified in Chapter 4. We do not discuss all MOAs and associated endpoints that could be elucidated; rather, we group the endpoints and assays by general approach, provide a few detailed examples, and include references to additional citations for in-depth reading on specific processes and procedures. To provide a reference as to how various endpoints can be used in ecological risk assessment, we present a series of case studies at the end of the chapter. In addition, as a general aid to the reader, we have identified those species or classes for which particular endpoints have or have not been utilized to a significant degree.

Comparison of Oviparous and Viviparous Reproduction and Development

This section will contrast the differences between mammalian and non-mammalian oviparous species in order to identify critical reproductive and developmental processes of susceptibility in non-mammalian vertebrates. For details on specific physiological functions, the reader is referred to Chapter 3.

The identifying characteristic of the oviparous group (i.e., egg-laying) has proven to be the most sensitive endpoint for several toxicants. For example, several pesticides and their metabolites, such as p,p'-dichlorodiphenyldichloroethylene (DDE), induce eggshell thinning in avian and reptilian species both in the field and in the laboratory at exposure levels considerably lower than levels that cause other toxicological responses. Other obvious qualitative differences in reproductive strategies between egg-laying and live-bearing animals that may be sensitive to toxicological insult include hatching, metamorphosis (in some fish and amphibian species), external fertilization (some species), sources of embryonic and juvenile nutrition (resulting in different kinetics of maternal deposition and exposure to contaminants), and degree of maternal care (lacking in some species). Some non-mammalian species also have different active steroids (e.g., 11-ketotestosterone), different receptors (e.g., multiple androgen receptors, cell membrane progestin receptors on the egg), and different synthetic or degradative pathways for hormones than do mammals (Borg 1994; Ankley, Mihaich et al. 1998). For example, in some fishes, 11-ketotestoserone, rather than testosterone, is the dominant androgen in males (Idler 1961,1976; Fostier et al. 1983).

There also are qualitative differences that are not absolute but that differ in degree, such as plasma hormone levels, degree of serum binding of the hormones, hormonal regulation of processes such as behavioral sex differentiation, and the effects of hormones or chemicals on gonadal differentiation.

Another major consequence of oviparity is production of the egg protein, vitellogenin, by the liver. Vitellogenin is transported in large quantities to the ovary, where it is added to the egg yolk prior to ovulation. The synthesis of this protein is regulated primarily by estrogens and hence serves as a useful biomarker of xenoestrogen exposure in many oviparous species, in particular in males where vitellogenin production is an abnormal process. Increased vitellogenin production has been reported from several locations: in the UK, in caged male fish in one river contaminated with alkylphenols from detergents used in wool processing; in male fish exposed to natural and pharmaceutical estrogens from sewage effluent; and in the U.S., in male fish in Lake Mead, Nevada, apparently also as a consequence of sewage exposure. Although vitellogenin production in males may not be considered adverse, high levels have been associated with reduced testis weight, and abnormal production of this protein challenges the metabolic reserve of the animal. Finally, in females, vitellogenin is not transported alone to the egg. It serves as a vehicle for maternal deposition of hormones, calcium, growth factors, and lipophilic toxicants. Maternal deposition can result in variations in exposure of the embryo to both endogenous and exogenous compounds throughout incubation, including early critical stages of organogenesis. In contrast, there is no preferential deposition of lipophilic chemicals in mammalian embryos at the beginning of development. Instead, because of their interaction with the maternal organism, mammalian embryos can be exposed to chemicals continuously or at discrete intervals during development. Also, there can be considerable exposure of juvenile mammals to maternally stored toxicants through lactation. For those species with external fertilization and development of the eggs and embryos, direct exposure to the progeny from chemicals in the environment can occur, whereas mammalian embryos can be exposed only indirectly through the maternal organism. For further consideration of chemical exposure differences in oviparous versus viviparous animals, see Chapter 2.

Although examples from the field have yet to be defined, toxicants that lower or elevate thyroid hormone levels likely would alter metamorphosis in anurans and fish species. The consequences of disrupting metamorphosis are particularly profound because this process is critical to adult differentiation and involves the nervous and gastrointestinal systems, the limbs, and the reproductive organs. In these species, sexual maturation and normal steroidal function may require prior action of thyroid hormones. For example, if xenobiotics alter levels or interactions among thyroid receptors, this may lead to altered expression of retinoic acid receptors (RARs/RXRs) and thereby change homeotic gene expression, resulting in effects on limb development. Exposure to retinoid analogues has been proposed as one possible mechanism responsible for limb defects in amphibians in north central and northeastern North America (Ankley and Giesy 1998).

It has been suggested that estrogens in the environment could alter gonadal sex determination. For example, in field studies, abnormal testicular development has been reported in birds (Fry 1995) and reptiles (Guillette et al. 1994) exposed to xenoestrogens and anti-androgens. In birds, feminization of the testis was seen after in ovo exposure to estrogenic/anti-androgenic insecticides, including organophosphates (parathion, azinphos), organochlorines (aldrin, endosulfan, lindane), pyrethroids, and 2,4-dichlorophenoxy acetic acid (2,4-D) and simazine (Lutz and Lutz-Ostertag 1975).

Some reptiles, on the other hand, are genetically indeterminate, with phenotype dependent upon incubation temperature (Raynaud and Pieau 1985). Many reptiles, including crocodile and turtle species, exhibit temperature-dependent sexual phenotype expression. For example, in the saltwater crocodile (*Crocodylus porosus*) and the American alligator (*Alligator mississippiensis*), embryos incubated at 30 °C develop ovaries and hatch as females, while those incubated at 32 °C develop testes and hatch as males (Raynaud and Pieau 1985; Smith and Joss 1994). Conversely, the red-eared slider turtle (*Trachemys scripta*) produces all female hatchlings at warmer temperatures (31 °C) and all male hatchlings at cooler temperatures (26 °C), while intermediate temperatures produce varying ratios of males to females (Bergeron et al. 1994). Lizards, geckos, and iguanas are variable, with temperature affecting sex ratios in some species but not in others (Raynaud and Pieau 1985). Those that are temperature-dependent show a higher proportion of males at lower temperatures. Some oviparous teleost fish also show temperature sensitivity, with males produced at higher temperatures (17 to 25 °C) and females at the cooler temperatures (11 to 19 °C) (Norris 1997; Van Der Kraak and Pankhurst 1996). The toxicological significance of temperature-dependent sex determination is that it is under hormonal control, and exposure of animals to chemical agonists or antagonists of these systems can affect these processes and change the temperature-specific responses (Ankley, Mihaich et al. 1998).

In addition to receptor binding and direct hormonal changes resulting in unwanted effects, changes in behavior, (i.e., maternal care or lack of it and nest abandonment) have been reported following exposure to organophosphate insecticides in the field (Fairbrother 1996). Although lactating dams also may display abnormal maternal behavior as a consequence of neurotoxicity or endocrine disruption, this effect would be more difficult to document in the wild for mammals, fish, and amphibians than it is for avian species because of the relatively conspicuous nature of many avian nesting habits.

Mechanisms of Toxicity

The above discussion suggests many potential mechanisms by which environmental chemicals may exert disruptive effects. As described in the present section, direct chemical modulation may occur at molecular sites but results in whole organism effects emanating from an integration of these lower-level changes. Thus, this section looks primarily at suborganismal effects, and subsequent sections consider methods for measuring effects that address whole organism questions.

While interspecies homology in signal transduction pathways and protein synthesis and processing provides a framework for extrapolating effects across species (Ankley et al. 1997; Ankley, Mihaich 1998; Van Der Kraak, Zacharewski et al. 1998), a better understanding of interspecific differences, such as those seen in protein homology and organization of regulatory sites on protein and deoxyribonucleic acid (DNA), is needed to quantify toxicodynamic differences. Cellular processes, such as growth, differentiation, and apoptosis, are controlled by multiple factors, and the importance of these factors varies across species. Developmental phenology also must be kept in mind, as there are windows of sensitivity to chemical modulators corresponding to times when specific organ systems are developing. Development represents a time of rapid proliferation and differentiation of tissues. Not only are cells dividing rapidly, but also they are turning into other types of cells and may migrate from one location to another within the developing organism. Chemicals that interfere with such processes, such as those that affect the cytoskeleton, can lead to developmental defects. As a consequence, the extent to which subcellular or cellular effects can be extrapolated across organisms requires an understanding of species-specific integration of organizational, structural, and functional processes.

Mechanisms of endocrine modulation

Environmental chemicals affect endocrine function through effects on hormone biosynthesis, transport, activity, and metabolism (Van Der Kraak, Zacharewski et al. 1998). Most of our current knowledge comes from information related to the effects of environmental chemicals on steroid and thyroid hormone physiology and resulting influences on growth, development, and reproduction. By comparison, our understanding of the adverse effects of xenobiotics on the synthesis and actions of peptide and catecholamine hormones is limited.

One of the most common mechanisms of endocrine modulation involves the ability of chemicals, other than the endogenous ligand, to bind to steroid hormone receptors and trigger responses that usually are less pronounced but are qualitatively similar to those induced by the hormone itself. Examples of environmental hormone agonists include various pesticides and phytoestrogens that stimulate estrogenic responses by interaction with the estrogen receptor.

Not all endocrine disrupters that interact with steroid hormone receptors function as agonists. Many compounds bind to the hormone receptor and act as antagonists, preventing receptor occupancy by the endogenous ligand or interfering with subsequent steps in the signaling pathway. Although these compounds bind to the receptor, they are not capable of producing the typical changes in gene transcription that normally result from the interaction of the steroid hormone with its receptor.

Endocrine function may be modulated by interference with the biosynthesis or degradation of steroid hormones. For example, exposure of white sucker (*Catostomus commersoni*) to bleached kraft pulp mill effluent leads to significant reductions in plasma sex steroid hormone levels (Van Der Kraak et al. 1992; McMaster et al. 1995). This can result from changes in availability of precursors such as cholesterol, or altered activity of enzymes involved in hormone synthesis or catabolism. For example, a large number of xenobiotics either induce or inhibit specific cytochrome P450-dependent monooxygenases, which are important enzymes in the production or degradation of many hormones. Although there has been considerable focus on xenobiotic modulation of steroid hormone biosynthesis in wildlife, relatively little is known regarding potential effects of environmental chemicals on peptide hormone synthesis or release. Yet, reduced plasma gonadotropin hormone II (GTH-II) levels in white sucker chronically exposed to pulp mill effluents suggest that protein hormone homeostasis may be a target of environmental chemicals (Van Der Kraak et al. 1992).

Once released into the bloodstream, a hormone may circulate freely if water soluble or bind to plasma proteins if lipid soluble. Thus, water soluble peptides and amines circulate freely while lipid soluble steroids and thyroid hormones are bound to plasma proteins. Plasma proteins are expressed in all vertebrates except for sex hormone binding globulin (SHBG), which appears to be absent in birds (Wingfield et al. 1984; Pasmanik and Callard 1986). Testosterone and 17β-estradiol bind to SHBG, glucocorticoids and progesterone bind to corticosteroid binding globulin (CBG or transcortin), and thyroid hormones and retinoids bind to thyroid-retinol binding globulin (TBG or transthyretin). Depending on the hormone, 98 to 99.9% of circulating hormone may be bound to these plasma proteins. Exposure to xenobiotics can influence the degree of hormone binding to plasma proteins and, therefore, affect free levels of hormones available to elicit effects. This most likely would involve decreased binding of the hormone to SHBG, CBG, or TBG. Chemicals acting as ligands for hormone receptors that are present in relatively higher concentrations in the bloodstream, such as compounds with estrogenic or androgenic activity, may also bind to SHBG and displace the endogenous hormones from plasma proteins. This would result in increased plasma clearance of the endogenous hormone and decreased plasma hormone concentrations. For example, a hydroxylated polychlorinated biphenyl metabolite was found to bind to the transthyretin-retinol plasma binding protein complex in rats, resulting in increased clearance of plasma retinol (Brouwer and Van Den Berg 1986). Moreover, SHBG may facilitate the transport of sex steroids into hormone responsive tissues through alterations of cell membranes (Joseph 1994; Hammond 1995) and also may do the same with environmental chemicals (Nakhla and Rosner 1996).

Potential effects of xenobiotics on hormone metabolism and excretion may arise after alterations in activities of specific enzymes mediating degradation. Induction or inhibition of monooxygenases has the potential to affect the metabolic degradation of numerous hormones. For example, 2,3,7,8-tetrachlorodibenzo-*p*-dioxin (TCDD) and related chemicals increase the metabolism of sex steroids by inducing monooxygenases responsible for their hydroxylation and subsequent excretion (Goldstein 1980). Similarly, induction or inhibition of phase II enzymes may influence hormone metabolism. For example, TCDD reduces plasma thyroid hormone (T_4) levels in mammals, partly through induction of uridine 5'-diphosphate-glucuronosyl transferase, which increases the biliary excretion of T_4-glucuronide (Bastomsky 1977).

Receptor mediated mechanisms

A major MOA of toxicants is by means of receptor signaling pathways. Receptors are proteins that bind to ligands with high affinity. Receptors serve the dual functions of recognition and transmission of a ligand signal leading to a biological response. Receptors fall into 2 broad categories, membrane or nuclear, which reflect their cellular localization and the general properties of the ligands (i.e., size and solubility) that they recognize. This separation of receptors into 2 categories is not all inclusive as other receptors including the aryl hydrocarbon receptor (AhR), and some steroid hormone receptors are cytoplasmic in origin but, once activated by a ligand, function in a manner similar to nuclear receptors. Chapter 3 provides an in-depth review of receptor function and physiology.

Signaling pathways occurring downstream of ligand-receptor binding are well conserved in vertebrates. Components of the signal transduction pathway down-stream of catecholamine and peptide hormone receptor binding (e.g., heterotrimeric G-proteins, adenylyl cyclase, intrinsic tyrosine kinase activity) appear to function in a similar manner across species (e.g., Chang and Jobin 1994; Van Der Kraak et al. 1998). Similarly, the general pattern of response in which binding to nuclear receptors or Ah receptors leads to gene activation and the transcription and translation of specific proteins is similar across species. Overall it seems that the variation in a receptor-mediated biological response across verte-brate classes is associated more with species-specific gene products than with marked divergence in the signaling pathways used by hormones. Estrogen-induced expression of vitellogenin in all egg-laying vertebrates and its absence in viviparous mammals illustrates this point.

Aryl hydrocarbon receptor

Because of the somewhat unique nature of the AhR, and the importance of this system in a number of well-documented impacts on wildlife populations, the toxic MOA associated with this receptor is covered in some detail. The AhR functions as a ligand-activated transcription factor and is a member of a growing family of regulatory proteins defined by the presence of a PAS (Per, AhR ARNT, Sim) domain, a stretch of 100 to 350 amino acids containing two 51-amino acid repeats that are involved in dimerization and, in the case of AhR, in ligand binding and hsp90 association. The AhR is involved in mediating the effects of TCDD and related compounds, including polychlorinated dibenzo-*p*-dioxins (PCDDs), polychlorinated dibenzofurans (PCDFs) and polychlorinated biphenyls (PCBs). An endogenous ligand for the AhR has yet to be isolated and identified from vertebrate tissue. It generally is accepted that modulation of gene transcription is a key step in media-tion of TCDD toxicity. At the molecular level, the mechanism by which TCDD up-regulates transcription of a battery of TCDD-inducible genes is well understood (Schmidt and Bradfield 1996). TCDD binds to the AhR, a basic helix-loop-helix (bHLH), PAS protein, facilitating dissociation of hsp90 and other AhR associated proteins (e.g., p37 and p43). The liganded AhR translocates to the nucleus, dimer-izes with another bHLH/PAS protein, ARNT, forming an active transcription factor that binds to enhancer elements (dioxin responsive elements, or DREs) in the 5' regulatory region of TCDD-inducible genes, some of which code for xenobiotic metabolizing enzymes (Sutter and Greenlee 1992). The most thoroughly character-ized of these genes, in terms of TCDD-inducibility, is cytochrome P4501A1 (CYP1A1), an enzyme that possesses ethoxyresorufin-*O*-deethylase activity. However, induction of neither CYP1A1 nor any other TCDD-inducible gene identified to date has been causally linked to TCDD toxicity. Nevertheless, the presence of the AhR seems to be required for manifestation of TCDD toxicity (Poland and Glover 1980; Poland and Knutson 1982; Fernandez-Salguerro et al. 1996). In fact, the AhR is necessary but not sufficient for all of the effects of TCDD (Birnbaum 1994).

Three potential mechanisms by which the AhR could play a role in causing TCDD toxicity are 1) genes directly linked to toxicity are transcriptionally inducible by the liganded AhR/ARNT complex but have not yet been identified; 2) genes directly linked to toxicity are transciptionally inducible by the liganded AhR complexed with an alternative dimerization partner but have not yet been identified, or 3) genes directly linked to toxicity are transcriptionally down-regulated by the liganded AhR, perhaps by competition for a limiting factor (such as ARNT) necessary for transcription of these genes. If either the first or third possibility is true, then expression of both the AhR and ARNT would be required for TCDD-responsiveness. It is also possible for the AhR to function as a transcriptional repressor causing decreased gene expression. Lastly, the AhR may function in the control of cellular proteins leading to altered phosphorylation and calcium mobilization. These latter effects are nonnuclear (Matsumura 1994).

Deoxyribonucleic acid damage

One of the best understood mechanisms of adverse reproductive and developmental effects involves compounds that react with DNA to induce mutagenicity and/or genotoxicity. Chemicals can be either directly or indirectly genotoxic. Direct-acting genotoxicants are highly reactive electrophiles that are able to form covalent bonds to individual nucleotides, leading to point mutations, and single- or double-strand breaks, depending on the type of adduct formed and the repair mechanisms that attempt to correct the damage. In addition, reactive oxygen species can lead to direct strand breaks. Reactive electrophiles, as well as active oxygen species (e.g., superoxide, singlet oxygen, hydroxyl radicals, nitric oxide), often are generated in the cell by various metabolic processes such as biotransformation reactions. Chemicals may interfere with normal processes that denature reactive oxygen species as electrophiles, resulting in an increased incidence of strand breakage. Point mutations can lead to altered expression of key genes. Mutations in genes that control cell cycle or express growth factors and/or their receptors have pleiotropic consequences. Gene expression also can be altered by changes in methylation patterns.

Genotoxic events require not only the direct insult to DNA but also fixation of the lesion through replication of the cell. Cells of developing organisms are dividing rapidly, thereby increasing the possibility of fixation of damaged DNA. Intense DNA damage often activates repair mechanisms with varying efficiencies. While a low level of point mutations may be tolerated if they have not occurred in critical genes, greater numbers of mutations as well as effects on higher orders of genetic structure, such as chromosomal alterations, usually have physiological consequences such as structural alterations or alterations in proliferation and differentiation, key determinants of normal development.

Posttranslational modifications

Alterations in second messengers and signal transduction can occur via posttranslational modifications of critical cellular proteins. Protein kinases specifically phosphorylate key regulatory molecules, thus controlling their activity. In contrast, there are phosphatases that remove phosphate groups, leading to modulation of activity. Exogenous chemicals can perturb these processes. It is interesting that certain receptor systems, although normally thought of as modulating gene expression, also can play a role in such phosphorylation events (e.g., nonnuclear role of AhR leading to tyrosine kinase activation; genistein as both a ligand for ER and as a tyrosine phosphatase). Glycosylation and myristylation are also important in enzymatic activity and can be modified by the action of exogenous chemicals.

Trafficking

Trafficking involves the synthesis of apoproteins and is the transport of product within the endoplasmic reticulum, movement toward and fusion with the Golgi apparatus, unidirectional movement of Golgi-derived vesicles, fusion at the plasma membrane, and release from the cell. Requirements include sufficient adenosine triphosphate levels in the cell (apoprotein synthesis, fusion of endoplasmic reticulum with Golgi), and proper directional movement of vesicles by cytoskeletal components (intermediate filaments and microtubules). Important enzyme functions include energy production, membrane modification and protein synthesis. Chemicals may perturb any of the above.

Enzyme inhibition and ion channel modulation

Enzyme inhibition and modulation of ion channels by xenobiotics can lead to a wide range of adverse effects in an organism. Interactions with enzymes and ion channels within the nervous system are of note because of the potential to disrupt behavioral responses associated with avoidance of noxious stimuli, attraction to food or mates, and social interactions (e.g., aggregation, territoriality, courtship, and parental care). Examples of toxic MOA and associated compound classes include organophosphorus (OP) and carbamate insecticides that inhibit acetylcholinesterase (AChE) (Fukuto 1990), dichlorodiphenyltrichloroethane (DDT)-type compounds that prevent the deactivation of axon sodium gates (Coats 1990), chlorinated alicyclics that bind at the picrotoxinin site in the δ-aminobutyric acid (GABA) chloride ionophore complex (Coats 1990), and pyrethroids that interfere with sodium gates (Bradbury and Coats 1988; Coats 1990). Enzyme inhibition also can cause structural malformations. For example, OPs cause micromelia in chicks by inhibition of kynurenin formamidase and wry neck and other skeletal malformations by inhibition of AChE (Misawa et al. 1982). Metalloenzymes such as superoxide dismutase or glutathione peroxidase require the incorporation of metals into the reactive site, but also can be inhibited from function in the presence of excess metals (Johnson and Fischer 1994). Furthermore, calcium homeostasis is essential for proper functioning of many homeostatic enzymes such as phosphates, phospholipases, proteases, and endonucleases.

While industrial organic chemicals usually are not designed to have biological activity, it is estimated that approximately 70% of these compounds may elicit acute, baseline toxicity through one or more narcosis modes of action (Bradbury 1995). These modes of action are thought to result from the partitioning of xenobiotics within currently unidentified ion channels and/or lipid-protein interfaces within axon membranes (Bradbury et al. 1989; Franks and Lieb 1990).

Cytotoxicity

Cytotoxic chemicals lead to cell death, thus preventing the proliferation necessary for embryonic development. Exposure early in development results in massive cell death, generally followed by degradative changes referred to as necrosis. Necrosis occurs as a result of activation of intracellular enzymes and denaturation of proteins due to the rupturing of intracellular lysosomes after the cell has died. Later in its development, or at lower doses, cytotoxicity can be manifested by subtle changes such as growth retardation or by significant increases in the occurrence of structural or functional abnormalities in the embryo. While the preimplantation mammalian embryo generally is insensitive to chemical effects, it is not known if a similar stage of insensitivity occurs within the embryonated egg of oviparous species. Different species may experience significant differences in embryonic exposures to cytotoxic agents, depending upon the timing of yolk-sac resorption relative to the stage of embryonic development.

Cytotoxicity may cause regenerative hyperplasia, triggered by differential cell death. Because cells in different points in their replication cycle have different sensitivities to cytotoxic agents, the restorative growth following cytotoxic insults results in a highly uncoordinated process of replication. This can lead to mistiming of inductive interactions and subsequent developmental defects, particularly if exposure occurs during the organogenesis phase. The developing embryo and metamorphosing amphibian relies on a programmed cycle of cell death known as apoptosis. Cytotoxic chemicals rarely, if ever, mimic the timing and selectivity of these preprogrammed events. In nearly all cases, regenerative hyperplasia results in deformities rather than the normal developmental or morphogenic cascade of events.

Behavior modification

Behavioral responses are based on an integration of underlying physiological and biochemical functions. Behavioral toxicity occurs when exposure to a chemical, or other stressful condition, induces changes in observable, recordable or measurable activities of a living animal that exceed the normal range of variability. Exposure to a xenobiotic could affect avoidance or preference to noxious stimuli, attraction to food or mates, feeding, and social interactions (e.g., aggregation, territoriality, courtship, and parental care). These effects can result from xenobiotic interactions with sensory reception (e.g., chemoreception, phototaxis), motor activity associated with the peripheral nervous system, or processing by the central nervous system (Rand 1985).

In addition to altering behavior by directly interacting with neurotransmitters, receptors, and neuronal structures, it is also possible for xenobiotics to interact with the endocrine system and modify behaviors by influencing signal reception and transmission, activation of neuron clusters, and differentiation of brain structure and function. Examples of endocrine influences on nervous system structure and/or function in birds, fish, and amphibians are available (e.g., Manns and Fritzsch 1992; Adkins-Regen et al. 1994; Weber and Spieler 1994; Hill et al. 1995). Metals and metalloids also can affect the nervous system directly; a number of metals have been reported to influence synaptic transmission or neurotransmitter release in the peripheral or central nervous system (Weber and Spieler 1994). Moreover, most organometals are developmental toxicants.

Techniques and Endpoints for Assessing Toxicity

Both in vitro (cellular and subcellular) and in vivo (whole organism) approaches are available for assessing the effects of xenobiotics on the reproduction and development of vertebrate species. There has been considerable discussion about whether adverse effects associated with a specific MOA such as disruption of endocrine systems should narrowly focus on subcellular/cellular effects associated, for example, with nuclear hormone receptor signaling pathways or should include all aspects of hormone production, release, transport, metabolism, binding, action, elimination, and, ultimately, adverse effects in the whole organism (Kavlock et al. 1996). Certain groups (e.g., Klotz et al. 1996; Shelby et al. 1996; EC 1997) subscribe to the view that effects associated with exposures to xenobiotic compounds is meaningful only if changes occur to the whole organism. We believe that comprehensive risk assessments require consideration of effects at all levels of biological organization. A summary of test methods, including endpoints studied, is provided below.

Subcellular/cellular endpoints

Normal cellular function involves the integration of multiple subcellular processes. These are modulated by endogenous regulators, and can be affected by xenobiotics—from actions on receptor biochemistry and signal tranduction processes to effects on gene expression and protein synthesis and activity. Collectively, these events affect normal cell functions through actions on aspects of growth and proliferation, differentiation, and programmed cell death. These endpoints can be used in defining the actions of xenobiotics. However, differences in the rates of growth, specificity of cell physiology, and the unique processes regulated by different cell types (enzyme activity, cell products) contribute to the need for species-specific test systems (Ankley, Mihaich et al. 1998). Other aspects of cell function, however, may represent common endpoints that are sensitive to perturbations by xenobiotics in all species. In particular, conserved functions include those affected by energy and ion/osmotic balance, such as cell growth and proliferation, pH, and cell volume.

Receptor and SHBG-binding assays are commonly used techniques for studying the potential for a chemical to mimic a hormone agonist or antagonist. The methods have been well described in a variety of texts (e.g., Norris 1997) and other publications and generally are applicable to both mammals and oviparous species. The basic approach is reviewed briefly below.

Binding assays may be evaluated in a cell-free environment (Shelby et al. 1996; Danzo 1997) and are applicable to a wide range of organisms. For example, cytosolic preparations appropriate for ER-binding assays have been made from spotted seatrout (*Cynoscion nebulosus*) (Thomas and Smith 1993) and turtle liver (Ho et al. 1988) and from alligator (Vonier et al. 1996) and chicken oviducts (Kon et al. 1980). The major limitation of cell-free receptor binding assays is that they assess only the ability of a chemical to bind a specific receptor and do not include any measure of its ability to pass through the cell membranes to contact the nuclear receptors or information about whether the receptor-ligand complex initiates messenger ribonucleic acid (mRNA) transcription.

An alternative approach to cell-free receptor binding is measurement in cell or tissue culture preparations. This maintains the integrity of the cell and therefore includes a measure of cell penetration as well as some information on cell responses. Certain cell lines have been developed and used to screen chemicals for receptor binding capacity in mammals (e.g., the MCF-7 cell line for ER) (Soto et al.1995); similarly, primary cell cultures can be derived from oviparous embryo fibroblasts and hepatocytes (Berry et al. 1990; White et al. 1994). Cell lines also have been transfected, e.g., with portions of the gene coding for the estradiol-inducible protein, pS2, linked to a luciferase enzyme reporter; luciferase activity is then measured as bioluminescence. Other enzyme activities indicative of cell proliferation also can be measured colorimetrically, thereby providing another alternative for simplifying the methodology (Holinka et al. 1989).

In addition to radioligand assays characterizing the affinity, specificity, and kinetics of receptor binding, various tissue and cell cultures have been used to measure downstream effects of receptor activation. For example, cDNA probes have been developed for the elucidation of synthesis of mRNA and other gene activation processes (Chen 1988).

Physiological/organismal endpoints

In retrospective analyses, in vivo information often is required as part of the process for determination of MOA and confirmation of the causative agent. In prospective toxicity analyses, in vivo studies also would be required to determine if the potential effects observed from in vitro assays were realized and to establish the dose-response relationship for developmental or reproductive effects of the toxicant. Results of in vitro assays enable the investigator to formulate a hypothesis and "tailor" in vivo testing towards the effects expected for a given MOA. It is critical to understand that extrapolation from in vitro to in vivo requires not only knowledge with regard to MOA but also a good understanding of the basic biology of the species of concern. Different species can display very disparate responses to the same cellular event. For example, plumage is sexually dimorphic in many passerines, with males being brightly colored as a result of testicular steroids (Norris 1997). In contrast, in many raptor species, the sexes appear similar despite the fact that males have serum concentrations of androgens comparable to passerines. The effects of hormonally active toxicants on sexual differentiation, seen in mammalian tests, are likely to differ greatly in oviparous species, where estrogens play a more dominant role (Fry 1995). In addition to significant among-class variations in oviparous animals, some aspects of sexually dimorphic reproductive and nonreproductive behaviors differ greatly between closely related genera. For example, even though the cellular and molecular mechanisms controlling development are identical, closely related sympatric species of anurans utilize mating calls that are quite distinct from one another (Norris 1997).

Standardized reproductive toxicity tests that have been developed for some oviparous species include endpoints relevant to populations such as

- lethality, which is assessed at all stages of development;
- growth, which is included in several standard assays; and
- reproductive fitness, which is assessed in the parents as age-specific fecundity, clutch size, eggshell quality, and age at maturity and in offspring as hatchability, teratogenesis, and development and maturation to reproductive maturity.

Standardized in vivo toxicity tests have been developed for several species of teleost fish, including carp (*Cyprinus carpio*), guppy (*Lebistes reticulatus*), rainbow trout (*Onchorhyncus mykiss*), sheepshead minnow (*Cyprinodon variegatus*), fathead minnow (*Pimephales promelas*), medaka (*Oryzias latipes*), and zebrafish (*Brachydanio rerio*) (USEPA 1972, 1982,1986, 1987, 1994a, 1994b). Similar types of avian reproductive toxicity tests are also in use for northern bobwhite quail (*Colinus virginianus*), mallards (*Anas platyrhynchos*), and Japanese quail (*Coturnix coturnix japonica*) (OECD 1984, 1996; Bennett and Ganio 1991). There are no standardized tests for reptiles; for amphibians, the frog embryo teratogenesis assay - *Xenopus* (FETAX) has been standardized (ASTM 1991). The endpoints measured vary among these whole animal tests, but most could be modified to include additional endpoints. It must be noted that standardization can be problematic as well, since tests generally are designed to respond to specific situations and may not always be appropriate. For example, many of the avian pesticide protocols were designed to identify the effects of highly bioaccumulative chemicals, and may not identify hazards posed by newer, biodegradable agents. For aquatic assays, chemicals with very low water solubility may be difficult to administer according to existing protocols. Additionally, behavioral aspects of mating, incubation, migration, and parental care are critical to reproductive success but are not evaluated in routine testing. Nor is metamorphosis specifically addressed, and it is not clear whether, in addition to perturbing the developmental changes occurring in organisms, chemicals might not affect the transition from aquatic to terrestrial breathing in amphibians.

Many of the endpoints measured in standardized whole animal tests have been evaluated in field situations, including those described in the case studies at the end of this chapter. For example, in Great Lakes fish and birds, early-life-stage mortality and abnormalities were critical endpoints. In studies of effects of pulp mill effluents and polycyclic aromatic hydrocarbons (PAHs) on fish, a number of endpoints related to the reproductive fitness of the parents were measured, including fecundity, age at maturity, and effects on viability of gametes and larval and juvenile stages. These types of endpoints were also measured in studies of selenium effects in fish and birds. Effects of contaminants on behavior, which are not often assessed in standardized tests, were observed following field exposures to organophosphorus pesticides.

Gonadosomatic index (GSI), the ratio of gonad weight to body weight, and related measures such as oviduct weight in birds, are commonly used endpoints in whole animal toxicity studies. Although GSI is not specific for any particular mechanism of toxicant action, it is a good predictor of reproductive success and may be linked to population-level impacts. Various measures of gonadal weight or gonadal growth were used in several case studies, including those on effects of pulp mill effluents in fish, effects of PAHs in fish, effects of pesticide residues on Lake Apopka alligators, and effects of oil on birds.

Alterations in sexual differentiation also may be a valuable endpoint, and their use is illustrated in the study of effects of estrogenic substances on fish in the UK, and in studies of the alligators from Lake Apopka, Florida. Alterations in secondary sexual characteristics also may be good endpoints of disrupted reproductive function. In relevant fish species, endpoints such as gonopodium alteration, size of breeding tubules, or color changes may be utilized (Turner 1942; Howell et al. 1980; Bortone et al. 1989; McMaster et al. 1995). In chickens, comb development, which is under the control of testosterone, has been used as a bioassay to detect natural and synthetic androgenic compounds or compounds that inhibit androgenic activity (Munson and Sheps 1958; Dorfman et al. 1966; Boris and Stevenson 1966; Gawienowski et al. 1977; Henessey et al. 1986). Sex reversal also has been a useful endpoint in studies with amphibians, such as the African clawed frog, *X. laevis* (Ramsdell et al. 1996). However, this latter endpoint can be problematic because environmental factors such as water temperature can affect sex determination in some species of fish, reptiles, and amphibians. Moreover, sexual differentiation mechanisms in fish are variable, and some species are normally sequential hermaphrodites; therefore, model species need to be well characterized with respect to normal patterns of sexual differentiation.

Final gamete maturation can be a useful measurement endpoint. This involves collecting eggs and sperm from exposed animals and assessing gamete quality. Endpoints such as gamete size, germinal vesicle breakdown (GVBD), ovulation, hydration, and lipid and oil droplet formation may be assessed with light microscopy or in vitro (Ghosh and Thomas 1995). Assays for GVBD have been used widely in fish and also would be appropriate for amphibians. In some species, treatment with gonadotropin-releasing hormone agonists may be required to obtain gametes. The disadvantage of this technique is that it would obscure possible disruption of the neuroendocrine axis or behavior; however, responsiveness to the treatment also may be used as a measure of reproductive function. The use of such endpoints is illustrated in the case studies on effects of PAHs and pulp mill effluents on fish.

Histopathology is used in addition to gross morphological observations for assessing toxicant effects on many of the endpoints discussed above. It played an important role in determining the cellular and tissue lesions of adult birds following exposure to selenium, in forming the basis for associations of effects in coastal ocean-bottom-dwelling flatfishes, in detection and progression of vascular lesions of early life stages of Great Lakes fishes, and in detection of cardiovascular toxicity in early life stages of birds in the Great Lakes. When reduction in gonadal size or quality is indicated by gross alteration, histopathology has been used to identify the cellular basis for the lesions and has led to diagnoses of cystic change, retarded gametogenesis, and alterations in sustentacular cells (e.g., Lake Apopka alligators, PAHs in fish, estrogenic substances in UK fish). This tool has particular importance in analysis of small organisms or in early life stages where other approaches may not provide the precision needed. Newer methods for tissue processing in water-miscible plastics such as glycolmethacrylate provide the added resolution needed for determination of specific cellular effects. In addition, freeze drying and glycolmethacrylate embedding of fresh specimens has provided a means to evaluate immunohisto- and cytochemical properties, enzyme histo- and cytochemiststry, and conventional light microscopy in serial sections (Teh and Hinton 1993).

Alterations in plasma hormone concentrations are another endpoint that can be assessed in whole animals in both laboratory and field situations (exemplified in the case studies of pulp mill effluents and PAHs on fish, petroleum effects on birds, and pollution effects on Lake Apopka alligators). Hormone alterations can be caused by changes in hormone synthesis, but also by changes in hormone secretion, metabolism, excretion, or binding to plasma proteins. Also, chemical exposure can lead to alterations in hypothalamic neuroendocrine and neurotransmitter functions that regulate secretion of pituitary hormones, thus altering normal reproduction and behavior. Other mechanisms can involve toxicity to endocrine cells and depletion of germ cells, resulting in reduced reproductive success. In many studies, sex steroids such as estradiol, testosterone, and maturation-inducing progestin steroids have been monitored. Additional hormones, such as thyroxin and cortiocosteroids, have been measured as well. Hypothalamic-pituitary function may be assessed by measuring plasma concentrations of pituitary hormones, such as gonadotrophins, prolactin, corticotropic hormones, releasing factors, and neurotransmitters. However, where hormones show structural diversity across taxonomic groups, antibodies (for radioimmunoassay or enzyme-linked immunosorbent assay) may be available only for a limited number of species. Hormones are more difficult to measure in small animals because of difficulties in obtaining sufficient volumes of blood, but they can be measured in whole body extracts, in composite samples, or in tissue explants. A more serious concern is that steroid concentrations fluctuate throughout the reproductive cycle; therefore, knowledge of normal endocrine cycles is needed before this endpoint can be applied in toxicological studies. In some species, including certain fish and birds, steroids also show diurnal rhythms (Rattner et al. 1984; Zohar et al. 1988; Matsuyama et al. 1990), so animals must be sampled at the same time of day for appropriate comparisons.

In addition to measuring endogenous hormone concentrations in animals, explants of tissue that have been exposed to xenobiotics (either by dosing the intact animal or by exposing tissue ex vivo) can be exposed to physiological stimuli, such as a releasing hormone, to determine tissue responsiveness. Gonadal steroidogenesis has been frequently used in laboratory exposure studies as an endpoint, both in whole animal tests and in vitro with isolated ovarian follicles (MacLatchy and Van Der Kraak 1995; McMaster et al. 1995; Alfonso et al. 1997).

Vitellogenin induction is an important endpoint that may be indicative of exposure to ER agonists in males of oviparous species and of reproductive function in females undergoing normal gonadal development. Vitellogenin induction has been used routinely as a diagnostic biomarker in fish, and has been used as an endpoint in birds (Robinson and Gibbins 1984), turtles, and amphibians (Palmer and Palmer 1995). Because vitellogenin structure differs among species, antibodies developed for one species in some instances may not cross-react with another, requiring species-specific vitellogenin purification and antibody development for measurement purposes.

In vivo techniques used to assess the potential neurotoxic effects of industrial organic chemicals currently are limited. It has been estimated that at least 5% of the 70,000 industrial chemicals in the U.S. are neurotoxic; however, few have been evaluated for neurotoxicity (NRC 1992). As a consequence, there has been a call to develop valid, sensitive, and reproducible methods to identify neurotoxic potential, to characterize the nature of neurotoxic effects and to determine associated mechanisms of action (Tilson 1996). The National Research Council observed that the greatest challenge in developing methods for neurotoxicology lies in integrating behavioral observations with neurormorphological, neurochemical, and neurophysiological alterations (NRC 1992). Because behavioral processes are an integration of biochemical, physiological, and morphological processes, behavioral observations provide a unique means to potentially link cellular and subcellular processes to ecological consequences (Little 1990). As a result, assays capable of identifying chemical-induced neurotoxicity using functional endpoints that are indicators of net sensory, motor, and integrative processes occurring in the central and peripheral nervous system will be an important component of future screening and diagnostic techniques (Tilson 1990).

From the standpoint of population relevance, there is concern about the exposure of embryos of all oviparous animals to chemicals. Such exposure occurs either indirectly, when females are exposed to chemicals and transfer them to the egg, or by direct exposure of the egg after release. Egg injection assays are used as an alternative to maternal transfer studies (Walker et al. 1992; Noesk et al. 1993). This method has some advantages in that it is less costly, may allow exposure to compounds that are metabolized in the adult, and can be used in species that cannot be propagated easily in the laboratory. Types of endpoints measured in embryos exposed in this fashion may include sex ratios, weights of gonads, oviducts or other sex organs, and gonadal histology. Other in ovo exposure techniques such as egg painting, spraying, or immersion have been used to assess toxicant impacts on development (e.g., Hoffman 1990; Bergeron et al. 1994; Gross and Guillette 1994).

Case Studies

The following case studies were developed to demonstrate various aspects of the toxic effects of contaminants on developmental and reproductive processes in oviparous fish and wildlife in field settings using measurements at many levels of biological organization. We emphasize examples of effects observed in either individuals or populations for which established or plausible toxic MOA exist. The case studies range in degree of completeness from those where MOAs and causative toxicants are reasonably well established (e.g., Lake Ontario lake trout sac-fry mortality) to those where MOAs are much more speculative with few (or no) established chemical stressors (e.g., amphibian deformities). The case studies illustrate the critical need to understand chemical MOAs in retrospective assessments and ways in which this understanding leads to the use of formal hypothesis testing and deductive logic in reaching a correct diagnosis (Fairbrother et al. 1997; Ankley and Giesy 1998).

The case studies were selected to represent the major classes of oviparous vertebrates (fishes, birds, amphibians, and reptiles), multiple MOAs and processes reflective of those described in early sections of this chapter, and a variety of chemical classes. However, the preponderance of examples are for fishes and birds. This does not reflect a lack of concern for amphibians or reptiles but the fact that very few ecotoxicological studies have been conducted with these classes, a clear shortcoming in terms of understanding the potential for environmental chemicals to cause developmental and reproductive effects (Ankley et al. 1997). Similarly, although a relatively wide range of toxic mechanisms are illustrated in the case studies, there are several key MOAs that are not. Adverse impacts resulting from overt genotoxicity are not represented, nor are effects due to changes in nutrient status and resulting bioenergetic alterations. Furthermore, the effects of xenobiotics on the general process of metamorphosis is not addressed in any of the case studies. Again, a lack of inclusion of case studies related to these MOAs does not imply that they are unimportant. Certainly all reproductive/developmental mechanisms and processes that have been demonstrated to be sensitive to contaminant-induced effects in laboratory studies have the potential to cause similar effects in the field, provided that adequate exposure occurs.

Chemicals in the case studies include PAHs, PCBs, PCDDs, PCDFs, organochlorine insecticides, selenium, organophosporus insecticides, and naturally occurring and synthetic estrogens. Not represented are metals such as methylmercury, organotins, cadmium, and lead. Likewise, pesticides representing classes other than insecticides, but of potential significance in terms of developmental/reproductive toxicity (e.g., the fungicide vinclozolin or the herbicide atrazine), are not included in any case study. Finally, virtually all the chemicals discussed are of high to moderate persistence and are not necessarily representative of the many readily degraded but potentially toxic compounds released into the environment.

Nevertheless, these case studies represent one of the most inclusive reviews of field-related incidences of reproductive and developmental effects of environmental contaminants on oviparous vertebrates. It is our hope that this review will illustrate the diagnostic and epidemiological methods used to correlate observed effects with potential causative agents. This discussion also illustrates how an understanding of the potential MOA for chemicals that disrupt development and reproduction can facilitate better diagnostic capabilities in the future and begin to allow forecasting of potential effects prior to release of chemicals into the environment.

Early-life-stage mortality—Great Lakes fish

Transfer of lipophilic, bioaccumulative developmental toxicants from maternal fish tissues to eggs prior to spawning is a potential hazard to early-life-stage survival of teleosts (Walker et al. 1994). In the Great Lakes, the eggs of various fish species are contaminated with low levels of PCDDs, PCDFs, and PCBs. This poses a potential hazard to the maintenance of sustainable wild fish populations because complex mixtures of the PCDDs, PCDFs, and planar PCBs in fish eggs have been shown to interact in an additive fashion, by way of an AhR-mediated mechanism, to cause fish early-life-stage mortality (Zabel, Walker et al. 1995; Walker et al. 1996; Cook et al. 1997). The mortality is characterized by a generalized, developmental cardiovascular anomaly known as blue sac syndrome (Spitsbergen et al. 1991; Walker and Peterson 1994; Elonen et al. 1998). This adverse effect has clear ecological relevance, since it may lead to decreased recruitment of yearling fish into feral populations, resulting in a decline in population size.

There has been a long history of problems maintaining naturally reproducing populations of lake trout (*Salvelinus namaycush*) in Lake Ontario and the other Great Lakes. Lake trout populations declined in the 1940s and by the mid-1950s were deemed essentially extinct throughout the Great Lakes, except for isolated populations in Lake Superior. Subsequent stocking of the Great Lakes with fin clipped yearling lake trout resulted in stocked fish reaching sexual maturity and producing fertilized eggs, but there was no recruitment of young into the population.

Mechanism of action

Various hypotheses have been put forward to account for the failure of lake trout recruitment, including low stocking densities, predation by and competition with other fish species, low genetic diversity, lack of homing to appropriate spawning sites, degraded spawning sites, and chemical contaminants. It was hypothesized that contamination of Lake Ontario lake trout eggs with complex mixtures of PCDDs, PCDFs, and PCBs may contribute to recruitment failure of lake trout by causing early-life-stage mortality. In support of this notion, salmonids express the AhR and its dimerization partner, ARNT, and therefore have the capacity to respond to TCDD and related chemicals. Also, PCDD, PCDF, and PCB congeners that are approximate stereoisomers of TCDD have been found in eggs of Lake Ontario lake trout and cause signs of toxicity in fish identical to those caused by TCDD (Walker and Peterson 1991; Zabel, Walker et al. 1995). Furthermore, among salmonid fish species studied thus far, lake trout are the most sensitive to TCDD-induced early-life-stage mortality, with an egg LC50 (concentration causing 50% mortality) in the environmentally relevant range of 65 pg TCDD/g egg (Spitsbergen et al. 1991; Walker et al. 1991).

It is known that TCDD induces oxidative stress in exposed organisms (Stohs et al. 1990). The mechanism, although unknown, might be related to the ability of TCDD to induce genes involved with oxidative metabolism and oxidative stress response, collectively referred to as the Ah gene battery (Nebert et al. 1990). The consequence of oxidative stress in the vasculature of the developing embryo of at least some fish species is aberrant induction of programmed cell death, or apoptosis. Japanese medaka embryos exposed to TCDD have tissue-specific patterns of apoptosis, and the initial site for this to occur is in the endothelial cells of the vasculature (Cantrell et al. 1996). Apoptosis in the endothelial cells of the vasculature leads to loss of functional integrity and may be mechanistically linked to TCDD-induced mortality (Cantrell et al. 1996, 1998). The simultaneous occurrence of CYP1A induction and apoptotic cell death in this tissue, as well as parallel dose-response curves for embryo mortality, CYP1A induction and apoptotic cell death in the medial yolk vein, and the ability to mitigate TCCD-induced apoptosis and embryo mortality of Japanese medaka with the P450 inhibitor piperonyl butoxide, suggests a possible mechanistic linkage among these events.

Lake trout and other fish species exposed to TCDD as embryos in laboratory toxicity studies exhibit ischemia, edema, hemorrhages, craniofacial malformations, and growth retardation (Walker et al. 1991; Walker and Peterson 1994; Henry et al. 1997; Elonen et al. 1998). These signs of toxicity are essentially identical to blue sac disease in salmonid fish species and are expressed at the sac-fry stage of development, culminating in mortality (Wolf 1954). The TCDD dose-response relationship for induction of CYP1A in vascular endothelium of lake trout sac fry is similar to that for sac-fry mortality (Guiney et al. 1997). This suggests that either CYP1A or AhR in the endothelium may be linked to the hemodynamic alterations that are hallmark signs of TCDD toxicity in fish. However, the precise mechanism by which TCDD causes developmental cardiovascular dysfunction remains to be determined. There are wide differences in susceptibility of different fish species to early-life-stage mortality caused by TCDD and related AhR agonists. Recognizing these differences is important in predicting which species are most vulnerable to adverse impacts in aquatic environments contaminated by PCDDs, PCDFs, and PCBs. The rank order of sensitivity, from most sensitive to least sensitive, based on the egg concentration of TCDD causing 50% lethality is lake trout > brook trout > rainbow trout = fathead minnow = channel catfish > lake herring = medaka > white sucker = northern pike = zebrafish (Walker and Peterson 1994; Henry et al. 1997; Elonen et al. 1998).

Toxicity of PCDDs, PCDFs, and planar PCBs to developing lake trout embryos also has been examined by the direct injection of environmentally derived mixtures into newly fertilized trout eggs (Wilson and Tillitt 1996). An organic extract was made from whole adult lake trout collected from Lake Michigan in 1988. Graded doses of the final extract were injected into eggs of hatchery reared rainbow trout (Wilson and Tillitt 1996) and lake trout (Tillitt and Wright 1997). The signs of TCDD-like toxicity (i.e., yolk-sac edema, craniofacial deformities, and hemorrhaging) occurred in a dose-dependent fashion in both species. It was concluded that the mixture of organic chemicals present in feral lake trout from Lake Michigan can induce AhR-related mortality in the developing embryos of trout and that this mixture of chemicals interacts in an additive manner. These and related studies (Zabel, Cook et al. 1995a, 1995b; Walker et al. 1996; Cook et al. 1997) support using the TCDD equivalence approach, which assumes AhR agonists interact additively, for characterizing the risk of early-life-stage mortality caused by mixtures of PCDDs, PCDFs, and PCBs in trout eggs.

Epidemiology

To conduct retrospective epidemiological assessments of the causative agents of early embryo mortality in Great Lakes fish, the TCDD toxicity equivalence approach in combination with dose response data for early-life-stage mortality in lake trout was applied to estimate historical concentrations of TCDD and related chemicals in Lake Ontario lake trout eggs (Cook et al. 1999). Specifically, congener-specific toxicity equivalency factors (TEFs) relative to TCDD (based on trout early-life-stage mortality) were used in conjunction with bioavailability estimates and an additive toxicity model in an epidemiological assessment of the historical effects of PCBs, PCDFs, and PCDDs on lake trout recruitment. Of pivotal importance for this assessment was the biota-sediment accumulation factor (BSAF) that related the concentrations of individual PCDD, PCDF, and PCB congeners in surface sediment to concentrations in eggs. The BSAF is defined as the concentration of a chemical in eggs on a lipid-normalized basis divided by its concentration in sediment on an organic-carbon-normalized basis (Ankley et al. 1992). In Lake Ontario, concentration profiles of TCDD and related chemicals from segments of dated sediment cores have been obtained. They show that increasing contamination of Lake Ontario sediments with PCDDs, PCDFs, and PCBs began in the 1930s, peaked around 1960, and declined thereafter. The BSAF and relative potency (TEF) values for individual congeners have been used to estimate lake trout egg concentrations of TCDD equivalents in Lake Ontario based on historical concentrations of the TCDD-like PCDD, PCDF, and PCB congeners in dated sediment cores. Analysis of these retrospectively estimated concentrations of TCDD equivalents in lake trout eggs suggests that PCDDs, PCDFs, and PCBs probably contributed to extinction of native lake trout in Lake Ontario prior to 1960 and to recruitment failure of stocked lake trout reintroduced into Lake Ontario since 1973.

Conclusions

Since the 1960s, TCDD toxicity equivalent concentrations in lake trout eggs have dropped significantly. It is estimated that, after the mid-1980s, lake trout sac fry mortality caused by TCDD-like chemicals likely no longer contributed to the recruitment failure of Lake Ontario lake trout. In support of this assessment, mortality associated with blue sac disease was observed for lake trout sac fry hatched from eggs collected from Lake Ontario lake trout during the period of 1977 to 1984 (Symula et al. 1990). However, in 1991, when the TCDD equivalents in Lake Ontario lake trout eggs were determined to be less than the no-observable-adverse-effects level (NOAEL) of 34 pg TCDD/g egg (Walker et al. 1991), there was no evidence of such mortality in Lake Ontario lake trout sac fry (Guiney et al. 1996). The extent to which recent exposures have resulted in decreased survival due to sublethal effects or AhR agonists occurring at egg concentrations of TCDD toxicity equivalents below the NOAEL for early-life-stage mortality of lake trout remains to be determined.

Great Lakes Embryo Mortality, Edema, and Deformities Syndrome

One of the best-studied examples of xenobiotic-induced impacts on oviparous populations is reproductive failures of colonial fish-eating water birds of the Great Lakes region. Severe declines in populations of these species, as well as of predatory avian species across the western world, were first seen during the 1940s and continued into the 1970s, at which time several organochlorine compounds, notably DDT and PCBs, were banned in North America. Initial investigations in the 1960s and 1970s attributed most of the population declines to DDT, which was shown to cause eggshell thinning in numerous avian species (Cooke 1973), and the subsequent recovery of populations of pelicans (*Pelecanus*), eagles (*Haliaeetus leucocephalus*), peregrine falcons (*Falco peregrinus*), and kestrels (*Falco sparverius*) across the continent after DDT was banned substantiated this hypothesis. In the Great Lakes region, however, developmental deformities and embryo mortality continue to be abnormally high among fish-eating birds in areas contaminated with PCBs.

Complex mixtures of chlorinated organic compounds also have been associated with abnormal behaviors in breeding herring gulls (*Larus argentatus*) and western gulls (*Larus occidentalis*) (as summarized by Fry 1995; Yamamoto et al. 1996). In Great Lakes herring gulls, a complex mixture of DDE, PCBs and other organochlorines was associated with decreased incubation attentiveness and defense of territories (Fox et al. 1978) and increases in female-female pairing. Temporal and geographical localization of female-female pairing, supernormal clutches of eggs and skewed sex ratios in western gulls (in the Channel Islands off the coast of California) and herring gulls (in Lakes Ontario and Michigan) were associated with reduced numbers of breeding males. One sample of gull embryos from Lake Ontario also showed a high incidence of testicular feminization (Fox et al. 1978). Ring doves (*Streptopelia risoria*) fed synthetic mixtures of organochlorine insecticides representative of those observed in impacted Great Lakes sites elicited behavioral responses consistent with those observed in herring gulls (McArthur et al. 1983).

In contrast to the DDT-induced syndrome, which was characterized by normal embryonic development with mortality induced by shell breakage, the PCB-induced syndrome comprises relatively normal shell development; in ovo mortality; subcutaneous, pericardial, and peritoneal edema; and deformities of the beak and limbs (Gilbertson et al. 1991). This syndrome, named Great Lakes Embryo Mortality, Edema, and Deformities Syndrome (GLEMEDS) (Gilbertson et al. 1991), proved to be strikingly similar to chick edema disease, which occurs in offspring of hens (*Gallus gallus*) exposed to PCDFs and PCDDs.

Extensive epidemiologic evidence indicates that GLEMEDS, like chick edema disease, is caused by PCDFs or PCDDs, as well as by those PCB congeners that are AhR agonists. Therefore the primary toxic mechanism leading to both syndromes is assumed to be binding of the contaminants to the AhR. Most of the mechanistic studies of AhR-mediated embryotoxicity in birds have been carried out in chickens, during the investigation of chick edema disease. It was shown that PCBs injected on day 0 of incubation caused embryolethality at 5 mg/kg (Aroclor 1254), that some mixtures were lethal at early stages of organogenesis (Aroclor 1242), and that post-hatching growth of chicks exposed before organogenesis was slower than that of chicks exposed later in development (Carlson and Duby 1973). When PCBs were administered in the diet, embryolethality initially occurred in the late stages of incubation; continued feeding led to earlier embryolethality (Platonow and Reinhart 1973), suggesting that bioaccumulation in the hens led to greater PCB deposition in the eggs. Malformations were not a prominent feature of chick edema disease, but they were reported in some studies with chicks (Tumasonis et al. 1973) as well as in ring doves (Peakall and Peakall 1973) fed Aroclor 1254.

Mechanism of action

The cascade of events leading from receptor binding of AhR agonists to the appearance of GLEMEDS is not known. Exposure to TCDD-like compounds alters the ratio of retinol to retinoyl palmitate in yolk (Spear et al. 1989). Since the retinoid receptor belongs to the superfamily of receptors that also includes the steroid receptors, and since cross-talk among these receptors and the AhR is known to occur (Birnbaum 1994), this is not entirely surprising. Appropriate levels of retinoids are essential for normal development in all vertebrates, and either excess or insufficient levels of retinoic acid lead to malformations. Since not all conversions among the various retinoic acid precursors can be carried out efficiently by the embryo, the alteration in ratios of retinol to retinoyl palmitate may be relevant to development of GLEMEDS.

Conclusions

Although it is thought that the various chemicals inducing GLEMEDS act by a single AhR-mediated mechanism, individual organochlorines vary markedly in potency across species within a vertebrate class (Kennedy et al. 1996), presenting challenges for quantitative among-species extrapolations. Moreover, GLEMEDS constitutes an interesting ecological risk management challenge, given that it causes marked effects at the individual level (deformities, lethality), while general population levels of most affected bird species continue to increase.

Eggshell thinning

Toxicant-induced eggshell thinning (ET), especially in predatory birds, can result in cracked or broken eggs and other adverse reproductive effects. During the period of use of DDT as an insecticide in North America, p,p' DDE-induced ET nearly resulted in the extinction of several avian species. Although ET can be produced by several chlorinated hydrocarbons and some metals as well as p,p' DDE, the critical cases of ET in wild birds are all related to an accumulation of p,p' DDE in the eggs (Ratcliffe 1967). Induction of ET by p,p' DDE has been observed in about 30 avian species and effects on reptilian eggshell formation have been reported as well (Ratcliffe 1967).

Feeding p,p' DDE to avian species for a few weeks can result in ET that persists for months or even years in susceptible species (Peakall et al. 1975).

Eggshell thinning does not occur in all birds. Avian species have been divided into 3 categories with respect to their sensitivity to p,p' DDE-induced ET:

- ET > 30 % (e.g., falcon, brown pelican),
- ET ranges from 5 to 15% (e.g., kestrel, Japanese quail), and
- insensitive species in which p,p' DDE does not induce ET (e.g., domestic fowl)

Although p,p' DDE exposures were accompanied by high levels of other pollutants, careful field studies indicated that p,p' DDE, rather than PCBs or dieldrin, was responsible for most, if not all, of the ET. These observations were verified in several laboratory studies using a variety of susceptible avian species. Although several toxicants (PCBs, lindane, dieldrin) also can induce ET, their effects typically differ from those induced by p,p' DDE, being less persistent and of lesser magnitude (up to 10%). For example, in some cases, ET lasts only for a few days after cessation of treatment. In addition to specific effects on ET, nonspecific toxicity, reduced food consumption, or reduced dietary calcium can induce ET. It is of interest that some of these effects of toxicants or dietary restriction of calcium can be reversed by estradiol administration.

Laboratory studies demonstrate transgenerational effects of estrogenic chemicals on ET (Gildersleeve et al. 1985; Bryan et al. 1989). In ovo treatment with diethylstilbestrol (DES), estradiol, or o,p' DDT (the estrogenic isomer) results in female progeny that lay eggs without shells on a regular basis. These females typically also have reproductive tract malformations (extraoviductal tissue).

Mechanism of action

The site of action of p,p' DDE on ET appears to be located in the mucosal cell of the shell gland of the oviduct, and the probable MOA involves altered calcium metabolism in the SG (reviewed by Lundholm 1984, 1994). Eggshell formation is a complex and lengthy physiological process. Vitellogenin (the egg yolk protein) is produced in the liver and is added to the egg in the ovary. After ovulation, the egg moves into the oviduct, where albumin is added over a 2- to 3-h period (in chickens), followed by 2 proteinaceous membranes and some calcium carbonate (1.5 to 2 h). Next, the water and electrolytes are secreted by the SG mucosa and absorbed into the albumin, adding 25% weight to the egg. Then the shell is secreted over a 13 to 15 hour period. The calcium of the hard shell is 98% calcium carbonate and is deposited in 2 layers, mammillary cones and palisades. Finally, in the vagina, the egg is covered by a 10 µm protein cuticle.

It appears that p,p' DDE acts directly within the SG mucosa. However, calcium homeostasis is a complex process that includes absorption from the gut, transport in blood, deposition into and out of bone, and transport into the SG. These physiological processes are regulated by several hormones including vitamin D_3, estradiol, progesterone, and parathyroid hormone. Such a complex process obviously provides many levels where a compound could disrupt eggshell formation by endocrine and nonendocrine mechanisms. For this reason, it is not surprising that the mechanism by which p,p' DDE induces ET is still the subject of investigation and some debate.

Within the SG, there are several key physiological events necessary for the formation of a normal eggshell:

- translocation of Ca^{+2} from the blood into the shell gland against a concentration and electrolyte gradient;
- secretion of HCO^{3-} by the shell gland;
- resorption of H^+ ions to form a carbonate ion CO_3^{-2};
- secretion of K^+ in exchange for H^+;
- absorption of Na^+, coupled to secretion of Ca^{+2}; and
- absorption of PO_3^{-3} from the fluid in the SG during shell formation to create a solution suitable for crystallization of calcite.

Hence, drugs like furosomide that alter sodium/potassium levels can result in ET.

Several mechanisms for the action of p,p' DDE on ET have been proposed. In some of the early work it was observed that dietary p,p' DDT inhibited carbonic anhydrase and reduced serum estradiol levels (Peakall 1970). However, because p,p' DDE injections also caused ET and reduced carbonic anhydrase without lowering serum estradiol, it was proposed that reduced serum estradiol was not responsible for p,p' DDE-induced ET. More recently, it was proposed that inhibition of Ca^{+2}-Mg^{+2}-ATPase activity in the SG by p,p' DDE was responsible for ET; however, Lundholm (1984) reported that this effect, although prominent in vitro, was only occasionally produced in vivo. Rather, Lundholm proposed that p,p' DDE in vivo altered stimulus-secretion mechanisms in the SG. Among the hormonal factors that he considered as possibly being influenced by p,p' DDE were progesterone interactions with its cellular receptor and the synthesis of prostaglandins in the SG. He reported that p,p' DDE reduced progesterone binding to its receptor and that p,p' DDT and DDE both reduced receptor binding of calmodulin, a polypeptide involved in calcium regulation that was proposed to bind to a segment of the progesterone receptor (PR). p,p' DDE was as potent as the calmodulin antagonists calmoduzaline and trifluroperazine in inhibiting binding of progesterone to the PR.

More recently, Lundholm (1994) showed that p,p' DDE-induced ET was accompanied by a reduction in prostaglandin synthesis. Lundholm (1994) proposed that the probable mechanism of action of p,p' DDE in ducks is inhibition of prostaglandin synthesis. He found that decreased PGE_2 blocks HCO_3^- transport resulting in reduced Ca^{+2} transport. He also demonstrated that reduced HCO_3^- transport did not result from an inhibition of HCO_3^--stimulated ATPase activity, as proposed by some. In support of his hypothesis, Lundholm (1994) found that indomethacin treatment (which inhibits prostaglandin [$PGF_{2\alpha}$ and PGE_2] synthesis in SG) at the onset of shell calcification reduced eggshell thickness by 21% 14 hours after treatment. At this time, the calcium level in the SG mucosa was increased by 153%.

Conclusions

Although the worst of the p,p' DDE-induced ET is behind us, at least in the U.S., the question arises as to whether any pesticides in current use present a similar hazard. In 1984 the U.S. Environmental Protection Agency (USEPA) reported that no other chemical has been shown to produce the degree and duration of ET induced by p,p' DDE. However, Schwarzbach et al. (1988) reported that dicofol caused ET in ring doves in the laboratory (fed 33 mg/kg in the diet) that was equivalent to or greater than the effect of p,p' DDE (37 mg/kg in the diet). These authors concluded that dicofol is as potent as p,p'-DDE in inducing ET and cautioned that "until field observations are conducted, confidence that dicofol does not present a hazard to avian reproduction would seem unwarranted." It is clear that we do not need to test all chemicals released to the environment for potential toxicity associated with ET.

Twenty-five years after restriction of the use of p,p' DDT we are still attempting to define the precise mechanism of its ET action. Although several nonendocrine mechanisms have been proposed, at present the data support the idea that p,p' DDE acts on ET through a direct endocrine mechanism in the SG mucosa.

Lake Apopka alligators

Several problems associated with eggs and the reproductive system of alligators (*Alligator mississippiensis*) have been seen in Lake Apopka, Florida. This 12,500 hectare lake is polluted with contaminants from agricultural activities around the lake, sewage treatment from a nearby town, and a major pesticide spill in 1980 at Tower Chemical Company, now a USEPA Superfund site. Although this spill reportedly resulted in contamination of the lake with dicofol, DDT, and sulfuric acid, some confusion exists about the actual contaminants released into the lake. For example, Rohm and Haas, the manufacturer of Kelthane® (dicofol), claimed in a letter to the USEPA that Tower Chemical was not manufacturing dicofol at the time of the spill (Rohm and Haas letter from A. Tilman to J. Loranger of the USEPA Aug 5, 1995). Shortly after this spill, there was a 90% decline in the alligator population that has persisted to the present.

The Tower Chemical spill, however, was not the only potential source of dicofol and DDT. In the late 1980s dicofol and DDT, a contaminant in dicofol at that time, were used on citrus in Florida, including the area around Lake Apopka. The estimated use in 1980 in the entire state was 389,000 kg applied as 42% active ingredient with 5.9% total DDT (22,980 kg) of which 1.6% was DDE (1,680 kg) (USEPA 1984). In 1982, the Florida Game and Freshwater Fish Commission analyzed fish for pesticide residues from Lake Apopka following reports that catfish contained sufficiently high DDT residues to raise potential public health concerns (Florida Game and Fish

Commission 1982, unpublished, Dingell-Johnson Project F-30-9. First Annual Performance Report). Survey results found that several fish (fillets, not whole fish) species had DDE + p,p'-dichlorodiphenylethane (DDD) levels in the .03 to .525 ppm range. It was concluded "that chronic exposure at these concentrations would likely result in adverse reproductive effects on birds," but there was no consideration of the potential effects of these contaminants in other animals until several years later (Florida Game and Fish Commission 1982, unpublished).

In 1989, Woodward et al. reported that Lake Apopka alligator egg clutch viability was depressed as a result of an embryonic mortality rate of 80% in the first month after fertilization (Woodward et al. 1989, 1993). In addition, juvenile alligators displayed a number of endocrine and morphological abnormalities (Guillette et al. 1994). Females displayed polyovular follicles with multinucleated oocytes. Males exhibited plasma testosterone levels one-third those of normal males in nearby Lake Woodruff, while phalli were 24% smaller than those of Lake Woodruff male alligators, and abnormal structures were observed within the seminiferous tubules of the testes. Even though serum testosterone levels were reduced by about two-thirds, ex vivo testicular testosterone production was not reduced in Lake Apopka juveniles compared to juvenile males from Lake Woodruff (Guillette et al. 1996). In contrast, testicular estradiol synthesis was elevated 3-fold in Lake Apopka males over males from Lake Woodruff. Not surprisingly, the alligators are not the only species that have been adversely affected in Lake Apopka. Freshwater turtles from this lake show developmental abnormalities of the gonads and altered levels of plasma sex steroids similar to those seen in alligators (Guillette 1995).

An analysis of the contaminant load in Lake Apopka alligator eggs indicated that these reptiles were exposed to several pesticides. p,p' DDE was the most abundant contaminant, occurring at levels of 5.8 mg/kg in alligator eggs (Guillette et al. 1994), above the threshold for decreased embryonic viability in avian eggs (Heinz et al. 1991). In addition to p,p' DDE, these eggs contained up to 1.8 mg/kg DDD, 0.02 to 1.0 mg/kg dieldrin, and up to 0.25 mg/kg chlordane.

Mechanism of action

Although the effects in Lake Apopka alligators are often attributed to "environmental estrogen" exposure, none of the egg contaminants display estrogenic activity in standard in vivo assays. Although some estrogenic activity has been reported from mammalian in vitro assays for these chemicals, the relevance of these effects is unclear because they occur only at concentrations so high that they are unlikely to be seen in exposed wildlife. The fact that p,p' DDE displays anti-androgenic activity in mammalian in vitro and in vivo assays led Guillette et al. (1995) and Kelce et al. (1995) to hypothesize that the small phallus size could result from an anti-androgenic effect of p,p' DDE in juvenile male alligators. On the other hand, Guillette et al. (1995) proposed that the increase in testicular abnormalities could result from an "estrogenic environment" created by p,p' DDE through inhibition of testosterone action via the androgen receptor and increasing synthesis of testicular estradiol. Recently it was reported that many of the contaminants identified in the eggs of alligators in Lake Apopka displayed weak but measurable affinity for the alligator estrogen and progesterone receptors when provided in combinations (Vonier et al. 1996). Based on these data, the authors proposed that the reproductive abnormalities in Lake Apopka alligators may be associated with the interaction of the contaminants with the estrogen or progesterone receptors.

In addition to anti-androgenic action, p,p' DDE disrupts endocrine function by way of multiple mechanisms, as do most endocrine-disrupting chemicals (Conney et al. 1967). This toxicant also stimulates sterol hydroxylases, increasing steroid hormone metabolism such that serum hormone levels are reduced, a fact that could explain the reduction in plasma testosterone in Lake Apopka juvenile male alligators in spite of normal ex vivo testosterone synthesis (Crain et al. 1998). No one has yet proposed a specific mechanism for the effects of the contaminants on the ovary, but occurrence of polyovular follicles with multinucleated oocytes is pathognomonic of in utero DES exposure in female mice, suggesting to us that an estrogenic effector may be present in the Lake Apopka ecosystem.

Conclusions

At present, laboratory studies testing the hypothesis that p,p' DDE alters reptilian reproductive function by acting in an antiandrogenic fashion are ongoing, as are studies on the reproductive effects of other contaminants. In addition to the reptilian populations, other classes of vertebrates (e.g., fish) in the lake are likely to have been adversely affected as well, although the effects are less well characterized at this time. In addition to a need for additional mechanistic work on the action of p,p' DDE and other pesticides on the fish and wildlife in this lake, there is a critical need for a broader contaminant analysis of the entire food chain, sediment, and water column in Lake Apopka. Such information is critical to strengthen the association between toxicant exposure and the reproductive effects seen in this lake.

Pulp mill effluents

There is substantial evidence that exposure to pulp and paper mill effluent leads to alterations in endocrine and reproductive function in fish (McMaster et al. 1995; Sandstrom 1996). Perhaps the most extensive of these studies are with white sucker (*Catostomus commersoni*) exposed to bleached kraft pulp mill effluent (BKME) in Jackfish Bay, Lake Superior. These fish exhibit a wide array of altered reproductive responses, including reduction in gonad size, delayed sexual maturation, and reduced expression of secondary sexual characteristics (Munkittrick et al. 1991; McMaster et al. 1991, 1995, 1996). White sucker of both sexes have reduced levels of the dominant sex steroid hormones throughout the period of gonadal development. Female white sucker exposed to bleached kraft pulp mill effluent and sampled during vitellogenesis (period of ovarian growth) exhibited reduced circulating levels of 17β-estradiol and testosterone (McMaster et al. 1991; Munkittrick et al. 1991) and reduced levels of testosterone and 17α,20β-dihydroxy-4-pregnen-3-one (17,20β-P) during the prespawning and spawning periods (McMaster et al. 1991; Van Der Kraak et al. 1992). Male white sucker exposed to BKME also have depressed circulating levels of testosterone, 11-ketotestosterone and 17,20β-P (McMaster et al. 1991; Munkittrick et al. 1991; Van Der Kraak et al. 1992). Assessment of white sucker collected during their spawning migration demonstrated that a number of sites within the reproductive-endocrine axis were affected by exposure to BKME (Van Der Kraak et al. 1992). Impacts included altered pituitary function (i.e., reduced circulating levels of GTH-II and a diminished GTH-II response to a superactive gonadotropin-releasing hormone [GnRH] analog), reduced ovarian steroid biosynthetic capacity (lower basal in vitro steroid production and a diminished response to human chorionic gonadotropin), and altered peripheral steroid metabolism (i.e., depressed levels of glucuronidated testosterone). Recent studies provided evidence of elevated apoptotic DNA fragmentation and increased expression of the 70-kDa heat shock protein in ovarian follicular cells from prespawning white sucker exposed to BKME (Janz et al. 1997). These responses coupled with the lower steroid production may contribute to the reduced ovarian size and delayed sexual maturity. Despite these reproductive responses, exposure to BKME affected neither fertility of the eggs or sperm obtained from fish reaching sexual maturity nor subsequent larval development (McMaster et al. 1995).

Depressed reproductive steroid serum concentrations also were found in longnose sucker (*Catostomus catostomus*) and lake whitefish (*Coregonus clupeaformes*) collected in the vicinity of Jackfish Bay (Munkittrick, McMaster et al. 1992; Munkittrick, Van Der Kraak et al. 1992b). Interestingly, the consequences of alterations in steroid hormone levels to whole-animal reproductive fitness varied greatly between these species. Bleached kraft pulp mill effluent exposure did not affect gonadal size in the longnose sucker. In contrast, lake whitefish from Jackfish Bay had limited ovarian and testicular development and > 90% of the fish were not developing gonads for the upcoming spawning season. While these observations provide strong evidence for species differences in responsiveness to BKME, the factors contributing to these differences are unknown.

Since initial work in the early 1990s, numerous studies have demonstrated that alterations in reproductive endocrine function and gonadal development in fish are common features of exposure to a number of pulp and paper mill effluents. Depressed steroid levels were demonstrated in white sucker at other mills (Hodson et al. 1992; Gagnon et al. 1994; Munkittrick et al. 1994; McMaster et al. 1995) and in other species at other sites (Adams et al. 1992; McMaster et al. 1995). A recent review of the in situ assessment of pulp mill effluents on reproductive parameters indicated that 8 of the 10 populations exhibited an increased age to sexual maturation and 4 of the 6 species studied had reduced gonadal size (Sandstrom 1996). These negative reproductive responses have not been seen universally. Studies on feral fish populations downstream of several modernized Canadian and Swedish pulp mills found no differences in gonadal development or circulating steroid-hormone levels (Kloepper-Sams et al. 1994; Lander et al. 1994; Swanson et al. 1994). There is also recent evidence that process changes within pulp mills can contribute to a recovery of reproductive function in fish (Munkittrick et al. 1997).

Mechanism of action

Laboratory tests using fathead minnows exposed over their life cycle to BKME have confirmed depression in sex steroid production, delay in sexual maturity, reduced egg production, and changes in secondary sexual characteristics (Robinson 1994). Separate studies on fathead minnows exposed to effluent from a different bleached kraft pulp mill showed impacts on time to maturation, egg production, and gender (Kovacs, Gibbons, Tremblay et al. 1995). Fish exposed to 20% effluent did not spawn, whereas fish exposed to 10% effluent experienced a 116-d delay in maturation. Female fish also showed male secondary sexual characteristics at effluent concentrations above 5%. The threshold for spawning and egg production was calculated to be 1.7% effluent (Kovacs, Gibbons, Tremblay et al. 1995). These studies provided evidence that the reproductive responses were directly associated with effluent exposure and were not simply the result of habitat alterations. Subsequent studies on exposure of fathead minnows to secondary-treated thermomechanical pulp mill effluent (Kovacs, Gibbons, Martel et al. 1995) and repeat studies on BKME effluent (Kovacs, Gibbons, Tremblay et al. 1995) following a series of mill upgrades and process changes (Kovacs et al. 1997) provided no evidence for reproductive impacts on fathead minnows in life-cycle studies. Again, these studies suggest that mill process changes can reduce the reproductive impacts of pulp mill effluents (Munkittrick et al. 1997).

There is evidence that plant-derived compounds may be responsible for some of the effects associated with pulp mill effluents. Bleached kraft pulp mill effluent contains the plant sterol β-sitosterol, which can affect a diversity of reproductive processes in vertebrates (Van Der Kraak, Munkittrick et al. 1998). Both injection (MacLatchy and Van Der Kraak 1995) or waterborne exposure of goldfish (Carassius auratus) to β-sitosterol (MacLatchy et al. 1997) decreased plasma testosterone and 11-ketotestosterone levels in males and testosterone and 17β-estradiol levels in females; β-sitosterol also caused a reduction in the capacity of goldfish ovarian follicles and testicular fragments to secrete testosterone and pregnenolone in vitro. Similar reductions in plasma testosterone and pregnenolone levels were observed in sexually immature rainbow trout exposed to waterborne β-sitosterol (Tremblay and Van Der Kraak, 1998). β-sitosterol has estrogenic properties as well; for example, it induced the expression of the vitellogenin gene in the liver of juvenile and methyl-testosterone-treated rainbow trout (Onchorhynchus mykiss) (Mellanen et al. 1996), bound in vitro to rainbow trout hepatic estrogen receptors (Tremblay and Van Der Kraak 1995), and induced vitellogenin synthesis in male goldfish (MacLatchy et al. 1995). Other studies have quantified high levels of estrogenic activity in waste streams from pulp mills (Zacharewski et al. 1995), which lend support to the hypothesis that estrogenic compounds contribute adverse reproductive responses. Female mosquito fish (*Gambusia affinis*) exposed to pulp and paper effluent from a Florida mill exhibited masculinization due to either aerobic/anaerobic bacteria transforming β-sitosterol into an androgenic hormone (Rosa-Molinar and Williams 1984) or by the actions of stigmastanol and its degradation products (Howell and Denton 1989). Certainly, a more comprehensive evaluation of the endocrine modulating activity of pulp mill effluents is warranted.

Conclusions

The chemicals responsible for the reproductive effects observed in fish downstream of pulp and paper mills are not known. Dioxins and related compounds are possible causative agents since they can be formed during the bleaching process and are known to alter reproductive function in fish (Walker and Peterson 1994) and mammals (Peterson et al. 1993). However, it seems unlikely that persistent bioaccumulative compounds such as the more chlorinated dioxins or furans are responsible for the altered reproductive responses downstream of pulp and paper mills. Canadian mills have eliminated much of the dioxins and furans from released effluents through high levels of chlorine dioxide substitution, the use of low-active chlorine multiples, efficient chemical mixing and secondary treatment (Servos et al. 1994; Servos 1996), but these technological improvements did not eliminate effluent effects on cytochrome P450-mediated monooxygenase (MO) induction or endocrine function (Munkittrick, Van Der Kraak et al. 1992a, 1992b; Munkittrick et al. 1994; Servos et al. 1994). In addition, white sucker exposed to effluent from mills that do not use chlorine bleaching have induced hepatic MO activity and depressed sex steroid levels (Munkittrick et al. 1994; McMaster et al. 1995). Sampling 10 d following a mill maintenance shutdown revealed recovery of biochemical responses to BKME as the level of MO induction was reduced and plasma sex steroid levels increased to levels near those found in reference fish (Munkittrick et al. 1992a). This pattern of response is unlike that observed with persistent bioaccumulative organochlorines and suggests that chemicals responsible for reproductive responses at Jackfish Bay were relatively short-lived and that continual exposure is needed for the endocrine responses to persist.

Vitellogenin induction in fish

In the mid-1980s, a research team led by John Sumpter sought to address causative factors underlying the observation of hermaphroditic fish caught by anglers from waters below sewage treatment plants in the UK. They hypothesized that one possible contributor to the phenomenon was the presence of estrogenic compounds in the effluents (Purdom et al. 1994). To assess this possibility they utilized a novel bioassay system, induction of vitellogenin, to screen effluents and various substances found in effluents for estrogenic activity in fish and fish cell lines (Sumpter 1985). Vitellogenin, the precursor to egg-yolk protein, is produced in the liver and released to the plasma of female egg-laying animals, including fish. It can be induced both in males and females by exposure to estrogen receptor agonists but is an exceptionally attractive biomarker in the former because it is an abnormal physiological condition.

There is evidence that inappropriate vitellogenesis in males can result in kidney damage, and it has been proposed that it can divert metabolic resources from growth and spermatogenesis (Herman and Kincaid 1988). The induction of vitellogenin in male rainbow trout following estrogen treatment has been correlated with decreased testicular growth, which generally is accepted as a negative reproductive consequence (Jobling et al. 1996). It also has been shown that vitellogenin induction in female fathead minnow exposed to ethynylestradiol is associated with a suppression of egg production (Laenge et al. 1997).

Mechanism of action

Using vitellogenin induction as an endpoint, the UK group performed a series of caged fish studies in which unexposed animals were held in different types of effluents for various periods of time. They found that a relatively large percentage of the effluents caused significant elevation of plasma vitellogenin concentrations (Purdom et al. 1994; Harries et al. 1996). An initial suspicion in the UK was that the estrogenic components in sewage effluents were alkylphenols derived from alkylphenol ethoxylates. Studies in various systems indicate that different alkylphenols are indeed ER agonists and have been shown to induce vitellogenin and inhibit testicular growth in fish (Jobling and Sumpter 1993; White et al. 1994; Jobling et al. 1996). However, in considering the potency of alkylphenols, it seemed that their concentrations in most effluents might be too low to cause the estrogenic responses observed. Recent work by the UK group, using an effects-based fractionation approach similar to that described by Burkhard and Ankley (1989), revealed that in many effluents the likely cause of the vitellogenin induction was the presence of natural and synthetic estrogens, such as estradiol, estrone, and ethynylestradiol (used in birth control pills), excreted by humans (Desbrow et al. 1996). Unlike the alkylphenols, the estrogens often were found at sufficient concentrations in effluents to induce the observed vitellogenin concentrations in plasma. Recent observations of abnormally high plasma concentrations of vitellogenin in fish from certain sites in North America (Bevans et al. 1996; Folmar et al. 1996) should further heighten interest in this particular MOA and associated causative factors.

Oil spill effects on birds

Oil spills during transportation or as industrial effluents can expose adult and juvenile birds either externally through oiling of plumage or internally by ingestion of contaminated food and water or through preening. Embryos may be exposed as well, primarily through absorption of petroleum hydrocarbons through the eggshell. Toxicological effects of oil are variable because of the heterogeneous nature of oils (e.g., crude oils versus distilled products such as gasoline, jet fuel, diesel fuel, kerosene, and fuel oils) and the weathering of oils once they are released into the environment. Toxicologically active components include PAHs, such as benzo-a-pyrene (BaP), and certain volatile organic compounds such as benzene, hexane, and toluene, which are highly toxic (Andrews and Snyder 1986) but tend to evaporate quickly following a spill. By contrast, the alkane and alkene portions of the oils are relatively nontoxic.

Birds that attempt to clean themselves by preening ingest oil from their feathers. It has been estimated that a moderately contaminated duck could ingest up to 10 g/kg body weight of oil over an 8-d period. Trivelpiece et al. (1984) demonstrated reduced survival of newly hatched Leach's storm-petrel (*Oceanodroma leucohoa*) chicks when adults were exposed orally to a single dose of 0.1 ml Purdhoe Bay crude oil during egg laying, well within the amount that would be ingested by preening.

Mechanism of action

Physiological and reproductive responses of birds exposed to oil were reviewed by Leighton (1983), Holmes (1984), Rattner et al. (1984), and Jessup and Leighton (1996). Observations that birds exposed to oil succumbed to mild cold stress more readily than nonexposed birds suggested that the adrenal gland may be affected. Measurements of severely diminished levels of plasma corticosterone in contaminated birds supported this hypothesis. Further support came from observations of extensive structural degeneration of the mitochondria in the adrenocortical inner zone cells, reduced production of corticosterone from adrenocortical tissue perfused ex vivo with medium containing adrenocorticotrophic hormone, and kinetic analysis of injected doses of tritiated corticosterone. In addition to changes in hormone synthesis and secretion by the adrenal gland, petroleum-induced changes in liver function may contribute to low plasma corticosterone levels. Hepatic MO activity can be elevated in petroleum-contaminated birds. It has been suggested that this induction may increase metabolic clearance of corticosterone.

Cavanaugh (1982) and Albers (1983) reviewed the effects of petroleum on reproduction. Low reproductive success in female mallards (*Anas platyrhynchos*) ingesting South Louisiana crude oil was evidenced by significantly fewer live ducklings hatched per breeding pair. Direct ovarian involvement, including suppression of follicular development, altered gonadal hormone levels, eggshell thinning, and decreased hatchability, has been observed. Prolactin concentrations of oil-exposed birds were significantly lower than controls throughout oviposition and incubation, although the pattern of a sharp decrease in circulating levels at onset of incubation still occurred (Cavanaugh et al. 1983). Similarly, levels of plasma estradiol, estrone, progesterone, and luteinizing hormone (LH) in females exposed to oil were reduced (Cavanaugh and Holmes 1987). It is likely that the oil acts directly on ovarian steroidogenesis, resulting in insufficient stimulatory feedback to the pituitary and a subsequent decline in circulating LH. The net result is a delay in ovarian maturation and a marked reduction in fertility.

Very small quantities of oil applied to the external surface of the shell of bird eggs result in high levels of embryo mortality (reviewed by Leighton 1983). Estimates suggest that 3.6 to 14.2 μL per egg would kill 25 to 100% of the embryos (differences result from using various types of oils in the studies). These amounts are well within the range of what an incubating adult bird could bring back to the clutch on oily plumage. The mechanisms involved in this toxicity are not known, although reduced gas exchange due to partial sealing of the shell by oil has been ruled out (Jessup and Leighton 1996).

Conclusions

Other effects of petroleum exposure not directly related to growth or reproduction have been summarized by Leighton (1983) and include hyperthermia resulting from fouling of plumage and damage to the kidneys and intestinal mucosa. Salt gland function may be affected by crude oil, resulting in problems with osmoregulation. Hemolytic anemia from damaged red blood cells has been reported in birds exposed to oils with high polynuclear aromatic content.

Impacts of oil in the environment have been demonstrated in adult and embryonic birds along many shorelines, from small to large colonies of breeding seabirds, exposed to spill events such as the *Exxon Valdez* spill in 1989. While reproductive and other toxic effects to individuals definitely occur, long-term population changes have not yet been documented.

Effects of polycyclic aromatic hydrocarbons on marine fish

Polycyclic aromatic hydrocarbons derived from fossil fuels and related petroleum products or from combustion of wood and other organic matter affect reproduction and development of marine fish. The impacts of PAHs on reproductive function in English sole (*Pleuronectes vetulus*), a common marine flatfish occurring on the west coast of the United States, have been examined in a series of studies on sole populations from contaminated embayments in Puget Sound, Washington. Various studies (Varanasi and Stein 1991) show that English sole from industrialized areas in Puget Sound take up PAHs. In comparison to other fish species, English sole generate high quantities of reactive PAH metabolites; the activities of hepatic xenobiotic metabolizing enzymes (i.e., MOs, epoxide hydrolases, and glutathione-S-transferases) in this species are consistent with higher activation and lower detoxication of PAHs than is typically seen in related flatfish species (Collier et al. 1992). These metabolites have been shown to bind to DNA in English sole, forming persistent adducts (Stein et al. 1994). Consequently, English sole are highly susceptible to the development of liver cancer and related lesions (Myers et al. 1994, 1997) and also appear to be prone to reproductive anomalies related to PAH exposure.

Various types of reproductive impairment have been documented in English sole residing in PAH-contaminated areas of Puget Sound. Female sole from areas with high concentrations of PAHs in sediment are less likely to produce vitellogenin (40 to 60% versus 80 to 90% in reference sites) and have lower plasma concentrations of 17β-estradiol than do sole from uncontaminated reference sites (Johnson et al. 1988; Collier et al. 1997). In epidemiological analyses of these data, exposure to PAHs, as indicated by concentrations of PAH metabolites in bile of individual fish, were identified as major risk factors for depressed plasma steroid concentrations and inhibited ovarian development. English sole from PAH-contaminated areas also display increased ovarian atresia, particularly of primary oocytes (Johnson et al. 1988, 1997). Increased atresia is associated with a trend toward reduced egg production and increased egg weight, although the magnitude of the effect is not marked (Johnson et al. 1997).

If English sole from PAH-contaminated sites do enter vitellogenesis successfully, they may still suffer from inhibited spawning ability and reduced viability of eggs and larvae (Casillas et al. 1991). When gravid English sole were brought into the laboratory and artificially induced to spawn, responsiveness to luteinizing-hormone-releasing hormone treatment was significantly lower in fish from PAH-contaminated sites; animals took longer to spawn and the proportion that failed to spawn was higher than was observed in reference animals. Eggs of sole from the PAH contaminated sites also exhibited a 50% reduction in fertilization success, while larval viability declined by approximately 15%. Exposure to PAHs in the water column (e.g., fluoranthene at 0.075 to 7.5 mg of PAH per liter of seawater) caused larvae to become disoriented and to exhibit signs of narcosis; mortality resulted at higher concentrations (Eddy et al. 1993).

Effects of PAHs on reproduction have not been studied as extensively in male as in female English sole, but there is evidence that males may also be susceptible to PAH-related reproductive dysfunction. Preliminary studies suggest that while testicular development in male fish from PAH contaminated sites appears to be normal, there is some reduction in circulating plasma concentrations of 11-ketotestosterone and testosterone in animals with particularly high concentrations of PAH metabolites in bile (S. Sol, NOAA, Seattle, WA, personal communication). Sperm quality also appears to be somewhat lower in male sole collected from spawning grounds near the heavily industrialized Elliott Bay than in males from a non-urban spawning ground, although additional data are needed to confirm this trend.

Polycyclic aromatic hydrocarbons also may affect the growth of juvenile English sole. In a recent laboratory study (Kubin 1997), exposure to PAH-contaminated sediment for 90 d led to increased variation in growth rates of young-of-the-year English sole and a trend toward reduced fish size at the highest exposure level. After 6 months of exposure to a sediment mixture containing 1.6% Eagle Harbor sediment, juvenile sole were about 15% smaller than sole reared on reference sediments.

At this time, however, there is no consistent evidence of suppressed growth in subadult sole collected from sites with high sediment PAH concentrations.

Reproductive abnormalities similar to those observed in female English sole have been reported in wild populations of several other fish species exposed to PAHs from industrial sources, although few species are as sensitive as sole. For example, depressed plasma vitellogenin levels and increased ovarian atresia have been observed in winter flounder (*Pleuronectus americanus*) from PAH-contaminated sites in Boston Harbor (Pereira et al. 1992; Johnson et al. 1994). Winter flounder from industrial areas in Boston Harbor also exhibited reduced egg size and 10 to 30% declines in fertilization success, percent hatch, percent viable hatch, larval yolk-sac lipid concentrations and yolk-sac size, and larval size in comparison to reference fish (Nelson et al. 1991; Perry et al. 1991). Reductions in gonadal growth, egg quality, and other indicators of reproductive function have been observed in field popula-tions of starry flounder (*Platichthys stellatus*) from San Francisco Bay (Spies and Rice 1988) and in white croaker (*Genyonemus lineatus*) from the Los Angeles area (Cross and Hose 1988) in conjunction with exposure to CYP1A-inducing compounds, including PAHs.

Effects similar to those observed in English sole have also been induced in the laboratory through controlled exposure to model PAHs. In Atlantic croaker (*Micropogonias undulatus*), for example, exposure to oil or naphthalene led to reductions in plasma estradiol concentrations and in vitro ovarian estradiol produc-tion, as well as reduced gonadal growth, increased ovarian atresia, and reduced spawning ability (Thomas 1988; Thomas and Budiantara 1995). A number of laboratory studies have documented the suppressive effects of exposure to PAHs, particularly petroleum hydrocarbons, on the viability and growth of larval and juvenile fish (e.g., Kuhnhold et al. 1978; Whipple et al. 1978; Moles and Rice 1983; Paine 1989; Carls and Rice 1990; Onuoha and Nwadukwe 1990; Vignier et al. 1992; Vogelbein et al. 1996; Moles and Norcross 1997). Effects observed include delayed or accelerated hatching, reduced larval size, slower developmental rates, and morphological abnormalities, including curvature of the notochord, and cardiovas-cular abnormalities.

There also is evidence from several studies that exposure to PAHs can impair the reproductive health of other male fish in ways similar to those observed in English sole. Male winter flounder exposed to mixtures of oil and sand exhibit changes in reproductive steroid metabolism, including reduced plasma glucuronides and reduced plasma 11-ketotestosterone and testosterone (Truscott et al. 1992; Idler et al. 1995). It has been suggested that androgen glucuronides can function as phero-mones in some fish, therefore these changes in androgen metabolism potentially could affect the breeding success of these animals. However, such impacts have not yet been demonstrated in field studies. Recently, Nagler and Cyr (1997) reported that exposure of male American plaice *(Hippoglossoides platessoides)* to marine sediments contaminated with a mixture of PAHs and PCBs resulted in a 30 to 50%

reduction in viable hatch in eggs fertilized with their sperm. However, in these studies, as in English sole, testicular growth in PAH-exposed males generally was not depressed.

Mechanism of action

A notable feature of this class of contaminants is that, unlike polychlorinated aromatic hydrocarbons that bioaccumulate in tissues, PAHs are metabolized extensively in vertebrates. It is often these metabolites, rather than the parent compounds, that exert toxic effects. Metabolites of PAHs can bind to proteins or DNA (Stein et al. 1994), inducing DNA damage or cytotoxic effects, such as oocyte atresia (Mattison et al. 1983). Exposure to PAHs also may cause the generation of reactive oxygen intermediates (DiGiulio et al. 1993; Livingstone et al. 1993), that can have a variety of detrimental effects on cell growth and function (Suzuki et al. 1997). Additionally, like TCDD and related compounds, PAHs can bind to the AhR and trigger induction of AhR responsive genes, including CYP1A1. Anti-estrogenic effects mediated by the AhR include increased oxidative metabolism of estradiol, down-regulation of ER levels, and inhibition of 17β-estradiol-induced gene expression (Santodonato 1997).

The mechanisms through which PAHs disrupt reproduction and development in English sole have not been fully elucidated, but it is likely that multiple modes of action are involved. First, these compounds appear to have anti-estrogenic effects that may be related to their potency as CYP1A1 inducers. Several recent studies indicate that CYP1A1 induction in sexually maturing fish may lead to disruption of estrogen-regulated reproductive processes. For example, CYP1A1-inducing compounds such as the PAH β-naphthoflavone have anti-estrogenic activity in both in vitro and in vivo laboratory exposure tests with rainbow trout (Anderson, Miller et al. 1996; Anderson, Olsen et al. 1996). Preliminary data show a similar effect in English sole; exposure to benzo[a]pyrene or PAH contaminated sediment extracts inhibited vitellogenin induction in juvenile English sole pretreated with estradiol, while concurrently inducing aryl hydocarbon hydroxylase activity in the liver, an indicator of CYP1A1 induction (Anulacion et al. 1997).

Reproductive impairment in English sole also may be mediated in part by effects of PAHs on steroid metabolism, clearance, or biosynthesis. Studies with trout have shown that CYP1A1-inducing PAHs may increase Phase II conjugation of 17β-estradiol and thus enhances its excretion rate (Forlin and Haux 1985). This mechanism also may be involved in reduction of estradiol concentrations in English sole, although the evidence is not entirely clear. Stein et al. (1991) found that injection of English sole with a sediment extract containing PAHs induced biliary excretion of 17β-estradiol, but they found no evidence of enhanced catabolism of 17β-estradiol. In vitro production of 17β-estradiol by ovarian tissue also appears to be reduced in sole collected from sites with high concentrations of PAHs (Johnson et al. 1993).

Polycyclic aromatic hydrocarbons also disrupt reproductive function through modulatory effects on estrogen-mediated neuroendocrine and endocrine processes in the hypothalamus and pituitary. Because estrogens and aromatizable androgens play an important role in regulating gonadotropin (GTH) synthesis and release (Pakdel et al. 1990, Xiong et al. 1994), effects of PAHs on plasma estradiol concentrations, or on binding of ER to estrogen response elements associated with the GTH genes, could inhibit gonadal development and spawning. The effects of PAH exposure on gonadotropin production and release have not been studied in English sole because of the lack of homologous assays for measurement of English sole GTHs. However, Thomas and Budiantara (1995) observed decreased secretion of pituitary gonadotropins in female Atlantic croaker exposed to diesel fuel oil and naphthalene in the laboratory during reproductive development. Consequently, it seems likely that PAHs also may affect pituitary function in English sole. The likelihood that the pituitary may be a target organ for PAHs is enhanced by the findings of Andersson et al. (1993) that CYP1A1 is inducible by model PAHs in the pituitary of sexually maturing rainbow trout.

Timing of exposure has a strong modulating influence on the actions of PAHs on reproductive function, including their potency in inhibiting vitellogenesis. Anderson, Miller et al. (1996) and Anderson, Olsen et al. (1996) observed that the anti-estrogenic potency of CYP1A1-inducing PAHs was reduced when administered to rainbow trout whose circulating plasma estradiol concentrations were high. This effect may be related to the well-documented suppressive effects of estradiol on CYP1A1 activity, which has been observed to occur normally during the reproductive cycle in female fish (Pajor et al. 1990; Gray et al. 1991). These findings are consistent with field observations (Johnson et al. 1988, 1994) that species such as English sole, which remain at contaminated near-shore sites throughout previtellogenesis and early vitellogenesis, appear to be more susceptible to the anti-estrogenic effects of PAHs than species such as winter flounder, that spend a portion of this part of their life cycle in uncontaminated waters offshore (Hanson and Courtenay 1996).

Conclusions

In summary, a variety of effects on reproduction and development have been observed in wild flatfish populations exposed to PAHs in the Puget Sound, as well as in flatfish and other marine fish species exposed to PAHs in the laboratory. The impacts include alterations in steroid hormone concentrations, reduced fertility and fecundity, larval deformities, and changes in juvenile growth rates. Current research suggests that PAHs interfere with normal reproduction and development through multiple mechanisms. Metabolites of PAHs can induce DNA damage or have cytotoxic effects on cells and tissues involved in reproduction. They can also interact with the Ah receptor and trigger the induction of AhR responsive genes, which may mediate anti-estrogenic effects such as down-regulation of ER levels and inhibition of 17β-estradiol-induced gene expression. Exposure to PAHs may also affect endocrine function, altering rates of steroid metabolism and biosynthesis or modulating neuroendocrine and endocrine processes in the hypothalamus and pituitary. Patterns of activation and detoxication of PAHs in exposed animals appear to be important determinants of their sensitivity to PAH-mediated reproductive and developmental effects and may help to identify those ecologically and commercially important species that are particularly at risk for PAH-related reproductive injury.

Health impacts similar to those found in English sole also have been observed in other marine fish in PAH-contaminated urban coastal waters throughout Europe and the U.S., suggesting that these compounds may be a widespread threat to fish species that rely on near-shore coastal areas for nursery and feeding grounds. Although at this point we have only a limited understanding of the ecological risks posed by PAHs, the evidence from laboratory and field studies shows clearly that they are capable of decreasing fisheries' productivity and, consequently, of reducing the capacity of fish populations to withstand the additional stresses associated with overharvest, habitat destruction, and changes in climatic conditions. As such, the potential impacts of PAHs on fish growth, survival, and reproduction should be taken into account in our management of marine resources.

Amphibian deformities

Many amphibian populations have experienced, or are currently experiencing, significant declines (NBS 1995). Habitat degradation almost certainly is one cause of these declines. However, a host of other causative factors including toxic chemicals and increases in ambient ultraviolet radiation (due to ozone depletion) have been hypothesized (Pechman and Wilbur 1994). In addition, there have been several recent reports of the occurrence of severely deformed frogs, representing several species, in abnormally large numbers at various locations across the U.S. and Canada (Tietge 1996; Oeullet et al. 1997). These deformities include missing limbs, supernumerary limbs, bony limb-like projections, and digit and musculature malformities, as well as eye and central nervous system abnormalities. It is uncertain, however, whether in some instances population declines could be linked to the occurrence of deformities and/or whether the same environmental stressors could be causing both phenomena. The possible contribution of chemical toxicants to either or both of these effects is particularly difficult to assess because, with the exception of the relatively well characterized teratogenesis assay, FETAX, with *X. laevis* (ASTM 1991), there has been little systematic ecotoxicology research with amphibians.

Mechanism of action

Despite this uncertainty, a plausible mechanistic argument can be made for the possible role of xenobiotics in causing the deformities in amphibians. Specifically, it is possible to induce many of the same types of malformations observed in frogs from the field through administration of excess retinoic acid (RA) to a variety of embryonic vertebrate systems, including amphibians (Scadding and Maden 1986; Mohanty-Hejmadi et al. 1992). Retinoic acid, as well as other endogenous metabolites of vitamin A, exert their influence through one or more receptor families (e.g., RAR, RXR) that are members of the well-conserved nuclear steroid-hormone receptor superfamily (Schena 1989). The RA system, under normal conditions, controls processes related to cellular differentiation, pattern development and the establishment of embryonic polarity (Wagner et al. 1990; Shimeld 1996). However, administration of excess RA during specific windows of development has been shown to cause defects in central nervous system and craniofacial structures, as well as abnormal limb pattern development, including supernumerary limbs and digits (Shenefelt 1972; Rutledge et al. 1994; Scott et al. 1994). Hence, it is reasonable to hypothesize that there may be environmental chemicals that act as agonists or antagonists of the RA receptors. A report by Harmon et al. (1995) indicates that metabolites of the widely used insecticide methoprene, an insect growth regulator, bind to at least one of the RA receptors (i.e., RXR) and activate gene transcription. While organism studies with the Northern leopard frog (*Rana pipiens*) have indicated that methoprene and its metabolites can cause significant developmental effects, these effects do not include limb malformations (Ankley, Tietge et al. 1998).

Conclusions

It is obvious that much more research is required before the possible role of chemicals at environmentally relevant concentrations in producing amphibian malformations can be assessed. Our main purpose for including this example is to illustrate the thought processes used in the identification of a plausible MOA around which hypotheses can be formulated and tested.

Organophosphorus ester insecticides

The effects of organophosphate (OP) insecticides on the reproduction and development of egg-laying vertebrates have been well documented in the laboratory but not necessarily in the field. The mechanisms by which OPs cause malformations have been identified (below), and numerous studies document the occurrence, at environmentally relevant exposures, of developmental toxicity (see review by Hoffman 1990; also: King et al. 1984; Bennett and Bennett 1990; Bennett et al. 1990, 1991; Solecki et al. 1996), disruption of egg laying (Stromborg 1977), transient decrease in LH secretion (Rattner et al. 1986) and damage to male reproductive organs (Maitra and Sarkar 1996). Other studies demonstrate no effects except those accounted for by decreased food intake and overt morbidity (Haegele and Tucker 1974; Stromborg 1986), suggesting that some OPs do not selectively affect reproduction, even at doses causing mortality. There are also numerous studies suggesting that OPs are more lethal in nestlings than in adult birds (e.g., Grue and Shipley 1984). Nevertheless, there are few field studies that demonstrate either reproductive or developmental consequences of OP exposure.

Organophosphorus esters comprise a large and varied class of pesticides consisting primarily of insecticides but also including a few herbicides and fungicides (Gallo and Lawryk 1991). The MOA of the latter are not well understood, but the acute toxicity of all OPs to both target and nontarget organisms results from inhibition of AChE. Because the nervous systems of all organisms from arthropods to mammals use AChE to modulate transmission of nerve impulses at the synapse, OPs are acutely neurotoxic to all vertebrates and invertebrates. There is, however, considerable variability in the acute toxicity of different OPs within species, and correspondingly great differences in the toxicity of single OPs exist across phyla and between classes of vertebrates. Structure-activity relationships have been elucidated for both target and nontarget species, allowing the design of insecticides that are selectively more toxic to insects than to vertebrates (e.g., Metcalf and Metcalf 1973).

Given the enormous structural diversity of the class, it is not surprising that individual OPs may exhibit toxic effects that are unrelated, or only marginally related, to AChE inhibition (Marquis 1986). Most of these effects are restricted to one or a few compounds and are of little general interest. There are, however, 2 irreversible forms of OP toxicity that occur frequently: organophosphate-induced delayed neuropathy (OPIDN) and developmental toxicity.

Organophosphate-induced delayed neuropathy

Organophosphate-induced delayed neuropathy is an irreversible paralysis that begins 5 to 30 d following acute exposure to some OPs. Symptoms result from axonal degeneration of peripheral nerves, and are typically most severe in the hind limbs. Both rodents and Japanese quail are resistant to OPIDN-induced paralysis; chickens are commonly used as a model species. Immature animals generally are resistant to classical OPIDN, although children have been affected (Smith and Spalding 1959). Organophosphate-induced delayed neuropathy is highly correlated with > 80% inhibition of an esterase designated neuropathy target enzyme (NTE); the normal function of this enzyme is not known (Aldridge and Barnes 1966; Johnson and Henschler 1975; Lotti 1992). All OPs that cause OPIDN inhibit AChE (directly or through metabolites), but the potential for an OP to cause OPIDN is not well correlated with its inhibition of AChE (Metcalf et al. 1988).

Developmental toxicity

In terms of developmental toxicity, OP ester insecticides are not, in general, mammalian teratogens (reviewed by Farage-Elawar and Francis 1990). In contrast, OPs induce 2 well-defined syndromes of malformations in birds (Roger et al. 1969). Type I defects include short limbs (micromelia), parrot beak, and abnormal feathering; these effects result from inhibition of kynurenine formamidase. Type II malformations include primarily skeletal defects such as leg malformations and wry neck and result from inhibition of AChE in the developing embryo. Malformations resembling the Type II avian group are also seen in frogs (*Rana perezi*) exposed to either methyl parathion (*O*-4-nitrophenyl *O,O*-dimethyl phosphorothionate) or the carbamate insecticide pirimicarb (2-(dimethylamino)-5,6-dimethyl-4-pyrimidinyl N,N-dimethylcarbamate) (Alvarez et al. 1995). Type II malformations result from the excess of acetylcholine present in the embryo due to inhibition of AChE; there is no known antidote. In contrast, Type I malformations can be mitigated by administration of nicotinamide for up to 4 d after injection of the OP. Such mitigation suggests that OP-induced inhibition of kynurenine formamidase leads to depletion of nicotinamide-adenine dinucleotide (NAD) over a period of days and that the depletion of NAD is the proximate cause of the malformations (Misawa et al. 1982).

Organophosphorus ester insecticides differ not only in the dose required to cause malformations, but also in which of the 2 malformation syndromes they cause (reviewed by Hoffman 1990). The structure-activity relationships governing developmental toxicity are not known, but they are not directly correlated with the inhibition of AChE. In fact, the teratogenic dose can be below that required for significant inhibition of AChE (Hoffman and Eastin 1981). Ducks, pheasants, and quail, as well as chickens, are susceptible to OP-induced malformations; no resistant species of birds have been identified.

There is considerable evidence from laboratory studies that OPs cause functional deficits as well as structural malformations. Roger et al. (1969) observed that O-4-nitrophenyl O-ethyl phenylphosphonothioate (EPN) did not cause either Type I or Type II malformations, but newly hatched chicks "lacked equilibrium and extended their heads backward." Similarly, in a series of studies on OPIDN (Farage-Elawar and Francis 1987, 1988a, 1988b), injection of certain OPs into hens' eggs after organogenesis resulted in newly hatched chicks that were seemingly normal but could not stand or hold up their heads. Severely affected chicks died. Administered shortly after hatching, these OPs caused gait alterations lasting up to 90 d of age. These effects did not correlate well with inhibition of either AChE or NTE. Moreover, although carbamate insecticides also exert their acute toxicity by inhibiting AChE, they do not appear to cause malformations in birds (Hoffman 1990). There is, however, some evidence that they can induce functional deficits such as paralysis when administered in ovo (Farage-Elawar 1989; Hoffman 1990). In sum, these studies suggest that inhibition of AchE—or some other function of certain AChE inhibitors—alters development of the nervous system and so causes functional effects even in the absence of structural defects.

High mortality (up to 25%) was documented among laughing gull (*Larus atricilla*) chicks in nests near cotton fields at 3 Texas locations treated with parathion or EPN (White, Mitchell, and Prouty 1983). Adult deaths also occurred, and both adult and juvenile mortality was accompanied by severe inhibition of brain AChE (> 50%). In a subsequent study, White and colleagues dosed one member of pairs of nesting birds with 6 mg parathion/kg body weight by gavage and observed the nesting behavior of the treated birds (relative to that of sham-dosed controls) at 10-min intervals over a 3-d period (White, Mitchell, and Hill 1983). Birds were dosed on their nests and then observed from a blind constructed before nest-building began. Thus the investigators were able to demonstrate that sublethal doses of parathion decreased nest attentiveness by exposed birds for up to 48 h following exposure. Because only one member of each pair was dosed, reproductive success of treated pairs relative to controls was not decreased: the undosed parent compensated by increased nest attentiveness. The authors point out, however, that this would be improbable in natural exposure situations, in which parents probably would be exposed at the same time (White, Mitchell, and Hill 1983). In a separately reported study, the same investigators dosed laughing gulls with 5 mg parathion/kg body weight and were unable to document behavioral changes (King et al. 1984).

Powell (1984) found no significant decrease in reproductive success of red-winged blackbirds (*Agelaius phoeniceus*) in fields treated with fenthion, other than a tran-sient decrease in growth of young nestlings. Unaffected reproductive endpoints included incidence of nest abandonment, clutch size, hatching success, and fledg-ling success. However, there was no significant inhibition of brain cholinesterases in the nestlings, suggesting that exposure levels were very low. Thus no conclusions can be drawn about the effects of significant exposures to fenthion on reproductive success in this species. Several other studies also suggest that many forms of insecticide application simply do not result in high enough exposure of birds to cause significant inhibition of AChE, even though adequate insect control was achieved (Lari et al. 1994; Fair et al. 1995).

Reproductive behavior

As summarized by Fairbrother (1996), there is conflicting evidence concerning the effects of AChE inhibitors on reproductive behavior in birds. For example, mallard (*Anas platyrhinchos*) and teal (*Anas discors*) hens were reported to abandon their nests following a single field exposure to methyl parathion, possibly as a result of reduced prolactin levels. White-throated sparrow (*Zonotrichia albicolis*) population reductions following field application of fenitrothion were attributed, in part, to territory abandonment, decreases in territory defense, clutch desertion and desul-tory incubation. Alternatively, no effects of methyl parathion on hatching and fledgling success were noted for red-winged blackbirds or European starlings (*Sturnus vulgarus*). The differences in results among these studies may be the result of interspecies variability or variation in exposures or habitats.

Conclusions

Laboratory data suggest that OPs, and perhaps other cholinesterase inhibitors, affect survival and reproduction in egg-laying vertebrates both by causing malfor-mations and by altering parental behavior. There are, however, few field studies confirming these effects.

The overall lack of field studies may result from the certainty imparted by labora-tory studies that define mechanisms as well as syndromes, i.e., a sense that the questions have already been answered. More probably, the scarcity of field studies reflects the difficulty of examining short-lived effects of biodegradable contaminants. Environmental pollution by the persistent organochlorines results in lifelong contamination of organisms, with long-term consequences on reproduction because physical and behavioral changes persist within the individual and the species. Effects of OPs, in contrast, are short-lived and most often result in death of a subset of the affected organisms, with full recovery of the others. Only if significant but sublethal exposure occurs (a narrow window for many of the highly toxic OPs, such as parathion or fenthion) will reproductive processes or behavior be affected, and the effects almost certainly will be transient. In many species, such short-lived effects probably can be compensated within the same breeding season (e.g., by laying replacement eggs). Moreover, unless intense observation is maintained, even relatively severe losses of young may be attributed to nonspecific factors, since reproductive success within a species varies naturally from year to year and site to site.

Thus, at first approximation, it is easy to conclude that OPs do not pose a serious threat to egg-laying vertebrates, despite their well-documented teratogenicity and the laboratory evidence for altered reproductive success. If one examines the implications of the studies by White, Mitchell, and Prouty (1983) and by White, Mitchell, and Hill (1983), however, it becomes clear that OPs can strongly affect the reproductive success of species or populations whose reproductive fitness is already limited (King et al. 1984). For example, reproduction of laughing gulls (*Larus atricilla*) in Texas cotton-growing areas was severely affected in 3 of 4 years by OP-induced mortality (White, Mitchell, and Prouty 1983). The investigators were motivated to carry out these studies by the at-risk status of laughing gulls in Texas; quite possibly, the difference between endangered colonies and thriving colonies of this species is the presence or absence of OPs.

Alterations in reproductive and developmental behavior

While the role of abnormal reproductive behavior in causing population-level responses in oviparous animals can be hypothesized, specific examples where behavioral alterations have been clearly identified as the primary cause of adverse population effects are scarce. The potential role of *o,p*-DDT and PCBs in altering male breeding behavior of colonial-nesting, fish eating birds, within adversely affected populations of the West coast of North America and the Great Lakes, are among the best documented cases for attempting to test such an hypothesis. However, the contribution of chemically suppressed breeding behavior to these population responses must be considered within the context that the reduction in functional breeding males may also have been due to additional anatomical and physiological effects associated with feminization of embryos and/or decreased survival of male juvenile birds (as summarized by Yamamoto et al. 1996).

Establishing indices diagnostic of behavioral effects on reproduction in the field is difficult. In part, this is because of the co-occurrence of natural and anthropogenic stimuli that can elicit subtle behavioral modifications. To some extent, behavioral alterations associated with insecticide and metalloid exposures can be related to AChE activity (e.g., Fairbrother 1996), residue levels in the central nervous system (e.g., Bradbury and Coats 1988; Blus et al. 1996), or pathological lesions (e.g., Heinz 1996). However, such relationships typically have been most credible in the context of acute exposures and behavioral alterations preceding death. Studies in which consistent behavioral alterations, and/or related anatomical or physiological modifications, can be elicited in controlled experiments using actual or synthetic mixtures representative of field exposures can contribute to establishing plausibility of cause and effect relationships. These approaches were employed to contribute to the weight-of-evidence concerning the role of *o,p* DDT and PCBs in feminization of herring and western gulls (as summarized by Fry 1995; Yamamoto et al. 1996). The concordance of these responses with studies using endogenous hormones (e.g., Adkins-Regen and Pniewski 1978; Adkins-Regen et al. 1994) are noteworthy.

Mechanism of action

A variety of chemicals cause alterations in behavior through effects on the nervous or endocrine systems. Several classes of pesticides are specifically designed to disrupt the nervous system. For example, OP and carbamate insecticides inhibit AChE activity (Fukuto 1990), DDT-type compounds prevent the deactivation of axon sodium gates (Coats 1990), chlorinated alicyclics bind at the picrotoxinin site in the GABA chloride ionophore complex (Coats 1990), and pyrethroid insecticides interfere with sodium gates (Bradbury and Coats 1988; Coats 1990). The role of specific classes of organophosphates in eliciting a delayed distal polyneuropathy has also been reported using the chicken as a model (Abou-Donia 1981). This effect appears to be strongly correlated with the inhibition of an enzyme activity termed neurotoxic esterase (Johnson 1977; Abou-Donia 1981).

While industrial organic chemicals are not designed, or intended, to have biological activity it is estimated that approximately 70% of these compounds may elicit acute toxicity through one or more narcosis modes of action (Bradbury 1995). These modes of action are thought to result from anesthetic-like mechanisms (Bradbury et al. 1989; Franks and Lieb 1990). Metals and metalloids also can directly affect the nervous system. For example, a number of metals have been reported to influence synaptic transmission or neurotransmitter release in the peripheral or central nervous system (Weber and Spieler 1994) and methyl mercury causes neuron degeneration and demethylation (Heinz 1996).

In addition to compounds that can cause behavioral modifications by directly interacting with neurotransmitters, receptors, and neuronal structure, it is also possible for xenobiotics to interact with the endocrine system and subsequently to alter behavior by influencing signal reception and transmission, activation of neuron clusters, and differentiation of brain structure and function. Examples of investigations of endocrine influences on nervous system structure and/or function in birds, fish, and amphibians are available (e.g., Manns and Fritzsch 1992; Adkins-Regen et al. 1994; Weber and Spieler 1994; Hill et al. 1995).

Epidemiology

Organism-level responses as a consequence of exposure to neurotoxic pesticides have been reported extensively for acute exposures. Acute organophosphate and carbamate exposures to birds and fish are characterized by signs of overstimulation of the parasympathetic autonomic nervous system, skeletal muscles, and to some extent, the central nervous system (e.g., see Drummond and Russom 1990; Bradbury et al. 1991; Fairbrother 1996). Exposures to organochlorine and pyrethroid insecticides in wildlife and fish include hyperactivity, spasms and convulsions (Bradbury et al. 1991; Blus et al. 1996). In birds, experimental studies and field observations indicate that methyl mercury intoxication is associated with weakness of the extremities, tremors, ataxia, and paralysis (Heinz 1996). Exposures of fish to narcotic industrial organic compounds depress spontaneous locomotor activity and responses to outside stimuli and cause rapid and shallow respiration (Drummond and Russom 1990).

There also are numerous laboratory investigations that have reported that sublethal exposures of organic chemicals and metals influence locomotion, ventilation, avoidance, startle response, forage, and reproductive behaviors (e.g., see Little and Henry 1990; Featherstone et al. 1993; Weber and Spieler 1994; Fairbrother 1996; Blus et al. 1996). Homing migration of adult salmon has been shown to depend on the olfactory sensory system, which can be sensitive to natural and anthropogenic compounds. Examples are available where amino acids (e.g., L-serine), inorganic mercury, and copper have been shown to alter impulses from olfactory nerves at sublethal concentrations. In turn, copper exposures have been shown to affect the attractiveness of home stream waters to migrating salmonids (summarized by Heath 1987).

Conclusions

While credible examples establishing causal relationships between xenobiotic exposures, acute or sub-chronic intoxication, and behavioral effects in the field are available (e.g., see Fairbrother 1996; Blus et al. 1996), there are only a limited number of controlled field studies or extensive field observations, in association with controlled experimental studies, to evaluate the plausibility of cause-effect relationships for reproductive or developmental behavioral alterations within impacted ecosystems. However, behaviors of oviparous animals are critical and underassessed endpoints and are potentially very sensitive to the effects of xenobiotics that can affect successful reproduction.

Selenium

Both field and laboratory exposures of fishes and birds have documented food-chain bioaccumulation of selenium and subsequent reproductive failure. Combustion of coal at electric generating stations has led to elimination of entire fish communities in major reservoirs in the eastern U.S. (Lemly 1995). Several areas in the western U.S. have had accumulation of high levels of selenium in rivers and wetlands that drain naturally seleniferous soils or receive agricultural drain water.

Selenium toxicosis was first recognized during the late 1930s as causing reduced growth and hatchability in chickens (*Gallus gallus domesticus*) at > 5 mg/kg in food (Poley et al. 1937; Poley and Moxon 1938; Poley et al. 1941). Shortly thereafter, in the early 1940s, selenium was identified as the causative agent for huge losses in the livestock industry of South Dakota, Wyoming, and adjacent states due to ingestion of seleniferous plants (Oldfield 1995). During the intervening years between the 1940s and 1980s, relatively little attention was paid to selenium poisoning in wetland or terrestrial systems (Eisler 1985; Ohlendorf 1989).

In 1983, incidents of mortality, congenital deformities, and reproductive failures in waterfowl and shorebirds were discovered at Kesterson National Wildlife Refuge in the San Joaquin Valley, California. Agricultural drain water was discharged to the wetlands through a series of drains that moved irrigation water through seleniferous soils and into the terminal ponds that formed the wetlands. Analysis of water for potential contaminants showed high concentrations of selenium (> 300 µg/L). A series of field and laboratory studies conducted by the U.S. Fish and Wildlife Service (USFWS) resulted in a very detailed description of the toxicological effects of selenium in birds, including food-concentration-response relationships, egg-concentration-response relationships, and an increased understanding of the mechanisms of action of selenium.

At Kesterson, embryos or chicks with developmental abnormalities were found in 19% of the bird nests examined, while 33% of the nests had at least 1 dead chick. Overall, 39% of the 578 nests monitored had a dead or deformed chick or embryo (Ohlendorf 1989). The lesions seen were consistent with those described from laboratory studies of mallards fed diets containing 8 to 25 mg/kg selenium (Heinz et al. 1987; Hoffman and Heinz 1988; Heinz et al. 1989). These included defects of the eyes (anophthalmia and microphthalmia), feet or legs (amelia, ectromelia, and ectrodactyly), beak (incomplete development of the lower beak), brain (hydrocephaly and exancephaly), and abdomen (gastroschisis). Defects in adult birds included wasting; gelatinous or serous fluid in the abdominal cavity, air sacs, and pericardial sac; swollen, nodular, pale livers with focal necrosis; and marked loss of feathers from the head (Ohlendorf 1996).

Because of the difficulty in accurately describing diets of free-ranging birds, researchers examined the relationship between egg selenium concentrations and embryotoxicity. This was done both on an individual basis (i.e., what is the probability of embryotoxicity in an individual egg at various egg selenium concentrations) and on a population basis (i.e., what is the probability that some individuals in a population would experience reduced hatchability given a particular mean egg selenium concentration for that population). Skorupa and Ohlendorf (1991) summarized mean egg selenium concentrations from 74 bird populations located in areas considered low in selenium. From this they concluded that a mean egg selenium concentration at or below 3 mg/kg is indicative of normal background concentrations for that population. However, individual eggs from low selenium areas may have as high as 8 mg/kg selenium without showing signs of reproductive toxicity. They further showed that there is a 40% probability of teratogenic effects in a population of birds that has a mean egg selenium concentration between 13 and 24 mg/kg selenium; at a mean egg selenium concentration of 25 to 48 mg/kg, nearly all populations of birds will have some malformations evident (Skorupa and Ohlendorf 1991). This is in agreement with experimental studies in mallards (*Anas platyrhnchos*) that suggested that the teratogenesis threshold lies between 12 and 27 mg/kg mean egg selenium concentration (Hoffman and Heinz 1988; Heinz et al. 1989).

Ohlendorf et al. (1986) studied the relationship between individual egg selenium concentrations and the predicted incidence of embryotoxicity or the probability of successful hatch of a clutch of eggs. Skorupa and Ohlendorf (1991) looked at the same relationship based on the population mean egg selenium concentration. Skorupa et al. (1996) summarized this information on embryotoxicity. The data show that mallard embryos are more sensitive to the effects of selenium than are black-necked stilts (*Himantopus mexicanus*) and that American avocets (*Recurvirostra americana*) are less sensitive still. They speculate that this is because of the evolution of avocets in more saline environments that also contain higher levels of sulfur. As selenium may readily substitute for sulfur in membranes and enzymes, whatever mechanism avocets have for depurating excess sulfur also may act on selenium. This assumes that these mechanisms are operative in the embryo (which experiences the high selenium concentrations in the egg) and ignores the fact that many stilt and duck populations occur in saline environments as well. Nevertheless, there is an obvious species-related difference in embryo sensitivity to egg selenium. Heinz (1996) and Skorupa et al. (1996) suggest that 15 mg/kg egg selenium is a threshold for embryotoxicity (approximately an EC_3 for the mallard).

Mosquitofish (*Gambusia affinis*) also were affected by high selenium at Kesterson. Selenium body burdens averaged 170 mg/kg. Studies performed using resident individuals inhabiting waters of the San Luis Drain, the source of Kesterson's water, showed that these fish produced an average of 12 fry per brood compared to the geographically relevant reference value of 25 fry per brood (Zahm 1986; Saiki and Ogle 1995).

Mechanism of action

Selenium is a micronutrient required by animals and plants for the basic functions of life. Selenium is biochemically active at dietary levels of less than 1 mg/kg food, and has a range of nutritional activity of about 100-fold. For example, dietary deficiency can occur at levels of 0.04 mg/kg, while toxicity can occur above 4 mg/kg. About 30 to 40% of the selenium in the body is contained in glutathione peroxidase, an enzyme important in preventing oxidation of cell membranes and subsequent structural damage and disruption of cellular integrity (Combs and Combs 1986). It is present as selenocysteine both within the peptide backbone of the enzyme and at its catalytic site. Other proteins also contain selenocysteine, although not in their peptide backbones nor as an active component of the protein. In addition, it is thought that selenium may substitute for sulfur or bind to sulfhydryl groups in other enzyme and membrane systems. Inactivation of sulfhydryl-containing proteins leads to a cessation of mitosis in a variety of cells, resulting in decreased cell division. Reduction in intracellular glutathione peroxidase results in an inhibition of DNA synthesis. In the presence of excess selenium, rapid division of embryonic cells would be arrested prior to completion of their developmental processes, resulting in the observed embryonic and neonatal deformities in fish and birds.

The MOAs of selenium are well conserved across all species and are not specific to oviparous vertebrates. Differences in observed effects between oviparous species such as birds and fish and mammalian species likely are due to exposure differences. In birds and fish, it is likely that incorporation within yolk of eggs forms the major route of exposure. Chicks and larvae may incorporate the metalloid from their diet (aquatic invertebrates to birds or larval fish), but the presence of necrotic embryos within exposed maternal mosquitofish and dead and malformed avian embryos suggests maternal transfer as the most likely route of exposure. Although embryonic and larval stages are particularly susceptible, diminished growth and reduced body weight are also seen in young birds and fish (Fairbrother et al. 1994). Cold winter temperatures also may increase the susceptibility of birds and fish to selenium poisoning (Sorenson 1991; Heinz and Fitzgerald 1993; Lemly 1993).

Epidemiology

During the late 1980s and early 1990s, 25 wildlife refuges in the western U.S. that receive agricultural drain water were surveyed to determine if selenium or other compounds were causing adverse effects to wildlife, similar to what had occurred at the Kesterson Wildlife Refuge. Most of these surveys measured concentrations of selenium in water, sediment, plants, invertebrates, and fish. Selenium concentrations in bird eggs also were measured at 15 of the sites. Detailed field observations, including nest monitoring (Lambing et al. 1988; Rinella and Schuler 1992) and estimation of numbers of breeding pairs, numbers of nests, nest success, embryo viability/deformities, and juvenile survival and growth measures (Nimick et al. 1996) were conducted at 3 of the sites. Concentration of selenium in food items (invertebrates and fish) exceeded dietary threshold values for reproductive impairment of mallards (e.g., reduced hatchability and increased incidence of teratogenesis) (Heinz et al. 1987, 1989) at several of the refuges. Selenium concentrations in some eggs from 12 of the 15 sites where selenium concentrations were measured exceeded the values known to cause embryo teratogenicity (Skorupa 1996). However, selenium concentrations in food items, environmental media, or eggs from any of the refuges surveyed were much lower than those measured at Kesterson, and no embryo mortality or teratogenicity was documented at any of the sites.

Conclusions

The laboratory data suggest that the avian embryo probably is very sensitive to the effects of selenium, more so than mammals. However, field epidemiological studies have demonstrated that it may be very difficult to detect ecologically relevant changes in avian reproduction when egg selenium concentrations are at, or only slightly above, threshold levels.

Summary and Conclusions

This chapter highlighted a number of issues relative to developmental and reproductive toxicology of contaminants in oviparous animals. Key differences between oviparous and viviparous reproductive strategies were identified in the "Comparison of Oviparous and Viviparous Reproduction and Development" section in the context of potential chemical sensitivity. Unique processes that could be affected by xenobiotics in oviparous animals include: fertilization (for those species utilizing external fertilization), nutrient production/utilization, egg shell production, hatching, and metamorphosis. It should be noted that several of these processes could be sensitive to xenobiotics because of true toxicological differences between oviparous and viviparous animals or because of exposure/toxicodynamic differences. For example, vitellogenesis and subsequent deposition of egg proteins could be significantly affected by xenobiotics that activate maternal ERs, thus causing developmental defects as a result of nutrient imbalances. However, there also are important toxicokinetic issues associated with this process, as vitellogenin production and deposition affect maternal transfer of contaminants to the potentially sensitive embryos. But, this latter issue probably should be viewed as more an exposure concern than a true difference in sensitivity of oviparous versus viviparous embryos (see Chapter 2).

The "Mechanisms of Toxicity" section provided a brief discussion of toxic MOAs, primarily at the molecular and subcellular levels. At these levels, there is a high degree of across-species similarity in toxic mechanisms. For example, a significant degree of similarity exists in the organization of the neuroendocrine system and hypothalamic-pituitary axis, synthesis of estrogens and androgens by the gonads, and the actions of hormones at target sites in terms of receptor binding, subsequent interaction with DNA, production of mRNA and, ultimately, protein expression. Similarly, the toxicity of planar organochlorines such as TCDD and structurally-similar PCDDs, PCDFs, and PCBs appears to be expressed via the AhR in virtually all vertebrates, with common across-species manifestations of toxicity (e.g., edema, delayed mortality). In general, the most divergence among vertebrate classes occurs downstream from basic molecular interactions at the cellular and organismal levels. For example, the role of estrogens/androgens on sexual differentiation in mammals versus birds is reversed. These types of among-species variations in responses, although of critical toxicological importance, are not necessarily inherent to any particular type of reproductive strategy (i.e., oviparity versus viviparity). A factor that often differs significantly across animal species is the dose needed to elicit toxicity via any given MOA. However, whether this is related to oviparous versus viviparous reproductive modes is unclear. For example salmonids, in particular lake trout, seem to be far more sensitive to TCDD than other vertebrate models, including mammals. But the relationship of this sensitivity to oviparity is uncertain. This type of observation highlights the need for a comprehensive understanding of among-species differences in biologically based, dose-response relationships.

Subcellular changes caused by toxicants become important in the context of ecological risk assessments only if they result in changes at the whole organism level. Of particular interest are those changes that directly or indirectly affect the reproductive process, growth, and development (see Chapter 4 for a discussion of potential ecological effects and Chapter 6 for the risk assessment process). The section entitled "Techniques and Endpoints for Assessing Toxicity" described methods for measuring changes at various biological levels of organization and methods to relate such changes to specific MOAs for particular chemicals. Emphasis was placed on methods that would be unique to oviparous species, such as measurements of egg formation or embryo hatchability. Indirect effects of chemical changes on reproduction also were considered, such as the role of sex steroids in development and expression of appropriate reproductive behaviors. This section pointed out that many of the changes in whole organisms in response to chemical exposure cannot yet be predicted with accuracy solely from studies at the cellular or molecular level; in addition, cross-species extrapolations cannot be conducted with certainty. Therefore, there remains a critical need for conducting whole-animal toxicity studies in the context of both prospective and diagnostic assessments.

The final section of the chapter presented a series of case studies concerning developmental and reproductive effects of contaminants in oviparous animals, illustrating practical application of many of the conceptual points covered in the previous sections. The case studies are summarized in Table 5-1. The first case study described a comprehensive diagnostic analysis in which early-life-stage mortality of lake trout was linked to AhR agonists, specifically PCBs, PCDFs, and PCDDs, including TCDD. Field observations of decreased lake trout populations, as well as observations of specific developmental abnormalities, were used as a basis

Table 5-1 Case studies of toxicological effects on reproduction and development of oviparous species

Case study	Observed field effect	Hypothesized MOA	Suspect chemicals	Subcellular	Cellular	Tissue/organ whole organism
Early life stage mortality in fish (Great Lakes lake trout)	Sac fry mortality, blue sac syndrome, edema, craniofacial malformations	AhR agonists	PCBs, PCDFs, PCDDs	AhR binding	Cardiovascular dysfunction	Developmental cardiovascular toxicity, pathology (consistent with field observations)
Great Lakes bird mortality/Deformities	Mortality, deformities (embryos), behavior (adults)	AhR agonists	PCBs, PCDF, PCDD	AhR binding	MO induction	Mortality, similar deformities in field and surrogate species (consistent with field effects), abnormal behavior
Eggshell thinning in birds	Eggshell thinning/reduced reproduction (Adults/ embryos)	Affects calcium deposition (via various mechanisms)	DDE, DDT, other organochlorines	Progesterone binding	Decreased prostaglandin synthesis, inhibition of Ca Mg ATPase	Egg shell thinning in lab models (consistent with field effects)
Alligator reproductive anomalies (Lake Apopka, FL)	Pathological effects, reproductive steroids (Embryos)	ER/AR agonists, antagonists,	Unknown (Organochlorine pesticides?)	ER, AR binding	Steroids	Gross pathological alterations
Pulp mill effluents (Canada)	Decreased gonad size/delayed maturity, steroid hormone changes (Adults)	ER, AR, AhR agonists, oxidative stress	Unknown (Phytosterols?)	ER, AhR binding,	Steroid biosynthesis, MO induction	Alterations in steroid biosynthesis, ovarian apoptosis, delayed sexual maturity, mortality
Hermaphroditic fish	Hermaphroditic fish (Adults)	ER agonists	Synthetic/natural estrogens, alkylphenols	ER binding	Vitellogenin induction	Testicular atrophy

Table 5-1 continued

Case study	Observed field effect	Hypothesized MOA	Suspect chemicals	Subcellular	Cellular	Tissue/organ whole organism
West coast fish	Pathological effects on reproductive organs/function (adults)	DNA damage, oxidative stress, ER agonists	PAHs	MO induction, glutathione reduction	Steroid biosynthesis	Pathology, MO induction, reduced growth, decreased gonad size/sexual maturity, poor gamete quality
Deformed amphibians	Deformed (limb, eggs) frogs (Embryos)	RA/RXR agonists	Unknown (Pesticides?)	Unknown	Gene transcription (Methoprene)	Limb malformations (with model retinoids)
Pesticide reproductive effects	Mortality (nestlings); death	Inhibition of AChE, kynurenine formamidase	OP insecticides	Enzyme inhibition	Muscle paralysis, suppression of cell division	Deformities, functional deficits/behavior
Selenium (Kesterson, CA and Belows Lake, NC)	Mortality, deformities, reproductive failure (Embryos)	Glutathione peroxidase modification	Selenium	Inhibition of DNA synthesis, reduced glutathione peroxidase activity	Suppression of mitosis	Terata, reduced hatachability; decreased fertility
Oil spills	Early embryo death	Effects on corticosteroids	Oil/PAHs	Increased cytochrome P450 activity	Reduced ovarian steroidogenesis	Reduced reproduction, embryo mortality, decreased circulating steroid hormones

for more mechanistic laboratory experimentation to focus on suspect chemicals. The experimentation was conducted at subcellular and whole-organism levels and indicated a high concordance between field observations and laboratory studies. Then, a complete exposure analysis, conducted with field-collected samples, was used to confirm the presence of AhR agonists at concentrations consistent with the original pathology observed in embryos from natural populations. This study clearly demonstrates how an understanding of MOA, linked to field and laboratory observations, can form the basis of a definitive diagnostic assessment.

The second case study also emanated from the Great Lakes, only in this example the concern was for mortality and deformities in fish-eating birds, such as terns and cormorants. Again, initial observations consisted of a relatively specific suite of symptoms in animals from the field (e.g., embryonic mortality, wasting, craniofacial abnormalities). Based on laboratory studies with other avian species (e.g., chickens) exposed to TCDD, it was hypothesized that a plausible MOA for toxicity in the Great Lakes birds was mediated through the AhR. In this case study, a dose-response analysis was somewhat more difficult than for the lake trout in that direct toxicology studies with the species of concern were difficult to conduct; consequently, a weight-of-evidence case for the effects, primarily of PCBs, on the birds needed to be constructed. This was achieved through collecting data at the subcellular, cellular, tissue, and whole-organism levels in the laboratory and comparing laboratory responses to field observations. As was true for the lake trout studies, described in the previous case study, the primary linchpin for this weight-of-evidence diagnostic analysis was an understanding of the toxic MOA.

The third case study described eggshell thinning, an endpoint of unique concern to oviparous, in particular avian, species. Despite a significant amount of evidence for the environmental contaminant responsible for this response (i.e., p,p DDE), and many years of research both at the subcellular and organismal levels, the exact molecular MOA involved remains uncertain. However, enough is understood about the process that predictions can be made as to those avian species most likely to experience eggshell thinning. Further, this presents an example where better definition of the MOA involved would greatly aid in prospective assessments of the potential risk of new chemicals that could elicit this response.

The next 2 case studies, Lake Apopka alligators and pulp mill effluents, both described situations where marked effects have occurred in field populations. However, there is uncertainty as to chemicals causing the effects, in large part because the mechanisms underlying the responses are unclear. The Lake Apopka case study described adverse effects in reptiles, a class of animals rarely considered in ecological risk assessments. Alligators in Lake Apopka, Florida have experienced population-level declines with concomitant changes in gonadal morphology and circulating sex steroids. However, the mechanism underlying these effects is not known; for this reason, chemicals responsible have been difficult to pinpoint. The pulp mill effluent case study described a suite of physiological responses (e.g., MO induction, abnormal gonadal histology, alterations in circulating sex steroids, etc.) in fish exposed to pulp and paper mill effluents; again, however, because the exact mechanisms underlying these responses are uncertain, identification of the chemicals responsible has not been achieved. Both these examples demonstrate the innovative use of subcellular responses to assess organism-level effects, but also demonstrate the critical importance of understanding MOA in the context of diagnostic assessments.

The case study on vitellogenin induction in fish described a situation in which the observed response in caged fish, i.e., vitellogenin induction, could be very closely related to a known mechanism: stimulation of the ER. Using this knowledge, it was relatively straightforward (albeit not simple) to design laboratory studies of subcellular responses and to identify the synthetic estrogens responsible for these effects. There remains uncertainty, however, as to the exact relationship between the original observations of concern at the organism level (i.e., hermaphroditic fish) and the vitellogenin biomarker. Thus, this case study highlights the necessity of understanding relationships between endpoints at different biological levels of organization.

The case study on effects of polycyclic aromatic hydrocabons on marine fish described the effects of PAHs on reproductive success of fish, primarily English sole. As in several previous case studies, initial observations of adverse effects emanated from field populations. Based on these observations, an exhaustive amount of integrative laboratory and field experimentation sought to identify contaminants of concern and to define mechanisms through which effects occurred. These studies assessed linkages between endpoints at the subcellular, cellular, tissue, and whole-organism levels to derive a weight-of-evidence case for PAHs as causative agents of reproductive impacts in the fish. The study illustrated a particularly difficult aspect of environmental diagnostics, as multiple MOAs likely were involved in producing toxicity.

Finally, several additional case studies were presented to illustrate the application of toxicological principles to the diagnosis and prediction of reproductive or developmental effects of contaminants in a wide variety of oviparous species. The case study on selenium illustrated how a naturally occurring substance can be concentrated through human-induced environmental alterations to cause developmental effects in species as diverse as fish and birds, resulting from effects on highly conserved processes such as enzyme-dependent aspects of cell division. The case study of petroleum toxicity is particularly interesting as an example of a toxicological mechanism restricted to oviparity; direct toxicity to the developing embryo occurred through uptake of a chemical across the eggshell, thus completely bypassing any parental toxicity. The two case studies organophosphorus ester insecticides and "alterations in reproductive and developmental behavior, describing the potential for adverse effects to occur in oviparous species based on laboratory studies, were included as an illustration of how toxicological principles can be used in a prospective assessment of potential risks of new environmental chemicals. And finally, an ongoing, incomplete diagnosis of the causes of amphibian deformities was presented as a case study to illustrate that these methods are still being used to help understand causes of environmental abnormalities, whether as a result of natural or human-induced stressors, chemical-based or not.

Taken as a whole, the information presented in this chapter should provide the reader with a general understanding of how chemicals potentially can alter the reproductive and developmental processes of animals. In particular, the reader should have an understanding of why egg-laying vertebrates may be expected to differ significantly from mammals in some, but certainly not all, aspects of the reproductive cycle and to some, but not all, classes of environmental contaminants. Given that our knowledge of the MOA of all chemicals in all species is far from complete, there remains a high degree of uncertainty in extrapolating information among species. However, as illustrated by the case studies, an understanding of a chemical MOA is essential to the diagnostic process and can significantly reduce uncertainty in prospective assessments. Sufficient knowledge currently exists to solve many environmental problems, and methods are available for the development of information to apply to other current and future environmental assessments. We encourage the reader to keep the information on comparative toxicology and reproductive physiology presented in this and the preceding chapter in mind while perusing the following chapters on the application of such information in an environmental context.

References

Abou-Donia MB. 1981. Organophosphorus ester-induced delayed neurotoxicity. *Annual Rev Pharmacol* 21:511–548.

Adams SM, Crumby WD, Greely Jr MS, Shugart LR, Saylor CF. 1992. Responses of fish populations and communities to pulp mill effluents: a holistic approach. *Ecotoxicol Environ Saf* 24:347–360.

Adkins-Regen EK, Mansukhani V, Stewart C, Thompson R. 1994. Sexual differentiation of brain and behavior in the zebra fish: Critical periods for effects of early estrogen treatment. *J Neurobiol* 25:865–877.

Adkins-Regen EK, Pniewski EE. 1978. Control of reproductive behavior by sex steroids in male quail. *J Comp Biochem Physiol* 89:61–71.

Albers PH. 1983. Effects of oil on avian reproduction: a review and discussion. In: The effects of oil on birds: physiological research, clinical applications and rehabilitation. Wilmington DE: Tri-State Bird and Research, Inc. p 78–97.

Aldridge WN, Barnes J. 1966. Esterases and neurotoxicity of some organophosphorus compounds. *Biochem Pharmacol* 15:549–554.

Alfonso LOB, Campbell PM, Iwama GK, Devlin RH, Donaldson EM. 1997. The effect of the aromatase inhibitor Fadrozone and two polynuclear aromatic hydrocarbons on sex steroid secretion by ovarian follicles of coho salmon. *Gen Comp Endocrinol* 106:169–174.

Alvarez R, Honrubia MP, Herraez MP. 1995. Skeletal malformations induced by the insecticides ZZ-Aphox and Folidol during larval development of *Rana perezi*. *Arch Environ Contam Toxicol* 28:349–356.

Anderson MJ, Miller MR, Hinton DE. 1996. In vitro modulation of 17β-estradiol-induced vitellogenin synthesis: Effects of cytochrome P4501A1 inducing compounds on rainbow trout *Oncorhynchus mykiss* liver cells. *Aquat Toxicol* 34:327–350.

Anderson MJ, Olsen H, Matsumura F, Hinton DE. 1996. In vivo modulation of 17β-estradiol-induced vitellogenin synthesis and estrogen receptor in rainbow trout (*Oncorhynchus mykiss*) liver cells by β naphthoflavone. *Toxicol Appl Pharmacol* 137:210–218.

Andersson T, Forlin L, Olsen S, Fostier A, Breton B. 1993. Pituitary as a target organ for the toxic effects of P4501A1 inducing chemcials. *Mol Cell Endocrinol* 91:99–105.

Andrews LS, Synder R. 1986. Toxic effects of solvents and vapors. In: Klaassen CD, Amdur MO, Doull J, editors. Casarett and Doull's toxicology: the basic science of poisions. 3rd edition. New York: MacMillian.

Ankley GT, Cook PM, Carlson AR, Call DJ, Swenson JA, Corcoran HF, Hoke RA. 1992. Bioaccumulation of PCBs from sediments by oligochaetes and fishes: comparison of laboratory and field studies. *Can J Fish Aquat Sci* 49:2080–2085.

Ankley GT, Giesy JP. 1998. Endocirne disruptors in wildlife: a weight-of-evidence perspecitve. In: Kendall R, Dickerson R, Suk W, Giesy J, editors. Principles and processes for evaluating endocrine disruption in wildlife. Pensacola FL: SETAC. p 349–367.

Ankley GT, Johnson RD, Detenbeck NE, Bradbury SP, Toth G, Folmar LC. 1997. Development of a research strategy for assessing the ecological risk of endocrine disrupters. *Rev Toxicol* 1:71–106.

Ankley G, Mihaich E, Stahl R, Tillitt D, Colborn T, McMaster S, Miller R, Bantle J, Campbell P, Denslow N, Dickerson R, Folmar L, Fry M, Giesy J, Gray LE, Guiney P, Hutchinson T, Kennedy S, Kramer V, LeBlanc G, Mays M, Nimrod A, Patino R, Peterson R, Purdy R, Ringer R, Thomas P, Touart L, Van Der Kraak G, Zacharewski T. 1998. Overview of a workshop on screening methods for detecting potential [anti-] estrogenic/androgenic chemicals in wildlife. *Environ Toxicol Chem* 17:68–87.

Ankley GT, Tietge JE, DeFoe DL, Jensen KM, Holcombe GW, Durhan EJ, Diamond SA. 1998. Effects of methoprene and ultraviolet light on survival and development of *Rana pipiens*. *Environ Toxicol Chem* 17:2530–2542.

Anulacion BF, Lomax DP, Bill DD, Johnson LL, Collier TK. 1997. Assessment of antiestrogenic activity in CYP1A induction in English sole exposed to environmental contaminants. Society of Environmental Toxicology and Chemistry (SETAC) 18th Annual Meeting; 16–20 Nov 1997; San Francisco CA. Pensacola FL: SETAC. p 137.

[ASTM] American Society of Testing and Materials. 1991. Standard guide for conducting the frog embryo teratogenesis assay-*Xenopus* (FETAX). Annual Book of ASTM Standards. Philadelphia PA: American Society of Testing and Materials. E 1429-91.

Bastomsky CH. 1977. Enhanced thyroxine metabolism and high uptake goiters in rats after single dose of 2,3,7,8-tetrachlorodibenzo-dioxin. *Endocrinology* 101:292–296.

Bennett JK, Bennett RS. 1990. Effects of dietary methyl parathion on northern bobwhite egg production and eggshell quality. *Environ Toxicol Chem* 9:1481–1485.

Bennett RS, Ganio LM. 1991. Overview of methods for evaluating effects of pesticides on reproduction in birds. Corvallis OR: U.S. Environmental Protection Agency. EPA 600/3-91-048.

Bennett RS, Williams BA, Schmedding DW, Bennett JK. 1991. Effects of dietary exposure to methyl parathion on egg laying and incubation in mallards. *Environ Toxicol Chem* 10:501–507.

Bergeron JM, Crews D, McLachlan JA. 1994. PCBs as environmental estrogens: turtle sex determination as a biomarker of environmental contamination. *Environ Health Perspect* 102:780–781.

Berry M, Metzger D, Chambon P. 1990. Role of the two activating domains of the oestrogen receptor in the cell-type and promoter-context dependent agonistic activity of the anti-oestrogen 4-hydroxytamoxifen. *EMBO Journal* 9:2811–2818.

Bevans HE, Goodbred SL, Miesner JF, Watkins SA, Gross TS, Denslow ND, Schoeb T. 1996. Synthetic organic compounds and carp endocrinology and histology in Las Vegas wash and Las Vegas and Callville Bays of Lake Mead, Nevada, 1992 and 1995. Washington DC: U. S. Department of Interior-U.S. Geological Survey. Water-Resources Investigations Report 96-4266.

Birnbaum LS. 1994. The mechanism of dioxin toxicity: relationship to risk assessment. *Environ Health Perspect* 102:157–167.

Blus LJ, Wiemeyer SN, Henny CJ. 1996. Organochlorine pesticides. In: Fairbrother A, Locke LN, Hoff GL, editors. Noninfectious diseases of wildlife. Ames IA: Iowa State University Press. p 61–70.

Borg B. 1994. Androgens in teleost fishes. *Comp Biochem Physiol* 109C:219–245.

Boris A, Stevenson RH. 1966. Further studies on a nonsteroidal anti-androgen. *Endocrinol* 78:549–555.

Bortone SA, Davis W, Bundrick CM. 1989. Morphological and behavioral characters in mosquitofish as potential bioindication of exposure to kraft mill effluent. *Bull Environ Contam Toxicol* 43:370–377.

Bradbury SP, Carlson RW, Henry TR. 1989. Polar narcosis in aquatic organisms. In: Cowgill UM, Williams LR, editors. Aquatic toxicology and hazard assessment: Twelfth Symposium. American Society for Testing and Materials Philadelphia PA. ASTM STP 1027. p 59–73.

Bradbury SP, Carlson RW, Niemi GJ, Henry TR. 1991. Use of respiratory-cardiovascular responses of rainbow trout (*Oncorhynchus mykiss*) in identifying acute toxicity syndromes in fish: Part 4. Central nervous system seizure agents. *Environ Toxicol Chem* 10:115–131.

Bradbury SP, Coats JR. 1988. Comparative toxicology of the pyrethroid insecticides. *Rev Environ Contam Toxicol* 108:133–177.

Bradbury SP. 1995. Quantitative structure activity relationships and ecological risk assessment: An overview of predictive aquatic toxicology research. *Toxicol Lett* 79:229–237.

Brouwer A, Van Den Burg KJ. 1986. Binding of a metabolite of 3,4,3',4' tetrachlorobiphenyl to transthyretin reduces serum vitamin A transport by inhibiting the formation of the protein complex carrying both retinol and thyroxine. *Toxicol Appl Pharmacol* 85:301–312.

Bryan TE, Gildersleeve RP, Wiard RP. 1989. Exposure of Japanese quail embryos to *o,p'* DDT has long-term effects on reproductive behaviors, hematology and feather morphology. *Teratol* 39:525–536.

Burkhard LP, Ankley GT. 1989. Identifying toxicants: NETAC's toxicity based approach. *Environ Sci Technol* 23:1438–1443.

Cantrell S, Joy-Schlezinger J, Stegeman JJ, Tillitt DE, Hannink M. 1998. Quantitation of programmed cell death in TCDD-treated medaka (*Oryzias latipes*). *Toxicol Appl Pharmacol* 148:24-34.

Cantrell S, Lutz LH, Tillitt DE, Hannink M. 1996. Embryotoxicity of TCDD: the embryonic vasculature is a physiologic target for TCDD-induced DNA damage and apoptotic cell death in medaka (*Oryzias latipes*). *Toxicol Appl Pharmacol* 141:23–34.

Carls MG, Rice SD. 1990. Abnormal development and growth reductions of pollock *Theragra chalcogramma* embryos exposed to water-soluble fractions of oil. *Fish Bull* 88:29–37.

Carlson RW, Duby RT. 1973. Embryotoxic effects of three PCBs in the chicken. *Bull Environ Contam Toxicol* 9:261–266.

Casillas E, Misitano DA, Johnson LL, Rhodes LD, Collier TK, Stein JE, McCain BB, Varanasi U. 1991. Inducibility of spawning and reproductive success of female English sole (*Parophrys vetulus*) from urban and nonurban areas of Puget Sound, Washington. *Mar Environ Res* 31:99–122.

Cavanaugh KP. 1982. The effects of South Louisiana and Kuwait crude oils on reproduction. In: Scanes CG, Ottinger MA, Kenney AD, Balthazart J, Cronshaw J, Jones IC, editors. Aspects of environmental endocrinology: practical and theoretical implications. Lubbock TX: Texas Tech University. Grad. Studies Paper No. 26. p. 371–377.

Cavanaugh KP, Godlsmith AR, Holmes WN, Follett BK. 1983. Effects of ingested petroleum on the plasma protein levels during incubation and on the breeding success of paired mallard ducks. *Arch Environ Contam Toxicol* 12:335–341.

Cavanaugh KP, Holmes WN. 1987. Effects of ingested petroleum on the development of ovarian endocrine function in photostimulated mallard ducks (*Anas platyrhynchos*). *Archives Environ Contam Toxicol* 16:247–253.

Chang JP, Jobin RM. 1994. Teleost pituitary cells: isolation, culture and use. In: Hochachka PW, Mommsen TP, editors. Biochemistry and molecular biology of fishes: analytical techniques. New York: Elsevier. p 205–213.

Chen TT. 1988. Investigation of effects of environmental xenobiotics to fish at sublethal levels by molecular biological approaches. In: Stegeman JJ, Moore MN, editors. Responses of marine organisms to pollutants. *Mar Environ Res* 24:333–337.

Coats JR. 1990. Mechanisms of toxic action and structure-activity relationships for organochlorine and synthetic pyrethroid insecticides. *Environ Health Perspect* 87:255–262.

Collier TK, Johnson LL, Myers MS, Stehr CM, Krahn MM, Stein JE. 1997. Fish injury in the Hylebos Waterway of Commencement Bay, Washington. Interpretive Report prepared for NOAA Damage Assessment Center.

Collier TK, Singh SV, Awasthi YC, Varanasi U. 1992. Hepatic xenobiotic metabolizing enzymes in two species of benthic fish showing different prevalences of contaminant-associated liver lesions. *Toxicol Appl Pharmacol* 113:319–324.

Combs Jr GF, Combs SB. 1986. The role of selenium in nutrition. New York: Academic. 532 p.

Conney AH, Welch RM, Kuntzman P, Burns JJ. 1967. Effects of pesticides on drug and steroid metabolism. *Clinical Pharmcacol Therapy* 8:2–10.

Cook PM, Endicott DD, Robbins J, Marquis P, Berini C, Libal J, Kizlauskis A, Walker MK, Zabel EW, Guiney PD, Peterson RE. 1999. Effects of chemicals with an Ah receptor-mediated mode of early-life-stage toxicity on lake trout reproduction and survival in Lake Ontario: retrospective and prospective risk assessments. *Environ Sci Technol* (In preparation).

Cook PM, Zabel EW, Peterson RE. 1997. The TCDD toxicity equivalence approach for characterizing risks for early life-stage mortality in trout. In: Rolland RM, Gilbertson M, Peterson RE, editors. Chemically induced alterations in functional development and reproduction of fishes. Pensacola FL: SETAC. p 9–27.

Cooke AS. 1973. Shell thinning in avian eggs by environmental pollutants. *Environ Pollut* 4:85–152.

Crain DA, Guillette Jr LJ, Pickford DB, Percival HF, Woodward AR. 1998. Sex steroid and thyroid hormone concentrations in juvenile alligators (*Alligator mississippiensis*) from contaminated and reference lakes in Florida, USA. *Environ Toxicol Chem* 17:446–452.

Cross JN, Hose JE. 1988. Evidence for impaired reproduction in white croaker (*Genyonemus lineatus*) from contaminated areas off southern California. *Mar Env Res* 24:185–188.

Danzo BJ. 1997. Environmental xenobiotics may disrupt normal endocrine function by interfering with the binding of physiological ligands to steriod receptors and binding proteins. *Environ Health Perspect* 105:294–301.

Desbrow C, Waldock M, Sheahan D, Blackburn M, Routledge E, Sumpter J, Brightly G. 1996. The identification of compounds causing endocrine disruption in fish in UK rivers. Society of Environmental Toxicology and Chemistry (SETAC) 17th Annual Meeting; 17–21 November 1996; Washington DC. p 36.

Di Giulio RT, Habig C, Gallagher EP. 1993. Effects of Black Rock Harbor sediments on indices of biotransformation, oxidative stress, and DNA integrity in channel catfish. *Aquat Toxicol* 26:1–22.

Dolwick KM, Swanson HI, Bradfield CA. 1993. In vitro analysis of Ah receptor domains involved in ligand-activated DNA recognition. *Biochemistry* 90:8566–8570.

Dorfman RI, Rooks WH, Jones HJB, Leman JD. 1966. Androgenic activity of highly purified 5-androstane and 5-androstane-17-ol. *J Med Chem* 9:930–931.

Drummond RA, Russom CL. 1990. Behavioral toxicity syndromes: a promising tool for assessing toxicity mechanisms in juvenile fathead minnows. *Environ Toxicol Chem* 9:37–46.

[EC] European Community. 1997. European workshop on the impact of endocrine disrupters on human health and wildlife. Report of proceedings for the workshop held on 3–4 December 1996, Weybridge, UK. Report reference EUR 17549. 125 p.

Eddy S, Misitano D, Casillas E. 1993. Relative sensitivity of two species of larval flatfish to a model mixture of PCBs and fluoranthene. Proceedings of 1993 Pacific Northwest Regional SETAC meeting, Newport, OR.

Eisler R. 1985. Selenium hazards of fish, wildlife, and invertebrates: a synoptic review. Washington DC: U.S. Fish and Wildlife Service. Contaminant Hazard Reviews No. 5, Biological Report 85 (1.5).

Elonen GE, Spehar RL, Halcombe GW, Johnson RD, Fernandez JD, Tietge JE Cook PM. 1998. Comparative toxicity of 2,3,7,8-tetrachlorodibenzo-*p*-dioxin to seven freshwater species during early life stage development. *Environ Toxicol Chem* 17:472–482.

Fair JM, Kennedy PL, McEwen LC. 1995. Effects of carbaryl grasshopper control on nesting kildeer in North Dakota. *Environ Toxicol Chem* 14:881–890.

Fairbrother A. 1996. Cholinesterase-inhibiting pesticides. In: Fairbrother A, Locke LN, Hoff GL, editors. Noninfectious diseases of wildlife. Ames IA: Iowa State University Press. p 52–60.

Fairbrother A, Fix M, O'Hara T, Ribic CA. 1994. Impairment of growth on immune function of avocet chicks from sites with elevated selenium, arsenic, and boron. *J Wildl Dis* 30:222–233.

Fairbrother A, Kapustka LA, Williams BA, Bennett RS. 1997. Effects-initated assessments are not risk assessments. *Human and Ecological Risk Assessment* 3:119–124.

Farage-Elawar M. 1989. Enzyme and behavioral changes in young chicks as a result of carbaryl treatment. *J Toxicol Environ Health* 26:119–131.

Farage-Elawar M, Francis BM. 1987. Acute and delayed effects of fenthion in young chicks. *J Toxicol Environ Health* 21:455–469.

Farage-Elawar M, Francis BM. 1988a. Effects of fenthion, fenitrothion, and desbromoleptophos on gait, acetylcholinesterase, and neurotoxic esterase in young chicks after in ovo exposure. *Toxicology* 49:253–261.

Farage-Elawar M, Francis BM. 1988b. Effects of three organophosphorus esters on the central and peripheral neurotoxic esterase and acetylcholinesterase. *Pestic Biochem Physiol* 31:175–181.

Farage-Elawar M, Francis BM. 1990. Developmental toxicity of organophosphorus ester insecticides. *Comm Toxicol* 4:15–38.

Featherstone D, Drewes CD, Coats JR, Bradbury SP. 1993. A non-invasive neurotoxicity assay using larval medaka. In: Gorsuch JW, Dwyer FJ, Ingersoll CG, LaPoint TW, editors. Philadelphia PA: Environmental toxicology and risk assessment 2nd Volume. American Society for Testing and Materials. ASTM STP 1216. p 275–288.

Fernandez-Salguerro PM, Hilbert DM, Rudikoff S, Ward JM, Gonzalez FJ. 1996. Aryl-hydrocarbon receptor-deficient mice are resistant to

2,3,7,8-tetrachlorodibenzo-*p*-dioxin-induced toxicity. *Toxicol Appl Pharmacol* 140:173–179.

Folmar LC, Denslow ND, Rao V, Chow M, Crain PA, Enblom J, Marcino J, Guillette Jr LJ. 1996. Vitellogenin induction and reduced serum testosterone concentrations in feral male carp *(Cyprinus carpio)* captured near a major metropolitan sewage treatment plant. *Envrion Health Perspect* 104:1090–1101.

Forlin L, Haux C. 1985. Increased excretion in thebile of 17β-H³-estradiol-derived radioactivity in rainbow trout treated with β-napthoflavone. *Aquat Toxicol* 6:197–208.

Fostier A, Jalabert B, Billard R, Breton B, Zohar Y. 1983. The gonadal steroids. In: Hoar WS, Randall DJ, Donaldson EM, editors. Fish physiology, Volume 9A. New York: Academic Press. p 277–372.

Fox GA, Gilman APG, Peakall DB, Anderka FW. 1978. Behavioral abnormalities of nesting Lake Ontario herring gulls. *J Widl Manage* 42:477–483.

Franks NP, Lieb WR. 1990. Mechanisms of general anesthesia. *Environ Health Perspect* 87:199–205.

Fry DM. 1995. Reproductive effects in birds exposed to pesticides and industrial chemicals. *Environ Health Perspect* 103 (Suppl. 7):165–171.

Fukuto TR. 1990. Mechanisms of action of organophosphorus and carbamate insecticides. *Environ Health Perspect* 87:245–254.

Gagnon MM, Dodson JJ, Hodson PV, Van Der Kraak G, Carey JH. 1994. Seasonal effects of bleached kraft mill effluent on reproductive parameters of white sucker (*Catostomus commersoni*) populations of the St. Maurice River, Quebec, Canada. *Can J Fish Aquat Sci* 51: 337–347.

Gallo MA, Lawryk NJ. 1991. Organic phosphorus pesticides. In: Hays Jr WJ, Laws ER, editors. Handbook of pesticide toxicology. San Diego: Academic Press. p 917–1123.

Gawienowski AM, Stadnicki SSS, Stacewicz-Sapuntzakis M. 1977. Androgenic properties of gibberellic acid in the chick comb bioassay. *Experimentia* 33:1544–1545.

Ghosh S, Thomas P. 1995. Antagonistic effects of xenobiotics on steroid-induced final maturation of Atlantic croaker oocytes in vitro. *Mar Environ Res* 39:159–163.

Gilbertson M, Kubiak T, Ludwig J, Fox G. 1991. Great Lakes embryo mortality, edema, and deformities syndrome (GLEMEDS) in colonial fish-eating birds: similarity to chick-edema disease. *J Toxicol Environ Health* 33:455–520.

Gildersleeve RP, Tilson H, Mitchell C. 1985. Injection of diethylstilbestrol on the first day of incubation affects morphology of sex glands and reproductive behavior of Japanese quail. *Teratology* 31:101–109.

Goldstein JA. 1980. Structure activity relationships for the biochemical effects and the relationship to toxicity. In: Kimbrough RD, editor. Volume 4, Topics in environmental health. Halogenated biphenyls, terphenyls, naphthalenes, dibenzodioxins and related products. North Holland NY: Elsevier. p 151–190.

Gray SE, Woodin BR, Stegeman JJ. 1991. Sex differences in hepatic monooxygenses in winter flounder (*Pseudopleuronectes americanus*) and scup (*Stenotomus chrysops*) and regulation of P450 forms by estradiol. *J Exp Zool* 259:330–342.

Gross TS, Guillette LJ. 1994. Pesticide induction of developmental abnormalities of the reproductive system of alligators (*Alligator mississippiensis*) and turtles (*Trachemys scripta*). Estrogens in the environment III: Global health implicaitons. Washington DC: National Institute of Environmental Health Sciences.

Grue CE, Shipley BK. 1984. Sensitivity of nestling and adult starlings to dicrotophos, an organophosphate pesticide. *Environ Res* 35:454–465.

Guillette LJ. 1995. Endocrine disrupting environmental contaminants and developmental abnormalities in embryos. *Human Ecol Risk Assess* 1:25–36.

Guillette LJ, Gross TS, Gross DA, Rooney AA, Percival HF. 1995. Gonadal steroidogenesis in vitro from juvenile alligators obained from contaminated or control lakes. *Environ Health Perspect* 103:31–36.

Guillette LJ, Gross TS, Masson GR, Matter JM, Percival HF, Woodward AR. 1994. Developmental abnormalities of the gonad and abnormal sex hormone concentrations in juvenile alligators from contaminated and control lakes in Florida. *Environ Health Perspect* 102:680–688.

Guillette LJ, Pickford DB, Crain DA. Rooney AA, Percival HF. 1996. Reduction in penis size and plasma testosterone concentrations in juvenile alligators living in a contaminated environment. *Gen Comp Endocrinol* 101:32–42.

Guiney PD, Cook PM, Casselman JM, Fitzsimons JD, Simonin HA, Zabel EW, Peterson RE. 1996. Assessment of 2,3,7,8-tetrachlorodibenzo-*p*-dioxin-induced sac fry mortality in lake trout (*Salvelinus namaycush*) from different regions of the Great Lakes. *Can J Fish Aquat Sci* 53:2080–2092.

Guiney PD, Smolowitz RM, Peterson RE, Stegeman JJ. 1997. Correlation of 2,3,7,8-tetrachlorodibenzo-*p*-dioxin induction of cytochrome P4501A in vascular endothelium with toxicity in early life stages of lake trout. *Toxicol Appl Pharmacol* 143:256–273.

Haegele MA, Tucker RK. 1974. Effects of 15 common environmental pollutants on eggshell thickness in mallards and Coturnix. *Bull Environ Contam Tox* 11:98–102.

Hammond GL. 1995. Potential functions of plasma steroid-binding proteins. *Trends Endocrinol Metab* 6:298–304.

Hanson JM, Courtenay SC. 1996. Seasonal use of estuaries by winter flounder in the southern Gulf of St. Lawrence. *Trans Am Fish Soc* 125:705–718.

Harmon JE, Boehm MF, Heyman RA, Manglesdorf DJ. 1995. Activation of mammalian retinoid X receptors by the insect growth regulator methoprene. *Proc Natl Acad Sci* 92:6157–6160.

Harries JE, Sheahan DA, Jobling S, Matthiessen P, Neal P, Routledge EJ, Rycroft R, Sumpte JP, Taylor T. 1996. A survey of estrogenic activity in United Kingdom inland waters. *Environ Toxicol Chem* 15:1903–2002.

Heath AG. 1987. Behavior and nervous system function. In: Heath AG, editor. Water pollution and fish physiology. Boca Raton FL: CRC Press. p 181–200.

Heinz GH. 1996. Mercury poisioning in wildlife. In: Fairbrother A, Locke LN, Hoff GL, editors. Noninfectious diseases of wildlife. Ames IA: Iowa State University Press. p 118–127.

Heinz GH. 1996. Selenium in birds. In: Beyer WN, Heinz GH, Redmon-Norwood AW. Environmental contaminants in wildlife: interpreting tissue concentrations. Boca Raton FL: Lewis. p 447–458.

Heinz GH, Fitzgerald MA. 1993. Overwinter survival of mallards fed selenium. *Archives Environ Contam Toxicol* 25:90–94.

Heinz GH, Hoffman DJ, Gold LG. 1989. Impaired reproduction of mallards fed an organic form of selenium. *J Wild Manage* 53:418–428.

Heinz GH, Hoffman DJ, Krynitsky AKJ, Weller DMG. 1987. Reproduction in mallards fed selenium. *Environ Toxicol Chem* 6:423–433.

Heinz GH, Percival HF, Jennings ML. 1991. Comtaminants in American alligator eggs from lakes Apopka, Griffin and Okeechobee, Florida. *Environ Monit Assess* 16:277–285.

Henessey JV, Glass AR, Barns S, Vigersky BA. 1986. Comparative antiandrogenic potency of spironolactone and cimetidine: Assessment by the chicken cockscomb topical bioassay. *Proc Soc Exper Biol Med* 1982:443–447.

Henry TR, Spitsbergen JM, Hornung MW, Abnet CC, Peterson RE. 1997. Early life stage toxicity of 2,3,7,8-tetrachlorodibenzo-*p*-dioxin in zebrafish (*Danio rerio*). *Toxicol Appl Pharmacol* 142:56–68.

Herman RL, Kincaid HL. 1988 Pathological effects of orally administered estradiol to rainbow trout. *Aquaculture* 72:165–172.

Hill J, Clarke JDW, Vargesson N, Jowett T, Holder N. 1995. Exogenous retinoic acid causes specific alterations in the development of the midbrain and hindbrain of the zebrafish embryo including positional respecification of the Mauthner neuron. *Mech Develop* 50:3–16.

Ho MS, Fehrer S, Yu M, Lian L, Press D. 1988. High-affinity binding of [^{3}H] estradiol-17b by an estrogen receptor in the liver of the turtle. *Gen Comp Endocrinol* 70:382–394.

Hodson PV, McWhirter M, Ralph K, Gray B, Thivierge D, Carey J, Van Der Kraak G, McWhittle D, Levesque R. 1992. Effects of bleached kraft mill effluent on fish in the St. Maurice River, Quebec. *Environ Toxicol Chem* 11:1635–1651.

Hoffman DJ. 1990. Embryotoxicity and teratogenicity of environmental contaminants to bird eggs. *Bull Environ Contam Toxicol* 115:39–89.

Hoffman DJ, Eastin Jr WC. 1981. Effects of malathion, diazinon, and parathion on mallard embryo development and cholinesterase activity. *Environ Res* 26:472–485.

Hoffman DJ, Heinz GH. 1988. Embryotoxic and teratogenic effects of selenium in the diet of mallards. *J Toxicol Environ Health* 24:477–490.

Holinka FC, Anzai Y, Hata H, Kimmel N, Kuramoto H, Gurpide E. 1989. Proliferation and responsiveness to estrogen of human endometrial cancer cells under serum-free culture conditions. *Cancer Res* 49:3297–3301.

Holmes WN. 1984. Petroleum pollutants in the marine environment and their possible effects on seabirds. In: Hodgson E, editor. Reviews in environmental toxicology. Amsterdam: Elsevier. p 251–317.

Howell WH, Denton TE. 1989. Gonapodial morphogenesis in female mosquitofish, *Gambusia affinis affinis*, masculinized by exposure to degradation products from plant sterol. *Environ Biol Fish* 24:43–51.

Howell WM, Blank DA, Bortone SA. 1980. Abnormal expression of secondary sex characters in a population of mosquitofish, *Gambusia affinis holbrooki*: Evidence for environmentally induced masculinization. *Copeia* 4:676–681.

Idler DR, Reinboth R, Walsh JM, Truscott B. 1976. A comparison of 11-hydroxytestosterone and 11-ketotestosterone in blood of ambisexual and gonochroistic teleosts. *Gen Comp Endocrinol* 30:517–521.

Idler DR, Schmidt PJ, Biety J. 1961. The androgenic activity of 11-ketotestosterone, a steroid in salmon plasma. *Can J Biochem Physiol* 39:317–320.

Idler DR, So YP, Fletcher GL, Payne JF. 1995. Depression of blood levels of reproductive steroids and glucuronides in male winter flounder (*Pleuronectes americanus*) exposed to small quantities of Hibernia crude, used crankcase oil, oily drilling mud and harbour sediments in the 4 months prior to spawning in late May–June. In: Goetz FW, Thomas P, editors. Proceedings of the 5th International Symposium on Reproductive Physiology of Fish; 2-8 July 1995; Austin TX: p 187.

Janz DM, McMaster ME, Munkittrick KR, Van Der Kraak G. 1997. Elevated ovarian follicular apoptosis and heat shock protein-70 expression in white sucker exposed to bleached kraft pulp mill effluent. *Toxicol Appl Pharmacol* 147:391–398.

Jessup DA, Leighton FA. 1996. Oil pollution and petroleum toxicity to wildlife. In: Fairbrother A, Locke LN, Hoff GL. Noninfectious diseases of wildlife. Ames IA: Iowa State Univ. Press. p 141–156.

Jobling S, Reynolds T, White R, Parker M. 1995. A variety of environmentally persistent chemicals including some phthalate plasticizers are weakly estrogenic. *Environ Health Perspect* 103:582–587.

Jobling S, Sheahan D, Osborne JA, Matthiessen P, Sumpter JP. 1996. Inhibition of testicular growth in rainbow trout *(Oncorhynchus mykiss)* exposed to estrogenic alkylphenolic chemicals. *Environ Toxicol Chem* 15:194–202.

Jobling S, Sumpter JP. 1993. Detergent components in sewage effluent are weakly estrogenic to fish: An in vitro study using rainbow trout (*Oncorhynchus mykiss*) hepatocytes. *Aquat Toxicol* 27:361–372.

Johnson LL, Casillas E, Collier TK, McCain BB, Varanasi U. 1988. Contaminant effects on ovarian development in English sole (*Parophrys vetulus*) from Puget Sound, Washington. *Can J Fish Aquat Sci* 45: 2133–2146.

Johnson LL, Casillas E, Sol SY, Collier TK, Stein JE, Varanasi U. 1993. Contaminant effects on reproductive success in selected benthic fish species. *Mar Env Res* 35:165–170.

Johnson LL, Nelson GM, Sol SY, Lomax DP, Casillas E. 1997. Fecundity and egg weight in English sole *(Pleuronectes vetulus)* from Puget Sound, WA: influence of nutritional status and chemical contaminants. *Fish Bull* 92:232–250.

Johnson LL, Stein JE, Collier TK, Casillas E, Varanasi U. 1994. Indicators of reproductive development in prespawning female winter flounder *(Pleuronectes americanus)* from urban and nonurban estuaries in the northeast United States. *Sci Total Environ* 141:241–260.

Johnson MA, Fischer JG. 1994. Role of minerals in protection against free radicals. *Food Technol* 48(5):112–120.

Johnson MK, Henschler D. 1975. The delayed neuropathy caused by some organophosphorus esters: mechanism and challenge. *CRC Crit Rev Toxicol*. June 1975. p 289–316.

Johnson MK. 1977. Improved assay of neurotoxic esterase for screening organophosphates for delayed neurotoxicity potential. *Arch Toxicol* 37:113–115.

Joseph DR. 1994. Structure, function, and regulation of the androgen-binding protein/sex hormone-binding globulin. *Vitam Horm* 49:197–280.

Kavlock RJ, Daston GP, DeRosa C, Fenner-Crisp P, Gray LE, Kaatari S, Lucier G, Luster M, Mac MJ, Maczka C, Miller R, Moore J, Rolland R, Scott G, Sheehan DM, Sinks T, Tilson HA. 1996. Research needs for the risk assessment of health and environmental effects of endocrine disruptors: a report of the USEPA-sponsored workshop. *Environ Health Perspect* 104 (Suppl. 4):715–740.

Kelce WR, Stone CR, Laws S, Gray LE, Kemppainen JA, Wilson EM. 1995. Persistent DDT metabolite *p,p'* DDE is a potent androgen receptor antagonist. *Nature* 375:581–585.

Kennedy SJ, Lorenzen A, Jones SP, Hahn ME, Stegeman JJ. 1996. Cytochrome P4501A induction in avain hepatocyte cultures: a promising approach for predicting the sensitivity of avian species to toxic effects of halogenated aromatic hydrocarbons. *Toxicol Appl Pharmacol* 141:214–230.

King KA, White DH, Mitchell CA. 1984. Nest defense behavior and reproductive success of laughing gulls sublethally dosed with parathion. *Bull Environ Contam Toxicol* 33:499–504.

Kloepper-Sams PJ, Swanson SM, Marchant T, Schryer R, Owens JW. 1994. Exposure of fish to biologically treated bleached kraft pulp mill effluent. 1. Biochemical and pathological assessment of rocky mountain whitefish (*Prosopium williamsoni*) and longnose sucker (*Catostomus catostomus*). *Environ Toxicol Chem* 13:1469–1482.

Klotz DM, Beckman BS, Hill SM, McLachlan JA, Walters MR, Arnold SF. 1996. Identification of environmental chemicals with estrogenic activity using a combination of in vitro assays. *Environ Health Perspect* 104:1084–1089.

Kon O, Webster RA, Spelsberg TC. 1980. Isolation and characterization of the estrogen receptor in hen oviduct: Evidence for two molecular species. *Endocrinol* 107:1182–1191.

Kovacs TG, Gibbons JS, Martel PH, O'Conner BI, Voss RI. 1995. Effects of a secondary-treated thermomechanical pulp mill effluent on aquatic organisms as assessed by short-term and long-term laboratory tests. *J Toxicol Environ Health* 44:485–502.

Kovacs TG, Gibbons JS, Martel PH, Voss RI. 1997. Improved effluent quality at a bleached kraft mill as determined by laboratory biotests. *J Toxicol Environ Health* 49:533–561.

Kovacs TG, Gibbons JS, Tremblay LA, O'Conner BI, Martel PH, Voss RI. 1995. The effect of a secondary-treated bleached kraft pulp mill effleuent on aquatic organisms as assessed by short-term and long-term laboratory tests. *Ecotoxicol Environ Safety* 31:7–22.

Kubin LA. 1997. A study of growth in juvenile English sole exposed to sediments amended with aromatic compounds. [M.S. Thesis] Bellingham WA: Western Washington State University.

Kuhnhold WW, Everich D, Stegeman JJ, Lake J, and Wolke RE. 1978. Effects of low levels of hydrocarbons on embryonic, larval, and adult winter flounder (*Pseudopleuronectes americanus).* Proceed. Conf. Assess. Ecol. Impacts Oil Spills. Keystone, CO, 14–17 June 1978. AIBS, p 677–711.

Laenge R, Schweinfurth H, Crowdace CP, Panter GH. 1997. Growth and reproduction of fathead minnow (*Pimephales promelas*) exposed to the synthetic steroid hormone ethynylestradiol in a life-cycle test. Seventh Annual Meeting of SETAC-Europe; 1997 Apr 6–10; Amsterdam, The Netherlands.

Lambing JH, Jones WE, Sutphin JW. 1988. Reconnaissance investigation of water quality, bottom sediment, and biota associated with irrigation drainage in Bowdoin National Wildlife Refuge and adjacent areas of the Milk River Basin, northeastern Montana, 1986–87. Denver CO: U.S. Geological Survey Water-Resources Investigations. Report 87-4243.

Lander L, Grahn O, Hardig J, Lehtinen K-J, Monfelt C, Tana J. 1994. A field study of environmental impacts of a bleached kraft pulp mill site on the Baltic sea coast. *Ecotoxicol Environ Safety* 27: 128–157.

Lari L, Massi A, Fossi MC, Casini S, Leonzio C, Focardi S. 1994. Evaluation of toxic effects of the organophosphorus insecticide azinphos-methyl on experimentally and naturally exposed birds. *Arch Environ Contam Toxicol* 26:234–239.

Leighton FA. 1983. The pathophysiology of petroleum oil toxicity in birds: a review. In: The effects of oil on birds: physiological research, clinical applications and rehabilitation. Wilmington DE: Tri-State Bird and Research, Inc. p 1–28.

Lemly AD. 1993. Metabolic stress during winter increases the toxicity of selenium to fish. *Aquat Toxicol* 27:133–158.

Lemly AD. 1995. A protocol for aquatic hazard assessment of selenium. *Ecotoxicol Environ Safety* 32:280–288.

Little EE, Henry MG, editors. 1990. Symposium on behavioral toxicology. *Environ Toxicol Chem* 9:91–119.

Little EE. 1990. Behavioral toxicology: Stimulating challenges for a growing discipline. *Environ Toxicol Chem* 9:1–2.

Livingstone DR, Lemaire P, Matthews A, Peters L, Bucke D, Law RJ. 1993. Pro-oxidant, antioxidant and 7-ethoxyresorufin O-deethylase (EROD) activity responses in liver of dab (*Limanda limanda*) exposed to sediment contaminated with hydrocarbons and other chemicals. *Mar Pollut Bull* 26:602–606.

Lotti M. 1992. The pathogenesis of organophosphate polyneuropathy. *Toxicology* 21:465–488.

Lundholm CE. 1984. Effect of DDE on the Ca metabolism of the duck eggshell gland and its subcellular fractions: relation to functional stage. *Comp Biochem Physiol* C78:5–12.

Lundholm CE. 1994. Changes in the level of different ions in the eggshell gland lumen following p,p'-DDE-induced eggshell thinning in ducks. *Comp Biochem Physiol* C109:57–62.

Lutz H, Lutz-Ostertag Y. 1975. Intersexuality of the genital system and "free-martinism" in birds (chickens, ducks, quail). In: Reinboth R, editor. Intersexuality in the animal kingdom. p 382–391.

MacLatchy D, Peters L, Nickle J, Van Der Kraak G. 1997. Exposure to β-sitosterol alters the endocrine status of goldfish differently than 17β-estradiol. *Environ Toxicol Chem* 16:1895–1904.

MacLatchy DL, Van Der Kraak GJ. 1995. The phytoestrogen β-sitosterol alters the reproductive endocrine status of goldfish. *Tox Appl Pharmacol* 134:305–312.

MacLatchy DL, Yao Z, Tremblay L, Van Der Kraak G. 1995. The hormone mimic β-sitosterol alters the reproductive status in goldfish. In: Goetz FW, Thomas P, editors. Proceedings of the 5th International Symposium on Reproductive Physiology of Fish. Austin, Texas 2–8 July 1995. p 189.

Maitra SK, Sarkar R. 1996. Influence of methyl parathion on gametogenic and acetylcholinesterase activity in the testis of white-throated munia (*Lonchura malabarica*). *Arch Environ Contam Toxicol* 30:384–389.

Manns M, Fritzsch B. 1992. Reinoic acid affects the organization of reticulospinal neurons in developing *Xenopus*. *Neuroscience Letters* 139:253–256.

Marquis J. 1986. Contemporary issues in pesticide toxicology and pharmacology. Basel, Switzerland: Karger Press.

Matsumura F. 1994. How important is the protein phosphorylation pathway in the toxic expression of dioxin-type chemicals? *Biochem Pharmacol* 48:215–224.

Matsuyama M, Adachi S, Nagahama Y, Matsumura S. 1990. Diurnal rhythm of oocyte development and plasma steroid hormone levels in the female red sea bream, *Pagrus major*, during the spawning season. *Aquaculture* 73:357–372.

Mattison DR, Nightingale MS, Shiromizu K. 1983. Effects of toxic substances on the female reproduction. *Environ Health Perspect* 48:43–52.

McArthur MLB, Fox GA, Peakall DB, Philogene BJR. 1983. Ecological significance of behavioral and hormonal abnormalities in breeding ring doves fed an organochlorine chemical mixture. *Arch Environ Contam Toxicol* 12:343–353.

McMaster ME, Van Der Kraak G, Portt CB, Munkittrick KR, Sibley PK, Smith IR, Dixon DG. 1991. Changes in hepatic mixed-function oxidase (MFO) activity, plasma steroid levels, and age at maturity of a white sucker (*Catostomus commersoni*) population exposed to bleached kraft pulp mill effluent. *Aquat Toxicol* 21:199–218.

McMaster ME, Van Der Kraak GJ, Munkittrick KR. 1995. Exposure to bleached kraft pulp mill effluent reduces the steroid biosynthetic capacity of white sucker ovarian follicles. *Comp Biochem Physiol* 112C: 169–178.

McMaster ME, Van Der Kraak GJ, Munkittrick KR. 1996. An evaluation of the biochemical basis for steroid hormone depressions in fish exposed to industrial wastes. *J Great Lakes Res* 22:153–171.

Mellanen P, Petanen T, Lehtimaki J, Makela S, Bylund G, Holmbom B, Mannila E, Oikari A, Santti R. 1996. Wood derived estrogens—a study in vitro with breast cancer cell lines and in vivo in trout. *Toxicol Appl Pharmacol* 136:381–388.

Metcalf RL, Francis BM, Metcalf RA, Farange-Elawar M, Hansen LG. 1988. Structure-activity relationships in the acute and delayed neurotoxicity of methyl- and ethylphosphonothionates. *Pest Biochem Physiol* 30:45–56.

Metcalf RA, Metcalf RL. 1973. Selective toxicity of analogs of methyl parathion. *Pest Biochem Physiol* 3:149–159.

Misawa M, Doull J, Uyeki EM. 1982. Teratogenic effects of cholinergic insecticides in chick embryos. III. Development of cartilage and bone. *J Toxicol Environ Health* 10:551–563.

Mohanty-Hejmadi P, Dutta SK, Mahapatra P. 1992. Limbs regenerated at site of tail amputation in marbled balloon frog after Vitamin A treatment. *Nature* 355:252–253.

Moles A, Norcross BL. 1997. Effects of oil-laden sediments on growth and health of juvenile fishes. *Can J Fish Aquat Sci* 55:605–610.

Moles A, Rice SD. 1983. Effects of crude oil on growth, caloric content, and fat content of pink slamon juveniles in seawater. *Trans Am Fish Soc* 112:205–211.

Munkittrick KR, McMaster ME, Portt CB, Van Der Kraak G, Smith IR, Dixon DG. 1992. Changes in maturity, plasma sex steroid levels, hepatic MFO activity and the presence of external lesions in lake whitefish exposed to bleached kraft mill effluent (BKME). *Can J Fish Aquat Sci* 49:1560–1569.

Munkittrick KR, Portt C, Van Der Kraak GJ, Smith I, Rokosh D. 1991. Impact of bleached kraft mill effluent on population characteristics, liver MFO activity, and serum steroid levels of a Lake Superior white sucker population. *Can J Fish Aquat Sci* 48:1371–138.

Munkittrick KR, Servos MR, Carey JH, Van Der Kraak GJ. 1997. Environmental impacts of pulp and paper wastewater: evidence for a reduction in environmental effects at North American pulp mills since 1992. *Water Sci Technol* 35:329–338.

Munkittrick KR, Van Der Kraak GJ, McMaster ME, Portt CB. 1992a. Response of hepatic MFO activity and plasma sex steroids to secondary treatment of bleached kraft pulp mill effluent and mill shutdown. *Environ Toxicol Chem* 11:1427–1439.

Munkittrick KR, Van Der Kraak GJ, McMaster ME, Portt CB. 1992b. Reproductive dysfunction and MFO activity in three species of fish exposed to bleached kraft mill effluent at Jackfish Bay Lake Superior. *Water Poll Res J Can* 27:439–446.

Munkittrick KR, Van Der Kraak GJ, McMaster ME, Portt CB, van den Heuvel MR, Servos MR. 1994. Survey of receiving water environmental impacts associated with discharges from pulp mills. 2. Gonad size, liver size, hepatic MFO activity and plasma sex steroid levels in white sucker. *Environ Toxicol Chem* 13:1089–1101.

Munson PL, Sheps MC. 1958. An improved procedure for the biological assay of androgens by direct application to the comb of baby chicks. *Endorinology* 62:173–187.

Myers MS, Johnson LL, Hom T, Collier TK, Stein JE, Varanasi U. 1997. Toxicopathic hepatic lesions in subadult English sole *(Pleuronectes vetulus)* from Puget Sound, Washington, USA: relationships with other biomarkers of contaminant exposure. *Mar Environ Res* 45:47–67.

Myers MS, Stehr CM, Olson OP, Johnson LL, McCain BB, Chan S-L, Varanasi U. 1994. Relationships between toxicopathic hepatic lesions and exposure to chemical contaminants in English sole *(Pleuronectes vetulus)*, starry flounder *(Platichthys stellatus)* and white croaker *(Genyonemus lineatus)* from selected marine sites on the Pacific Coast USA. *Environ Health Perspect* 102:200–215.

Nagler JJ, Cyr DG. 1997. Exposure of male American plaice (*Hippoglossoides platessoides*) to contaminated marine sediments decreases the hatching success of their progeny. *Environ Toxicol Chem* 16:1733–1738.

Nakhla AM, Rosner W. 1996. Stimulation of prostate cancer browth by androgens and estrogen through intermediacy of sex hormone-binding globulin. *Endocrinol* 137:4126–4129.

[NBS] National Biological Service. 1995. Our living resources: a report to the nation on the distribution, abundance, and health of U.S. plants, animals, and ecosystems. Department of the Interior, Washington, DC.

Nebert DW, Petersen DD, Fornace AJ. 1990. Cellular response to oxidative stress: The [Ah] gene battery as a paradigm. *Environ Health Perspect* 88:13–25.

Nelson DA, Miller JE, Rusanowsky D, Greig RA, Sennefelder GR, Mercaldo-Allen R, Kuropat C, Gould E, Thurberg FP, Calabrese A. 1991. Comparative reproductive success of winter flounder in Long Island Sound: A three-year study (biology, biochemistry, and chemistry). *Estuaries* 14:318–331.

Nimick DA, Lambing JH, Palawski DU, Malloy JC. 1996. Detailed study of selenium in soil, water, bottom sediment, and biota in the Sun River irrigation project, Freezout Lake Wildlife Management Area, and Benton Lake National Wildlife Refuge, west-central Montana, 1990-92. U.S. Geological Survey Water-Resources Investigations Report 95-4170, Denver, CO.

Norris DO. 1997. Vertebrate endocrinology. Third edition. New York: Academic Press. 634 p.

Nosek JA, Sullivan JR, Craven SR, Gendron-Fitzpatrick A, Peterson RE. 1993. Embryotoxicity of 2,3,7,8-tetrachlorodibenzo-*p*-dioxin in the ring-necked pheasant. *Environ Toxicol Chem* 12:1215–1222.

[NRC] National Research Council. 1992. Committee on Neurotoxicology and Models for Assessing Risk. Environmental Neurotoxicology. Washington DC: National Academy Press.

[OECD] Organization for Economic Cooperation and Development. 1984. OECD Guidelines for testing of chemicals. Guideline 206 Avian reproduction test. Paris, France: OECD.

[OECD] Organization for Economic Cooperation and Development. 1996. Report of the SETAC/OECD workshop on avian toxicity testing. OEC/GS(96)166. Paris, France: OECD.

Ohlendorf HM. 1989. Bioaccumulation and effects of selenium in wildlife. In: Jacobs LW, editor. Selenium in agriculture and the environment. Madison WI: American Society of Agronomy and Soil Science Society of America. SSSA Special Publication No. 23.

Ohlendorf HM. 1996. Selenium. In: Fairbrother A, Locke LN, Hoff GL. Noninfectious diseases of wildlife, 2nd Edition. Ames IA: Iowa State University Press. p 128–140.

Ohlendorf HM, Hoffman DJ, Saiki MK, Aldrich TW. 1986. Embryonic mortality and abnormalities of aquatic birds: apparent impact of selenium from irrigation drain water. *Sci Total Environ* 52:49–63.

Oldfield JE. 1995. Serendipity. *Chemtech* 25:52–55.

Onuoha GC, Nwadukwe FO. 1990. Influence of liquid petroleum refinery effluent on the hatching success of *Clarias gariepinus* (African mud fish) eggs. *Environ Ecol* 8:1201–1206.

Ouellet M, Bonin J, Rodrique J, DesGranges J, Lair S. 1997. Hindlimb deformities (ectromelia, ectrodactyly) in free-living anurans from agricultural habitats. *J Wildlife Dis* 33:95–104.

Paine MD. 1989. Sublethal effects of Hibernia crude oil on capelin (*Mallotus villosus*) embryos. *Rapp P-V Reun Ciem* 191:493.

Pajor AM, Stegeman JJ, Thomas P, Woodin WW. 1990. Feminization of the hepatic microsomal cytochrome P450 system in brook trout by estradiol, testosterone, and pituitary factors. *J Exp Zool* 253:51–60.

Pakdel F, LeGac F, Le Goff P, Valataire Y. 1990. Full-length sequence and in vitro expression of rainbow trout estrogen receptor cDNA. *Mol Cell Endocrinol* 71:195–204.

Palmer B, Palmer SK. 1995. Vitellogenin induction by xenobiotic estrogens in the red-eared turtle and African clawed frog. *Environ Health Perspect* 103:19–25.

Pasmanik M, Callard G. 1986. Characteristics of a testosterone-estradiol binding globulin (TEBG) in goldfish serum. *Biol Reprod* 35:838–845.

Peakall DB. 1970. *p,p'*-DDT effect on calcium metabolism and concentration of oestradiol in the blood. *Science* 168:592–594.

Peakall DB, Miller DS, Kinter WB. 1975. Prolonged eggshell thinning caused by DDE in the duck. *Nature* 254(5499):421.

Peakall DB, Peakall ML. 1973. Effect of polychlorinated biphenyl on the reproduction of artificially and naturally incubated dove eggs. *J Appl Ecol* 10:863–868.

Pechman JHK, Wilbur HM. 1994. Putting declining amphibian populations in perspective: Natural fluctuations and human impacts. *Herpetologica* 50:65–84.

Pereira JJ, Ziskowsky J, Mercaldo-Allen R, Kuropat C, Luedke D, Gould E. 1992. Vitellogenin in winter flounder (*Pleuronectes americanus*) from Long Island and Boston Harbor. *Estuaries* 15:289–297.

Perry DM, Hughes JB, Hebert AT. 1991. Sublethal abnormalities in embryos of winter flounder, *Pseudopleuronectes americanus* , from Long Island Sound. *Estuaries* 14:306–317.

Peterson RE, Theobald HM, Kimmel GL. 1993. Developmental and reproductive toxicity of dioxins and related compounds: cross-species comparisons. *Crit Rev Toxicol* 23:283–335.

Platonow NS, Reinhart BS, 1973 . The effects of polychlorinated biphenyls (Aroclor 1254) on chicken egg production, fertility and hatchability. *Can J Comp Med* 37:341–346.

Poland A, Glover E. 1980. 2,3,7,8-tetrachlorodibenzo-*p*-dioxin: segregation of toxicity with the Ah locus. *Mol Pharmacol* 17:86–94.

Poland A, Knutson JC. 1982. 2,3,7,8-tetrachlorodibenzo-*p*-dioxin and related halogenated aromatic hydrocarbons: examination of the mechanism of toxicity. *Ann Rev Pharmacol Toxicol* 22:517–554.

Poley WE, Moxon AL. 1938. Tolerance levels of seleniferous grains in laying rations. *Poultry Sci* 17:72–76.

Poley WE, Moxon AL, Franke KW. 1937. Further studies of the effects of selenium poisoning on hatchability. *Poultry Sci* 16:219–225.

Poley WE, Wilson WO, Moxon AL, Taylor JB. 1941. The effect of selenized grains on the rate of growth in chicks. *Poultry Sci* 20:171–179.

Powell GVN. 1984. Reproduction by an altricial songbird, the red-winged blackbird, in fields treated with the organophosphate insecticide fenthion. *J Appl Ecol* 21:83–95.

Purdom CE, Hardiman PA, Bye VJ, Eno NC, Tyler CR, Sumpter JP. 1994. Estrogenic effects of effluents from sewage treatment works. *Chem Ecol* 3:275–285.

Ramsdell HS, Blandin DA, Schmechel TR. 1996. Developmental effects and biochemical markers of alkylphenol exposure in frog larvae. Society of Environmental Toxicology and Chemistry (SETAC) 17th Annual Meeting; 17–21 Nov 1996; Washington, DC. Pensacola FL: SETAC. p 140.

Rand GM. 1985. Behavior. In: Rand GM, Petrocelli SR, editors. Fundamentals of aquatic toxicology. Washington DC: Hemisphere Publishing Corporation. p 221–263.

Ratcliffe DA. 1967. Decrease in eggshell weight in certain species of birds of prey. *Nature* 215:208–210.

Rattner BA, Clarke RN, Ottinger MA. 1986. Depression of plasma luteinizing hormone concentration in quail by the anticholinesterase insecticide parathion. *Comp Biochem Physiol* 83C:451–453.

Rattner BA, Eroschenko VP, Fox GA, Fry DM, Gorsline J. 1984. Avian endocrine responses to environmental pollutants. *J Exper Zool* 232:683–689.

Raynaud A, Pieau C. 1985. Embryonic development of the genital system. In: Gans C, Billett F, editors. Biology of the reptilia, Volume 15: Development. New York: Wiley and Sons. p 149–300.

Rinella FA and Schuler CA. 1992. Reconnaissance investigation of water quality, bottom sediment, and biota associated with irrigation drainage in the Malheur National Wildlife Refuge, Harney county, Oregon, 1988–89. Denver CO: U.S. Geological Survey Water-Resources Investigations. Report 91-4085.

Robinson GA, Gibbins AMV. 1984. Induction of vitellogenesis in Japanese quail as a sensitive indicator of the estrogen-mimetic effect of a variety of environmental contaminants. *Poultry Sci* 63:1529–1536.

Robinson RD. 1994. Evaluation and development of laboratory protocols for predicting the chronic toxicity of pulp mill effluents to fish. [Ph.D. Dissertation] Guelph, Ontario, Canada: University of Guelph.

Roger JC, Upshall DG, Casida JE. 1969. Structure-activity and metabolism studies on organophosphate teratogens and their alleviating agents in developing hen eggs with special reference to Bidrin. *Biochem Pharmacol* 18:373–392.

Rosa-Molinar E, Williams CS. 1984. Notes on the fecundity of an arrhenoid population of mosquito fish, *Gambusia affinis holbrooks. Northeast Gulf Sci* 7:121–125.

Rutledge JC, Shourbaji, AG, Hughes, LA, Polifka JE, Cruz YP, Bishop JB, Generoso WM. 1994. Limb and lower-body duplications induced by retinoic acid in mice. *Proc Natl Acad Sci* 91:5436–5440.

Saiki MK, Ogle RS. 1995. Evidence of impaired reproduction by western mosquitofish inhabiting seleniferous agricultural drain water. *Trans Am Fish Soc* 124:578–587.

Sandstrom O. 1996. In situ assessments of pulp mill effluent impact on life history variables in fish. In: Servos MR, Munkittrick KR, Carey J, Van Der Kraak G, editors. Environmental effects of pulp mill effluents. Boca Raton FL: St. Lucie Press. p 449–457.

Santodonato J. 1997. Review of the estrogenic and antiestrogenic activity of polycyclic aromatic hydrocarbons: relationship to carcinogenicity. *Chemosphere* 34:835–848.

Scadding SR, Maden M. 1986. The effects of local application of retinoic acid on limb development and regeneration in tadpoles of *Xenopus laevis. J Embryol Exp Morphol* 91:55–63.

Schena M. 1989. The evolutionary conservation of eukaryotic gene transcription. *Experientia* 45:972–983.

Schmidt JV, Bradfield CA. 1996. Ah receptor signalling pathways. *Ann Rev Cell Dev Biol* 12:55–89.

Schwabe CW, Riemann HP, Franti CE. 1977. Epidemiology in veterinary practice. Philadelphia PA: Lea & Febiger. 313 p.

Schwarzbach SE, Shull L, Grau CR. 1988. Eggshell thinning in ring doves exposed to *p,p'*-dicofol. *Arch Environ Contam Toxicol* 17:219–227.

Scott WJ, Collins MD, Ernst AN, Supp DM, Potter SS. 1994. Enhanced expression of limb malformations and axial skeleton alterations in legless mutants by transplacental exposure to retinoic acid. *Develop Biol* 164:277–289.

Servos MR, Huestis SY, Whittle DM, Van Der Kraak G, Munkittrick KR. 1994. Survey of receiving water environmental impacts associated with discharges from pulp mills. 3. Polychlorinated dioxins and furans in muscle and liver of white sucker (*Catostomus commersoni*). *Environ Toxicol Chem* 13:1103–1115.

Servos MR. 1996. Origins of effluent chemicals and toxicity: recent research and future directions. In: Servos MR, Munkittrick KR, Carey J, Van Der Kraak G, editors. Environmental effects of pulp mill effluents. Boca Raton FL: St. Lucie Press. p 159–166.

Shelby MD, Newbold RR, Tully DB, Chae K, Davis VL. 1996. Assessing environmental chemicals for estrogenicity using a combination of in vitro and in vivo assays. *Environ Health Perspect* 104:1296–1300.

Shenefelt R. 1972. Morphogenesis of malformations in hamsters caused by retinoic acid: Relation to dose and stage of treatment. *Teratology* 5:103–118.

Shimeld SM. 1996. Retinoic acid, HOX genes and the anterior-posterior axis in chordates. *BioEssays* 18:613–616.

Skorupa J. 1996. Avian selenosis in nature: a phylogenetic surprise and its implications for ecological risk assessment. Poster presentation at Society of Environmental Toxicology and Chemistry 17th Annual Meeting; 17–21 Nov 1996; Washington DC.

Skorupa J, Ohlendorf HM. 1991. Contaminants in drainage water and avian risk thresholds. In: Dinar A, Zilberman D, editors. The economics and management of water and drainage in agriculture. Boston: Kluwer Academic.

Skorupa J, Morman SP, Sefchick-Edwards JS. 1996. Guidelines for interpreting selenium exposure of biota associated with nonmarine aquatic habitats. Sacramento CA: U.S. Fish and Wildlife Service National Irrigation Water Quality Program. Sacramento Field Office.

Smith CA , Joss JMP. 1994. Uptake of super(3)H-estradiol by embryonic crocodile gonads during the period of sexual differentiation. *J Exp Zool* 270:219–224.

Smith HV, Spalding JMK. 1959. Outbreak of paralysis in Morocco due to *ortho*-cresyl phosphate poisoning. *The Lancet* 5 Dec 1959. p 1019–1021.

Solecki R, Faqi AS, Pfeil R, Hilbig V. 1996. Effects of methyl parathion on reproduction in Japanese quail. *Bull Environ Contam Toxicol* 57:902–908

Sorenson EMB. 1991. Metal poisoning in fish. Boca Raton FL: CRC Press. 383 p.

Soto AM, Sonnenschein C, Chung KL, Fernandez MF, Olea N, Olea Serrano F. 1995. The E-SCREEN assay as a tool to identify estrogens: an update on estrogenic environmental pollutants. *Environ Health Perspect* 103:113–122.

Spear PA, Bourbonnais DH, Peakall DB, Moon TW. 1989. Dove reproduction and retinoid (vitamin A) dynamics of adult females and their eggs following exposure to 3,3',4,4'-tetrachlorobiphenyl. *Can J Zool* 67:908–911.

Spies RB, Rice JDW. 1988. Effects of organic contaminants on reproduction of starry flounder, *Platicthys stellatus*, in San Francisco Bay, California. II. Reproductive success of fish captured in San Francisco Bay and spawned in the laboratory. *Mar Biol* 98:191–200.

Spitsbergen JM, Walker MK, Olson JR, Peterson RE. 1991. Pathologic alterations in early life stages of lake trout, *Salvelinus namaycush*, exposed to 2,3,7,8-tetrachlorodibenzo-*p*-dioxin as fertilized eggs. *Aquat Toxicol* 119:41–72.

Stein JE, Hom T, Sanborn HR, Varanasi U. 1991. Effects of exposure to a contaminated-sediment extract on the metabolism and disposition of 17β-estradiol in English sole (*Parophrys vetulus*). *Comp Biochem Physiol* 99C:231–240.

Stein JE, Reichert WL, Varanasi U. 1994. Molecular epizootiology: Assessment of exposure to genotoxic compounds in teleosts. *Environ Health Perspect* 102:19–23.

Stohs SJ, Shara MA, Alsharif NZ, Wahba ZZ, Al-Bayati ZA. 1990. 2,3,7,8-Tetrachlorodibenzo-*p*-dioxin-induced oxidative stress in female rats. *Toxicol Appl Pharmacol* 106:126–135.

Stromborg K. 1986. Reproduction of bobwhites fed different dietary concentrations of an organophosphate insecticide, methamidiphos. *Arch Environ Contam Toxicol* 15:143–147.

Stromborg KL. 1977. Seed treatment of pesticide effects on pheasant reproduction at sublethal doses. *J Wildlife Manage* 41:632–641.

Sumpter JP. 1985. The purification, radioimmunoassay and plasma levels of vitellogenin from the rainbow trout, *Salmo gairdneri*. In: Lofts B, Holmers WH, editors. Trends in comparative endocrinology. Hong Kong: Hong Kong University Pr. p 355–357.

Suter GW. 1993. Ecological risk assessment. Boca Raton FL: CRC Press/Lewis Publishers. 538 p.

Sutter TR, Guzman K, Dold KM, Greenlee WF. 1991. Targets for dioxin: genes for plasminogen activator inhibitor-2 and interleukin 1-β. *Science* 254:415–418.

Suzuki YJ, Forman HJ, Sevanian A. 1997. Oxidants as stimulators of signal transduction. *Free Radic Biol Med* 22:269–285.

Swanson SM, Schryer R, Shelast R, Kloepper-Sams PJ, Owens JW. 1994. Exposure of fish to bleached kraft mill effluent. 3. Fish habitat and population assessment. *Environ Toxicol Chem* 13:1497–1507.

Symula J, Meade J, Skea JC, Cummings L, Colquhoun JR, Dean HJ, Miccoli J. 1990. Blue-sac disease in Lake Ontario lake trout. *J Great Lakes Res* 16:41–52.

Teh SJ, Hinton DE. 1993. Detection of enzyme histochemical markers of hepatic preneoplasia and neoplasia in medaka (*Oryzias latipes*). *Aquat Toxicol* 24:163–182.

Thomas P. 1988. Reproductive endocrine function in female Atlantic croaker exposed to pollutants. *Mar Environ Res* 24:179–183.

Thomas P, Budiantara L. 1995. Reproductive life history stages sensitive to oil and naphthalene in Atlantic croaker. *Mar Environ Res* 39:147–150.

Thomas P, Smith J. 1993. Binding of xenobiotics to the estrogen receptor of spotted seatrout: A screening assay for potential estrogenic effects. *Marine Environ Res* 35:147–151.

Tietge JE. 1996. Summary of the workshop on Central North American amphibian deformities. Duluth MN: U.S. Environmental Protection Agency.

Tillitt DE, Wright PJ. 1997. Dioxin-like embryotoxicity of a Lake Michigan lake trout extract to developing lake trout. Abstract to Dioxin 97 Conference, Indianapolis, IN, 25–29 August 1997.

Tilson HA. 1990. Behavioral indices of toxicity. *Toxicol Pathol* 18:96–104.

Tilson HA. 1996. Setting exposure standards: A decision process. *Environ Health Perspect* 104:401–405.

Tremblay L, Van Der Kraak G. 1995. Interactions of the environmental estrogens nonylphenol and β-sitosterol with liver estrogen receptors in fish. In: Goetz FW, Thomas P, editors. Proceedings of the 5th International Symposium on Reproductive Physiology of Fish; 2–8 July 1995; Austin TX. 202 p.

Tremblay L, Van Der Kraak GJ. 1998. Use of a series of homologous in vitro and in vivo assays to evaluate the endocrine modulating action of β-sitosterol in rainbow trout. *Aquatic Toxicol* 43:149–162.

Trivelpiece W, Butler RG, Miller DS, Peakall DB. 1984. Reduced survival of chicks of oil-dosed adult Leach's storm-petrel. *Condor* 86:81–82.

Truscott B, Idler DR, Fletcher GL. 1992. Alteration of reproductive steroids of male winter flounder (*Pleuronectes americanus*) chronically exposed to low levels of crude oil in sediments. *Can J Fish Aquat Sci* 49:2190–2195.

Tumasonis CF, Bush B, Baker FD. 1973. PCB levels in egg yolks associated with embryonic mortality and deformity of hatched chicks. *Arch Environ Contam Toxicol* 1:312–324.

Turner CL. 1942. A quantitative study of the effects of different concentrations of ethynyl testosterone and methyl testosterone in the production of gonopodia in females of *Gambusia affinis*. *Physiol Zool* 15:263–280.

[USEPA] U.S. Environmental Protection Agency. 1972. Recommended bioassay procedure for brook trout *Salvelinus fontinalis* (Mitchill) partial chronic tests. Duluth MN: USEPA.

[USEPA] U.S. Environmental Protection Agency. 1982. User's guide for conducting life-cycle chronic toxicity tets with fathead minnows (*Pimephales promelas*). Duluth MN: USEPA. EPA 600/8-81-011.

[USEPA] U.S. Environmental Protection Agency. 1984. Dicofol: special review petition document 2/3. Washington DC: USEPA Office of Pesticides and Toxic Substances.

[USEPA] U.S. Environmental Protection Agency. 1986. Ecological risk assessment. Hazard evaluation dividison standard evaluation procedure. Washington DC: USEPA. EPA 540/19-83-001.

[USEPA] U.S. Environmental Protection Agency. 1987. Guidelines for the culture of fathead minnow (*Pimephales promelas*) for use in toxicity tests. Duluth MN: USEPA. EPA/600/3-87-001.

[USEPA] U.S. Environmental Protection Agency. 1994a. Short-term methods for estimating the chronic toxicity of effluents and receiving waters to freshwater organisms. 3rd ed. Cincinnati OH: USEPA. EPA 600/4-91-002.

[USEPA] U.S. Environmental Protection Agency. 1994b. Short-term methods for estimating chronic toxicity of effluents and receiving waters to marine and estuarine organisms. Cincinnati OH: USEPA. EPA 600/4-91-003.

Van Der Kraak G, Munkittrick KR, McMaster ME, MacLatchy DM. 1998. A comparison of bleached kraft pulp mill effluent, 17β-estradiol, and β-sitosterol effects on reproductive function in fish. In: Kendall RJ, Dickerson RL, Suk WA, Giesy JP, editors. Principles and processes for evaluating endocrine disruption in wildlife. Pensacola FL: SETAC. p 249–265.

Van Der Kraak G, Munkittrick KR, McMaster ME, Portt CB, Chang JP. 1992. Exposure to bleached kraft pulp mill effluent disrupts the pituitary-gonadal axis of white sucker at multiple sites. *Tox Appl Pharmacol* 115:224–233.

Van Der Kraak G, Parkhurst NW. 1996. Temperature effects on the reproductive performance of fish. In: McDonald DG, Wood CM, editors. Global warming—implications for freshwater and marine fish. Society for Experimental Biology Seminar Series 61. Cambridge UK: Cambridge University Press. p 159–176.

Van Der Kraak G, Zacharewski T, Janz D, Sanders B, Gooch J. 1998. Comparative endocrinology and mechanisms of endocrine modulation in fish and wildlife. In: Kendall RJ, Dickerson RL, Suk WA, Giesy JP, editors. Principles and processes for evaluating endocrine disruption in wildlife. Pensacola FL: SETAC. p 97–119.

Varanasi U, Stein JE. 1991. Disposition of xenobiotic chemicals and metabolites in marine organisms. *Environ Health Perspect* 90:93–100.

Vignier V, Vandermeuler JH, Fraser AJ. 1992. Growth and food conversion by Atlantic salmon parr during 40 days exposure to crude oil. *Trans Am Fish Soc* 12:322–332.

Vogelbein WK, Williams CA, Van Veld PA, Unger MA. 1996. Acute toxicity resistance in a fish population with a high prevalence of cancer. Proc. SETAC 17th Ann. Meeting, Washington, DC. p 67.

Vonier PM, Crain DA, McLachlan JA, Guillette LJ, Arnold SF. 1996. Interaction of environmental chemicals with the estrogen and progesterone receptors from the oviduct of the American alligator. *Environ Health Perspect* 104:1318–1322.

Wagner M, Thaller C, Jessell T, Eichele G. 1990. Polarizing activity and retinoid synthesis in the floor plate of the neural tube. *Nature* 345:819–822.

Walker MK, Cook, PM, Batterman AR, Hufnagle LC, Peterson, RE. 1994. Translocation of 2,3,7,8-tetrachlorodibenzo-*p*-dioxin from adult female lake trout (*Salvelinus namaycush*) to oocytes: Effects on early life stage development and sac fry survival. *Can J Fish Aquat Sci* 51:1410–1419.

Walker MK, Cook PM, Butterworth BC, Zabel EW, Peterson RE. 1996. Potency of a complex mixture of polychlorinated dibenzo-*p*-dioxin, dibenzofuran, and biphenyl congeners compared to 2,3,7,8-tetrachlorodibenzo-*p*-dioxin in causing fish early life stage mortality. *Fundam Appl Toxicol* 30:178–186.

Walker MK, Hufnagle Jr LC, Clayton MK, Peterson RE. 1992. An egg injection method for assessing early life stage mortality of polychlorinated dibenzo-*p*-dioxins, dibenzofurans, and biphenyls in rainbow trout (*Oncorhynchus mykiss*). *Aquat Toxicol* 22:15–38.

Walker MK, Peterson RE. 1991. Potencies of polychlorinated dibenzo-*p*-dioxins, dibenzofurans, and biphenyl congeners, relative to 2,3,7,8-tetrachlorodibenzo-*p*-dioxin, for producing early life stage mortality in rainbow trout (*Oncorhynchus mykiss*). *Aquat Toxicol* 21:219–238.

Walker MK, Peterson RE. 1994. Aquatic toxicity of dioxins and related chemicals. In: Schecter A, editor. Dioxins and health. New York: Plenum Press. p 309–346.

Walker MK, Spitsbergen JM, Olson JR, Peterson RE. 1991. 2,3,7,8-Tetrachlorodibenzo-*p*-dioxin toxicity during early life stage development of lake trout (*Salvelinus namaycush*). *Can J Fish Aquat Sci* 48:875–883.

Weber DN, Spieler RE. 1994. Behavioral mechanisms of metal toxicity in fishes. In: Nalins DC, Ostrander GK, editors. Aquatic toxicology. Molecular, biochemical and cellular perspectives. Boca Raton FL: Lewis Publishers. p 421–467.

Whipple JA, Yocum TG, Smart DR, Cohen, MH 1978. Effects of chronic concentrations of petroleum hydrocarbons on gonadal maturation in Starry Flounder, *Platichthys stellatus*. Proceedings from Conference on Assessing Ecological Impacts of Oil Spills. 14–17 June 1978; Keystone CO: AIBS. p 757–806.

White DH, Mitchell CA, Hill EF. 1983. Parathion alters incubation behavior of laughing gulls. *Bull Environ Contam Toxicol* 31:93–97.

White DW, Mitchell CA, Prouty RM. 1983. Nesting biology of laughing gulls in relation to agricultural chemicals in south Texas, 1978–81. *Wilson Bull* 95:540–551.

White R, Jobling S, Hoare SA, Sumpter JP, Parker, JP. 1994. Environmentally persistent alkylphenolic compounds are estrogenic. *Endocrinology* 135:175–182.

Wilson PJ, Tillit DE. 1996. Rainbow trout embryotoxicity of a complex contaminant mixture extracted from Lake Michigan lake trout. *Marine Environ Res* 42:129–134.

Wingfield JC, Matt KS, Farner DS. 1984. Physiological properties of steroid hormone-binding proteins in avain blood. *Gen Comp Endocrinol* 53:281–292.

Wolf K. 1954. Progress report on blue sac disease. *Prog Fish-Cultur* 16:51–59.

Woodward AR, Jennings ML, Percival HF. 1989. Egg collecting and hatch rates of American alligator eggs in Florida. *Wildlife Soc Bull* 17:124–130.

Woodward AR, Jennings ML, Percival HF, Moore CT. 1993. Low clutch viability of American alligators on Lake Apopka. *Fl Sci* 56:52–63.

Xiong F, Suzuku K, Hew CL. 1994. Control of teleost gonadotropin gene expression. In: Farrel AP, Randall DJ, editors. Fish physiology, Volume 13. San Diego CA: Academic Press. p 135–160

Yamamoto JT, Donohoe RM, Fry DM, Golub MS, Donald JM. 1996. Environmental estrogens: implications for reproduction in wildlife. In: Fairbrother A, Locke LN, Hoff GL, editors. Noninfectious diseases of wildlife. Ames IA: Iowa State University Press. p 31–51.

Zabel EW, Cook PM, Peterson RE. 1995a. Potency of 3,3',4,4',5-pentachlorobiphenyl (PCB 126), alone and in combination with 2,3,7,8-tetrachlorodibenzo-*p*-dioxin (TCDD), to produce lake trout early life stage mortality. *Environ Toxicol Chem* 14:2175–2179.

Zabel EW, Cook PM, Peterson RE. 1995b. Toxic equivalency factors of polychlorinated dibenzo-*p*-dioxin, dibenzofuran, and biphenyl congeners based on early life stage mortality in rainbow trout (*Oncorhynchus mykiss*). *Aquat Toxicol* 31:315–328.

Zabel EW, Walker MK, Hornung MW, Clayton MK, Peterson RE. 1995. Interaction of polychlorinated dibenzo-*p*-dioxin, dibenzofuran, and biphenyl congeners for producing rainbow trout early life stage mortality. *Toxicol Appl Pharmacol* 134:204–213.

Zacharewski TR, Berhane K, Gilesby BE, Burnison BK. 1995. Detection of estrogen and dioxin-like activity in pulp and paper mill black liquor and effluent using in vitro recombinant receptor/reporter gene assays. *Environ Sci Technol* 29:2140–2146.

Zahm GR. 1986. Kesterson Reservoir and Kesterson National Wildlife Refuge, California USA. History, current problems, and management alternatives. In: McCabe RE, editor. Transactions of the North American wildlife and Natural Resources Conference, No. 51. 1986 Mar 21–26; Reno NV. Washington DC: Wildlife Management Institute. p 314–329.

Zohar Y, Pagelson G, Tosky M. 1988. Daily changes in reproductive hormone levels in the female gilthead seabream *Sparus aurata* at the spawning period. In: Reproduction in fish. Basic and applied aspects in endocrinology and genetics. *Les Colloques de l'INRA*. 44:119–125.

CHAPTER 6

Using Reproductive and Developmental Effects Data in Ecological Risk Assessments for Oviparous Vertebrates Exposed to Contaminants

James Clark, Kenneth Dickson, John Giesy, Robert Lackey, Ellen Mihaich, Ralph Stahl, Maurice Zeeman

Ecological risk assessment (ERA) is a tool often used to support the risk-based, decision-making process. An essential step in the process of applying reproductive and developmental effects data within a risk assessment context is developing an understanding of how the information is relevant to the risk-based policy question at hand. This chapter addresses how reproductive and developmental effects data from studies of contaminants can be used in ERAs in which oviparous vertebrates are primary resources of ecological and/or policy concern. We also discuss how a number of important policy, technical, and procedural topics are addressed in the course of framing and implementing the risk assessment. One of our objectives is to determine if the current ERA paradigm needs modifications to address unique risks of contaminants to oviparous vertebrates. A secondary objective is to provide background information on ERA to toxicologists, physiologists, ecologists, chemists, and modelers who are knowledgeable about oviparous vertebrates, illuminating how the results of their disciplines may be used in this process.

Historically, some adverse effects on oviparous vertebrates studied under field conditions have been attributed to exposure to synthetic chemicals, leading to the hypothesis that oviparous vertebrates are unique and may require special consideration in the ERA process. A critical examination of the species that have been adversely impacted by exposure to chemical stressors indicates that both viviparous and oviparous species have been affected. The commonalties among affected species primarily are factors of their life histories other than reproductive strategy. Often, affected species are either aquatic or feed on aquatic organisms. Generally, organ-

CHAPTER PREVIEW

Reproductive and Developmental Effects of Contaminants in Oviparous Vertebrates. Richard T. Di Giulio and Donald E. Tillitt, editors.
©1999 Society of Environmental Toxicology and Chemistry (SETAC). ISBN 1-880611-37-6

isms are at greatest risk when they are at the top of the food chain, which tends to maximize exposure to persistent contaminants. Reproductive strategy per se is less of a risk factor. As described in other sections of this book, oviparous and viviparous species with "r-selected" and "K-selected" life histories have been affected by exposures to chemicals. The affected species also are at risk due to some common use patterns and properties of compounds that have caused most of the adverse, population-level effects. Chemicals that are used over wide geographic areas, such as organophosphate pesticides, or those that are common constituents in industrial waste streams, such as polycyclic aromatic hydrocarbons (PAHs), have been implicated in a number of studies. However, the most common attributes of compounds reported to cause adverse effects are their hydrophobic properties, environmental persistence, and bioaccumulative potential.

General background on ecological risk assessment

In the 1990s, ERA became the dominant, decision-making tool supporting incorporation of ecological consequences into the resolution of some types of environmental policy questions, but it is neither a new concept nor a relatively refined tool in ecological decision-making (Regens 1995). Risk assessment has been used most effectively in casualty assessments surrounding the incidence of unexpected events (i.e., automobile, health, and life insurance, flood management, and nuclear accidents). Specifically, risk assessment is used to estimate the likelihood of an event occurring that is clearly recognized as adverse. This concept has been adapted for use in decision-making with regard to ecological issues. For example, ERA may be applied to estimate the likelihood of a specific, adverse event occurring within a specified magnitude of effect, such as determining the likelihood that a species will go extinct. A key requirement is that the ecological consequence of exposure to a stressor is adverse by definition, which enables the analyst to conduct the risk assessment. In classical risk assessment, the assumption of what is adverse is relatively easy to justify: a nuclear accident is universally accepted as adverse, as is an automobile fatality, a skiing injury, a heart attack, or personal loss due to fire or theft. It has been difficult to achieve scientific and political consensus regarding an analogous list of adverse ecological events.

Ecological risk assessment has enjoyed widespread support and has become a common analytical tool in environmental policy analysis (Molak 1996), although the use of ERA continues to be controversial (O'Brien 1995; Pagel 1995). Like all analytical techniques used to assist policy analysis or decision-making, ERA has strengths and weaknesses. Ecological risk assessments are used appropriately in some circumstances but not in others (Mazaika et al. 1995). The emerging consensus appears to be that ERA will be useful in decision-making for at least a certain class of policy questions: those dealing with the effects of chemicals, especially where there is a reasonably clear legislative or policy basis for defining what is ecologically "adverse."

To be technically tractable and credible, the risk "problem" must be defined in fairly narrow policy and scientific terms (Friant et al. 1995). Often, issues related to ecological resources that are affected or utilized by diverse public and private interests quickly become complicated because of competing priorities, multiple-use strategies, and differentials between rates of resource recovery or renewal for each of the uses. Even if an issue can be defined in fairly narrow terms, the analysis may be technically and scientifically quite complex and may require sophisticated and detailed ecological information.

Most often the issues are brought together by a legislative or policy mandate. The risk problem then becomes relatively simple analytically, e.g., one chemical is the stressor causing effects on a few biological components; the effects, if present, are adverse by definition. In the absence of such a legislative or policy mandate, policy decisions related to "ecological problems" for multiple use resources appear to be simply too complicated to be addressed by traditional risk assessment methods. At best, the approaches resort to arguable assumptions about societal values and preferences or technical simplification that shrouds the essence of the decision or policy issue (Menzie 1995; O'Brien 1995; Power and Adams 1997; Lackey 1997). The traditional definition of risk, as the probability of occurrence of a defined, adverse event, often is relaxed in practice to merely predicting the likely extent of change for an ecosystem component exposed to a stressor of concern.

However, assessing the risk of contaminants on oviparous vertebrates is an area in which the decision tool, ERA, and the biota of concern, oviparous vertebrates, appear to be well matched. The efforts of this workshop are focused on effects driven by chemical stressors. In real-world applications, ecological risk assessors and managers often are confronted with the single, multiple, and cumulative effects of stressors originating from chemical, biological, and physical changes operating in the environment (Figure 6-1). This chapter will reflect the workshop focus on contaminant issues and their impact on oviparous vertebrates. Readers are reminded that in some, if not most, instances, contaminant stressors may not be among the primary risk factors that are critical to the survival or sustainability of oviparous vertebrates. Stresses that result from large-scale changes due to natural occurrences, such as floods or fires, or habitat alteration associated with agriculture, forestry, and urbanization, may be more significant than chemical effects when evaluating issues at the ecosystem scale. Therefore, approaches described herein, using a rather narrow context of contaminant effects on oviparous verte-

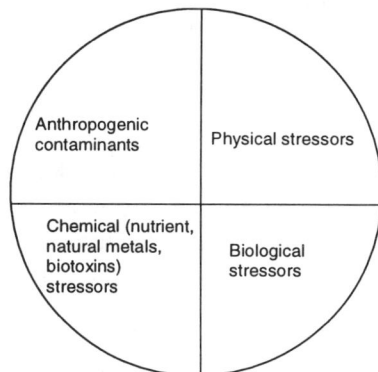

Figure 6-1 Relative contribution of stressors to ecological risk

brates, may not be directly applicable to issues framed with a much broader, and perhaps more ecologically relevant, perspective.

Nevertheless, practitioners of ERA have developed practical and quantitative approaches for assessing contaminant issues. Scientists and managers can use risk assessment tools to address the basic questions regarding potential exposures to toxic chemicals and to discern the potential for, and magnitude of, ecological change. These types of efforts have advanced our understanding of the fate and effects of chemicals released into the environment and of the dynamic biological and ecological interactions that affect the extent of impact chemicals can have. This chapter synthesizes those understandings and demonstrates how they are relevant to assessing ecological effects when information is available regarding reproductive and developmental effects of contaminants on oviparous vertebrates.

Ecological risk assessment is not the only tool that might be used to address questions regarding effects of environmental contaminants on fish and wildlife species. Others have proposed approaches based on use of the precautionary principle, analysis of ecological consequences, or economic analyses applied to ecological and other tradeoffs. Current applications of ERA provide a scientific and quantitative basis to evaluate the potential for and magnitude of adverse impacts, with some specific limitations and incorporating some uncertainties. This chapter will discuss some of these shortcomings.

Ecorisk Approaches Applied to Contaminant Effects on Oviparous Vertebrates

The U.S. Environmental Protection Agency (USEPA) has developed a framework for ERA (USEPA 1992) along with general guidance for its implementation (USEPA 1998). The conceptual ecological risk paradigm and basic approaches described in these documents have received additional elaborations and/or clarifications in peer reviewed literature (Bartell et al. 1992; Calabrese and Baldwin 1993; Suter 1993; Zeeman and Gilford 1993; Karr 1995; Mazaika et al. 1995; Molak 1996). This cumulative base of information is sufficient to address the broad array of ecological risk issues with respect to potential reproductive and developmental effects on oviparous vertebrates from exposures to contaminants.

The framework has been designed and applied with sufficient robustness to allow risk assessments to be conducted for a variety of conditions, such as transient or long-term exposure events, contaminant exposures from any number of environmental sources or pathways, and plant and animal receptors with a variety of life-history strategies. However, obtaining the relevant and appropriate exposure and response information required to execute the types of analyses laid out in these guidelines can be a major obstacle in attempting to conduct an appropriate ERA. Only when the physical, chemical, and biological data utilized in the ERA are obtained from laboratory or field studies that closely represent the conditions addressed in problem formulation can the ecological risks be expressed with significant degrees of certainty.

The various stages of organizing, collecting, analyzing, and communicating information as part of the ERA process, along with additional notations of specific considerations that might be undertaken when addressing contaminant risks to oviparous vertebrates, are presented in Table 6-1. The discussions below briefly describe this process and elaborate on ways in which the activities specifically can support the assessment of reproductive and developmental health issues.

Table 6-1 Stages of ERA and associated activities with considerations specific to oviparous vertebrates

Stage of ERA	Conventional activities	Specific oviparous vertebrate considerations
Problem formulation	Frame issue with risk manager/stakeholders; Identify assessment/measurement endpoints; Identify data needs; Develop conceptual model.	Determine how special status, commercial/ sport value considerations apply; Recognize unique exposure routes.
Exposure assessment	Identify relevant exposure pathways; Quantify oncentrations in media.	Quantify specific exposure routes for each life stage; Recognize potential for food-web exposures, especially persistent contaminants.
Hazard assessment	Assess relevancy of available toxicity data to assessment/ measurement endpoints; Evaluate dose-response relationships; Express toxicity relative to ecological receptors of concern.	Address specific roles of energy allocation and metabolism at various life stages; Take into account unique reproductive cycles, life-history strategies; Recognize contaminant stress in role of energy allocation/metabolism in oviparous vertebrate life history.
Risk characterization	Determine where and when contaminant and receptor co-occur in environment; Express contaminant risks relative to other ecological stresses.	

Problem formulation

An important early step in the ERA process is the problem formulation phase (USEPA 1992, 1998). Problem formulation establishes an understanding of the need for, and the extent of, an ERA. It evolves from iterative discussions between the risk managers requesting an assessment and the risk assessors that will perform the scientific aspects of ERA needed for risk management purposes. Elements that may be discussed include deciding what ecological components could be impacted and which ones may need to be protected; the manner in which to express or measure potential effects in ecologically significant terms; and what types of organisms, populations, communities, and ecosystems could be affected. It is during this phase of assessment that decisions to focus on oviparous vertebrates will be made because of their value to the risk managers (reflecting societal priorities), their vulnerability to the types of chemicals under consideration, or their ecological significance.

Role of risk managers, stakeholders, and policy guidance

When evaluating contaminant risks adverse to the development and/or reproduction of oviparous vertebrates, risk managers are the individuals assigned to consider the risks to those environmental components that are being or could be impacted by the contaminant-use decisions that they have made or could make. Risk managers could include corporate decision-makers considering the development of a specific product by their company, government regulators considering registration of such products, project directors tasked to clean up hazardous waste sites, or state administrators considering the ecological risks of granting air or water permits.

Assessment of the risk to oviparous vertebrates in the environment is only one of several factors that are typically considered in risk management decisions and environmental policy analyses. Other inputs such as tradeoffs between benefits and costs, conflicting societal values and preferences, or technical feasibility factors also may be considered. Other stakeholders, such as affected individuals (e.g., fishermen, landowners), state fish and wildlife officials, public interest groups, etc., also may seek to have input into and influence over some of these decisions.

Defining and improving the quality of interactions between those involved in the ERA process (the scientific risk assessors, the risk managers, and any stakeholder) is currently an actively evolving area (O'Brien 1995; Mazaike et al. 1995; Karr 1995). The social, legislative, and scientific guidance for what risk managers should consider in the environment, and the type and level of ecological change that may be accepted without being labeled and adverse change, appears highly variable (USEPA 1994, 1995d, 1997; SETAC 1997). The legislative and regulatory guidance provided to define adverse ecological change varies considerably. Some laws provide general directions on what to consider regarding protection of ecological resources from reduction, degradation, or loss in quality, quantity, or utility (Zeeman and Gilford 1993). Other laws set specific considerations and technical approaches for establishing water quality criteria that protect groups of aquatic organisms from identified priority-pollutant chemicals (USEPA 1986; Hudiburgh 1995). A few laws provide specific guidance to risk managers in defining harm and what the focus of restoration efforts should be, e.g., the Endangered Species Act (USEPA 1995d).

Deciding what to protect (and to what level)

Part of the discussions during problem formulation is deciding which oviparous vertebrates to protect from adverse developmental and reproductive effects. There are a number of reasons for focusing on oviparous vertebrates. These species can be readily observed in their habitats and some are important for economic and/or recreational purposes. Therefore, they may be highly noticed and valued by the public and natural resource managers. Any mortality, noticeable reduction in population abundance, or absence from certain environments is likely to be considered an important, adverse effect by the public and natural resource managers. In addition, some stakeholders view representatives of such highly visible and valued species as sentinels that integrate adverse effects to other species that may be masked by the dynamics of an ecosystem. Further, oviparous vertebrates are considered by some to be important sentinels for assessing larger, overall ecological effects of chemicals that may pose threats to human health (NRC 1991).

Decisions by risk managers as to what levels of chemical effects are unacceptable are likely to be somewhat variable because there is little technical or practical guidance available on this subject. Unlike human-health risk management, in which society appears to have prioritized risks of cancer at specific levels of incidence, management of ecological risks has not resulted in any such societal consensus. Scientifically, it appears to have become tractable to state that adverse ecological effects at the population level and above will be the principle areas of concern when dealing with organisms in the environment (Suter 1993; Tiebout and Brugger 1995: USEPA 1998). However, this may not be the case for highly endangered species where effects on the individual can have a disproportionate impact on a population when the number of individuals is low.

Using higher levels of ecological organization as the risk management basis, it is reasonable that defining a level of unacceptable reproductive or developmental effects likely will depend upon the specific life-history and population-dynamic characteristics of the species being considered. For example, fish that develop quickly and also spawn large numbers of eggs may be able to withstand certain localized chemical impacts, e.g., decreased egg or sperm viability. A species that requires a long time to reach adulthood and then spawns only a limited number of young may be at higher risk. Also, such long-lived species often face the increased opportunity to become exposed to chemicals, especially from persistent and bioaccumulative contaminants that might build up in body tissues over their lifetime.

Using ecologically significant measures of effect

Ecological significance commonly is applied as a subjective term because it depends on what the risk managers and risk assessors conclude are the most germane attributes. Some risk assessors may consider any incidence of bird, fish, or amphibian mortality caused by chemical contamination to be significant. A risk manager surely will ask what significance such contaminant mortality would have on the exposed populations community, or ecosystem compared with the other, ongoing stresses such as harvesting, disease, predation; this is the "so what" scenario.

A case has been made that perhaps only a certain proportion of species in an aquatic or terrestrial environment, i.e., 95%, would need to be protected (USEPA 1986;

OECD 1995). However, some might also consider subtle impacts on individual organisms as ecologically significant, e.g., effects upon behavior, development, or reproduction in a vertebrate species that is valued for various societal reasons, such as cardinals to watch, trout to catch, or frogs as an indicator of environmental problems.

This workshop has tried to develop a scientific rationale for focusing ecological concerns on contaminant impacts to reproductive and developmental processes of oviparous vertebrates. Therefore, it seems reasonable to raise ecological considerations if a chemical has been shown to cause these types of effects in the laboratory or if that chemical can be found in the environment. Ecological risk assessment is the tool to organize this information to address social and ecological priorities and to determine the likelihood that the potential for contaminant effects will be realized for ecological resources of concern.

Selecting sentinel species, surrogate species, bioindicators

As mentioned above, many oviparous vertebrates often become candidate ecological sentinels because they are sensitive to the adverse effects of chemical contaminants released to aquatic or terrestrial environments. For example, many consider salmon or trout as ideal sentinels for certain aquatic environments, hawks or other raptors as sentinels for specific terrestrial environments, and certain endangered or threatened species as representatives of larger issues surrounding preservation of unique habitats that may need special consideration. Being highly visible and/or valued often is a prerequisite for attaining sufficient attention by the public, elected officials, or organizations interested in protecting natural resources.

If oviparous sentinel species are highly visible, valued, or threatened, they may not be considered appropriate for laboratory studies. However, it is crucial in assessing risks to understand environmental exposure and dose-response relationships using typically measured endpoints such as mortality or adverse changes in development or reproduction. In many cases, suitable surrogate species will be selected to assess the possible risks of chemical contaminants to oviparous vertebrates using biological endpoints that would help express the potential for chemicals to cause adverse ecological effects.

For example, surrogate species often are selected for laboratory tests of short-term or longer-term toxic effects of specific chemicals (Smrchek et al. 1993; Zeeman and Gilford 1993; Zeeman 1997). Surrogate species are selected for these tests because they usually are amenable to laboratory culturing and because many of these species also are suitably sensitive to chemical hazards (e.g., rainbow trout are frequently sensitive to chemical toxicity and thus often are used as surrogates for all trout species, as well as for broader classes of aquatic vertebrates). Surrogate species can provide for a relatively quick, inexpensive, and reliable method to initiate the assessment of the acute and longer-term toxicity of many chemical contaminants.

Models to frame/analyze issues

The problem formulation phase of the ERA includes developing conceptual models as one way to help frame and analyze issues (USEPA 1992, 1998; Molak 1996). The type and complexity of models developed depend a great deal upon the type and complexity of the ecological problem that the risk manager has posed. Traditionally, ERAs are conducted with datasets representing 5 levels of data richness or complexity:

- screening (based on limited acute data or structure activity relationships),
- basic acute and chronic estimation studies (these include developmental and reproduction data),
- multigenerational (full life cycle, chronic effects data),
- population, and
- field.

Ecological risk assessments can range from being fairly simple and straightforward to being complex retrospective assessments. Simple, prospective assessment of the effects of one chemical product on a surrogate species may be achieved with only laboratory test data. On the other hand, complex mixtures of several different types of chemical contaminants may be assessed in an existing field situation. Such a study could assess the potential for adverse impacts over many years, incorporating information on several oviparous vertebrates as well as the invertebrates and plants that are important to the subject animals' survival in that environment.

Exposure-response assessment

Contaminant risks commonly are expressed as a probability function of exposure and resultant magnitude of effects (USEPA 1992; Suter 1993; Molak 1996). The relatively direct, uncomplicated relationships established under laboratory test conditions can provide useful insight into the potential for environmental risks. Exposure-response relationships established from field studies, although more policy relevant, carry varying degrees of uncertainty, depending on the preponderance of evidence supporting diagnosis of individual causative agents. Discussions in previous sections of this book have demonstrated the complex physical, biological, and chemical processes that can affect both exposure conditions and toxic responses of oviparous vertebrates to chemical exposure under laboratory or field conditions.

Applying these types of information to contaminant effects requires considerable evaluation as to the applicability of available data to the specific task at hand. Challenges include extrapolating data from laboratory to the field or between various taxonomic groups, clearly linking cause-and-effect information, developing a clear understanding of the toxic mode of action under consideration, and determining the ecological relevance of potential effects. The technical basis for setting a basic exposure-response relationship can become the key issue for accurately representing risks of chemical exposure to wildlife resources.

When the dose-response information necessary to conduct an ERA is focused specifically on contaminant effects on reproduction or development of oviparous vertebrates, several lines of evidence can be used. Depending on how the specific issue has been framed in the problem formulation phase, the results of laboratory tests that measure survival, growth and reproduction can be used to estimate hazard. This information may be assessed in terms of likelihood of observing impacts among exposed individuals or converted to population-level responses based on assessments of population dynamics. In this application, incremental increases in exposure are assessed in terms of the potential to affect the reproductive performance or developmental rate (efficiency) of individuals, or increased exposure is evaluated as to how it affects the ability of a population to sustain itself. Additional detail on these applications is provided in specific sections below.

Exposure

Obtaining an accurate estimate of contaminant exposure requires consideration of a wide variety of physical, chemical, and biological factors that may influence the magnitude and duration of contaminant exposure of oviparous vertebrates. The multiple life-stages characteristic of oviparous vertebrates may necessitate a more detailed and complex assessment of potential exposure pathways and vulnerable periods of exposure. In particular, some oviparous species are at the upper levels of trophic food webs. Therefore, analyses of exposures through food sources is an important consideration, especially when the contaminants under consideration are persistent and bioaccumulative chemicals.

Routes of exposure, contaminant bioavailability, and assimilation by receptor organisms can vary widely among species and habitats. Understanding the role of these factors is essential in quantitative risk assessments, since the actual duration and intensity of exposure are central factors in calculating risk. Even when duration and intensity of exposure can be quantified with confidence, low-level exposures that occur at sensitive life stages or during key developmental stages may pose significantly greater risks than do longer duration exposures to greater concentrations that might occur during less vulnerable life stages. Complete and accurate characterizations of exposure may be necessary if greater attention to the uncertainties in the risk assessment are required to resolve a risk management question.

Conceptual model of exposure/bioavailability

A number of concepts and factors must be considered when characterizing the exposure of ecological resources to environmental contaminants. Exposure of oviparous vertebrates is particularly complex because there are many potentially significant pathways to address (Figure 6-2). Bioavailability of contaminants is controlled by a number of environmental fate factors that operate in the various media where contaminants occur (i.e., air, water, soil/sediments, and food). Properties of a contaminant interact with properties of the environment to determine the status of the contaminant, including compartmentalization, chemical

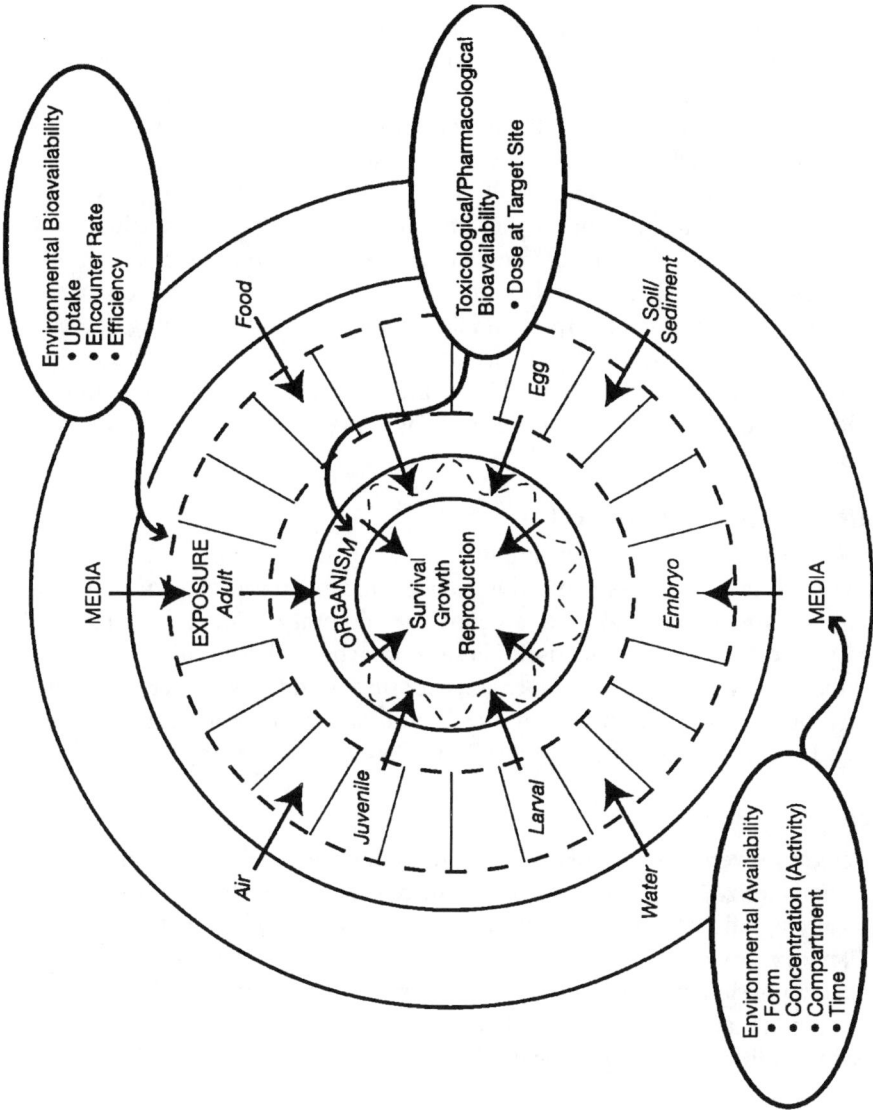

Figure 6-2 Considerations that link environmental exposures with effects of contaminants on oviparous vertebrates

form, concentration (activity), and persistence. In Figure 6-2 this is termed "environmental availability;" it is dependent on the chemical/physical properties of the compound and the environment to which it is released.

Oviparous vertebrates may be exposed incidentally or continually during various stages in their life cycle. Each stage may differ in encounter rates, uptake rates, and transfer efficiency. In all situations, exposures occur at the interface between the media and the life-cycle stages. Environmental bioavailability is the result of environmental availability and life-stage interactions. Toxicological/pharmacological bioavailability is the fraction of the dose absorbed/adsorbed by the life stages of the organism that reaches the target site that initiates biological response. The absorbed/adsorbed dose is a function of the activities specific to the organism's various life stages that can influence the effective intracellular concentrations. Thus, quantification of exposure must be conducted with significant understanding of physical, chemical, and biological dynamics. Since effects on survival, growth, and reproduction are key endpoints from an ERA perspective, understanding how the dose affects each life stage and the consequences of the effects on these ultimate measures of effect is critical. Figure 6-2 illustrates conceptually the kinds of information needed to assess exposure and effects of any contaminant to any oviparous vertebrate.

Next steps in exposure assessment

Rather than continuing to rely on simplistic assessments of exposure for any but the most simple screening assessments, the science is moving towards quantitative exposure assessments that take into account the temporal and spatial variability of exposure conditions for a population. These approaches utilize multimedia, probabilistic distributions of exposure and exposure conditions, rather than single point estimates meant to represent average conditions or values indicative of the upper end of the exposure range (Moore 1996; Power 1996; Cowan et al. 1995).

Effects

As covered by several sections of this book, reproductive and developmental effects can be characterized as occurring within acute or chronic time frames, leading to biologically significant impacts on survival, growth, or reproduction. Other sublethal changes may lead to similar adverse effects by indirect means, such as changes in behavior, physiology, or biochemistry. The key to incorporating indirect effects information into an ERA is to establish a meaningful link between the measured change in condition (e.g., physiological, biochemical, behavioral) to an ecologically relevant change.

The means by which biologically significant effects will be differentiated from ecologically significant changes is an important challenge that risk assessors must work through with risk managers during the problem formulation stage. When the risk assessment is targeted at higher levels of ecological organization (e.g., community, ecosystem), these distinctions can be particularly troublesome. Risk assessors must conduct comprehensive evaluations with sufficient completeness to ensure that measured effects can be placed into an ecological perspective relative to the temporal and spatial patterns of change that might be expected within the normal range of ecological variability. Obviously, large-scale, irreversible effects pose significantly greater ecological risks than more localized, transient changes. Changes in the reproductive capacity or incidence of developmental irregularities assessed at the population or community level of organization must be investigated with sufficient detail to allow perspective on ecological significance, providing a basis for meaningful, risk-based decision-making.

Application of Ecological Risk Assessment to Oviparous Vertebrates

General data considerations in ecological risk assessment

While many commonly used measurement endpoints are not specific to oviparous vertebrates, there are some endpoints that are particularly useful for these types of animals (Table 6-2). This list is intended to represent some, but not all, of the measurements that may be important. Depending on how the risk problem has been formulated, not all of these measurements will be needed in ERA. From an ERA perspective there is little that is unique about oviparous vertebrates, and few, if any, unique measurement endpoints would need to be selected to address most contaminant effects issues. In general, it is useful to have information on both inter- and intraspecies variability in response for the species of interest as well as estimates of the concentrations, duration, and timing of exposure. Though the specific responses to stressors might vary between oviparous and viviparous species, the fundamental concepts are not different. Both types of reproductive strategies provide opportunities to study a range of possible exposure-response endpoints that could be measured under both laboratory and field conditions. All of this information should be extrapolatable to real-world conditions and related to ecologically relevant effects endpoints.

Currently, effects on survival, growth, or development to maturity and reproductive performance are the most common endpoints considered when assessing contaminant risks for oviparous vertebrates. This information, which is based on the responses of individuals and specifically on the energetics of individual organisms, can be used to support population-level assessment endpoints. In aggregate, these types of data can provide estimates of rates of response that can be measured and expressed as ecologically relevant endpoints. The fundamental toxicological explanation behind their utility is the understanding that organisms have a limited amount of energy available to them. It takes energy to actively avoid a contaminant and to either excrete or detoxify contaminants (Forbes and Calow 1996). If indi-

Table 6-2 Example measurement endpoints that might be used to characterize contaminant effects on oviparous vertebrates and their relevance in ERA

Endpoint	Y/N
I. Survival of adults (Mortality)	Y
II. Fitness of adults	Y
Neuro-endocrine function	Y
Immune function	Y
Behavior	?
II. Growth of adults (energetics)	Y
Size	Y
Weight	Y
Lipid content	N
III. Reproductive output	Y
Fertility	Y
Fecundity	Y
Mortality of embryos	Y
Deformities (Teratogenicity)	Y
IV. Survival, growth, development of immature life stages	Y
Survival	Y
Size/weight	Y
Deformities	Y

vidual organisms need to expend energy to resist a stressor, chemical or otherwise, then less energy would be available for other functions (Calow 1994). These types of measurement endpoints are referred to as scope for growth and scope for reproduction.

Other less direct endpoints also can be measured as indicators of the potential for contaminants to adversely affect survival or reproduction. These endpoints could include behavioral changes or altered immune response that might affect the fitness of individuals or populations (Zeeman 1996). Various physiological or biochemical measures of toxicity or tissue damage could be used as measurement endpoints, as long as they are linked to the primary measurement of effect to individuals or population fitness.

The unique characteristic of oviparous vertebrates is that the embryo develops outside the body of the parent and is enclosed in a protective covering. However, there are many other important aspects of life history and physiological function that can affect how they respond to contaminant exposures and must be considered (Table 6-3). These may lead to greater or lesser exposure to stressors, depending on the specific nature of the species and stressor of interest.

Table 6-3 Characteristics that may influence the responsiveness of organisms to environmental contaminants

Characteristic	Y/N
Occurrence in higher trophic levels	Y
"r" or "K" reproductive strategy	N
Chemical tolerance	N
Immune function	N
Endocrinology	N
Exposure pathways, duration, magnitude	Y
Energy metabolism	N
Developmental stages	?
Behavior	?
Nerve function	N
Physiology	N
Biochemistry	N
Nutrition	N
Habitat	?

Issues in applied ecological risk assessment

In the application of ERA, it is important to consider that all types of information (Tables 6-2 and 6-3) may or may not be needed for making a final risk decision. Some types of needed data may not be available, requiring risk assessors to extrapolate data from other species, endpoints, or chemicals. Such extrapolation may lead to decisions based on a high degree of technical uncertainty. Alternatively, the lack of data, or availability of only highly uncertain information, may delay the decision-making process until a better scientific basis can be developed. Decisions regarding ecological risks that are perceived to be small in magnitude, localized, or of less concern to society will not require as much information and certainty as ecological risks that are perceived to be of greater severity, to be widespread in nature, or to involve oviparous vertebrates afforded special protection by law or policy.

Ecological risk assessment is practiced as an iterative and hierarchical evaluation process, especially when conducted as a prospective assessment (Zeeman and Gilford 1993). There is a premium on making accurate, scientifically defensible decisions in the most rapid, cost-effective manner possible. A tiered approach, which builds in multiple decision points for analysis of increasingly refined exposure and hazard estimates, has been an effective way to implement risk assessments. At all levels of the process, the probability of adverse consequences is expressed as a function of the degree of overlap between exposure and hazard.

In early tiers, this comparison is conducted by computing a hazard ratio, or hazard quotient, which is the quotient of some integrative measure of exposure divided by hazard. The estimate of exposure could be from a mathematical model or empirical data. Exposure estimates could be expressed as concentrations in various environmental matrices (e.g., air, water, food) or concentrations in the receptor's whole body or in specific tissues. Hazard is a measure of the magnitude of adverse effect, based on toxicity for chemical contaminants, associated with varying degrees of exposure and can be expressed as the concentration in environmental media, diet, or tissues that results in some defined level of effect. A dose or concentration that does not cause significant adverse effects is referred to as a reference dose or reference concentration and serves as a benchmark against which various exposure scenarios are evaluated.

The tiered approach begins by comparing the most relevant, readily obtained estimates of exposure and hazard. A hazard ratio of greater than 1.0 is used to indicate that there is a basis to express a level of concern (Rodier and Mauriello 1993), which might be sufficient to take some action or to justify further, more detailed evaluation of the stressor. This first phase of the risk assessment, screening, generally is based on purposefully conservative scenarios and parameter estimates that would be significantly protective of the ecological resources under consideration. Thus, exceeding a ratio of 1.0 in this tier does not necessarily mean that there would be risk but, rather, that there is reason to consider action or to further refine the risk assessment and reduce uncertainty. Generally, this is the point at which more detailed information on special exposure and effects conditions relative to oviparous vertebrates would begin to be utilized.

As currently practiced, the ERA process is dynamic and flexible. It often relies on professional judgment for assessing the relevancy of various studies and input data and for extrapolating laboratory data to field populations. This is true whether the assessment is for oviparous vertebrates or another ecological receptor. In this process, all relevant information eventually may be considered. However, there is a hierarchy or utility of the information, which can be useful for the ERA (Table 6-4).

If information that can be directly applied to the ecologically relevant processes of survival, growth, development, and reproduction of individuals is available, decisions on the probability of a contaminant affecting sustainability of populations generally are possible. The information that would be considered most relevant and given the greatest weight is empirical field data for which appropriate dosimetry and confounding factors are fully characterized. Similarly, multi-generation laboratory studies that measure a range of indicators of population sustainability and are conducted under a dosimetry regime expected in the field would receive greater weight. If these types of data are not available, information less directly applicable to the scenario under consideration is combined to provide insight into the probability of adverse population effects. The ERA process is sufficiently flexible to include a range of specific endpoints for oviparous vertebrates.

Table 6-4 Hierarchy of data potentially useful in ERA data generation and use that is currently feasible or practiced

Exposure data	Unique to O/V	Hazard data	Unique to O/V
Chemical identification name class	N	Acute toxicity Survival data Behavioral data	N
Physical chemical data Kow Koc Henry's constant	N	Chronic toxicity Survival data Growth data Developmental data Reproductive data Early life stage test Embryo-larval test	N
Persistence data hydrolysis photolysis biodegradability	N	Whole life-cycle toxicity Survival data Growth data Developmental data Reproductive data	N
Information on environmental releases chemical production use patterns	N	Population models Productivity Trophic dynamic	N
Fate models partitioning multi-media fate site specific	N	Field studies Mesocosm Controlled release Monitoring	N
Predicted environmental concentrations generic site/use specific	N		
Bioavailability bioconcentration biomagnification bioaccumulation	N		
Measured body burden whole body target tissues egg residue Lab studies Field studies Environmental monitoring	N		
Cheaper/faster analytical chemistry for exposure characterization Air Water Soil Sediments Food	N	Develop or improve surrogates available for testing Amphibians Reptiles Passerine birds	Y
		Better characterization of baseline developmental/ reproductive stages	Y
		Multigenerational toxicity tests P_1-F_1-F_2	N

Information that provides the most direct estimates of ecologically relevant endpoints would be given the greatest weight in ERAs. In this section, we provide a narrative of some examples of decisions that often are made relative to the relevance or utility of information. The problem formulation stage should delineate the extent to which oviparous vertebrates have special exposure vectors and sensitivity, information that dictates specific data acquisition tasks.

One of the greatest limitations to ERA, as with any other environmental decision-support tool, is a lack of information on the species of interest. When necessary, surrogate species can be used to estimate the hazard. Similarly, measured concentrations of exposure and dose are preferable to those estimated from models. In all cases, assumptions must be made about the extent of correspondence between the exposure conditions of the receptors of concern and the exposure-response information available to support the risk assessment. Such assumptions introduce uncertainty in the conclusions of the risk assessment. The more assumptions that are made in the assessment, the less certain the conclusions. In some cases the risk assessor may decide that there is so little information available that no estimates of risk should be made. This may result in a decision being made based on other criteria or in the decision being delayed until additional information has been generated.

Selection of exposure or toxicity data from surrogate species has been addressed in a variety of ways. Suter (1993) discusses the range of sensitivity among various taxa and the various attempts that have been made to extrapolate toxicity data among aquatic species, along with a caution that it is necessary to understand a chemical's mode of action to make appropriate extrapolations. Calabrese and Baldwin (1993) discuss derivation and extrapolation of toxicity reference values (TRVs) for birds and mammals and the accompanying cautions regarding extrapolations to other species. Because of the elaborate phamacokinetic and biochemical interactions that take place when a chemical elicits reproductive and developmental effects in oviparous vertebrates, selection and application of surrogate-species data must be conducted with considerable caution and enlightenment.

In addition, the route and timing of exposure need to be considered in evaluating the utility of various types of hazard data that may be used in the risk assessment process. For instance, egg injection studies can be conducted with most oviparous vertebrates. These types of studies have the advantage of precisely controlling the dose delivered to the embryo and may be appropriate for some persistent compounds that have their greatest effects directly on the developing embryo. However, risks from contaminants that affect reproduction by way of effects on the adult or from those contaminants that are metabolized in such a way that they are not deposited in the egg would not be accurately represented if based on egg injection studies. In that case, egg injection studies would be given less weight in the risk assessment process.

Other types of laboratory-derived data from biochemical or physiological studies pertinent to reproductive and developmental processes can serve to augment the ERA. Physiological data, life history, behavior, home range, breeding conditions, and a host of other types of ancillary information can be useful in the context of ERA (USEPA 1993b, 1993c; Pulliam 1994). Such data are particularly useful if they are mechanistic in nature, are applicable to physiological processes that are most sensitive to the type of stressor under consideration, or are unique to the receptor of concern. Even so, supportive data generally are not the key information upon which the ecological risk is based. For that reason, the need to generate these types of information de novo should be discussed early on and agreed to among those involved with the ERA and risk management decision (USEPA 1993a, 1995d).

Data analyses and applications

As noted earlier, the ecological risk assessor usually must rely on a limited set of data and is compelled to modify those data so that they are applicable to the hazard or exposure assessment at hand. It is important to emphasize that this situation is not unique to oviparous vertebrates but is commonly encountered in the field of ERA and environmental toxicology (Mayer and Ellersieck 1986; Calabrese and Baldwin 1993). Sometimes the only information available to the ecological risk assessor may be an LC50, LD50, or EC50, and this information usually comes from surrogates rather than from the species of concern (Zeeman and Gilford 1993; Stahl 1997). With the exception of the EC50, these effects levels are based on mortality and not necessarily on reproductive or developmental effects (Calow 1993). Thus, the ecological risk assessor may apply an uncertainty factor to the mortality data (LC50) and utilize this adjusted dataset in the ERA (Calabrese and Baldwin 1993) to assess the potential for long-term reproductive or developmental effects from lower-level exposures. Next, the ecological risk assessor may need to adjust the data if the study was conducted with a surrogate species (Calow 1993, 1994). At least these 2 types of data evaluations are needed for most ERAs. For more detailed reviews of toxicological data analyses, readers are referred to Calabrese and Baldwin (1993), Suter (1993), and Calow (1993).

Other important analyses of laboratory toxicity test data include deriving estimates of effects at the 20, 10, or 5% levels (i.e., the LC20, LC10, and LC5), generally to obtain estimates potentially relevant to the threshold for observing population effect levels (Barnthouse 1993). Even though there may not be quantitative responses observed at the population level to support selection of these lower-effects levels in the risk assessment, such analyses nonetheless can provide information useful in the ERA (Barnthouse 1993; Suter et al. 1987). Often these types of estimates are utilized with full recognition of the high degree of uncertainty surrounding their derivation.

In addition to short-term studies, analyses of effects of chronic exposure also produce information that can be useful for ERA (Suter et al. 1987). Chronic toxicity tests conducted in the laboratory usually incorporate exposure concentrations that tend to be more reflective of environmental exposure levels, cover a greater proportion of the animal's life stage, and, in some cases, expose sensitive life stages to the contaminant of concern. For these reasons, there generally is a lesser degree of critical analysis, estimation, and extrapolation associated with use of chronic toxicity data (Suter 1996). Chronic studies that include reproductive and developmental endpoints (e.g., egg production, hatching, survival) are some of the most useful datasets for estimating ecological risk at both the individual and population levels (Barnthouse 1993).

Empirical effects levels such as the lowest-observed-effect level, lowest-observed-adverse-effect level (LOAEL), no-observed-effect level (NOEL), and the no-observed-adverse-effect level (NOAEL) are key for ERA purposes because they can become benchmarks against which potential ecological risk can be estimated (Wagner and Lokke 1991; Suter 1993, 1996). These threshold levels of effects generally are measured directly, but other estimates of low level exposures that cause minimal or no effects (e.g., maximum-acceptable-toxicant concentration [MATC]) can be estimated based on the concentration-response or dose-response data available from chronic toxicity tests (Calabrese and Baldwin 1993).

In ERA, laboratory data are useful in the risk estimation phase, in which a quantitative expression of the magnitude and likelihood of adverse toxicological effect is developed. The ecological risk estimate derived from these data may be a point estimate or single value or a range of values, depending on the type of ERA conducted (Calabrese and Baldwin 1993). Another technique for estimating the potential risk is through the use of error propagation models (e.g., Monte Carlo simulation), which can enhance the robustness of the estimate under some conditions (Pastorok, Butcher et al. 1996). However, because the state of the science for probabilistic estimations has not yet fully matured for ERA, these types of estimates have yet to be used routinely (Suter 1993). It is expected that their development and use will increase rapidly in the next several years (Pastorok, Nielsen et al. 1996).

Ecological risk can be estimated with the hazard quotient (HQ) approach (Suter 1993; Tiebout and Brugger 1995), which has a long history of use in aquatic toxicology (Barnthouse et al. 1982; Zeeman and Gilford 1993) and is relatively well understood by both lay and technical groups. In practice, a NOAEL, LOAEL, or MATC from field or laboratory studies can become a TRV for anHQ analyses (Suter 1996). The ratio of the expected exposure to the TRV results in an HQ.

One additional important task in the risk characterization phase is to identify the uncertainties in the ERA, summarized in Table 6-5. Uncertainty is what we do not know or are unsure of and is not necessarily derived from the same sources as experimental error or variability (Stahl and Clark 1998). In the uncertainty analysis phase, the ecological risk assessor should document clearly the variability in available data, those areas in the risk assessment where data are lacking, and ways in which the absence of these data may impact the outcome of the ERA. It is important for the ecological risk assessor to point out these missing data to the risk manager, but it is not necessary that they also provide solutions to the problem. In many cases the risk manager will want to understand whether the uncertainties are significant enough to warrant further investigation or data collection. If so, then the assessment returns to the problem formulation phase, or, if the data required are easily collected and interpreted, they may be obtained without going through the major phases of the ERA framework.

Table 6-5 Summary of uncertainties encountered in the ERA of contaminants and oviparous vertebrates

Type of uncertainty	Level		
	Low	Medium	High
Extrapolation of laboratory data in non-oviparous surrogate to oviparous vertebrates of concern			x
Extrapolation of non-reproductive endpoint data to estimate reproductive endpoint or effects level		x	
Importance of maternal exposure vs. exposure to eggs or other sensitive developmental stages		x	
Understanding importance of paternal exposure on the egg and other sensitive life stages			x
Extrapolating maternal exposure scenarios to potential effects on the egg		x	
Ability to generalize effects between terrestrial oviparous vertebrates and aquatic species (extrapolating either way)			x
Understanding basic developmental and reproductive processes and linkage to potential ecological risks	x		
Understanding basic biochemical and physiological processes and their linkage to potential ecological risks			x
Extrapolation of laboratory-derived data to field situations			x

Derivation and use of relevant effects estimates from field studies

The use of effects data derived from field studies rather than to laboratory-derived data in many cases gives the ecological risk assessor information that is more applicable to a site-specific situation (Ma et al. 1991, 1996; Maughan 1993; Keddy et al. 1994). Field studies can provide direct evidence of reproductive and developmental effects, some of which can be readily observed or measured (Herricks et al. 1989; Giesy et al. 1994; Guillette et al. 1994; Hose and Guillette 1995). However, some of the effects on reproduction and development are not so pronounced and may require in-depth studies. In addition, it is difficult to diagnose the specific causative stressor, or suite of stressors, responsible for the observed effects. These types of efforts often utilize a preponderance of laboratory and field information to assess risks from contaminants found in the environment. Perhaps the most difficult aspect of using field studies is the level of effort or cost required to obtain scientifically suitable, reproducible results. In most cases, field studies need to be of sufficient length to allow researchers to understand the natural variability in a population; depending on the oviparous vertebrate involved, this can take several years (Kendall and Lacher 1994). Given the policy constraints placed on making timely decisions, seldom do risk assessors and risk managers have the option of undertaking such lengthy studies.

Types of Data/Information Needed or Useful to Assess Ecological Risks

For several reasons, ERAs should be conducted for populations of organisms rather than for individuals. The most common experimental unit upon which effects measurements are made is the individual organism. However, when statistical approaches are applied, the endpoints become population-level estimators. This is useful because ERAs seek to assign probabilities of response to given exposures, and those responses are expressed as rates. Mortality, which is a population-level phenomenon, is a crucial measurement used in ERA. Some responses of individuals such as fecundity and fertility of eggs can be expressed as rates, while others such as lethality cannot. In ERA applications, the information on population-level effects is most closely aligned with current ecological risk management policies under which contaminant releases are evaluated. Populations also are the level of ecological organization at which the technical capabilities of applied ecology are most closely linked to societal values. As one goes from population to higher levels of organization (e.g., food web to ecosystem to landscape scales) or from population to lower levels of biological organization (e.g., individual organisms to cellular level to subcellular level) it becomes increasingly difficult and complex to assess contaminant effects in terms that are relevant to societal values and policy directives (Figure 6-3).

At higher levels of organization, the potential number of stressors may be greater than those acting on a population, since the number of physical, chemical, and biological factors increases as one considers wider temporal and spatial scales associated with these larger constructs. Technically, it is more difficult to obtain credible measures of ecological effect at these higher scales. This leads to increased variability in measures of effect at the higher levels of organization and greater uncertainty regarding their predictive utility in supporting decision-making. Predicting ecological effects from studies conducted at levels of organizations below the individual organism also presents difficulty. Cellular and subcellular studies

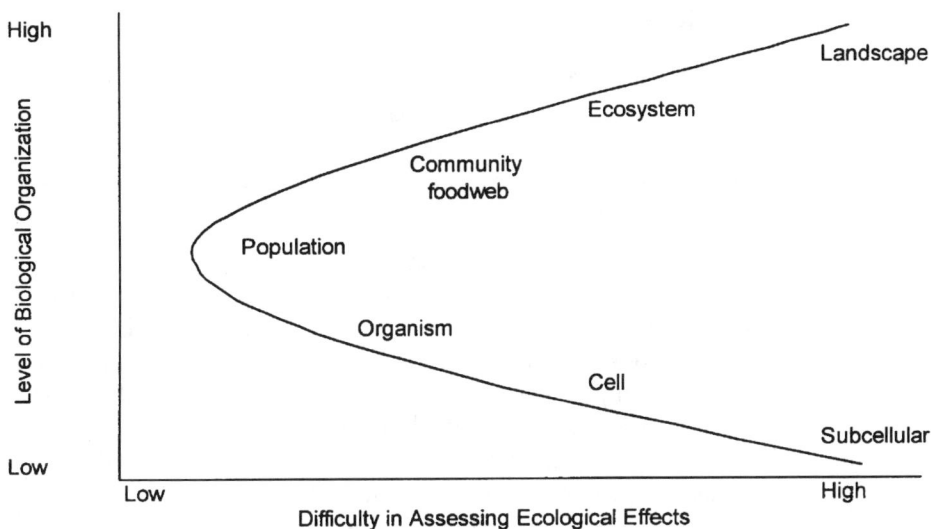

Figure 6-3 Comparison of degree of difficulty in assessing ecological effects from chemical effects data collected at different levels of biological organization

tend to be less relevant to the policy and societal values that are used to frame ERA questions than are studies conducted at higher organizational levels. Often measurements at lower levels of organization are easier to obtain and to replicate than are data collected at the higher levels. Although they may provide less variable information, these types of data contain residual uncertainty as to their utility in assessing risks relevant to ecological questions.

Historically, ecological risks have been assessed using data from various levels of ecological organization. Only recently has the practice of ERA been critically evaluated with sufficient detail and breadth of experience to allow development of a consensus regarding appropriate approaches (Suter 1993; Molak 1996; USEPA 1998). Although population-level approaches have become the most useful techniques for ERAs in contemporary practice (as depicted in Figure 6-3), there are historical, political, legislative, and technical reasons to consider conducting risk assessments at either the individual, population, food web, ecosystem, or higher levels of organization. In particular, the special status and value of some oviparous vertebrates and the technical tools available to deal with assessing contaminant effects provide opportunities for selecting any of the levels of organization, depending on the specific risk management problem to be addressed.

Risks to individuals

When is risk to individuals relevant?

Gould (1995) emphasizes evolution and natural selection acting on individuals. He said, "natural selection acts only for the benefit of individuals in reproductive success . . .Well designed organisms and balanced ecosystems are side consequences."

Because much of the available laboratory test data are generated at the level of individual organisms, assessing contaminant risks to individuals can be a relatively straightforward and rapid evaluation of the magnitude of effects that might be expected under various exposure scenarios. Because of the volume and complexity of information required to implement quantitative risk assessments at higher levels of organization, individual-level ERAs frequently are undertaken as the basis of screening-level, early tier risk assessments. In the process of emphasizing risks of impaired reproduction and development for each individual, the assessment need not become more complicated by addressing dynamic ecological processes that tend to mitigate the importance of individuals in ecological systems. If risk issues cannot be quantified or resolved with sufficient detail to remove a contaminant from further consideration, then ERAs commonly proceed to higher levels of organization.

In some cases assessing the ecological risks of contaminant exposure to individual oviparous vertebrates may be required when the receptors have been designated a species of special concern (e.g., rare/threatened/endangered species). For such species, risks must be managed to ensure full reproductive capacity and developmental potential of all exposed individuals. Even when a species is not of special concern, oviparous vertebrates often are the larger, more visible, and commonly more publicly valued natural resources that might be considered when assessing contaminant risks. Their commercial significance, recreational importance, and charismatic appeal to a wide public audience often make these fish, reptiles, amphibians, and birds highly valued receptors that frequently receive high degrees of protection relative to other ecological receptors. Thus, risk managers may decide to aggressively manage risks, regardless of the ecological significance of any reproductive or developmental change among a few exposed individuals. Risk assessors may not be able to justify setting exposures at risk levels that pose any population-level effect, knowing that the residual exposures may be sufficient to impair the reproduction or development of a few maximally exposed individuals. This type of conservative risk management approach frequently is supported by the concept that protection of all individuals in a sentinel or surrogate species will ensure the sustainability of all higher levels of ecological organization.

How does one assess risks to individuals?

In the contemporary practice of ERA, exposure assessments for individuals among a receptor species quickly moves towards defining conditions for maximally exposed individuals. Exposure conditions can be characterized over a continuum of space and time if one knows the range of territory utilized by individuals over daily, seasonal, and lifelong time frames. Often the default assumptions built into screening-level risk assessments consider limited daily or seasonal movement, low diversity of diet, and homogeneous, upper-range exposure conditions in soil, sediment, water, or air. This process of simplification greatly facilitates the estimation of the magnitude of exposure. If done within some practical limits of realism, this approach can set logical upper bounds on exposure conditions for individual organisms. Of course, exposures can be estimated more quantitatively for individuals for which there has been sufficient characterization of the quantities of environmental contaminants associated with various locations, animal activities, and exposure pathways.

An added advantage to focusing risk assessments on individuals is that much of the laboratory-derived toxicity-effects data are readily extrapolatable to risk management issues. Exposure assessments still have to take into account differences between lab and field exposure conditions (e.g., bioavailability, ionized versus unionized forms, oral versus dermal exposure). When exposure and effects can be compared quickly to existing laboratory data, the risk assessment process can be reduced to a simple mathematical exercise of generating hazard quotients. These are derived from division of environmental exposure concentrations by benchmark exposure values thought to be below levels of concern for receptor species (NOEL, TRVs, or other threshold effects estimates) under laboratory test conditions. Exceedences of effects thresholds for any individuals are deemed unacceptable risks in this approach, ensuring full reproductive and developmental capacity among the exposed receptors.

In nature, the importance of the individual to the sustainability of ecological structure and function is minimized by factors such as intra- and interspecific competition, changes in the extent or quality of habitat unrelated to contaminant effects, or natural cycles in meteorological conditions. Making a decision to manage contaminant risks to the extent that full reproductive and developmental capacity of all individuals in a receptor species will be protected often would not be cost-effective, practical, or necessary to sustain the receptor population. However, it may be warranted or required by regulatory directives or stakeholder values, since oviparous vertebrates continue to be placed at the top of the ecological hierarchy established through public values. Indeed, adverse toxicological effects observed in the field among individuals of oviparous vertebrates tend to raise public and scientific concerns about the status of the local ecosystems. Such observations can serve as an indicator of underlying contaminant effect issues.

Characterizing risks to individuals

Characterizing and communicating contaminant risks to individual organisms requires that risk assessors and risk managers appreciate the differences between ecologically significant changes and biologically significant changes. Even when there are regulatory or stakeholder directives to assess contaminant risks at the level of individual organisms, the risks should be put into an ecological context and a societal perspective. To do so would involve discussing with the stakeholder/ manager the temporal and spatial significance of the contaminant risks, the extent of contaminant effects relative to baseline variability in reproduction and developmental processes, and the severity of the potential contaminant-related effects relative to other physical and biological stresses acting upon the receptors. It is resolutely important that the basic exposure-and-effects assumptions, various extrapolations, and underlying association between the risk management issue and the risk assessment endpoints be fully explained when assessing ecological risks at the level of individual organisms. There should be transparent links between risks to individuals and the issues that drive risk management actions to protect these resources at risk.

Risks to populations

When is risk to populations relevant?

Field and laboratory studies seek to provide stressor-response information for a particular receptor and stressor and, by implication, the potential for impacts on other ecological receptors likely to be exposed to the same stressor. Increasingly, ecological risk assessors and risk managers have begun to focus their efforts on potential ecological risks at the population level and above (Croonquist and Brooks 1991; Sorensen 1996). However, ecological risk assessors and risk managers should recognize that legislation, public values, economics, or other issues may compel that the focus of the assessment be at the level of the individual (e.g., threatened or endangered species) or possibly at levels of organization above the population, including ecosystem or landscape levels. Therefore, with respect to any ecological receptor, the decision to focus the ERA at the population level may not be driven by scientific reasons alone (USEPA 1995d; SETAC 1997).

How does one assess risks to populations?

Exposure assessment

With oviparous vertebrates, an exposure assessment at the population level could be approached in a variety of ways. One step, exposure-pathway analysis, is directed at understanding the total input of a particular contaminant or stressor and ways in which that exposure occurs (Ma et al. 1991; Pascoe et al. 1996). It is particularly important to understand complete exposure pathways for the egg, embryo, and juvenile stages (see Figure 6-2) because these tend to be potentially most at risk in the context of reproductive and developmental effects (Pastorok, Butcher et al.

1996). Exposure information must be relevant to the sensitive life stages, particularly if these life stages have 1) a high probability of contact or co-occurrence with the stressor, 2) unique metabolic capabilities or lack thereof (Stahl and Kocan 1986), and 3) specific dietary preferences or feeding rates (Cripe et al. 1986). Understanding bioavailability in the context of the ERA is an important concept because the amount of a substance measured in soil, food, water, sediment, or animal tissue may not be relevant to the exposure most likely to result in impacts on reproduction and development. Without understanding bioavailability, the ecological risk assessor is likely to over- or underestimate exposure (Hamelink et al. 1994), thus increasing the uncertainty in the assessment.

Another important step in understanding exposure at the population level for purposes of ERA is through the analysis of food webs (Paine 1966; Pulliam 1994; USEPA 1995a, 1995b, 1995c; Pascoe et al. 1996). There are 2 important aspects to this analysis. The first is to understand the relative contribution of the contaminant to the overall exposure to various stresses, primarily focusing on adults. The second, given the life history strategies of oviparous vertebrates, is to understand the relative contribution of the contaminant to overall stresses in the egg, embryo, juvenile, or other sensitive life stages (Pascoe et al. 1996). Having both aspects, along with additional information on bioavailability and exposure scenarios where the sensitive life stages are likely to encounter the contaminant, can reduce the uncertainty in the overall assessment.

Effects endpoints

Unless the field and laboratory studies used in the ERA measured reproductive and developmental effects, they may not be relevant to population-level impacts (Barnthouse et al. 1986; Barnthouse 1993). Various approaches to population-level impacts have been taken, depending on the type of ecological receptor and/or stressor. Abundance is one common data point that is useful in understanding the particular population being studied (Gray 1989). Other approaches have utilized microcosm-type study designs to understand the complex nature of population change resulting from exposure to a selected contaminant (Croonquist and Brooks 1991; Bengtsson et al. 1985; Graney et al. 1995). Still other common measures of population-level effects and endpoints include fecundity, age-class distributions, sex ratios, recruitment, and rates of increase (Pulliam 1988, 1994; Walthall and Stark 1997). For ERA purposes, much of the information from population-level effects can be relevant as supportive evidence of potential or actual risks, as well as providing TRVs for estimating the potential ecological risk.

Another approach to measuring impacts at the population level is through population models. These have been described in Chapter 4 and previously by others (Emlen 1989; Barnthouse 1992, 1993). In ERA, these models can be helpful in estimating potential impacts on the receptor of concern using various default and/

or literature-based inputs and, where receptor-specific data are available, in estimating potential risks to a particular population (Emlen 1989; Akcakaya 1994).

Unfortunately, there are a number of potential problems associated with population-level studies and their use in ERA, regardless of whether these studies are specific to oviparous vertebrates or not. Background "noise" or normal fluctuation in a population can be substantial (Pulliam 1988; Norris and Georges 1993) and can make it difficult for the ecological risk assessor to differentiate contaminant-related impacts from those occurring under normal circumstances. The abundance, survival, reproduction, and recruitment of a receptor may change because of factors other than exposure to the stressor of interest (Herricks et al. 1989). Underlying this change could be 1) a pre-existing loss of reproductive capacity unrelated to contaminant stress (Pulliam 1988), 2) the inability of the population to adjust to physical/ chemical changes in the environment (Krebs 1972; Beeby 1993), and 3) changes in competition and predation (Connell 1961a, 1961b). In addition, there are seasonal and other cyclical changes in abundance that are influenced by the availability of food and habitat or that result from natural diseases.

An example of the difficulty in understanding population-level effects can be found in the work of Dobson and Hudson (1994). These workers found that 80% of passerine birds die before reaching age 1, and more than 70% of the variation noted in British bird survival could be attributed to temperature variation in just 2 months of the year. More quantitatively, a 1 °C change in temperature translated to a 3 to 4% change in survival in the case of British birds. Given this information, it could be difficult for an ecological risk assessor to estimate potential contaminant risk for these British birds since such a risk could be completely masked by the normal fluctuation in the population.

Another difficulty in estimating risks at the population level, whether for oviparous vertebrates or not, is that organisms vary in exposure and susceptibility to toxicants depending on age, locale, or habitat (Bowers 1994). In many cases, the ecological risk assessor may assume knowingly or unknowingly, that such variations are minimal or do not occur. Because of variations in rates of reproduction and survival, individuals contribute differently to long-term population dynamics and regulation, a fact that contradicts the implicit assumption in population-level risk assessment that all the individuals in the population respond to the stressor in the same way. Nevertheless, ecological risk assessors can draw upon demographic models where inputs to the model can be changed on the basis of field or laboratory effects, expected effects, or other points and obtain, iteratively, a possible range of potential outcomes (Akcakaya 1994). Another important point is that habitat selection can vary because not all individuals of the population will be capable of occupying all of the best habitat, particularly during breeding, spawning, and nesting periods (Neuhold 1986; Gray 1989; Suter 1990). For oviparous vertebrates, there will be some individuals that will not be able to compete as readily as others. The ability to reproduce and survive may be related to the specific habitat individuals are capable of occupying. In ERA, habitat may be just as important a variable in population effects as is exposure to contaminants.

Risks to communities

When is risk to communities relevant?

Communities are assemblages of interacting populations and have been used by ecologists as a construct to explain classical ecological hierarchy (i.e., organism, population, community, ecosystem and landscape levels). Examples of these constructs are benthic invertebrate community, fish community, and bird community). Such assemblages are artificial constructs because fish, bird, reptile communities do not interact only with like kinds but interact also with other biotic, physical, and chemical components of ecosystems during various life-cycle stages.

Assessing the ecological risks of contaminant exposures to oviparous vertebrates and subsequent effects on communities of aquatic and terrestrial organisms depends on the roles the affected populations play in physical, chemical, and biological interactions with the environment. If a population plays a keystone role in a food web and contaminants affect the population's structure and/or function, a potential exists for other populations to be affected. An example would be a contaminant's impact on the fecundity of a top-level fish predator that exerts top-down control on populations of prey species. A population decrease of such a top-level predator may release prey populations to increase in numbers, altering the structure of the fish and other aquatic populations via a cascade of interactions. Thus it appears that from an ERA perspective, the concept of community may be less useful than the concept of risk to the structure and function of food web components, which is a more scientifically tractable approach.

How does one assess risks to communities?

The major tool for assessing the potential effects of contaminants on oviparous vertebrates and subsequent impacts on food chains or food webs is trophic dynamic models. Examples of measures of effects or assessment endpoints can include allocation of numbers, biomass, growth among components of the food web, population attributes of components (e.g., age structure, abundance, sex ratios, mortality rates), or energy transfer and efficiency. Comprehensive assessment approaches such as the Index of Biotic Integrity, initially developed by Karr et al. (1986) and subsequently adapted for a variety of uses, begin to address the range of approaches that can be used to assess community impacts.

Ranking contaminant stress versus other stressors

The structure and function of food chains or food webs and their components are the consequence of not only biotic interactions but also of the combined interactions of the physical, chemical, and biotic aspects of the environment. Thus, an evaluation of the roles contaminants play in impairing food webs must include consideration of other factors affecting food webs. Figure 6-4 illustrates factors that control the biological integrity of aquatic and terrestrial food webs and ecosystems, respectively. The fate and bioavailability of contaminants in the aquatic environment are influenced primarily through the water/sediment quality component in Figure 6-4. However, the structure and function of aquatic food webs can be affected by stressors that change the energy available to food web organisms (e.g., allochthonous carbon from watershed impacted by land use), habitat limitations, amount of water available, and exotic species that compete with food-web organisms. Figure 6-4 also contains the major stressors of terrestrial systems. Plants and animals are exposed to contaminants in water, soil, and air. Habitat alteration and loss associated with land use changes can be major stressors.

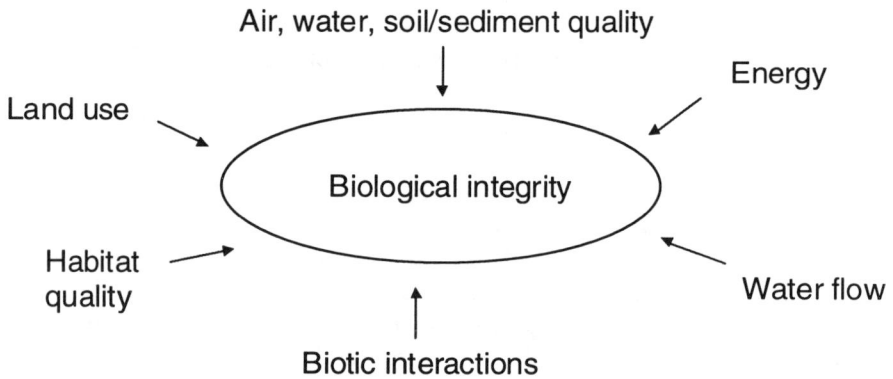

Air, water, soil/sediment quality

Land use

Energy

Biological integrity

Habitat quality

Water flow

Biotic interactions

Figure 6-4 Factors controlling the biological integrity of aquatic and terrestrial systems (Redrawn from Karr 1987 ©Springer-Verlag New York, Inc.)

Risks to ecosystems

Food webs operate in the context of ecosystems. Ecosystems incorporate the physical, chemical and biotic components of the environment and are a construct to help define and delimit operational units of the environment. From an ERA perspective, ecosystems are of concern because of their structural and functional attributes. They contain valued species and perform processes important to humans and the functioning of the earth. The magnitude of the services provided are starting to be appreciated and quantified (Costanza et al. 1997). An example would be a wetland ecosystem that serves as a habitat for a wide variety of species of interest to society and that also performs a variety of process such as carbon fixation, flood storage, water purification, or nutrient attenuation.

When is the ecosystem level of risk assessment relevant?

Assessing the impacts of contaminants on development and reproduction of oviparous vertebrates on ecosystems can be relevant if results of ERAs conducted at the food web level illustrate a potentially significant impact on biodiversity and/or productivity, both ecosystem-level assessment endpoints that people care about. For example, if contaminants reduce the population of a key avian predator that is important in keeping the population of an avian prey species relative stable, controlling factors on the prey species' population are released. The prey population may expand, but most likely it will experience more chaotic population fluctuations. If the prey population plays a significant role in the pollination process for certain plant species, chaotic pollination success may result for the plant species. Thus an important ecological process and one valued by humans could be adversely impacted.

Risks to ecological landscapes

Landscapes are a high level of ecological organization and can be thought of in several contexts. From a human perspective, a landscape can be visualized as the patchwork of land uses on the surface of the earth as viewed from an airplane, aerial photograph, or satellite image. The landscape to a salamander is much different and consists of the land uses associated with its home range. Clearly, changes in the amount, position, quality, and proximity of land uses/habitat have direct ecological implications to animals and plants.

If viewed from a human perspective of landscape, other stressors such as land-use modifications by humans, natural catastrophe (e.g., hurricanes, earthquakes, or fires), disease outbreaks, chemical contaminants (e.g., air pollutants, metals, or pesticide effects on vegetation) all appear to have greater implications for creating risks to landscapes than do contaminants that affect only oviparous vertebrates. Birds, fish, reptiles, and amphibians are components of the landscape but do not appear to be significant determinants of the physical, chemical, and biological aspects of nature at the landscape scale.

Summary and Conclusions

The objective of this chapter on ecological risk assessment is to determine if the currently operable ERA paradigm needs modifications to address unique risks of contaminants to oviparous vertebrates and, if so, to explore ways to adapt the paradigm to make better assessments of risks. A secondary objective is to provide background information on ERA to toxicologists, physiologists, ecologists, chemists, and modelers knowledgeable about oviparous vertebrates, illuminating how the results of their disciplines could be used. Based on deliberations at the workshop, the risk assessment workgroup reached the following conclusions:

1) The current ERA paradigm is sufficiently robust to characterize and quantify oviparous vertebrates/contaminant effects for species with a diversity of life histories, life-cycle stages, reproductive strategies, and exposure-pathway vulnerabilities.

The ERA paradigm can be used, and is being used, to assess risks associated with developmental and reproductive issues associated with oviparous vertebrates and does not need to be modified. A diversity of assessment endpoints can be addressed using the paradigm such as ecosystem level endpoints (e.g., nutrient cycling), population-level endpoints (e.g., salmon population attributes) and organismal-level endpoints (e.g., beak deformities in birds).

2) Risk management questions that are societal-based drive the risk assessment problem formulation, which drives data and analytical needs.

A fundamental part of the ERA paradigm is the interaction of risk assessors with risk managers to collaboratively develop assessment endpoints upon which risk decisions can be made. Assessment endpoints need to be ecologically important and reflective of things people care about.

3) In general, ERA endpoints will focus on population level effects because this level is relevant to current policy and societal priorities and the scientific/technical analysis is generally tractable.

An analysis of levels of biological organization from subcellular to landscape level demonstrated that the population level is not only the most scientifically tractable but also has the greatest ability for assessing ecological risks that are relevant to the societal values and policy directives that frame resource management issues. This is not to say that ERAs cannot be made at other levels of biological organization. However, risk assessments conducted at other levels generally yield greater technical and social uncertainty and less relevancy to the scientific aspects of ecological issues.

4) Given the life histories and vulnerable life stages of oviparous vertebrates, there are unique toxicological and physiological data and contaminant-fate data that are given more weight in, and are more relevant to, ERA.

ERAs use information on survival, growth, and reproduction and data on environmental contaminant-exposure concentrations to estimate risks to populations. Toxicologists and physiologists working with oviparous vertebrates need to link contaminant effects on life stages to these endpoints. However, basic knowledge of mechanisms of contaminants on physiology, pharmacology, and toxicology of life stages is useful in decreasing uncertainty in risk assessment by contributing to the weight of evidence relating contaminant exposure to effects.

5) Oviparous vertebrates that occupy higher trophic levels are particularly vulnerable to persistent chemicals through the food web.

Examination of case studies presented at the workshop consistently illustrated that adverse developmental and/or reproductive effects were present if the oviparous vertebrate was a top predator and the contaminant was persistent and concentrated in the food web.

6) In ERA of oviparous vertebrates, there exists an operational hierarchy of data in terms of utility.

The best data consist of exposure and effects information on the population of interest. However, these data rarely exist in sufficient quality or quality. In many risk assessments a complete set of data is not needed because the contaminant exposure concentration is significantly below estimated or measured effects concentrations. However, if the contaminant exposure concentration approximates the estimated or measured effects concentration, then definitive data with low uncertainty are needed on exposure and effects.

Recommendations

The following 6 areas of research needs should receive priority. These areas have been selected to be of the most direct value in advancing our ability to perform ERAs of chemical contaminants on oviparous vertebrates.

1) Development of cheaper, quicker, more accurate analytical methods for determining chemical contaminants in water, tissues (including eggs), and soil and sediments;

2) Identification of surrogates and the development of developmental/reproduction test methods for under-represented oviparous vertebrates, particularly amphibians, reptiles, and passerine birds, and baseline effects data on contaminants;

3) Development of improved methods for multigenerational studies in oviparous vertebrates and baseline effects data on chemical contaminants;

4) Continued development of population models for finer resolution of life-stage exposure-response dynamics;

5) Development of trophodynamic models appropriate for oviparous vertebrates; and

6) Developing diagnostic tools to differentiate the effects of chemical contaminants from other stressors (e.g., exotic species, habitat alteration).

References

Akcakaya HR. 1994. RAMAS/metapop: Viability analysis for stage-structured metapopulations (version 1.0). Setauket NY: Applied Biomathematics.

Barnthouse LW. 1992. The role of models in ecological risk assessment: a 1990's perspective. *Environ Toxicol Chem* 11:1751–1760.

Barnthouse LW. 1993. Population level effects. In: Suter GW, editor. Ecological risk assessment. Boca Raton FL: Lewis. p 247–274.

Barnthouse LW, DeAngelis DL, Gardner RH, O Neill RV, Suter GW, Vaughan DS. 1982. Methodology for environmental risk analysis. Oak Ridge TN: Oak Ridge National Laboratory. ORNL/TM-8167.

Barnthouse LW, O'Neill RV, Bartell SM, Suter GW. 1986. Population and ecosystem theory in ecological risk assessment. In: Poston TM, Purdy R, editors. Aquatic toxicology and environmental fate. 9th Volume. Philadelphia PA: ASTM. STP 921. p 82–96.

Bartell SM, Gardner RH, O'Neill RV. 1992. Ecological risk estimation. Chelsa MI: Lewis.

Beeby A. 1993. Applying ecology. London: Chapman and Hall.

Bengtsson G, Gunnarsson T, Rundgren S. 1985. Influence of metals on reproduction, mortality and population growth in *Onychiurus armatus* (Collembola). *J Appl Ecol* 22:967–978.

Bowers MA. 1994. Use of space and habitats by individuals and populations: Dynamics and risk assessment. In: Kendall RJ, Lacher TE, editors. Wildlife toxicology and population modeling. Integrated studies of agroecosystems. Boca Raton FL: Lewis. p 109–122.

Calabrese EJ, Baldwin LA. 1993. Performing ecological risk assessments. Chelsea MI: Lewis.

Calow P. 1993. Handbook of ecotoxicology. Volume 1. Oxford, UK: Blackwell.

Calow P. 1994. Handbook of ecotoxicology. Volume 2. Oxford, UK: Blackwell.

Connell JH. 1961a. The influence of interspecific competition and other factors on the distribution of the barnacle *Chthamalus stellatus*. *Ecology* 42:710–723.

Connell JH. 1961b. Effects of competition, predation by *Thais lapillus*, and other factors on natural populations of the barnacle *Balanus balanoides*. *Ecol Monogr* 31:61–104.

Costanza R, d'Arge R, de Groot R, Farber S, Grasso M, Hannon B, Limburg K, Naeem S, O'Neill RV, Paruelo J, Raskin RG, Sutton P, van den Belt M. 1997. The value of the world's ecosystem services and natural capital. *Nature* 387:253–260.

Cowan CE, Mackay D, Feijtel TCJ, van de Meent D, DiGuardo A, Davies J, Mackay, N. 1995. The multimedia fate model: a vital tool for predicting the fate of chemicals. Pensacola FL: SETAC.

Cripe GM, Hansen DJ, Macauley SF, Forester J. 1986. Effects of diet quantity on sheepshead minnow (*Cyprinodon variegatus*) during early life-stage exposures to chlorpyrifos. In: Poston TM, Purdy R, editors. Aquatic toxicology and environmental fate. 9th Volume. Philadelphia PA: ASTM. STP 921. p 450–460.

Croonquist MJ, Brooks RP. 1991. Use of avian and mammalian guilds at indicators of cumulative impacts in riparian-wetland areas. *Environ Manag* 15:701–714.

Dobson A, Hudson P. 1994. Assessing the impact of toxic chemicals: temporal and spatial variation in avian survival rates. In: Kendall RJ, Lacher TE, editors. Wildlife toxicology and population modeling. Integrated studies of agroecosystems. Boca Raton FL: Lewis. p 85–98.

Emlen JM. 1989. Terrestrial population models for ecological risk assessment: a state-of-the-art review. *Environ Toxicol Chem* 8:831–842.

Forbes VE, Calow P. 1996. Costs of living with contaminants: implications for assessing low-level exposures. *BELLE Newsletter* 4:1–8.

Friant SL, Bilyard GR, Probasco KM. 1995. Ecological risk assessment—is it time to shift the paradigm? *Human Ecol Risk Assess* 1:464–466.

Giesy JP, Ludwig JP, Tillitt DE. 1994. Deformities of birds in the Great Lakes region: assigning causality. *Environ Sci Technol* 28:128A–135A.

Gould SJ. 1995. Spin doctoring Darwin. *Natural History* 104:6–9, 70–71.

Graney RL, Giesy JP, Clark JR. 1995. Field studies. In: Rand GM, editor. Fundamentals of aquatic toxicology: effects, environmental fate, and risk assessment. 2nd ed. New York: Hemisphere. p 257–305.

Gray JS. 1989. Effects of environmental stress on species rich assemblages. *Biol J Linnean Soc* 37:19–32.

Guillette LJ, Gross TS, Masson GR, Matter JM, Percival HF, Woodward AR. 1994. Developmental abnormalities of the gonad and abnormal sex hormone concentration in juvenile alligators from contaminated and control lakes in Florida. *Environ Health Perspect* 102:680–687.

Hamelink JL, Landrum PF, Bergman HL, Benson WH. 1994. Bioavailability. Physical, chemical and biological interactions. Boca Raton FL: Lewis.

Herricks ED, Schaeffer DJ, Perry JA. 1989. Biomonitoring: Closing the loop in the environmental sciences. In: Levin SA, Harwell MA, Kelly JR and Kimball KD, editors. Ecotoxicology: problems and approaches. New York: Springer-Verlag. p 351–366.

Hose J, Guillette LJ. 1995. Defining the role of pollutants in the disruption of reproduction in wildlife. *Environ Health Perspect* 103:Supplement 4:87–91.

Hudiburgh Jr GW. 1995. The Clean Water Act. In: Rand G, editor. Fundamentals of aquatic toxicology—effects, environmental fate, and risk assessment. 2nd ed. Washington DC: Taylor and Francis. p 717–733.

Karr JR. 1987. Biological monitoring and environmental assessment: a conceptual framework. *Environ Manage* 11:249–256.

Karr JR. 1995. Risk assessment: we need more than an ecological veneer. *Human Ecol Risk Assess* 1:436–442.

Karr JR, Fausch KD, Angermeier PL, Yant PR, Schlosser IJ. 1986. Assessing biological integrity in running waters: a method and its rationale. Illinois Natural History Survey Special Publication 5. Champaign IL: Illinois Natural History Survey.

Keddy C, Greene JC, Bonnell MA. 1994. A review of whole organism bioassays for assessing the quality of soil, freshwater sediment, and freshwater in Canada. The National Contaminated Sites Remediation Program, Ecosystem Conservation Directorate, Scientific Series No. 198, Ottawa, Ontario.

Kendall RJ, Lacher TE, editors. 1994. Wildlife toxicology and population modeling. Integrated studies of agroecosystems. Boca Raton FL: Lewis.

Krebs CJ. 1972. Ecology. The experimental analysis of distribution and abundance. New York: Harper and Row.

Lackey RT. 1997. If ecological assessment is the answer, what is the question? *Human Ecol Risk Assess* 3:921–928.

Ma W, Denneman W, Faber J. 1991. Hazardous exposure of ground-living small mammals to cadmium and lead in contaminated terrestrial ecosystems. *Arch Environ Contam Toxicol* 20:266–270.

Ma W. 1996. Lead in mammals. In: Beyer WN, Heinz GH, Redmon-Norwood AW, editors. Environmental contaminants in wildlife: interpreting tissue concentrations. Boca Raton FL: Lewis. p 281–296.

Maughan JT. 1993. Ecological assessment of hazardous waste sites. New York: Van Nostrand Reinhold.

Mayer FL, Ellersieck MR. 1986. Manual of acute toxicity: interpretation and database for 410 chemicals and 66 species of freshwater animals. Washington DC: U.S. Department of the Interior, Fish and Wildlife Service. Resource Publication No. 160

Mazaika R, Lackey RT, Friant SL, editors. 1995. Ecological risk assessment: use, abuse, and alternatives. Amherst MA: Amherst Scientific.

Menzie CA. 1995. The question is essential for ecological risk assessment. *Human Ecol Risk Assess* 1:159–163.

Molak V. 1996. Fundamentals of risk analysis and risk management. New York: CRC/Lewis.

Moore DRJ. 1996. Using Monte Carlo analysis to quantify uncertainty in ecological risk assessment: are we guilding the lily or bronzing the dandelion. *Human Ecol Risk Assess* 2: 628–633.

Norris RH, Georges A. 1993. Analysis and interpretation of benthic macroinvertebrate surveys. In: Rosenberg DM, Resh VH, editors. Freshwater biomonitoring and benthic macroinvertebrates. New York: Chapman and Hall. p 234–286.

[NRC] National Research Council. 1991. Animals as sentinels of environmental health hazards. Washington DC: National Academy Pr.

Neuhold JM. 1986. Toward a meaningful interaction between ecology and aquatic toxicology. In: Poston TM, Purdy R, editors. Aquatic toxicology and environmental fate. 9th Volume. Philadelphia PA: ASTM. STP 921. p 11–21.

[OECD] Orgainization for Economic Cooperation and Development. 1995. Guidance document for aquatic effects assessment. Paris, France: OECD.

O'Brien MH. 1995. Ecological alternatives assessment rather than ecological risk assessment: considering options, benefits, and dangers. *Human Ecol Risk Assess* 1:357–366.

Pagel JE. 1995. Quandaries and complexities of ecological risk assessment: viable options to reduce humanistic arrogance. *Human Ecol Risk Assess* 1:376–391.

Paine RT. 1966. Food web complexity and species diversity. *Amer Natural* 100:65–75.

Pascoe GA, Blanchet RJ, Linder G. 1996. Food chain analysis of exposures and risks to wildlife at a metals-contaminated wetland. *Arch Environ Contam Toxicol* 30:306–318.

Pastorok RA, Butcher MK, Nielsen RD. 1996. Modeling wildlife exposure to toxic chemicals: trends and recent advances. *Human Ecol Risk Assess* 2:444–480.

Pastorok RA, Nielsen RD, Butcher MK. 1996. Future directions in modeling wildlife exposure to toxic chemicals. *Human Ecol Risk Assess* 2:570–579.

Power M. 1996. Probability concepts in ecological risk assessment. *Human Ecol Risk Assess* 2:650–564.

Power M, Adams SM. 1997. Assessing the current status of ecological risk assessment. *Environ Manage* 21:825–830.

Pulliam HR. 1994. Incorporating concepts from population and behavioral ecology into models of exposure to toxicants and risk assessment. In: Kendall RJ, Lacher TE, editors. Wildlife toxicology and population modeling: integrated studies of agroecosystems. Boca Raton FL: Lewis. p 13–26.

Pulliam HR. 1988. Sources, sinks and population regulation. *Amer Naturalist* 132:652–661.

Regens JL. 1995. Ecological risk assessment: issues underlying the paradigm. *Human Ecol Risk Assess* 1(4):344–347.

Rodier DJ, Maurillo DA. 1993. The quotient method of ecological risk assessment and modeling under TSCA: a review. In: Landis W, Hughes J, Lewis M, editors. Environmental toxicology and risk assessment. Volume 1. Philadelphia PA: ASTM. STP 1179. p 80–91.

[SETAC] Society of Environmental Toxicology and Chemistry. 1997. Concepts in ecological risk management. A discussion document from MERMD (The Multistakeholder Ecological Risk Management Dialog Group) and SETAC (Society for Environmental Toxicology and Chemistry). Prepared for the SETAC Workshop on Ecological Risk Management, Williamsburg, VA, July 25–27, 1997.

Smrchek J, Clements R, Morcock R, Rabert W. 1993. Assessing ecological hazard under TSCA: methods and evaluation of data. In: Landis W, Hughes J, Lewis M, editors. Environmental toxicology and risk assessment. Volume 1. Philadelphia PA: ASTM. STP 1179. p 22–39.

Sorensen MT. 1996. Annotated reference compilation, 1995 update: conducting ecological risk assessments at hazardous waste sites. *Human Ecol Risk Assess* 2:608–626.

Stahl Jr RG. 1997. Invited debate: Can mammalian and non-mammalian "sentinel species" data be used to evaluate the human health implications of environmental contaminants? *Human Ecol Risk Assess* 3:329–335.

Stahl Jr RG, Clark JR. 1998. Uncertainties in the risk assessment of endocrine modulating substances in wildlife. In: Kendall RJ, editor. Principles and processes for evaluating endocrine disruption in wildlife. Pensacola FL: SETAC. p 431–448.

Stahl Jr RG, Kocan RM. 1986. Influence of age on patterns of uptake and excretion of polycyclic aromatic hydrocarbons in the rainbow trout embryo. In: Poston TM, Purdy R, editors. Aquatic toxicology and environmental fate. 9th Volume. Philadelphia PA: ASTM. STP 921. p 287–303.

Suter GW. 1996. Toxicological benchmarks for screening contaminants of potential concern for effects on freshwater biota. *Environ Toxicol Chem* 15:1232–1241.

Suter GW. 1993. Ecological risk assessment. Boca Raton FL: Lewis.

Suter GW. 1990. Endpoints for regional ecological risk assessments. *Environ Manage* 14: 9–23.

Suter GW, Rosen AE, Linder E, Parkhurst DF. 1987. Endpoints for responses of fish to chronic toxic exposures. *Environ Toxicol Chem* 6:793–809.

Tiebout HM, Brugger KE. 1995. Ecological risk assessment of pesticides for terrestrial vertebrates: Evaluation and application of the U.S. Environmental Protection Agency's quotient method. *Conserv Biol* 9(6):1605–1618.

[USEPA] U.S. Environmental Protection Agency. 1986. Quality criteria for water. Washington DC: USEPA, Office of Water. EPA/440/5-86/001.

[USEPA] U.S. Environmental Protection Agency. 1992. Framework for ecological risk assessment. Washington DC: USEPA, Office of Research and Development. EPA/630/R-92/001.

[USEPA] U.S. Environmental Protection Agency. 1993a. A review of ecological assessment case Studies from a risk assessment perspective. Washington DC: USEPA, Risk Assessment Forum. EPA/630/R-92/005.

[USEPA] U.S. Environmental Protection Agency. 1993b. Wildlife exposure factors handbook. Volume I. Washington DC: USEPA, Office of Research and Development. EPA/600/R-93/187a.

[USEPA] U.S. Environmental Protection Agency. 1993c. Wildlife exposure factors handbook. Volume II. Washington DC: USEPA, Office of Research and Development. EPA/600/R-93/187b.

[USEPA] U.S. Environmental Protection Agency. 1994. Managing ecological risks at EPA. Washington DC: USEPA, Office of Research and Development. EPA/600/R-94/183.

[USEPA] U.S. Environmental Protection Agency. 1995a. Trophic level and exposure analyses for selected piscivorous birds and mammals. Volume I, Analyses of species in the Great Lakes Basin. Washington DC: USEPA, Office of Science and Technology.

[USEPA] U.S. Environmental Protection Agency. 1995b. Trophic level and exposure analyses for selected piscivorous birds and mammals. Volume II, Analyses of species in the conterminous United States. Washington DC: USEPA, Office of Science and Technology.

[USEPA] U.S. Environmental Protection Agency. 1995c. Trophic level and exposure analyses for selected piscivorous birds and mammals. Volume III: Appendices. Washington DC: USEPA, Office of Science and Technology.

[USEPA] U.S. Environmental Protection Agency. 1995d. Ecological risk: a primer for risk managers. Washington DC: USEPA, Office of Prevention, Pesticides and Toxic Substances. EPA 734-R-95-001.

[USEPA] U.S. Environmental Protection Agency. 1997. Priorities for ecological protection: an initial list and discussion document for EPA. Washington DC: USEPA Office of Research and Development. EPA/600/R-97/002.

[USEPA] U.S. Environmental Protection Agency. 1998. Guidelines for ecological risk assessment. Washington DC: USEPA Office of Research and Development. EPA/630/R-95/002F.

Wagner C, Lokke H. 1991. Estimation of ecotoxicological protection levels from NOEC toxicity data. *Water Res* 25:1237–1242.

Walthall WK, Stark JD. 1997. Comparison of 2 population-level ecotoxicological endpoints: the intrinsic (rm) and instantaneous (ri) rates of increase. *Environ Toxicol Chem* 16:1068–1073.

Zeeman M. 1996. Comparative immunotoxicology and risk assessment. In: Stolen JS, Fletcher TC, Bayne CJ, Seacombes CJ, Zelikoff JT, Twerdok LE, Anderson DP, editors. Modulators of immune responses: the evolutionary trail. Fair Haven NJ: SOS Publications. p 317–329.

Zeeman M. 1997. Aquatic toxicology and ecological risk assessment: USEPA/OPPT perspective and OECD interactions. In: Zelikoff JT, editor. Ecotoxicology: responses, biomarkers and risk assessment. Fair Haven NJ: SOS Publications. p 89–108.

Zeeman M, Gilford J. 1993. Ecological hazard evaluation and risk assessment under EPA's Toxic Substances Control Act (TSCA): an introduction. In: Landis W, Hughes J, Lewis M, editors. Environmental toxicology and risk assessment. Volume 1. Philadelphia PA: ASTM. STP 1179. p 7–21.

Reproductive and Developmental Effects of Contaminants in Oviparous Vertebrates: Workshop Summary, Conclusions, and Recommendations

William H. Benson, Richard T. Di Giulio, Donald E. Tillitt, Linda Birnbaum, Peter deFur, Jay Gooch, Ellen M. Mihaich, Charles Tyler

Oviparous vertebrates generally occupy important niches in aquatic as well as terrestrial systems, and reproductive and developmental effects on these species can be of relatively great ecological significance. Because these organisms have critical windows of development, they may be particularly susceptible to toxic perturbation resulting from episodic exposure to environmental contaminants. In addition, oviparous vertebrates that occupy higher trophic levels may be especially vulnerable to persistent organic chemicals by means of trophic transfer. In the context of oviparous vertebrates, the workshop focused on the processes that control contaminant exposure, basic physiological processes that control reproduction and development, toxicological mechanisms of contaminant exposure, and ecological ramifications. In addition, the workshop included critical discussions of laboratory and field approaches for determining exposure and effect, and the integration of this information into ecological risk assessments focused on reproductive and developmental contaminants. The common theme that surfaced through deliberations that took place at the workshop was that the life-cycle characteristics of oviparous vertebrates were the single most significant factor in defining an organism's vulnerability to the reproductive and developmental effects of contaminants. Moreover, the overriding demographic attributes of life-history strategies influence life-cycle characteristics, chemical exposure, and toxicity in a way that impacts factors that are critical to determining the risks associated with chemical exposure. This final chapter integrates the research gaps and recommendations identified during the workshop into a single set of workshop conclusions and recommendations.

CHAPTER PREVIEW

Reproductive and Developmental Effects of Contaminants in Oviparous Vertebrates. Richard T. Di Giulio and Donald E. Tillitt, editors.
©1999 Society of Environmental Toxicology and Chemistry (SETAC). ISBN 1-880611-37-6

Exposure of fish and wildlife to environmental contaminants may yield effects varying from acute lethality to sublethal changes in reproduction and development. The quantity and quality of offspring produced are critical components of population fitness and affect population dynamics. However, the emphasis on population-level effects, which is the focus of the current ecological risk assessment paradigm, has its foundation in individual-level effects. Chemical safety testing protocols and subsequent regulatory decisions designed to protect the environment are almost entirely based on exposure and effects measured in individuals. There are numerous biochemical and physiological mechanisms by which chemicals can impact reproductive success and/or developmental processes. These mechanisms include diversion of maintenance energy to biotransformation processes or tissue repair, impaired protein synthesis, and interference with gonad maturation and gamete production. These mechanisms are investigated at the organismal or molecular level of biological organization. The challenge to the scientific and regulatory communities is to understand or predict the potential for deleterious effects at higher levels of biological organization, namely population-level effects. Modeling efforts may provide an opportunity to link disturbances in reproductive and developmental processes measured at the individual level of organization to population-level effects. This connection may occur because the measurable endpoints of reproductive and developmental effects observed in organisms may be either directly used or proximately linked to population models. Therefore, reproductive and developmental events as they relate to chemical exposure and toxicity are of primary importance in the assessment of ecological risk.

The reproductive physiology of birds, fishes, reptiles, and amphibians is very diverse ranging from the production of yolk-laden eggs fertilized externally to the production of small eggs lacking yolk that are fertilized internally and require maternally supplied nutrition prior to parturition. Oviparous vertebrates are those animals that produce eggs that develop external to the female's body and are nourished by the maternal nutrients provided prior to fertilization. In the context of oviparous vertebrates, the workshop focused on bioavailability and toxicokinetic processes influencing exposure, basic physiological processes that control reproduction and development, toxicological mechanisms, and ecological ramifications. In addition, the workshop included critical discussions of laboratory and field approaches for determining exposure and effect, and the integration of this information into ecological risk assessments focused on reproductive and developmental toxicants.

Oviparous vertebrates generally occupy important niches in aquatic as well as terrestrial systems, and reproductive and developmental effects on these species can be of great ecological significance. Oviparous vertebrates that occupy higher trophic levels may be especially vulnerable to persistent organic chemicals by means of transfer through the food web. Additionally, changes in trophic complexity strongly impact dietary exposures of chemicals to higher trophic levels. With respect to nonpersistent environmental chemicals, these organisms have critical windows of development and may be particularly vulnerable to the impact of episodic exposures.

Oviparous vertebrates that are susceptible to reproductive and developmental toxicants include many species of great social value, i.e., commercial, recreational, and aesthetic value. There exists a rich scientific database for birds and fishes, the classes of vertebrates most often utilized as surrogate species for environmentally based chemical hazard and risk assessments. Recently, there has been considerable concern regarding the potential impacts of environmental contaminants upon amphibians and reptiles, 2 less-studied vertebrate classes. These vertebrate classes constitute the largest component of vertebrate biomass in almost any tropical and temperate terrestrial ecosystem. Furthermore, amphibians and reptiles may serve as especially good models to predict the impact of chemical contaminants because of the great diversity of their morphologies, habitats, and life-history strategies, and because of their relative abundance and vulnerability.

Workshop Objectives

The workshop brought together experts from the fields of reproductive and developmental biology, toxicology, chemistry, ecology, and risk assessment. The participants were divided into a number of workgroups based upon their area of expertise and affiliation (i.e., business, academia, government, non-government organization). This was done in an attempt to enhance communication and provide a forum for transfer of information between researchers and decision-makers from regulatory, business and academic communities. As discussed in Chapter 1, in order to prepare the participants for an active exchange of information and data, each workgroup was charged with discussing and communicating the state of the science of specific topic areas. The first goal was to identify critical areas that must be considered in order to advance our understanding of the potential impact of reproductive and developmental toxicants in oviparous vertebrates inhabiting aquatic and terrestrial ecosystems. A second goal was to identify major information gaps and suggest improved approaches for understanding impacts and assessing risks. Lastly, each workshop participant considered the scope of the problem with respect to what is known and what we need to know about reproduction and development in oviparous vertebrates. Through consideration of these topics and issues, each workgroup also developed research recommendations.

During the plenary sessions that were conducted throughout the workshop, informal presentations were made by the chemistry, physiology, toxicology, ecology, and risk assessment workgroups. A number of important research questions concerning the reproductive and developmental effects of contaminants in oviparous vertebrates were identified. It was recognized that there were great similarities between mammals and oviparous vertebrates with regard to the endocrine system and respective control mechanisms. Neuroendocrine physiology, for example, was identified as a critical area of basic science that will prove increasingly important in evaluating reproductive and developmental effects of contaminants in oviparous vertebrates. It was considered that critical issues related to neurosensory function and behavior will be as important to address as measures of performance (e.g., skeletal growth, gonadal development, gamete release, utilization of lipids). The more integrative a physiological endpoint, the more it is useful in assessing organismal- and population-level responses. It was recognized that population-level responses related to mortality, growth, reproduction, and development are considered valued endpoints when conducting ecological risk assessments. However, individual-level effects are used in ecological risk assessments because they may be the only available endpoints indicative of chemical exposure and/or effect. While individual-level effects may not, currently, be scalable to population-level effects, it was recognized that a greater degree of physiological and toxicological certainty was associated with individual-level effects and that such responses are of great value in the conduct of ecological risk assessments. Additionally, it is recognized that by the time population-level effects would be observed, it is too late and detrimental impact to overall environmental quality has already occurred; the use of individual-level effects may provide a predictive means to assess ecological risk.

Workshop theme: life-cycle vulnerability

By bringing together experts from the fields of reproductive and developmental biology, toxicology, chemistry, ecology, and risk assessment, it was possible to assess the state of the science on the reproductive and developmental effects of contaminants in oviparous vertebrates. Although this workshop was successful in advancing the state of the science, it was even more successful with respect to integrating a broad base of experience and viewpoints to advance the understanding of basic chemical and physiological processes, toxicological effects and mechanisms, ecological principles, and ways in which the science relates to risk assessment. The common theme or conclusion that surfaced through the workshop was that the life-cycle characteristics of oviparous vertebrates were the most significant factor when defining an organism's vulnerability with respect to exposure pathways, and to critical windows of physiological processes related to toxic perturbations (Figure 7-1). Oviparous vertebrates exhibit complex life-cycle stages that represent transitional periods in which new functions emerge. The transitional life stages are often the critical windows for chemical pertubation. The complex life-cycle stages, typical of oviparous vertebrates, also greatly complicate the analysis of population-level effects. Moreover, the demographic attributes of life-history strategies influence life-cycle characteristics, chemical exposure, and toxicity in a manner that impacts characteristics that define the risk to chemical exposure. Life-history characteristics are an important determinant of population-level changes in response to stressors, including contaminant exposure. Therefore, thorough knowledge of the life history

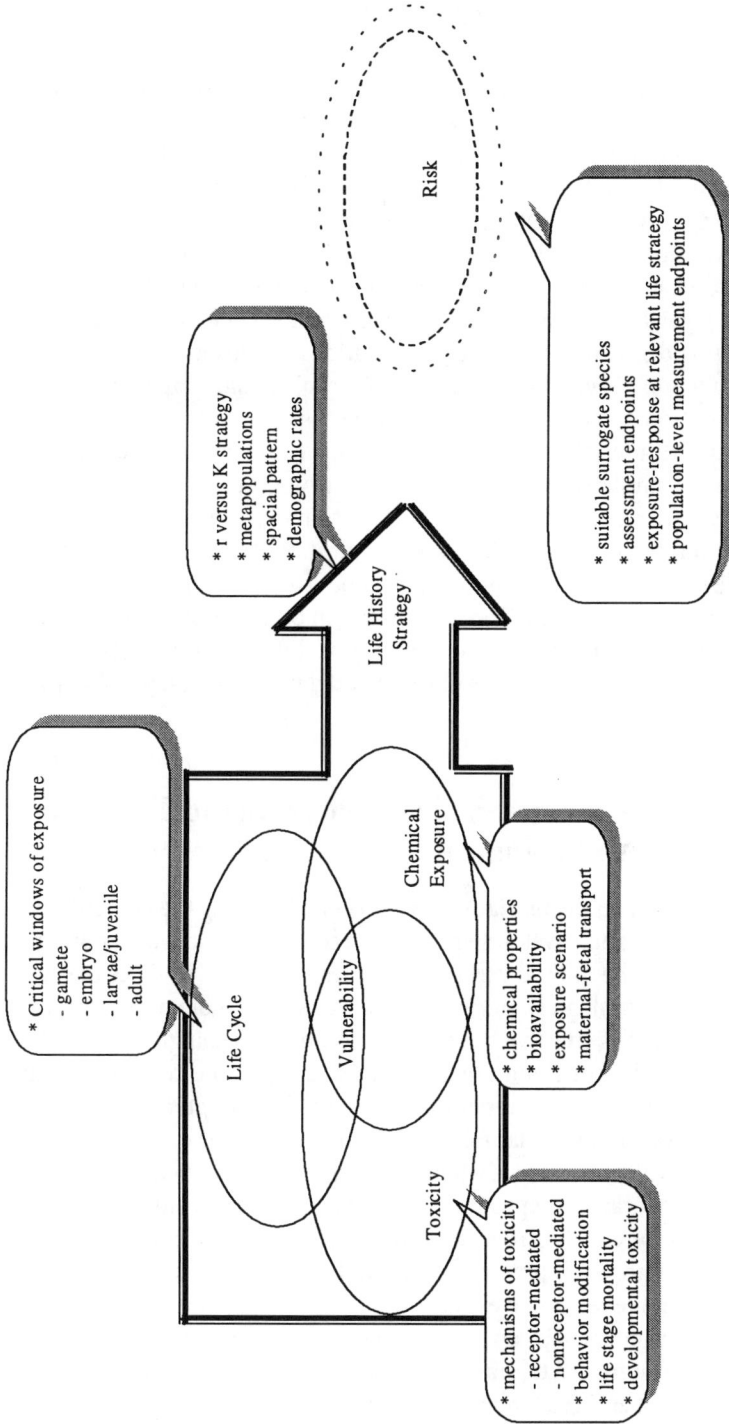

Figure 7-1 Representation of life-cycle vulnerability and life history as common themes in addressing reproductive and developmental effects of contaminants in oviparous vertebrates

of a species is essential to understanding population-level responses to chemical stress.

Workshop participants realized that the current status for evaluating the environmental impact of chemicals on reproductive and development processes was in an early state of understanding, particularly with regard to the availability and useful application of meaningful biological responses as tools for the conduct of ecological risk assessments. Although information exists regarding the development and reproduction in oviparous animals, little is known about the physiological processes controlling them, especially neural control mechanisms. There is a need to gain more information regarding the ranges of general physiological functions in response to natural and anthropomorphic perturbations. Additionally, maintenance of genetic/phenotypic variation is vital for the evolutionary capacity of a species, but the variation makes it difficult to ascribe "normal" values and determine "normal" variation. Extrapolating between populations within species adds a further dimension of complexity. In addition, the physiological performance of an individual can vary under normal conditions of variation in the natural environment. In studies of feral populations, there are the additional issues of tolerance and adaptation that deviate from baseline or normal values to permit an animal to compensate for a particular stressor. The interpretation of deviations in biological responses as toxic effects or compensatory responses is difficult without strong fundamental knowledge of the basic reproductive and developmental physiology of oviparous vertebrates.

Integration of Basic Chemical and Physiological Process with Ecological Risk Assessment

Clearly, more information on the fundamental processes affecting the vulnerable processes in reproduction and development is required to establish what "normal" is and to appreciate what the normal variability in these processes is. Once the mechanisms affecting physiological processes are understood, biological responses or biomarkers can be used as effective tools for the conduct of ecological risk assessments. As indicated throughout the text, there are numerous reasons why the workshop focused on oviparous vertebrates. As addressed in Chapter 3, the life cycle of oviparous vertebrates has particular nuances that increase their exposure to contaminants during critical life stages which, in turn, may make them more vulnerable to developmental and reproductive effects. Exposure to environmental contaminants may interfere with gamete motility and fertilization, embryonic hatching, larval or juvenile development, or the adult stage depending upon the species. An additional issue related to exposure, which is of significance with oviparous vertebrates, is maternal transport of chemical contaminants to developing gonads/embryos. Maternal-fetal transport represents a period of high vulnerability for the subsequent embryo/offspring. Although environmental toxicology and chemistry has traditionally addressed long term, low-level exposure to persistent organic chemicals, there is a growing need to understand the consequences of short-term episodic exposure to chemicals. Because of critical windows of development, oviparous vertebrates may be particularly sensitive to episodic discharge of environmental contaminants. For example, exposure at an early life stage may lead to alterations in key developmental processes, as well as increased susceptibility to chemical insults as adults. Therefore, it would be pertinent to concentrate research efforts on elucidating the physiological mechanisms effecting these processes. The processes and transitions of particular relevance include cleavage, embryogenesis, hatching, metamorphosis, and sex determination/differentiation. Sexual determination and differentiation represent periods of particular sensitivity to chemical perturbations. There is no doubt that sex determinations in many oviparous animals are more plastic than in mammals and therefore are probably more vulnerable to perturbations.

With respect to episodic exposure and critical windows of vulnerability, mobile animals may experience varying degrees of exposure as they move among areas with varying contaminant levels. Exposure of the mature adult could disrupt not only normal reproductive physiology but also mating and migratory behavior. Migratory patterns themselves greatly influence exposures during critical windows of vulnerability. Many species that migrate spend some portion of their early life in quasi-separate nursery grounds only to intermix later in life with individuals from other nursery areas. Immigration and emigration of young among nursery areas and intermixing of sub-populations makes site-specific contaminant effects difficult to determine.

There is a hierarchy in biological control systems, and the reproductive axis is regulated by the hypothalamus and pituitary of the central nervous system. There are numerous mechanisms by which environmental contaminants can interfere with higher-level control systems and it is likely that chemical perturbations that affect these control systems will have greater effects and generally more global effects on the animal. In this regard, another key factor is that the rate of response to a perturbation will relate to how that signal is communicated and to what organ/tissue it is communicated. The duration of the chemical insult may also affect the outcome of the organism. For example, a short-lived exposure may cause an acute abnormal response as the animal adapts to alteration in its environment, but then the animal's physiology returns to normal as the insult is removed, with no long-term, detrimental consequences. However, if the exposure is longer lived, the chemical insult may have more serious, long-term implications on developmental and/or reproductive process. Again, the issue of adaptive responses further emphasizes the need to understand the physiological mechanisms affecting reproductive and developmental processes and transition periods to enable the use of biological tools that are indicative of effects that will impact at the population level.

There are potentially useful measurement endpoints that are specific to oviparous vertebrates (see Chapter 6). Depending on how the ecological risk assessment strategy has been formulated, not all of the endpoints outlined may be needed. A critical examination of the species that have been adversely impacted by exposure to hydrophobic persistent and bioaccumulative chemicals indicates that both viviparous and oviparous species have been affected. The major commonalities among affected species are primarily factors of their natural histories other than reproductive strategy. Therefore, from an ecological risk assessment perspective there is little that is unique about oviparous organisms, and few, if any, truly unique measurement endpoints would need to be selected to address most contaminant effects issues. In general, it is useful to have information on the species of interest and estimates of inter- and intra-species variability and concentration duration, and timing of exposure, all of which should be interpretable and related to ecologically relevant endpoints. As mentioned previously, effects on survival, growth, or development to maturity and reproductive performance are the most valued endpoints considered when assessing contaminant risks.

There are several reasons why there are attempts to conduct ecological risk assessments for populations of organisms rather than for individuals. As mentioned above, the most common experimental unit on which effects measurements are made is the individual organism. When statistical approaches are applied, however, the endpoints become population-level estimators. Unless the field and laboratory studies used in ecological risk assessment measure reproductive and developmental effects, they may not be relevant to population-level impacts. Various approaches to population-level impacts have been taken depending on the target organisms of interest and/or stressors. One approach to measuring impacts at the population level is through the use of population models. These models can be useful in estimating potential impacts of concern using various default and or literature-based inputs and, if organism-specific data are available, in estimating potential risks to a particular population. If information that can be directly applied to the ecologically relevant processes of survival, growth, development, and reproduction of individuals is available, decisions on the probability of a chemical affecting the sustainability of populations is generally possible. The information that would be considered to be most relevant and given the greatest weight would be empirical field data for which appropriate dosimetry and confounding factors were fully characterized. Similarly, multigeneration laboratory studies that measure a range of indicators of population sustainability and are conducted under a dosimetry regime expected in the field would receive greater weight. While attempts are made to predict the potential for deleterious effects of contaminants at the population level, most of what is known about the linkages between chemical exposure and biological effect has been derived from the study of growth and reproduction of individual organisms of one species exposed at one time to one contaminant. Chemical stresses, however, vary in spatial extent, recurrence, interval and intensity. As the spatial scale of chemical exposure changes, other characteristics of contaminant stress tend to change. Cumulative impacts, where individually minor changes become significant as they aggregate through time and space, are also important to examine at the landscape scale.

Major Research Gaps

Through consideration of the topic areas discussed, numerous gaps in our knowledge of oviparous vertebrates, as pertaining to reproductive and developmental effects of contaminants, were identified, including the following:

- More information on the fundamental processes affecting development and reproduction of oviparous vertebrates is required to establish what "normal" is and to further understand the consequences of variability in these processes.
- Understanding of chemical deposition in oviparous vertebrates as it relates to reproductive biology and the relationship of exposure data to mechanistic studies of effect is needed.
- More basic physiological information and understanding of toxicological mechanisms are needed to allow extrapolation between species and across classes.
- Development of diagnostic tools to differentiate the effects of chemical contaminants from other stressors (e.g., exotic species, habitat alteration) is necessary.
- Development of more integrative approaches such as measures of performance (e.g., growth) is needed to assess reproductive and developmental effects of environmental contaminants.
- Development of a better understanding about exposure of developing eggs by means of maternal-fetal transport of chemical contaminants is needed.
- Development and use of bioenergetics-based models are needed to simulate the combined effects of chemical and non-chemical stressors and linkage of bioenergetics and physiologically based toxicokinetic (PB-TK) models. Additionally, further development of PB-TK models for birds, in particular, is needed.
- Identification of surrogate species and the development of appropriate model systems are needed to evaluate the reproduction and development in underrepresented oviparous vertebrates, particularly amphibians, reptiles, and passerine birds.
- Individual-level endpoints predictive of population-level effects that may be used to conduct ecological risk assessments are needed.
- Further development of population models for greater resolution of life-stage exposure-response dynamics is needed.

- Interpretation of field results has been difficult, inconsistent, and sometimes controversial, and there are a number of significant gaps that need to be addressed in several key areas:
  1) better understanding of reference site selection and the impacts of temporal and spatial variability,
  2) definition of sample size requirements,
  3) evaluation of effect of sampling bias and handling stress, and
  4) interpretation and use of field residue information.
- Development of trophodynamic models for oviparous vertebrates is needed.
- Definition of the differences between statistical significance and biological significance and what role, if any, genetic variation plays in the responses that have been reported in field studies is needed.
- Development of improved methods for conduct of mutligenerational studies in oviparous vertebrates is needed.
- Improvement of quantitative determination of exposure-response relationships is needed.

Conclusions and Recommendations

While serious gaps in our knowledge of the reproductive and developmental effects of contaminants in oviparous vertebrates were recognized, specific conclusions and recommendations were developed.

- Life-cycle characteristics of oviparous vertebrates are the most significant factor in defining an organism's vulnerability with respect to exposure pathways, and to critical windows of physiological processes related to contaminant exposure.

It is the transitional stages in which new functions emerge that are often the critical windows for chemical perturbation. These are times when the expression of the genetic potential for development may be disrupted. Sometimes these effects are seen in the immediate term; other times they may not be manifest until sometime later in the animal's life cycle.

- The demographic attributes of life-history strategies influence life-cycle characteristics, chemical exposure, and toxicity in a way that impacts factors critical to determining the risks associated with chemical exposure.

Life-history theory focuses on the constraints among demographic variables and the manner in which these constraints shape strategies for dealing with different kinds of environments. The consequences of contaminant effects can vary greatly on the life-history strategy of the species of interest. While fish, birds, reptiles, and amphibians obviously differ in the specifics of their life styles, there are attempts to use a common life-history-strategy framework to allow for cross-taxa comparisons.

- Significantly more attention must be given to obtaining information on the basic reproductive and developmental physiology of oviparous vertebrates.

Little is known of the normal hemostatic range of physiological processes controlling reproduction and development, and therefore it is difficult to interpret data as normal and compensatory responses versus abnormal responses indicative of contaminant stress. This is particularly true for neural control mechanisms that can be subject to chemical perturbation. Likewise, research efforts must be made in the development of model systems with respect to surrogate and sentinel species to assess the natural range of variability with respect to physiological responses.

- Many receptor systems are common across vertebrate groups.

General models can be drawn up for ligand-receptor interactions because many receptor systems are common across vertebrate groups. Recent advances in molecular biology have provided new insights into the fundamental processes of inter- and intra-cellular communication in both normal physiological processes and in response to chemical insult. There is a need to understand cell signaling, propagation of signals within the cell, and resulting changes in gene regulation following chemical exposure. Models of receptor-ligand interactions have been developed that are useful in understanding how chemicals interact with specific receptors and mediate their effects. Modeling systems for ligand-receptor interactions hold promise for identifying classes of chemicals that are potentially damaging to physiological processes mediated by specific endocrine signals.

- Maternal-fetal transfer in oviparous animals represents a period of high vulnerability for the subsequent embryo/offspring.

In oviparous animals the resulting egg must contain everything to sustain the subsequent embryo during the first period of its life. What is supplied by the female into the egg, e.g., the amounts of the various nutrients, maternal mRNA transcripts, associated contaminants, essentially direct the subsequent course of development for that offspring. This is a major point of vulnerability, especially if vitellogenesis takes place over a long period of time (e.g., 9 months in trout) and exposure is persistent, as the chemical is likely to accumulate in the developing embryo. If the chemical is lipophilic, molecules like vitellogenin are likely to facilitate their accumulation in the oocyte. Furthermore, depending on partitioning in the oocyte, it is problematic that stored contaminants will be released as the energy and protein reserves are being mobilized for the developing embryo at a highly vulnerable life stage.

- Attention to development of model systems for amphibians and reptiles is needed.

In terrestrial environments of temperate, subtropical, and tropical regions of the world, amphibians and reptiles dominate vertebrate faunas in terms of biomass, abundance, and ecological diversity. Because of these attributes, amphibians and reptiles may be especially good models to predict the impact of environmental chemicals; however, appropriate model systems do not exist to evaluate the reproductive and developmental effects of environmental contaminants on amphibians and reptiles.

- Oviparous vertebrates that occupy higher trophic levels are particularly vulnerable through the food web to persistent chemicals.

Examination of case studies consistently illustrates that adverse developmental and/or reproductive effects were present if the oviparous species was a top predictor and the contaminant was persistent and accumulated in the food web. Additionally, changes in trophic complexity strongly impact dietary exposures of chemicals to higher trophic levels.

- Greater attention must be given when evaluating the potential impact of nonpersistent environmental contaminants on reproduction and development in oviparous vertebrates.

Many oviparous vertebrates occupy higher trophic levels and, therefore, are particularly vulnerable to persistent chemicals. Because these organisms have critical windows of development, they are susceptible to toxic perturbation by the episodic discharge of environmental contaminants. Such contaminants need not be persistent to have significant detrimental reproductive and developmental effects.

- Development appropriate bioassays and test methodologies to assess a broader range of environmental chemicals and endpoints is needed.

To date, testing protocols are designed to evaluate exposure and effects of chemicals on individual animals. There is a need to continue present efforts to develop robust means to evaluate links between perturbation in reproductive and developmental processes measured at the individual level of biological organization. Furthermore there is need to develop multigeneration approaches to evaluate potential reproductive and developmental effects in oviparous vertebrates.

- Methodological approaches to provide a comprehensive evaluation of the influence of natural stressors is essential when addressing issues involving chemical stress.

Information on the number of individuals of each sex and age class suffering lethal and sublethal effects must be combined with data on population abundance and distribution, sex and age composition, proportion of nonbreeding adults, reproductive potential, rates of emigration and postfledging survival, relationships of natality and mortality to density, and the influence of natural stressors on population numbers.

- Ecological risk assessment endpoints, in general, focus on population level effects because this level of response is relevant to current policy and societal priorities.

An analysis of levels of biological organization from molecular to landscape level demonstrated that the population level is not more scientifically manageable but has the greatest ability of assessing ecological risks. This is not to say that ecological risk assessments cannot be made at other levels of biological organization; however, uncertainty associated with risk assessment at other levels is believed to be greater.

- In ecological risk assessment there exists an operational hierarchy of data in terms of utility.

The best data consist of exposure and effects information on the population of interest. However, these data rarely exist, and in many risk assessments are not needed because the contaminant exposure concentration is significantly below estimated or measured effects concentrations. If the contaminant exposure concentration approximates the estimated or measured effects concentration, definitive data with low uncertainty are needed on exposure and effects.

- The current ecological risk assessment paradigm is sufficiently robust to characterize and quantify contaminant effects for oviparous vertebrates.

The current ecological risk assessment paradigm is sufficient to associate risks associated with developmental and reproductive effects associated with oviparous vertebrates and requires no modification. A diversity of assessment endpoints such as organism-, population-, and ecosystem-level endpoints can be addressed utilizing the traditional paradigm.

- Essential need exists for integrative approaches to assess the reproductive and developmental effects of environmental contaminants on oviparous vertebrates.

Multidisciplinary approaches using population ecology, physiology, toxicology, and chemistry are essential to understanding the many nuances and processes involved with assessing the reproductive and developmental effects of environmental contaminants. Ecological risk assessments use information on mortality, growth, and reproduction and data on or estimates of environmental contaminant exposure concentrations to estimate risks to populations. Toxicologists and physiologists working with oviparous vertebrates need to link contaminant effects on life stages to these ecologically relevant endpoints. However, knowledge of mechanisms of contaminants on physiology, pharmacology, and toxicology of life stages is useful in decreasing uncertainty in risk assessment by contributing to the weight of evidence relating contaminant exposure to effects.

Summary

In summary, the workshop was successful in advancing the state of the science, as well as in bringing together a broad base of experience and viewpoints to advance integration of approaches to understanding basic chemical and physiological processes, toxicological effects and mechanisms, ecological principles, and risk assessment issues related to reproductive and developmental effects of environmental contaminants. It was clear that the demographic attributes of life-history strategies influence life-cycle characteristics, chemical exposure, and toxicity in a manner affecting characteristics that define the risk to chemical exposure.

Traditional approaches of toxicity testing methods and biological and chemical monitoring have contributed greatly to our current knowledge of reproductive and developmental effects and to our ability to assess and regulate the impacts of environmental contaminants. It was the intent of the workshop organizers to focus on reproductive and developmental effects of contaminants on those oviparous vertebrates that hold promise for use in ecological risk assessments. By providing this focus, the organizers intended to bring additional clarity to areas in which new research can provide the tools needed to better manage and protect ecological resources. A mechanistic understanding of environmental fate and reproductive and developmental effects of contaminants is attainable, but it must be considered a long-term goal. Continued progress toward this goal will enable researchers to demonstrate the link between contaminant exposure, basic physiological processes, toxicological effects, ecological consequences, and assessment of risk. Additional fundamental knowledge is essential to explain the mechanisms of reproductive and developmental effects and the processes governing the chemical fate and bioavailability of causative agents. As costs associated with compliance to regulations increase, regulators, as well as the regulated community, must make the most effective use of funds allocated for preventing or reducing deleterious impacts on environmental quality. Decisions based on a more complete mechanistic understanding of cause-and-effect relationships will be less likely to vary with technological trends and will be more defensible, while allowing economic growth to occur in a climate of ecologically meaningful environmental stewardship and maintenance of environmental quality.

Abbreviations

α-MSH	α-melanocyte stimulating hormone
β-LPH	β-lipotropin
AChE	Acetylcholinesterase
ACTH	Adrenocorticotropic hormone
AHH	Aryl hydrocarbon hydroxylase
AhR	Aryl hydrocarbon receptor
APE	Alkyl phenol ethoxylate
AR	Androgen receptor
ARNT	AhR nuclear translocator
ATP	Adenosine triphosphate
BaP	Benzo-a-pyrene
BAF	Bioaccumulation factor
bHLH	Basic helix-loop-helix
BKME	Bleached kraft pulp mill effluent
BMF	Biomagnification factor
BSAF	Biota-sediment accumulation factor
cAMP	Cyclic 3',5'-adenosine monophosphate
CBG	Corticosteroid binding globulin
cDNA	Carrier deoxyribonucleic acid
CRF	Corticotropin-releasing factor
CRH	Corticotropin releasing hormone
DBT	Dibutyltin
DCB	2,2'-dichlorobiphenyl
DDD	Dichlorodiphenlyethane
DDE	Dichlorodiphenyldichloroethylene
DDT	Dichlorodiphenyltrichloroethane

DES	Diethylstilbestrol
DHT	Dihydrotestosterone
DNA	Deoxyribonucleic acid
DRE	Dioxin response element
EGF	Epidermal growth factor
EMF	Egg-to-mother concentration factor
EPN	O-4-nitrophenyl O-ethyl phenylphosphonothionate
ER	Estrogen receptor
ERA	Ecological risk assessment
ERE	Estrogen response element
EROD	Ethoxyresorufin-O-deethylase
ET	Eggshell thinning
FF-MAS	Follicular fluid meiosis activating substance
FOM	Final oocyte maturation
FSH	Follicle stimulating hormone
GABA	δ-aminobutyric acid
GHRH	Growth hormone releasing hormone
GLEMEDS	Great Lakes embryo mortality, edema, and deformities syndrome
GnRH	Gonadotropin-releasing hormone
GR	Glucocorticoid receptor
GSI	Gonadosomatic index
GTH	Gonadotropin
GTH-I	Gonadotropin I
GTH-II	Gonadotropin hormone II
GTP	Guanosine triphosphate
GVBD	Germinal vesicle breakdown

HCB	Hexachlorobenzene
HDL	High-density lipoproteins
HEOD	A stable metabolite of aldrin and dieldrin
HOC	Hydrophobic organic chemicals
HQ	Hazard quotient
HRE	Hormone response element
HxCDD	1,2,3,6,7,8-hexachlorodibenzodioxin

| IGF | Insulin-like growth factor |

K_{aw}	Air-water partition coefficient
K_{oa}	Octanol-air partition coefficient
K_{ow}	Octanol-water partition coefficient

LDL	Low-density lipoprotein
LH	Luteinizing hormone
LOAEL	Lowest-observed-adverse-effect level
LP	Lipoprotein
LRS	Lifetime reproductive success
LXRa	Liver X-receptor a

MAP	Mitogen activating protein
MATC	Maximum acceptable toxicant concentration
MBT	Monobutyltin
MFO	Mixed function oxygenase
MIS	Maturation-inducing steroid
MOA	Mode of action
MR	Mineralocorticoid receptor
mRNA	Messenger ribonucleic acid

NAD	Nicotinamide-adenine dinucleotide
NO	Nitric oxide
NOAEL	No-observed-adverse-effect level
NOEL	No-observed-effect level
NP2EO	Nonylphenol di-ethoxylate
NTE	Neuropathy target enzyme
OP	Organophosphate or Organophosphorus ester insecticide
OPIDN	Organophosphate-induced-delayed-neuropathy
PAH	Polycyclic aromatic hydrocarbon or polyaromatic hydrocarbon
PAS	Per, AhR, ARNT, Sim
PBTK	Physiologically based toxicokinetic
PCB	Polychlorinated biphenyl
PnCDF	2,3,4,7,8-pentachlorodibenzofuran
PR	Progesterone receptor
QSPR	Quantitative structure-property relationship
RA	Retinoic acid
RAR	Retinoic acid receptor
RXR	Retinoid X-receptor
SG	Shell gland
SHBG	Sex hormone binding globulin
SMR	Standard metabolic rate
SPMD	Semipermeable membrane device
TBG	Thyroid-retinol binding globulin
TBT	Tributyltin
TCB	2,2',5,5' tetrachlorobiphenyl

TCDD	Tetrachlorodibenzo-*p*-dioxin or dioxin
TCP	2,4,6-trichlorophenol
TEBG	Testosterone-estrogen binding globulin
TEF	Toxic equivalency factor
TR	Thyroid receptor
TRH	Thyrotropin releasing hormone
TRV	Toxicity reference value
TSH	Thyroid stimulating hormone
UDP-GT	Uridine 5'-diphophate-glucuronosyltransferase
USEPA	U.S. Environmental Protection Agency
USFWS	U.S. Fish and Wildlife Service
VD3	Vitamin D3 receptor
VLDL	Very low density lipoprotein
VTG	Vitellogenin
WSF	Water soluable fraction
YOY	Young-of-the-year

Index

A

Absorption
  metabolic, 10, 25–26
  physical, onto settling particles and
    colloids, 31
Abundance monitoring, 238
Accessory sex organ development, 129, 178
Accessory sexual structures, 173
Accipiter nisus
  DDT and metabolites in, 56, 59, 60, 249–250
  elimination of contaminants into eggs, 60
Acclimatory adaptation, 167
Acetylcholinesterase (AChE), 292, 324–325, 326,
  328
Activin, 173
Adaptation
  compensatory mechanisms, 116
  physiology of, 167–168
  redundancy in control mechanisms, 116, 167
Adelie penguin, maternal transfer, 67
Adenylate cyclase, 139
Adenylyl cyclase, 290
Adipose fat. *See* Lipid reserves
Administered dose, *vs.* absorbed dose, 25–26
ß-Adrenergic receptors, 128, 137
Adrenocorticotropic hormone (ACTH), 123, 132
African clawed frog. *See* Frog embryo teratogen-
  esis assay - *Xenopus*; *Xenopus laevis*
Age
  of female, and egg quality, 191
  population distribution of, 240
  senescence, 154–164
Agelaius phoeniceus, 55, 326
Agonists
  definition, 125
  environmental pollutants as, 126, 136
Agricultural runoff, 21, 253–255
  effects of selenium, 257, 330–333
  effects on Florida alligators, 309–311
AHH. *See* Aryl hydrocarbon hydroxylase
Aimophila aestivalis, population model, 264
Air-water partition coefficient (K_{aw}), 23
Albumin, 143, 150
Aldosterone, vulnerable functions of, 131
Aldrin. *See* Dieldrin/aldrin
Alewife, maternal transfer, 67
Alkyl phenol ethoxylates (APE), 21
Alkyl phenols, 22, 315
Alligator. *See* Alligator mississippiensis

Alligator mississippiensis
  DDT and metabolites in, 250, 309–311, 337–
    338
  effects of contaminants on thyroid function,
    157
  sex determination, 286
Allochemics, 119
Allomones, 119
Allylisopropylacetamide, biotransformation by
  birds, 58
Altricial hatching, 159, 161, 166, 193
Aluminum, 258
Ameiurus nebulosus, body size and contaminant
  effects, 237
American avocet. *See* Recurvirostra americana
American kestrel. *See* Falco sparverius
American plaice. *See* Hippoglossoides platessoides
American robin. *See* Turdus migratorius
Amino acid derivatives, 121, 122, 123
Amitrol, 13
Amphibian(s)
  DDT in, 250
  detoxification potential, 253
  effects of organophosphates, 324–325
  frog embryo teratogenesis assay - *Xenopus*, 297,
    322
  larval period, 230
  life cycle, 230
  life history strategies, 233, 235–236
  malformations in, 225, 286, 322–323, 336
  metamorphosis, 146, 162, 170, 230
  newts, 262
  population modeling, 246, 262
  population responses, 267–268
  secondary sex characteristics, 180
  sex differentiation, 177, 298
  spermatogenesis, 187–188
Anadromous fish, 29
Anas platyrhynchos
  effects of oil spills, 316
  methylmercury in, 61
  selenium in, 61, 331
Anas rubripes, DDE in, 59
Androgens
  control mechanisms, 173
  immune system function and, 170
  molecular structure, 120
  response to stress, 169
  vulnerable functions of, 129
Aniline, biotransformation in fish, 52
Antagonists

Reproductive and Developmental Effects of Contaminants in Oviparous Vertebrates. Richard T. Di Giulio and Donald E. Tillitt, editors.
©1999 Society of Environmental Toxicology and Chemistry (SETAC). ISBN 1-880611-37-6

half-life, 20
physicochemical properties and fate, 15
production and usage, 19, 20
Toxicity. *See* Endpoints; Mechanisms of toxicity;
    Modes of action
Toxicity reference values (TRV), 380, 382, 387
Toxicity tests, 1, 296–297
Toxicokinetics, 10, 26, 72, 78–81, 82
Trachemys scripta, 255, 286–287
Trafficking, 292
Transcriptional regulation, 126–127, 133, 134,
    295
Transcription factors, 127, 154–155
Transformation of chemicals. *See* Degradation
    processes; Environmental fate and
    transport
Transforming growth factor α, 131, 134
Transforming growth factor ß, 131, 135
Transport
    of contaminants. *See* Environmental fate and
        transport
    of hormones, 142–144, 289
    of signaling molecules, 142–144
Transthyretin, 143, 151
Tree swallows. *See Tachycineta bicolor*
Tributyltin (TBT)
    environmental fate, 21
    in fish
        biotransformation, 51, 52
        elimination, 48
        uptake rate, 46
    production and usage, 21
Tributyltin chloride, trans-chorionic permeabil-
    ity, 45
Tributyl tin oxide, 13
1,2,3-Trichlorobenzene, 46
1,2,4-Trichlorobenzene, 46
Trichlorobiphenyls, 16
2,4,6-Trichlorophenol (TCP), 48
Trichogaster cosby, vitellogenesis, 182
Triflularin, 13
Trifluoro-nitromethyl-phenol, 253
17α,20ß-21-Trihydroxyprogesterone, 120, 130
Triiodothyronine, 44, 123, 130
Trophic level, vulnerability of a species and, 248,
    389, 394, 404, 414
Trophic transfer, 31–32
    in intermediate levels, 55
    productivity of system and, 31
Turbot. *See Scophthalmus maximus*

Turdus migratorius, 55
Turtle dove. *See Streptopelia risoria*
Turtles. *See* Reptiles
Two-compartment, open model, 72–74
2,4-D, 286
Tyrosine, 123
Tyrosine kinase activity, 135, 136, 292

U

Ultraviolet radiation, 322
Uncertainty factors, 265, 380, 383. *See also*
    Extrapolation
"Unit world," 23
Up-regulation, 125, 184
Uptake
    continuous, with nonpersistent chemicals, 252
    diffusion factors, and eggs, 69
    direct, in fish, 44, 45–46
    models for, 71–72, 74–76, 77, 81–82
Uria aalge, effects of oil spill, 258–259
U.S. Environmental Protection Agency (EPA),
    framework for ecological risk assessment,
    366–375, 394

V

Vapor pressure, 24
Vasopressin, 123
Very low-density lipoproteins (VLDL)
    in birds, 59, 182
    role in contaminant transfer, 42–45
Vinclozin, 14
Vitamin A, 150
    deficiency, 155
    metabolism of, and embryonic development,
        155
    retinoids derived from. *See* Retinoids
Vitamin D, 121
Vitamin derivatives, 121, 124
Vitellogenesis, 83, 174, 181–183
    in amphibians, 59
    in birds, 59
    in fish, 41–45, 147–149, 314–315
    hormone activity and, 129, 131, 147–149, 176,
        333
    in reptiles, 59
Vitellogenin (VTG)
    as biomarker for estrogens, 183
    in birds, 59, 182
    in fish, 161, 183, 314–315, 335, 338
    induction of, as an endpoint, 299, 315
    in oviparous *vs.* viviparous species, 285–286